Shedding of Plant Parts

PHYSIOLOGICAL ECOLOGY

A Series of Monographs, Texts, and Treatises

EDITED BY

T. T. KOZLOWSKI

University of Wisconsin
Madison, Wisconsin

T. T. KOZLOWSKI. Growth and Development of Trees, Volumes I and II – 1971

DANIEL HILLEL. Soil and Water: Physical Principles and Processes, 1971

J. LEVITT. Responses of Plants to Environmental Stresses, 1972

V. B. YOUNGNER AND C. M. McKELL (Eds.). The Biology and Utilization of Grasses, 1972

T. T. KOZLOWSKI (Ed.). Seed Biology, Volumes I, II, and III – 1972

YOAV WAISEL. Biology of Halophytes, 1972

G. C. MARKS AND T. T. KOZLOWSKI (Eds.). Ectomycorrhizae: Their Ecology and Physiology, 1973

T. T. KOZLOWSKI (Ed.). Shedding of Plant Parts, 1973

In Preparation

ELROY L. RICE. Allelopathy

SHEDDING
OF PLANT PARTS

Edited by T. T. KOZLOWSKI

Department of Forestry
School of Natural Resources
The University of Wisconsin
Madison, Wisconsin

1973

ACADEMIC PRESS   New York and London
A Subsidiary of Harcourt Brace Jovanovich, Publishers

ACADEMIC PRESS, INC.
111 Fifth Avenue, New York, New York 10003

United Kingdom Edition published by
ACADEMIC PRESS, INC. (LONDON) LTD.
24/28 Oval Road, London NW1

Library of Congress Cataloging in Publication Data

Kozlowski, Theodore Thomas, DATE
        Shedding of plant parts.

        Includes bibliographies.
        1.   Abscission (Botany)   I.   Title.
QK763.K69   581.3'1   73-5302
ISBN 0−12−424250−2

Contents

1. Extent and Significance of Shedding of Plant Parts

T. T. Kozlowski

2. Anatomical and Histochemical Changes in Leaf Abscission

Barbara D. Webster

3. Physiological Ecology of Abscission

Fredrick T. Addicott and Jessye Lorene Lyon

4. Internal Factors Regulating Abscission

Daphne J. Osborne

9. Shedding of Reproductive Structures in Forest Trees

G. B. Sweet

10. Anatomical Changes in Abscission of Reproductive Structures

Roy K. Simons

11. Chemical Thinning of Flowers and Fruits

L. J. Edgerton

12. Chemical Control of Fruit Abscission

William C. Cooper and William H. Henry

List of Contributors

Numbers in parentheses indicate the pages on which the authors' contributions begin.

FREDRICK T. ADDICOTT (85), Department of Botany, University of California, Davis, California

G. A. BORGER (205), Department of Biological Sciences, University of Wisconsin Center-Marathon County, Wausau, Wisconsin

W. R. CHANEY (149), Department of Forestry and Conservation, Purdue University, Lafayette, Indiana

WILLIAM C. COOPER (475), U. S. Horticultural Research Center, Agricultural Research Service, U. S. D. A., Orlando, Florida

L. J. EDGERTON (435), Department of Pomology, Cornell University, Ithaca, New York

G. C. HEAD* (237), Formerly, East Malling Research Station, Maidstone, Kent, England

WILLIAM H. HENRY (475), Citrus Research Investigations, Agricultural Research Station, Plant Science Research Division, Orlando, Florida

E. G. KIRBY (295), School of Forestry, University of Florida, Gainesville, Florida

T. T. KOZLOWSKI (1), Department of Forestry, School of Natural Resources, University of Wisconsin, Madison, Wisconsin

JESSYE LORENE LYON** (85), Department of Agronomy and Range Science, University of California, Davis, California

W. F. MILLINGTON (149), Department of Biology, Marquette University, Milwaukee, Wisconsin

DAPHNE J. OSBORNE (125), Agricultural Research Council, Unit of Developmental Botany, Cambridge, England

---

* Present address: World Health Organization, Geneva, Switzerland.
** Present address: Department of Plant Sciences, University of California, Riverside, California.

Roy K. Simons (383), Department of Horticulture, University of Illinois, Urbana, Illinois

Robert G. Stanley (295), School of Forestry, University of Florida, Gainesville, Florida

G. B. Sweet (341), Forest Research Institute, Roto Rua, New Zealand

Barbara D. Webster (45), Department of Agronomy and Range Science and Department of Vegetable Crops, University of California, Davis, California

Preface

The fact that plants recurrently shed cells, tissues, and organs throughout their lifetime has momentous implications to the plants themselves,
other plants, lower and higher animals, and the environment. Of para-
mount importance are the many influences on man, both helpful and
harmful, of the shedding of plant parts. Such considerations as well as
the rapid accumulation of important research results accented the need
for collating in one source the current state of knowledge and opinion
on the anatomical, physiological, and ecological features of shedding of
vegetative and reproductive parts of plants.

This book was planned as a text or reference for upper level under-
graduate students, graduate students, investigators, and growers of plants.
The subject matter overlaps into a variety of disciplines and should be
of interest to agronomists, plant anatomists, arborists, biochemists,
ecologists, entomologists, foresters, horticulturists, landscape architects,
meteorologists, plant pathologists, plant physiologists, and soil scientists.
Since much practical information is included on chemical inhibition and
stimulation of shedding of flowers and fruits, the work should be
particularly useful to those who grow flowers and fruits as crops or for
esthetic reasons.

The book is both authoritative and comprehensive. It encompasses
both natural and induced shedding and not only deals with true abscission (e.g., physiological changes causing cytolysis and weakening of
cells at a distinct separation layer) but also with the shedding of plant
parts due to mechanical factors and shedding as a result of death and
withering. The opening chapter outlines the extent of shedding of plant
cells, tissues, and organs and summarizes biological and economic implications of such shedding (e.g., production of organic matter, drought
resistance of plants, control of insects and diseases, control of species
composition, harvesting of crops, defoliation for military purposes, and
effects on plant growth). Separate chapters follow on anatomical and
histochemical changes in leaf abscission, the physiological ecology of
abscission, internal regulation of abscission, shedding of shoots and
branches, shedding of bark, shedding of roots, shedding of pollen and
seeds, shedding of reproductive structures of forest trees, anatomical
changes in abscission of reproductive structures, chemical thinning of
flowers and fruits, and chemical control of fruit abscission.

I am indebted to each of my colleagues who accepted invitations to write chapters. To these distinguished investigators I express my sincere thanks for their scholarly contributions. Mr. W. J. Davies and Mr. P. E. Marshall were most helpful in the Subject Index preparation as well as in other ways.

T. T. Kozlowski

. 1 .

Extent and Significance
of Shedding of Plant Parts

T. T. Kozlowski

I. Introduction

One of the most universal characteristics of plants is the more or less recurrent natural shedding of both vegetative and reproductive portions of their plant body. Such losses of plant parts occur as cells, tissues, or whole organs. Plant parts may be lost by formation of and separation at an abscission layer, by mechanical factors, by a combination of these, or by death and withering. In true abscission, physiological changes lead to cytolysis which weakens cells of the abscission layer until the parts fall of their own weight or because of some external force such as .wind. Cytolysis does not occur in mechanical separation.

Monocotyledons, plants with perennating organs below ground (e.g., potato), and many annual plants are considered to be nonabscising because they lose leaves and stems by withering and decay rather than by

1

formation of a discrete abscission layer. By contrast, in abscising plants such as deciduous trees, a natural rejection mechanism operates which involves physiological changes leading to formation of an abscission layer at which separation occurs (Osborne, 1968). But even in abscising plants (e.g., deciduous trees) some plant parts, such as small roots, are periodically lost through death and disintegration.

Abscission involves both separation and protection (Figs. 1 and 2). Separation occurs by dissolution of one or more layers of the cell wall (Fig. 3). For example, the middle lamella between 2 layers of cells dissolves but the primary wall does not (Fig. 3B), both the middle lamella and primary wall between 2 cells dissolve (Fig. 3C), or all the cells of one or more layers dissolve (Fig. 3D) (Addicott and Lynch, 1955). The first two types of separation are characteristic of leaf abscission of woody

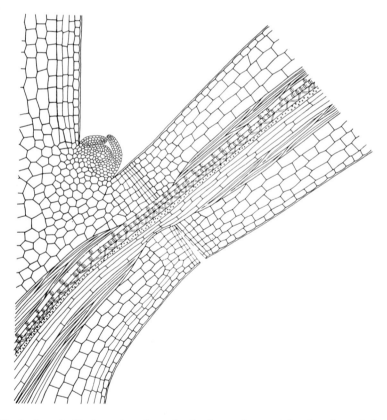

FIG. 1. Longitudinal section through the base of a petiole of a dicotyledonous leaf showing separation beginning at the abscission layer. (Photograph courtesy of F. T. Addicott.)

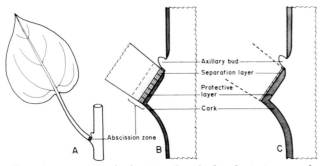

FIG. 2. Abscission zone of a leaf. A, Leaf with the abscission zone located at the base of the petiole. B, Layers of the abscission zone shortly before leaf abscission. C, Layers of the abscission zone after leaf abscission has occurred. (From Addicott, 1970; with permission of McGraw-Hill.)

plants. The third type occurs in leaf abscission of some herbaceous species and in certain kinds of dehiscence. Protection of the scar may involve suberization and lignification, as well as development of protective layers. After separation of a leaf occurs, cell division continues on the stem side of the abscission layer to produce a corky protective layer (Addicott, 1965). Separation by abscission may also involve varying degrees of mechanical disruption of cell walls.

In addition to natural losses of portions of the plant body, losses of plant parts also occur as a result of unusually severe environmental

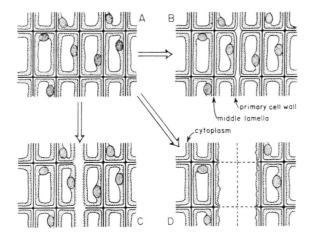

FIG. 3. Portion of an abscission layer showing various types of dissolution. A, Before dissolution begins. B, Dissolution of middle lamella only. C, Dissolution of middle lamella and primary cell wall. D, Dissolution of whole cells. (Adapted from Addicott, 1954.)

stresses (e.g., drought, frost), insect attacks, disease, damage, or mechanical breakage (see Chapter 3). Furthermore, various economic and aesthetic considerations often provide the impetus for artificially inducing leaf fall or removal of twigs, branches, bark, roots, flowers, and fruits. Hence, because of recurrent natural shedding and cumulative, externally induced losses of its plant body, the dry weights of many mature plants represent only a fraction of the photosynthate that was actually converted into structural tissues during development.

II. Shedding of Various Plant Parts

Vegetative parts which are lost by abscission include buds, branches, prickles, cotyledons, leaves, leaflets, leaf stalks, stipules, bark, and roots. Among abscising reproductive plant parts are inflorescences, pedicels, flowers, buds, calyxes, corollas, stamens, styles, ovaries, fruits, fruit hulls, strobili, scales of strobili, and seeds. Dehiscence of anthers and fruits is often regarded as a specialized form of abscission (Addicott, 1965). In this section I will briefly discuss some aspects of shedding of vegetative and reproductive plant parts.

A. LEAVES

In simple leaves abscission occurs at or near the base of the petiole. In compound leaves separate abscission zones form at the bases of leaflets, petiolules, and, also, petioles. In a study by Moline and Bostrack (1972), leaf fall in both *Acer negundo* and *Fraxinus americana* was the result of orderly and fragmentary abscission, with leaflets falling before petioles. The sequence of abscission was from terminal to basal leaflets. In general, abscission of leaflets occurred similarly to that of petioles. An exception was absence of a protective layer proximal to the separation layer of leaflets which fell at the end of the growing season. However, if leaflets were shed prematurely because of injury, a protective layer was formed.

Leaves of most deciduous trees of the temperate zone form an abscission layer during the season in which they expand and fall in autumn of the same year. However, some marcescent species (e.g., several species of *Quercus*) retain leaves through the winter and shed them the following spring. In *Quercus palustris* and *Q. coccinea*, abscission layers of leaves do not form by the end of summer, but are first confirmed in early winter. Leaf shedding begins the following March (Hoshaw and Guard, 1949). In *Quercus velutina* some leaves form well-developed separation layers and abscise in the autumn but the majority are retained on the tree until the following spring. In both cases separation occurs from mechanical breakage through the thin walls of the recently divided cells of the separation layer. No dissolution phenomena are involved (Marvin, 1964). Closely related marcescent species often shed their leaves at different times. For example, leaves of *Quercus rubra*, with large vascular strands with heavy cell walls, are shed later than those of *Q. coccinea* with small vascular strands (Berkley, 1931).

1. Variations in Leaf Abscission

Considerable insight into quantitative and seasonal aspects of leaf shedding may be gained from studies of litter production in forest stands because leaves invariably are the major litter component. According to Rodin and Bazilevich (1967), leaf fall contributes 40 to 65% of total litter fall in mature deciduous forests of the temperate zone and 75 to 85% in young forests. In a study of 8 widely separated forest stands including gymnosperm and angiosperm types, leaves comprised 60 to 76% of the litter, branches 12 to 15%, bark 1 to 14%, and fruits 1 to 17% (Bray and Gorham, 1964). By comparison, in a hardwood forest in New Hampshire leaves, branches, stems, and bark contributed 49, 22, 14, and <2%, respectively, of above-ground litter. Other deciduous structures (e.g., bud scales, flowers, fruits, etc.) contributed 10.9% (Gosz *et al.*, 1972).

Viewed broadly, the amount and seasonal distribution of leaf shedding are exceedingly variable and influenced largely by such factors as species, environment (e.g., climate and latitude, altitude, exposure, soil fertility, soil moisture, temperature), plant density, and time (e.g., seasonal variation, annual variation, age of plants) (Bray and Gorham, 1964; Gosz *et al.*, 1972).

In the northern hemisphere about 17% more litter was produced annually by gymnosperm than by angiosperm trees (Table I). However, detailed studies show considerable variation among and within species in leaf and litter production. This may be predicted from the following values of mean leaf crop in tons per hectare per year for closed canopy angiosperm forests: 3.7 for 11 *Quercus* sites; 3.0 for 3 *Fagus* sites; 2.9 for 2 *Salix* sites; 2.6 for 2 *Alnus* sites; 2.5 for 2 *Fraxinus* sites; 2.5 for 3 *Populus* sites; and 2.4 for 8 *Betula* sites (Bray and Gorham, 1964). *Quercus* sites produced large amounts of litter because of high leaf production and slow leaf decay. Leaf crops within a species often vary greatly with stand density. For example, Bray and Gorham (1964) gave the following values for leaf crops for 3 *Quercus* forests: *Quercus ellipsoidalis–Q. alba*

TABLE I

LITTER PRODUCTION BY EVERGREEN GYMNOSPERMS AND DECIDUOUS ANGIOSPERM
TREES IN THE NORTHERN HEMISPHERE[a]

Litter	Gymnosperms[b]	Angiosperms
Total	3.7	3.2
Leaf	2.6	2.4
Other		
By difference	1.1	0.8
Observed	0.7	0.7

[a] From Bray and Gorham (1964).
[b] Data given in metric tons/ha/year.

closed forest, age 52 years, 5.2 tons/ha/year; *Q. ellipsoidalis* open forest, age 43 years, 2.1 tons/ha/year; *Q. ellipsoidalis–Q. macrocarpa* savannah, age 50 years, 0.8 tons/ha/year.

Much more litter is produced in warm than in cold regions. Bray and Gorham (1964) gave the following values for average annual litter production (metric tons/hectare) for 4 major climatic zones: Arctic–Alpine, 1.0; Cool Temperate, 3.5; Warm Temperate, 5.5; Equatorial, 10.9. These differences reflected not only the high temperatures and long growing seasons of the warmer zones but also the greater amount of insolation during photosynthetic activity.

The amount of leaf fall of a given species or individual tree often varies from year to year (Fig. 4). In a stable forest in southern Canada the annual production of leaves varied from 2.8 to 3.2 tons/ha in 4 different years (Table II). The fall of branches and stem components was much more erratic and averaged 1.3 metric tons/ha, with a range of 0.5 to 3.2 tons. These trends were rather similar to those found by Sykes and Bunce (1970) who studied seasonal and annual fluctuations in a mixed deciduous forest in England during 3 consecutive years. The amount of nonwoody litter varied less between years (362 gm/m^2 to 393 gm/m^2) and collection sites than the amount of woody litter, which varied from 106 gm/m^2 to 162 gm/m^2 in the 3 years under study.

Annual variations in leaf fall of a given stand or tree often reflect climatic variations (see Chapter 3). In addition, because of competition within plants in vegetative and reproductive growth, reciprocal relationships may be expected between the amount of leaves and seeds, fruits, and cones which are found in litter. During a year of heavy fruiting or seed production, shoot growth and leaf crops are reduced (Kozlowski, 1971b). For example, a heavy crop of apples was followed by a reduction in shoot growth of 9 to 53% the next year, depending on root stock

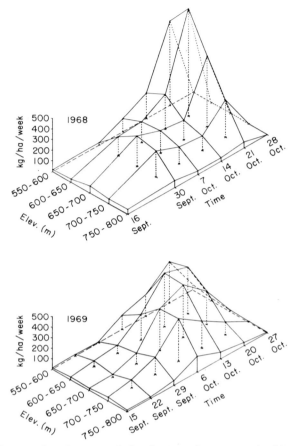

FIG. 4. Elevational and temporal distribution of sugar maple (*Acer saccharum*) leaf fall during September and October of 1968 and 1969 in a northern hardwood forest in New Hampshire. Values are in kg/ha/week. (From Gosz *et al.*, 1972.)

(Rogers and Booth, 1964). Deblossomed apple trees have been shown to produce more and larger leaves than are produced by fruiting trees (Maggs, 1963). Avery (1969) found that decrease in leaf size was an early effect of fruiting in apple trees. Similar relations have been shown for forest trees. Morris (1951) reported a reduction in leaf production of *Abies balsamea* trees during years of large seed crops. Furthermore, leaf production in individual trees was reduced in the portions of the crown that had abundant conelets. Tappeiner (1969) found that needle length and shoot elongation of *Pseudotsuga menziesii* were greatly reduced in good seed years. Apparently reproductive structures, especially fruits, at certain stages of development are powerful sinks which mobilize

TABLE II

LITTER PRODUCTION IN GLENDON HALL FOREST, TORONTO, CANADA[a]

	Litter fall (metric tons/ha/year)				
	1957	1958	1960	1961	Mean
Leaf, including bud scales, fruit	2.8	3.2	3.2	3.1	3.1
Stem, including bark	0.6	3.2	0.5	0.8	1.3
	3.4	6.4	3.7	3.9	4.4

[a] From Bray and Gorham (1964).

large amounts of carbohydrates from vegetative tissues (Kozlowski and Keller, 1966; Kozlowski, 1971a). For example, developing fruits of *Prunus persica* and *Prunus americana* compete very successfully for current photosynthate with nearby expanding foliage. Even when ripe these stone fruits are the major carbohydrate sinks (Kriedemann, 1968). Developing strobili of gymnosperms also are powerful sinks which preferentially mobilize currently produced carbohydrates (Dickmann and Kozlowski, 1968, 1970).

2. Seasonal Variations in Leaf Fall

In cool temperate forests leaf fall of most deciduous angiosperms is almost wholly an autumn phenomenon which follows leaf senescence (Figs. 5 and 6). Sykes and Bunce (1970), for example, showed that more than four-fifths of seasonal leaf fall in a mixed deciduous forest occurred from September to November. It should be remembered, however, that leaves of deciduous trees may be lost at any time during the growing season in response to injury, disease, or unfavorable environmental conditions (Kozlowski, 1971a).

In warm temperate forests the fall of litter occurs throughout the year, but superimposed on this pattern are maxima which often are associated with local climatic conditions (Fig. 7). In eastern Australia peaks of litter fall in spring and early summer are associated with increases in rainfall and temperature. In western Australia most leaf fall of *Eucalyptus marginata* and *E. diversicolor* occurs during the warm and dry months of January to March. In New Zealand introduced gymnosperms tend to shed most needles during the autumn months (March to May). Leaf fall tends to be much more continuous throughout the year in tropical forests than in those of temperate zones (Figs. 5–8). Nevertheless, there is considerable variation in annual leaf shedding in tropical climates because

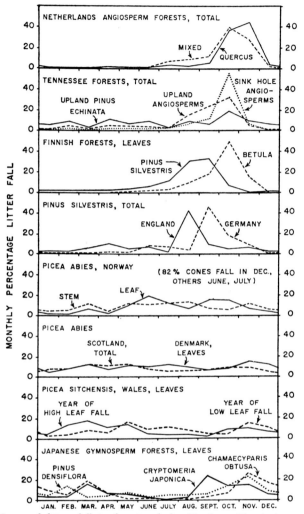

FIG. 5. Seasonal litter fall in forests of the Northern Hemisphere. (From Bray and Gorham, 1964.)

of variations in heredity and seasonal distribution of rainfall. Even in the uniform, hot, and wet climate of Singapore (approximately 100 inches of rain/year) there is wide species variation in the annual pattern of leaf abscission. Holttum (1931, 1940), for example, classified deciduous species of Singapore into those with (1) annual leaf periods (e.g., *Kigelia pinnata, Hymenaea courbaril*), (2) leaf periods > 12 months (e.g., *Cedrela glaziovii, Koompassia malaccensis*), (3) leaf

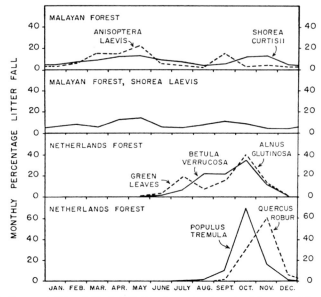

FIG. 6. Seasonal litter fall of individual species in mixed forests. (From Bray and Gorham, 1964.)

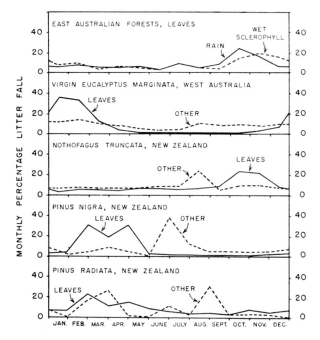

FIG. 7. Seasonal litter fall in forests of the Southern Hemisphere. (From Bray and Gorham, 1964.)

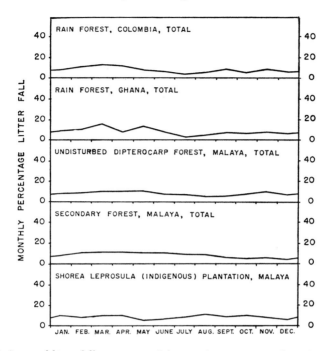

FIG. 8. Seasonal litter fall in equatorial forests. (From Bray and Gorham, 1964.)

periods between 6 and 12 months (e.g., *Cassia fistula, Ficus variegata*), and (4) irregular leaf periods (*Hevea brasiliensis, Mangifera indica*).

Kozlowski (1971a) stated that in a given species the leaves of adult trees usually abscise earlier in the year than leaves of juvenile plants. Whereas leaf abscission was well underway in 50-year-old *Robinia pseudoacacia* trees by mid-November near Paris, 10-year-old trees of this species had not begun to shed their leaves (Trippi, 1963).

B. Shoots

Shedding of portions of branches or whole ones occurs commonly and is shown by the fact that branches may comprise up to 30% of forest litter. Möller *et al.* (1954) found the annual loss of branches of *Fagus sylvatica* trees to be 0.8% of the total volume of stem and branches, or 12.6% of the annual increment in stems and branches. Loss of branches may result from mechanical damage (e.g., from strong winds), death and decay, desiccation or shading, or normal abscission. The shedding of branches of plants of arid regions is a common drought-resistance adaptation (Orshan, 1954, 1963, 1972) (see Section III,B).

Abortion of shoot tips by an abscission process is a common phenomenon in many sympodial species which do not form true terminal buds (Millington, 1963; Romberger, 1963). For example, expansion of *Citrus* shoots occurs in recurrent flushes from false terminal buds. Shedding of branches sometimes up to one inch (2.5 cm) or more in diameter by an abscission process occurs in some woody plants (van der Pijl, 1952; Licitis-Lindbergs, 1956). It is especially common in the tropical genera *Albizzia, Casuarina, Xylopia, Sonneratia,* and *Persea* (Kozlowski, 1971b). Death of lower branches without formation of an abscission layer is well known in forest trees. The degree of such natural pruning varies among species, with some (e.g., *Pinus palustris*) losing their lower branches much more readily than others (e.g., *Pinus coulteri*). The speed with which the lower stem sheds branches is sometimes used by foresters as an index of shade tolerance, since death of lower branches appears to be a response to shading (Baker, 1950).

Very slow shedding of lower branches occurs in many trees in forest stands. Such natural pruning of branches involves sequential death of branches, their separation, and healing over of the wound by cambial growth. Death of branches usually is preceded by a decrease in xylem increment in the branch near the main stem. After a branch finally dies it is weakened by attacks of saprophytic fungi and insects. Eventually the branch is shed. It may be broken of its own weight or shed as a result of the action of wind, rain, or whipping by adjacent trees. The rate of natural branch pruning varies among species and regions. It generally occurs faster in warm humid regions than in cool dry ones because fungi which weaken branches are more active in the former. Natural pruning is also stimulated by high stand density since lower branches die faster when trees are grown closer together (Hawley and Smith, 1954). The slow losses of branches by natural pruning contrast sharply with the rapid seasonal shedding of branches in many desert shrubs during the dry season (Orshan, 1954).

C. Roots

The shedding of root tissues is traceable to loss of the cortex, shedding of periderms in roots undergoing secondary thickening, death of whole roots, and loss of root hairs and root cap cells (Kozlowski, 1971b). In roots which do not undergo secondary thickening, the cortex may persist for many years (e.g., monocotyledons, pteridophytes, many herbaceous dicotyledons). However, whenever secondary thickening of roots occurs, as in woody angiosperms and gymnosperms and in many herbaceous plants, or when internal periderms form, the cortex is shed (Eames and MacDaniels, 1947). Nevertheless, root losses are appreciable in plants without secondary thickening since many of their roots die each year, and even in surviving roots, root cap cells are lost and root hairs are constantly shed and replaced.

Weaver and Zink (1946) placed many small bands around individual roots of 10 native prairie grasses and studied root longevity for 3 years thereafter. In *Agropyron cristatum* 97% of the banded roots survived the first summer and 75% survived the first year. Root losses occurred gradually in all species studied, and after 3 growing seasons survival of banded roots in 4 species was as follows: *Bouteloua gracilis,* 45%; *B. curtipendula,* 14%; *Andropogon scoparius,* 10%; and *Stipa spartea,* 10%. The losses, however, were not serious since roots were replaced much faster than they were lost. The average number of roots produced by individual plants varied from 175 to 882 at the end of the third summer. Mortality (as percent of the total number of roots) generally amounted to only 2 to 8% over the 3-year period. Stuckey (1941) reported that whereas some species of grasses produced a new root system annually, others had perennial roots and produced relatively few new ones each year.

Woody plants have large perennial roots as well as many small roots which die back. Longevity of these small rootlets varies markedly. Childers and White (1942) noted that many small roots of certain varieties of *Malus* survived for only about a week. By comparison, Bosse (1960) observed that in seedlings of other *Malus* varieties the small rootlets were shed in 80–100 days after their formation. Many small rootlets of *Picea excelsa* lived for several years, and about 20% were still alive after 4 years. Nevertheless, there was continuous root mortality and about 10% of the small roots died during the first year (Orlov, 1960). In the temperate zone, most shedding of rootlets occurs during the winter (Bode, 1959; Voronkov, 1956). There has been considerable controversy about whether the shedding of small roots constitutes normal physiological abscission or is caused by environmental stresses and various pests. Since root growth depends on a continuous supply of leaf products it appears likely that environmental stresses, fungus diseases, insect defoliation, or other agents which reduce leaf metabolism, or cause premature defoliation, or both, may lead to root mortality (Kozlowski, 1971b).

It should be remembered that even in healthy growing roots the root hairs are lost periodically. Some persist only for hours and others for several days or weeks. As old root hairs die and are shed new ones form regularly behind the apex of an elongating root. Hence, the zone of active root hairs migrates and remains at a more or less fixed distance behind the tip of the root.

D. REPRODUCTIVE STRUCTURES

Reproductive structures such as pollen, floral parts, seeds, fruits, and strobili are shed periodically.

1. *Pollen*

Plants produce and shed tremendous quantities of pollen. However, the amounts shed vary greatly with species, age of the plant, and time. For an individual species the amount of pollen produced and shed differs greatly from year to year and from plant to plant. Some idea of the extent of pollen production by a single plant may be gained from Faegri and Iverson (1964) who reported that in hemp one anther may produce 70,000 pollen grains and an individual shoot as many as 500 million grains. Pollen production by some trees is especially voluminous. For example, a 10-year-old branch system of *Betula, Quercus,* or *Picea* may produce over 100 million pollen grains, and of *Pinus* more than 300 million.

The production of pollen in different parts of trees varies greatly and this is especially marked in gymnosperms where production of staminate strobili is highly localized in the tree crown. In *Pinus* the pollen-producing staminate strobili form at the base of the current year's growth in the lower portion of the crown. The staminate strobili of *Picea* occur along twig growth of the previous year. In *Abies* the staminate strobili develop in the spring from buds formed along 1-year-old twigs. In *Abies balsamea* staminate strobili occur in the zone almost 10 to 15 ft below the top of the crown whereas ovulate strobili are found primarily in the top 4 to 5 ft of the crown.

Species vary in the time and duration of pollen shedding, but pollen may be dispersed in a few hours in some species (e.g., *Quercus prinoides*) and in a matter of days in others (e.g., *Pinus sylvestris*) (Sharp and Chisman, 1961; Dorman and Barber, 1956; Sarvas, 1955a,b, 1962). Male flowering and pollen shedding are strongly influenced by temperature, humidity, and wind (Kozlowski, 1971b). In Finland often only one warm day after a cold spring is required for pollen shedding. Large diurnal variations in pollen shedding are also related to temperature and humidity changes. Most pollen is shed during the middle of the day when temperature is high and humidity low. At night when the temperature is low and relative humidity approaches 100%, pollen shedding is negligible (Sarvas, 1962).

2. *Flowers, Fruits, and Seeds*

Following successful pollination of flowers, the rate of growth of the ovary increases and floral parts such as stamens and pistils generally are shed. Subsequent shedding of floral structures may involve abscission of inflorescences, whole flowers, or parts of flowers. Cell division usually does not occur before the shedding of petals and the abscission layer is poorly differentiated. Loss of petals occurs as a result of softening of the middle lamella (Esau, 1965). The formation of floral cup and style abscission zones is discussed by Lott and Simons (1964, 1966, 1968a,b) and in Chapter 10 of this volume.

Abscission of fruits may occur at various stages of development. The abscission of immature fruits is well known and is a matter of concern for growers of apples, pears, and citrus fruits. Apple fruits tend to be shed in 3 waves: (1) "early drop" between initiation of ovary enlargement and beginning of endosperm development, (2) "June drop" of young apples, and (3) preharvest drop.

In a study by Erickson and Brannaman (1960) at Riverside, California, abscission of various reproductive structures (buds, flowers, fruits) of Washington Navel orange occurred rather continuously from February to July. About half of the buds were shed before they opened. Total abscission during the six months amounted to 198,274 structures (96,343 buds; 33,235 flowers; 68,696 fruits). Most abscission of reproductive structures took place when fruits were very small. The mature fruit crop amounted to only 0.2% (419 mature fruits/tree), emphasizing the tremendous losses by abscission.

The premature abscission of fruits and strobili of forest trees is also well known. For example, Williamson (1966) reported that about 90% of the acorns of *Quercus alba* abscised from early May to the middle of July (during the period of pollination, ovule development, and fertilization). In *Pinus radiata*, Sweet and Thulin (1969) observed about half of all the ovulate strobili aborted during the first year following pollination.

The proportion of the total plant body which is lost by premature abscission and in the harvested reproductive structures differs greatly among species, within a species, and from year to year. Nevertheless, over the life span of the perennial woody plant the weight of shed reproductive structures is immense. For example, during a 32-year period Rhode Island Greening apple trees produced an average of 11.4 lb of mature apples per tree (Shaw, 1934). When the losses due to premature shedding in the bud, flower, or immature fruit stages are added to this amount it is clear that the cumulative total loss is very sizable. Weight losses due to shedding of reproductive structures in forest trees are also appreciable.

For example, in a study by Ovington (1963) the dry weight of the spring crop of catkins in *P. tremuloides* was about one-fourth as great as that of the entire annual leaf crop.

The marked annual variations in amount of reproductive structures shed by trees are correlated to a considerable degree with inherent patterns of periodicity of reproductive growth. In a number of orchard trees, alternate or biennial bearing is characteristic of individual trees and branches as well. Some trees have large fruit crops one year and a small one or none the next year. Within a tree, some branches often produce flowers and fruits whereas others do not. In the following year, the previously fruiting branches remain vegetative and the previously vegetative branches form reproductive structures (Davis, 1957).

Production and shedding of reproductive structures of forest trees are much more irregular in forest trees than in orchard trees. The amount of seed produced by forest trees varies greatly with species, tree age, and competition among trees, and it varies from year to year. Whereas *Tsuga heterophylla* is a prolific seeder, *Abies grandis* and *Pinus monticola* are light seeders. Some species (e.g., *Nyssa aquatica, Magnolia grandiflora, Celtis laevigata*) tend to produce good seed crops annually; others produce good seed crops at fairly regular intervals (e.g., *Fraxinus* spp., *Acer saccharum*) or at irregular intervals (e.g., *Fagus grandifolia, Juglans nigra, Liriodendron tulipifera*). There often is wide variation in periodicity of reproductive growth of different species of the same genus. For example, *Quercus phellos* tends to have good seed crops annually whereas *Quercus alba* has them at irregular intervals of 4 to 10 years. *Populus trichocarpa* generally has good seed crops annually, but *P. grandidentata* and *P. tremuloides* have good seed crops at 4- to 5-year intervals (Kozlowski, 1971b).

Under the severe competition which occurs in forest stands, most reproductive structures are produced by vigorous trees. For example, practically all of the strobili of *Pinus ponderosa* and *P. lambertiana* were produced by dominant trees, with only about 1 to 1.5% borne on codominant trees. Only an insignificant number of strobili were produced by trees of intermediate and suppressed crown classes. Dominant trees of *Abies concolor* bore 88% of the strobili produced whereas codominant trees had only 12% (Fowells and Schubert, 1956).

3. *Plant Age and Shedding of Reproductive Structures*

In woody plants the proportion of the plant body which is regularly shed in the form of reproductive structures varies with plant age. During ontogenetic development woody plants undergo a phase change from a juvenile stage during which they do not flower to an adult stage during which they achieve the capacity for reproductive growth. Once the juvenile stage is passed, the capacity for flowering is retained until late in the life of the plant. Nevertheless, as mentioned, many adult trees do not flower annually because environmental stresses control the initiation of floral primordia after the adult stage has been reached. The length of the juvenile, nonreproductive stage varies considerably in different species (Sax, 1962; Kozlowski, 1971a).

Aging (reduction in vigor and associated changes which occur as trees increase in size and complexity) has predictable effects on reproductive growth and, consequently, on the amounts of reproductive parts shed. The capacity for reproductive growth in trees is greatest during middle age and wanes thereafter. However, the initial age and duration of reproductive capacity vary considerably. Whereas *Juniperus virginiana* begins to produce large amounts of seed when 10 years old and has an optimum seed-bearing age of 25 to 75 years, *Picea engelmannii* begins to produce seeds in quantity at about 20 years and is most prolific at an age of 200 to 250 years (Kozlowski, 1971a).

III. Importance of Natural and Induced Shedding of Plant Parts

Natural and induced shedding of plant parts may be considered extremely beneficial to very harmful depending on many factors. Removal of plant parts often is economically beneficial, e.g., in improving harvesting of plant products, but may be harmful to the plant or ecosystem. For example, artificial defoliation of tropical forests has been considered of great benefit in military campaigns because it denies cover to the enemy, but some biologists have considered the effects of such defoliation on the ecosystem to be harmful or catastrophic.

Many values can be ascribed to natural shedding of plant parts. For example, it provides for regeneration and extension of ranges of species by dissemination of spores, pollen, seeds, and fruits. It accounts to a large degree for drought resistance of many plants in arid environments. As a process of self-pruning it removes injured, diseased, or senescent parts. Such self-pruning reduces competition for water and nutrients by removing the less vigorous leaves, branches, and fruits (Addicott, 1965). By thinning reproductive structures it improves the quality of residual fruits. In forest trees the natural pruning of branches results in knot-free wood. Natural shedding by plants also provides litter which becomes a major component of soil organic matter. Induced shedding also has beneficial effects in facilitating harvesting of plant crops, influencing the quality of harvested crops such as edible fruits, and handling of nursery stock. On the other hand, natural and induced abscission, especially premature abscission, often are very harmful. Among the deleterious effects of loss of plant parts are inhibition of plant growth and killing of plants, inhibition of seed germination by the physical and chemical effects of litter, loss of nutrients from ecosystems, maintenance of undesirable species composition, and loss of useful plant products such as wood and edible fruits. It should be remembered, however, that a distinction between natural and induced abscission is not always easy to make. For example, abscission often is accelerated by environmental stresses of varying degree, and whether such abscission is natural or induced has been a subject for lively debate.

In this section I will discuss some of the major effects of natural and induced shedding of vegetative plant parts. For a discussion of the importance of induced shedding of reproductive organs and tissues the reader is referred to Chapters 11 and 12.

A. PRODUCTION OF ORGANIC MATTER

Soil organic matter is produced by decomposition of shed plant parts and by synthesis of microbial cells which die and are decomposed by other microorganisms. Soil organic matter provides important beneficial effects on soil structure and plant growth. It affects soil fertility both directly and indirectly. The fact that soil organic matter plays an important role in cation exchange in soils is shown by the decrease in exchange capacity following destruction of the organic fraction with hydrogen peroxide. For example, Bartlett *et al.* (1939) noted up to a 90% loss of exchange capacity following treatment of Maryland soils with hydrogen peroxide. The soil organic matter also affects fertility directly because it contains nutrients which become readily available through activity of microbial organisms. It furnishes requirements for processes of the nitrogen cycle and fuel for the necessary microbial transformations (Broadbent, 1953). Thus, the total soil organic matter is an important factor in determining the amount of nitrogen a soil supplies for plant growth (Ensminger and Pearson, 1950). Much of the phosphorus absorbed by plants is also derived from organic forms (Bower, 1949). However, the soil organic fraction is not an important direct source of nutrients other than nitrogen and phosphorus. Nevertheless other nutrients are held in an exchangeable state by organic colloids. Furthermore, the availability of such elements as iron, manganese, and sulfur is affected by organic matter and soil remains (Broadbent, 1953).

The continual shedding of plant parts to produce litter is not always beneficial to plant growth. For example, the presence of a layer of leaf litter often interferes with regeneration of plants because of unfavorable physical characteristics and release of toxic chemical compounds.

Mineral soil often is a better medium than the litter layer for seed germination and seedling growth because of high infiltration capacity, good aeration, and close contact between the imbibing seed and soil particles. In addition, mineral soil warms faster than loose organic matter and offers little resistance to root penetration. By comparison, litter and duff tend to inhibit root penetration, prevent seeds from making good soil contact, and shade out the small germinants (Place, 1955). Many species cannot become established because their roots cannot penetrate a thick litter layer before the surface layers dry out. Hence, regeneration often is better on thin than on thick litter (Phelps, 1948). Winget and Kozlowski (1965a) found that forest litter was a barrier to regeneration of *Betula alleghaniensis*. Rootlets of seedlings germinating on humus or sandy soil readily penetrated the rooting medium and the seedlings grew upright. In contrast, the roots of germinants in litter tended to grow horizontally over the leaf mat. Often the rootlets were exposed and seedling stems were prostrate. Sometimes seedlings were overturned by the levering action of the young root against the leaf surface, causing rapid desiccation of these plants.

Leaves and other litter components may be decomposed by weathering and by soil microorganisms to liberate various allelopathic chemicals which adversely affect adjacent plants. Among the injurious effects of decomposition products of plants are delay or complete inhibition of seed germination, stunted growth, injury to roots, deranged nutrient absorption, chlorosis, wilting, and killing of plants (Patrick, 1971). Allelopathic substances may affect plants directly or indirectly by being decomposed into secondary products which are the effective agents (Tukey, 1969). Among the naturally occurring compounds which appear to inhibit seed germination and growth of neighboring plants are phenolic acids, coumarins and quinones, terpenes, essential oils, alkaloids, and organic cyanides. Naturally occurring allelopathic chemicals are ecologically important because they influence succession, dominance, vegetation dynamics, species diversity, structure of plant communities, and productivity (Whittaker, 1970). In addition to being released from shed tissues of plants, allelopathic substances may also be released from intact plants by leaching, volatilization, excretion, and exudation.

Several investigators have reported marked inhibitory effects of specific plants on seed germination and growth of adjacent plants. The toxic leachates of leaves of *Artemisia absinthium* (Funke, 1943), *Encelia farinosa* (Gray and Bonner, 1948), and *Ailanthus* (Mergen, 1959) are well known. In California the annual vegetation adjacent to naturalized stands of *Eucalyptus camaldulensis* is greatly inhibited. Annual herbs rarely survive to maturity where there is an accumulation of *Eucalyptus* litter. Del Moral and Muller (1970) identified several volatile and water-soluble toxins in *Eucalyptus* tissues, including the terpenes, cineole, and α-pinene.

Allelopathic chemicals sometimes are considered beneficial since they appear to contribute to survival of certain desert shrubs by suppressing development of competing vegetation (Kozlowski, 1972). For example, the presence of *Salvia mellifera* inhibited seed germination and prevented establishment of *Adenostema* seedlings under it (Went, 1952). In dry regions of India, *Prosopis juliflora* forms dense thickets of shrubs to trees. The ground underneath becomes covered with a thick layer of leaf litter and very few plants come up within the *Prosopis* community. Inhibitors present in the leaves inhibit seed germination and growth of seedlings. The inhibitory influence of the allelopathic chemicals is increased by the physical effects of the leaf litter (Lahiri and Gaur, 1969).

B. DROUGHT RESISTANCE

One of the most important drought resistance mechanisms of plants involves reduction in transpiring surface. Thus, natural shedding of leaves and other plant parts reduces the transpiring surface area during a period of soil drying and often prevents dehydration of plants to lethal levels. Leaf shedding generally begins with the oldest leaves and progresses to young ones, and water often is translocated from old to young leaves before the former are shed (Kozlowski, 1964a, 1972). Leaf shedding during droughts occurs more commonly in angiosperms than in gymnosperms. In many plants, tissues other than leaves are also shed during drought. For example, Orshan (1954) classified drought-resistant plants of arid zones into the following 4 types:

(1) Leaf shedders—plants which seasonally shed only leaves and inflorescences

(2) Branch shedders—plants which seasonally shed parts of their branches

(3) Whole shoot shedders—plants which seasonally shed all above-ground parts

(4) Annuals—plants which seasonally were entirely renewed

In desert plants the leaves often abscise, are reduced to stipules, or are replaced by brachyblasts. A number of plants have both long and short shoots. Seasonal dimorphism results from shedding and growth of various types of branches and leaves. For example, small summer leaves often replace the large winter leaves. Desert and semidesert shrubs may reduce the weight of their young shoots, through abscission, by a third to well over three-fourths. For example, shedding of leaflets while retaining petioles, reduced the transpiring surface of *Zygophyllum* by 87% (Zohary and Orshan, 1954). In New Mexico, shrubs such as *Ephedra trifurca* survive as leafless green twigs. Plants of Indian semideserts (e.g., *Euphorbia royaleana*) also have no leaves (Sarup, 1952).

The capacity for shedding of leaves in response to drought varies widely among species. In Brazil the leaves of practically all deciduous trees of the "Caatinga" abscise with onset of a drought and are replaced when the drought ends (Alvim, 1964). In Trinidad many deciduous trees have a long leafless period as a result of progressive abscission during periods of increasing soil water deficits. Other species, including *Hymenea courbaril, Platyminium trinitatis, Copaifera officinalis,* and *Albizzia caribaea* have a leafless period of only a few days or weeks at the beginning of the dry season, and they leaf out again by the time the dry season is at its height. Still other species lose only some of their leaves during the dry season and essentially lose none of their leaves during the wet season (Beard, 1946).

The reduction of transpiring surface by shedding of leaves has been considered to be the most important factor in survival of many desert plants (Orshan, 1954). In the arid parts of southwestern United States *Fouquieria splendens, Encelia farinosa,* and *Larrea tridentata* drop their leaves during periods of soil drying (McCleary, 1968). In Israel, leaf shedding desert genera include *Noea, Haloxylon, Anabasis,* and *Zygophyllum,* among others. Seasonal leaf shedding during the dry season is also characteristic of such Mediterranean plants as *Poterium spinosum, Cistus salvifolius, Thymus capitatus,* and *Artemisia monosperma.*

Increases and decreases in the transpiring plant surfaces occur through vigorous growth when water is readily available and by gradual leaf shedding and dieback of branches during drought. Under Mediterranean climates such changes are seasonal and vary in different species. For example, leaf shedding predominates in *Zygophyllum domosum.* However, in such articulated plants as *Haloxylon articulatum* and *Anabasis articulata,* which bear no leaves, transpiring area is reduced by dying and shedding of the green stem cortex. In addition, these plants shed branches (Orshan, 1963, 1972).

Appreciable photosynthesis of branches and bark often occurs in drought deciduous plants of arid regions. In the Sonoran desert, for example, several species (e.g., *Canotia holocantha, Holocantha emoryi,*

Koeberlinia spinosa, and *Dalea spinosa*) essentially lack functional leaves but they have rigid, thorny photosynthesizing branches (Jaeger, 1955). In Israel, the young green branches of *Retama raetam* and *Calligonum comosum* also synthesize carbohydrates (Fahn, 1964). Similarly in the southwestern desert of the United States the palo verde (*Cercidium floridum*) is leafless during most of the year but its green branches are very efficient in carrying on photosynthesis (Adams *et al.*, 1967). Adams and Strain (1969) showed that the stems of *Cercidium* contributed more than 40% of the total photosynthesis of leafy trees. After a severe drought, plants without leaves exhibited low photosynthetic capacity, but their rate of CO_2 uptake increased after autumn and winter rains. In fact, by midwinter the rate of CO_2 uptake by a leafless plant was 86% of that in a plant with leaves growing in an area where it had not been subjected to extreme drought.

C. Control of Insects and Diseases

Natural or induced shedding of plant parts may either increase or decrease the incidence of disease and insect attack. A general response to premature defoliation of trees is loss of vigor which often is followed by increased susceptibility to attack by secondary insects and by fungi which cause diseases (Houston and Kuntz, 1964; Kozlowski, 1969; Kulman, 1971). However, many cases have been documented in which natural or induced shedding of senescent, diseased, or injured plant parts which might become diseased has prevented the spread of infections. For example, the spread of South American leaf blight caused by *Dothidella ulsi*, can be checked in dense stands of *Hevea brasiliensis* by inducing rapid defoliation with herbicides at a critical stage of development of the disease near the center of the outbreak of the disease. Inducing rapid defoliation of trees with herbicides or by stem girdling has been shown to control tsetse flies (*Glossina* spp.) which thrive only in shady habitats (Osborne, 1968). Killing of leaves of potato vines with chemical desiccants prevents infection by late blight and prevents spread of virus diseases (Addicott and Lynch, 1957).

D. Control of Species Composition

Induced defoliation with herbicides has been practiced successfully for many years to control species composition of forest stands. For example, in mixed forests hormone-type herbicides are used selectively to foster growth of pines by inducing defoliation and thereby eliminating competition of brush and broad-leaved trees of low economic value. This practice has been particularly useful in the southern United States where, over a 20-year period, angiosperm trees replaced pines on some 13 million acres of land (Hayward, 1957). "Conifer release" is most often accomplished with aerial foliage applications of low volatile esters of 2,4-D,[1] 2,4,5-T,[1] or mixtures of these (Kozlowski, 1960; Arend and Roe, 1961). When skillfully applied at proper dosages, the broad-leaved trees in a mixed stand are defoliated and often killed, whereas the pines survive and subsequently grow vigorously. To release pines by herbicide defoliation a complete kill of all competing broad-leaved trees is not necessary or even desirable. The overall objective of the practice is usually to temporarily suppress competition so as to provide the pines with sufficient opportunity to overtop the broad-leaved trees. Ordinarily the light intensity requirements of competing broad-leaved trees are lower than those of the pines for maximum physiological activity and growth. Hence, under natural conditions growth of pines in mixed stands is checked, or more often the pines are eliminated, by the more shade-tolerant angiosperms (Kozlowski, 1949; Kramer and Kozlowski, 1960).

The successful use of defoliating herbicides for conifer release requires considerable skill. There is some hazard of volatility and wind drift of the herbicides, especially in farming areas. In a 5- to 10-mile/hr wind, drift of hormone-type herbicides has killed trees a quarter of a mile away.

Unfortunately, many examples of successful use of defoliating herbicides in conifer release have suggested to potential users that such chemicals are completely selective and not toxic to gymnosperms. Actually considerable damage has been caused to gymnosperms as a result of overdosage, herbicide application at the wrong time of year, or application to young gymnosperm seedlings. The selectivity of herbicides used in conifer release depends primarily on the physiological state of the trees to be released. If they are still growing vigorously they may be injured or killed by even light applications of herbicides. For example, in the north central United States, foliage sprays of 1 and 2 lb acid equivalent of 2,4-D and 2,4,5-T at 2.5 and 5 gal/acre caused injury to new growth of *Picea glauca* until July 15, *Pinus resinosa*, *P. strobus*, and *Picea abies* until August 1, and *Pinus banksiana* and *P. sylvestris* until August 15 (Arend and Roe, 1961). Hence, spraying should be limited to the latter part of the growing season when gymnosperm trees have hardened off but broad-leaved trees are still sensitive to the herbicides.

The hormone-type herbicides are particularly toxic to young gymnosperm seedlings and affect all classes of foliar appendages, including

[1] 2,4-Dichlorophenoxyacetic acid and 2,4,5-Trichlorophenoxyacetic acid, respectively.

cotyledons, primary needles, and secondary needles. Seedling responses include chlorosis; curling, shrivelling, or fusion of cotyledons; distortion and growth inhibition of foliar appendages; needle dieback; and death of plants (Sasaki and Kozlowski, 1968). Even if seedlings are not killed, the inhibition of growth of cotyledons and primary needles by herbicides lowers the growth rate and planting quality of gymnosperm seedlings (Kozlowski and Sasaki, 1968, 1970; Sasaki and Kozlowski, 1968, 1969, 1970; Kozlowski and Borger, 1971; Wu *et al.*, 1971; Wu and Kozlowski, 1972).

E. Harvesting of Crops

It often is useful or imperative to stimulate premature abscission or death of foliage by hand, animal defoliation or, much more commonly, by chemical defoliation or desiccation. In this chapter a distinction is made between chemical defoliation (in which physiological changes leading to abscission are accelerated) and chemical desiccation (in which the plant tissues are passive and separation does not occur at a discrete abscission layer). Desiccants appear to kill or denature protoplasm and destroy selective permeability of cell membranes, thereby allowing rapid water loss and desiccation. In desiccation, tissues at some distance from the point of application are not affected by the desiccant.

Induced premature defoliation and desiccation have played important roles in facilitating harvesting of certain crops; permitting accurate scheduling of harvests; curing of hay; swathing of grains; shocking of corn; and preparation of nursery stock for digging, shipping, and storage (Addicott and Lynch, 1957; Addicott, 1965).

Chemical harvest aids are widely used in mechanical harvesting of such crops as cotton, ramie, and hops. For example, chemicals are used in harvesting more than three-fourths of the cotton acreage in the United States (Walhood and Addicott, 1968). When cotton is harvested by machine, a dense leaf canopy prevents good spindle contact with the bolls. The moist leaves also cause lint staining and are a source of trash. For these reasons defoliation of cotton is necessary. It also helps to prevent lodging, to increase light penetration and rapid opening of mature bolls, to allow drying of the early opening bottom bolls so as to prevent boll rot, and control insects, especially when it can reduce overwintering populations of the boll weevil and pink bollworm. In defoliated plants, the moisture content of cotton on the stalk decreases to a safe picking level sooner in the day so harvesting can begin earlier and last longer.

The successful use of defoliating chemicals depends on such factors as variety, maturity (mature leaves defoliate easily, juvenile leaves poorly or not at all), leaf surface condition, thorough coverage of the leaves with the defoliating chemical, and climate (high temperatures generally are favorable, wind is unfavorable). The major adverse effect of premature chemical defoliation of cotton is immature fibers. For this reason chemical defoliants generally are not applied until one-half to three-fourths of the bolls have opened.

Many chemicals with defoliating properties have been tested but most of these have not been approved for field use because of high cost or inadequate performance. The approved list of chemical defoliants for cotton now includes only a few dusts and sprays. The dusts provide better dispersal than the sprays into dense cotton foliage. However, dusts require the presence of dew on leaves and therefore are impractical in arid regions such as the high plains of Texas or desert valleys of California. Approved dust defoliants include sodium chlorate, tributyl phosphorotrithioate, and tributyl phosphorotrithiorite. Among approved spray defoliants are ammonium nitrate, anhydrous ammonia, sodium chlorate, tributyl phosphorotrithioate and tributyl phosphorotrithioite. Formulations, dosages, and guides for application are given by Walhood and Addicott (1968) and Hoover and Walhood (1969).

Defoliants applied at correct dosages appear to be nonspecific and have in common the capacity to injure leaves with little or no injury to the abscission zone. The injury to leaves sets in motion the sequential physiological events (see Chapter 4) which culminate in cytolysis and separation at the abscission layer. Excessive application of defoliants often kills cells of the abscission zone and, as does heavy frost, may actually prevent abscission.

The principal objective of applying chemical desiccants is to accelerate dehydration (curing) of leaves. Sometimes the term "spray-curing" is used for this practice. Usually all aerial tissues, except seeds, are killed by desiccants in a manner similar to that following application of contact herbicides. The effects of chemical desiccants and defoliants overlap somewhat since high dosages of defoliants may cause some desiccation and low dosages of desiccants may induce defoliation. However, the response also varies with species. For example, leaves of cereal grains do not abscise and only a few leaves abscise in tomato and potato plants. Among approved desiccants in common use on cotton are ammonium nitrate, arsenic acid, and paraquat (1,1-dimethyl-4,4,4-bipyridinium) (Hoover and Walhood, 1969).

As with chemical defoliation, the effects of applying chemical desiccants are variable. Use of chemical desiccants is an accepted practice with only few crops and is valuable only under limited conditions for various reasons. Chemical desiccation has been found particularly advantageous with: moisture-stressed cotton of the high plains of Texas and Oklahoma, rice, small seeded legumes (including alfalfa, clover, birdsfoot trefoil), potato, and seed grasses (Addicott and Lynch, 1957). Removal of tomato leaves with sodium chlorate has only slight effect on picking time, yield, and rate of fruit ripening. Most of the differences due to treatment are unimportant (MacGillivray, 1960).

1. Defoliation of Nursery Stock

Many young woody ornamental plants and fruit trees retain their leaves beyond the growing season. This poses a serious problem in storage or shipping. Leaves must often be removed, particularly to avoid losses from fungus diseases. Nurserymen have used several methods of defoliation such as hand or mechanical stripping of leaves. Since these methods are slow, expensive, and often injurious to plants, much interest has been shown in chemical defoliation of nursery stock in the field (Tischler et al., 1950; Pridham, 1951, 1952). Chadwick and Houston (1948) tested many defoliating chemicals and found sprays of Naccanol NR (sodium salts of alkylated benzene sulfonic acid) satisfactory for defoliating several kinds of nursery stock. Success varied with species as well as with the chemical and dosage used. The earlier the chemical defoliant was applied in the autumn, the higher was the dosage required for satisfactory defoliation. Pridham (1952) found that Naccanol NR and Endothal (disodium 3,6-endohexahydrophthalate) caused rapid and satisfactory defoliation of ornamental nursery stock in the field. These chemicals were more effective than cyanamide dust, ammonium thiocyanate solution, or potassium thiocyanate solution. Kofranek and Leiser (1958) tested several chemicals and found that Vapam (sodium methyl dithiocarbamate) and Folex (tributyl phosphorotrithioite), in particular, were effective defoliants.

2. Chemical Debarking

In addition to mechanical debarking of harvested trees, chemical debarking of standing trees has been practiced, especially with pulpwood species (Wilcox, 1954a,b; Wilcox et al., 1956; Myers and Braga, 1969). According to Cook (1954), chemical debarking can provide inexpensive, efficient, and easy removal of bark from most species of northern hardwoods in the United States except ash (Fraxinus) and birch (Betula). Hence, production costs can be decreased by saving space during transportation of pulpwood and by transporting drier and lighter wood (McIntosh, 1951). Woodfin (1963) found that chemically debarked *Pinus ponderosa* trees weighed a third less than green untreated wood. Chemical debarking also decreases possibility of decay and insect attack in storage areas.

Many chemicals including, among others, 2,4-D, 2,4,5-T, Ammate, sodium arsenite, ammonium fluoride, and copper chloride have been used in chemical debarking. The best effects are obtained with a chemical that has a fast toxic action. All arsenic compounds are highly effective provided they are not diluted below 10% concentration of arsenic. Sodium arsenite has been used with most success (Wilcox *et al.*, 1956).

The mechanism of chemical debarking involves killing of tissues of cambial cells and their derivatives, preferably while they are still thin-walled. This is followed by cellular collapse, causing spaces to form between the bark and wood. Freezing and thawing during the winter cause further separation. Following chemical treatment, partially collapsed immature vessels and differentiating phloem elements often are seen.

Among factors influencing chemical debarking are species, season of the year, tree vigor, composition of the chemical used, available moisture, wetting agents, and time of standing after treatment before felling and peeling (Huber, 1948; Wilcox *et al.*, 1956). Some idea of species variation in debarking response to sodium arsenite may be gained from Tables III and IV. Vigorous, dominant trees with very active cambial zones have a longer peeling season than intermediate or suppressed trees.

TABLE III

PEELING CLASS OF ARSENIC-TREATED TREES BY TREATING SEASON[a]

	Presap peeling[b]	In sap-peeling season			Postsap peeling	Mean
		Early	Mid	Late		
Gymnosperms	74.9[c]	43.5	49.5	52.8	57.3	54.1
Angiosperms	74.1	55.2	51.1	53.7	74.2	55.2
All species	74.7	52.0	51.1	53.4	66.1	54.8

[a] From Wilcox *et al.* (1956).

[b] The data are based on a resistance to peeling scale based on the following: 0, Bark off on the standing tree; 1, Bark falls off in felling the tree; 2, Bark comes off in skidding the tree; 3, Bark comes free without pressure when using peeling spud; 4, Better than sap peeling; 5, Average sap peeling condition; 6, Harder than sap peeling; 7, Difficult to peel; 8, Bark must be scraped from wood; and 9, Will not peel.

[c] The values in the table are computed on the average of the peeling figure for all bolts of a tree multiplied by ten.

TABLE IV
PEELING CLASSES OF DIFFERENT SPECIES[a,b]

Species	Peeling class	No. of trees
Quercus rubra	30.8	6
Ulmus americana	34.5	10
Tsuga canadensis	34.9	230
Populus tremuloides, P. grandidentata	37.9	196
Liriodendron tulipifera	38.0	10
Fagus sylvatica	41.6	136
Picea rubens	48.0	213
Acer saccharum	50.6	214
Betula alleghaniensis	51.8	231
Acer rubrum	52.2	273
Abies balsamea	58.0	303
Betula papyrifera	70.2	212
Fraxinus sp.	75.1	44
		2078

[a] From Wilcox et al. (1956).
[b] Values are four-year averages for treatment with sodium arsenite at midpeeling season.

Chemical treatment varies greatly with season of application (Table III). Successful debarking of *Pinus ponderosa* with 40% sodium arsenite is possible throughout the year, except when wood and bark are frozen, provided the interval between treatment and peeling is extended for at least one growing season. For example, Woodfin (1963) found that trees treated between April and early July were successfully debarked in 4 to 6 months; winter-treated (September to February) trees were successfully debarked after a 7- to 12-month interval.

F. DEFOLIATION FOR MILITARY PURPOSES

In the tropics visibility is obscured by high tree density as well as by the presence of vines, epiphytes, buttressed roots, and fallen trees. The profusion of vegetation inhibits troop movements and visibility in horizontal, oblique, and vertical planes. For these reasons herbicides have been used in military operations to defoliate vegetation with the view of permitting easier detection of personnel and equipment and reducing possibility of ambush.

There has been lively controversy about the ecological importance of herbicide defoliation in moist tropical forests (Tschirley, 1969; Westing,

1971a), and a growing body of evidence indicates that serious environmental deterioration may be the result. For example, fear has been expressed that the adverse effects of defoliation may include invasion of economically undesirable plants, loss of nutrient capital, laterization of soils, erosion, flooding, killing of soil microorganisms, and elimination of some animals by affecting susceptible links in important food chains (Galston, 1967, 1971; Orians and Pfeiffer, 1970; Whiteside, 1970; Westing, 1971a,b).

The specific effects of herbicide defoliation on species composition and regeneration in the moist tropics vary somewhat as shown by their effects on dense upland forests and on mangrove swamps in Vietnam. The former consist of several strata, up to 130 ft high, and a multitude of species of dicotyledonous trees, lianas, herbs, monocotyledons, and ferns. The dominant family is the Dipterocarpaceae but members of Lythraceae, Leguminosae, Guttiferae, and Meliaceae also are important. By comparison, the mangrove forests are simple and consist of relatively few species of small trees, 10 to 50 ft high. In Vietnam, *Avicennia marina* is the pioneer species of the mangrove type. Within 5 to 6 years, species of *Rhizophora, Bruguiera,* and *Cerciops* may be found and at 20 years *Rhizophora* and *Bruguiera* predominate.

The species composition of moist tropical forests is materially altered by induced defoliation with herbicides. In South Vietnam, for example, repeated defoliation of high forest with phenoxy herbicides caused replacement of forest trees with the grass, *Imperata cylindrica,* and bamboos (*Bambusa* sp.). Once bamboos become established they remain for a long time. Their predecessors, the dicotyledonous forest trees, generally can reinvade the area only when bamboos flower and die. Such reinvasion may not occur until many years after herbicide treatment, since bamboo flowers only at very long intervals. For example, bamboo flowers in from 7 to 13 years in southern Brazil and up to 30 years in other regions (Alvim, 1964; Walter, 1962). Recovery of forest lands is rendered difficult not only because the invading bamboo and grasses are difficult to eradicate, but also because of loss of nutrients.

Mangrove forests are particularly sensitive to phenoxy herbicides. Most of the relatively few species which comprise the mangrove forest are killed by a single herbicide spraying (*Nipa fruticans* is an exception). According to Westing (1971b), defoliation appears to preclude recolonization by any new plant community, mangrove or otherwise, for at least 6 years and probably many more. Ecologists are particularly concerned about loss of mangrove forests because they are a vital link in the chain of requirements for completion of the life cycle of a variety of aquatic animals, including shellfish and migratory fish. The mangrove swamps provide abundant nutrients as well as breeding and nursery grounds for aquatic animals. Furthermore, mangrove forests play an important role in prevention of erosion and stabilization of shorelines.

Moist tropical forests are closed and efficient but fragile ecosystems. In contrast to temperate forests, the bulk of the cycling nutrient capital is in the plants themselves, particularly in leaves and twigs. Since leaf abscission in undisturbed moist tropical forests occurs throughout the year, the litter decomposes rapidly and nutrients are reabsorbed by plants. However, if massive defoliation is induced by herbicides, the litter will decompose rapidly but reabsorption of nutrients by the physiologically inactive forest trees will be severely curtailed. Hence, the nutrients will be lost by leaching and runoff during heavy tropical rains (Westing, 1971a). Such nutrient losses are well known after clearing of moist tropical forests (Nye and Greenland, 1960).

G. Growth of Plants

Premature defoliation inhibits both vegetative and reproductive growth, with the effects varying with such factors as severity of defoliation, the defoliating agency, time of defoliation, and physiological preconditioning of plants. Defoliation may cause bud mortality and shoot dieback; inhibition of shoot growth, cambial growth, root growth, and reproductive growth (Kozlowski, 1971a,b; Kulman, 1971); epicormic sprouting (Houston and Kuntz, 1964); and plant mortality (Kulman, 1971). Furthermore, defoliation causes increased susceptibility to secondary insects and diseases.

Late season defoliation of woody plants generally does not have obvious inhibitory effects on growth of many species during the current year. Nevertheless, after seasonal meristematic activity stops in above-ground tissues defoliation often decreases current year root growth, which continues for a longer time than shoot growth, and is likely to have important inhibitory effects on shoot growth in the subsequent year (Kozlowski, 1964a, 1968a,b, 1971a; Kozlowski and Keller, 1966). Current year root growth of apple trees was influenced by photosynthetic products of foliage after annual shoot growth ceased (Priestley, 1964).

As long as fully expanded leaves remain alive and healthy, they are capable of contributing to the reserve carbohydrate pool. In general, accumulation of carbohydrates occurs during the latter part of the growing season when meristematic activity declines. The importance of food reserves in plants is shown by their mobilization during growth. For example, Krueger (1967) demonstrated that carbohydrates stored by 1-year-old *Pseudotsuga menziesii* shoots were mobilized rapidly from April to early June by expanding shoots. Direct evidence of the use of prior year photosynthate in shoot growth is available for both angiosperms and gymnosperms. Quinlan (1969) exposed leaves of apple trees to $^{14}CO_2$ in the autumn (October) after seasonal shoot elongation had stopped. In the following spring (late May) ^{14}C was detected in new leaves and expanding internodes, indicating mobilization of stored carbohydrates. When old needles of *Pinus resinosa* were exposed to $^{14}CO_2$ late in the growing season, some of the ^{14}C was fixed in reserves. During the next growing season a portion of the ^{14}C was mobilized and used in shoot expansion (Schier, 1970; Gordon and Larson, 1970). In growth of gymnosperm shoots, reserve carbohydrates may be mobilized from old needles as well as twigs and stems. For example, Kozlowski and Clausen (1965) and Clausen and Kozlowski (1967) recorded a significant decrease in dry weight of old needles of *Pinus resinosa* as shoots elongated in the growing season, indicating mobilization of reserves. Schier (1970) found that 2-year-old needles of *Pinus resinosa* contributed about 85% of autumn [^{14}C] photosynthate to growth of various tissues in the following spring, and Splittstoesser and Meyer (1971) noted that reserve carbohydrates in old needles of *Taxus media* contributed importantly to shoot expansion in the spring. These observations emphasize that late season defoliation may ultimately have important inhibitory effects on growth but these are not obvious for a long time after the occurrence of defoliation.

There is also evidence that late season and winter photosynthesis in gymnosperms has important effects on growth even before the next growing season. Winget and Kozlowski (1965b) found that cambial growth of the evergreen *Tsuga canadensis* continued later into the autumn than was the case with several species of deciduous trees in the northern conifer–hardwood subformation of Wisconsin. Sweet and Wareing (1968) compared seasonal rates of growth and dry matter production in *Larix leptolepis, Pinus contorta,* and *P. radiata* seedlings in Wales. The deciduous *Larix* had a higher rate of dry weight increment than either species of *Pinus* until the time of leaf abscission, and this was accompanied by greater height and diameter increment. However, between the time of leaf fall in *Larix* and the end of the growing season, the species of *Pinus* increased in dry weight by more than 25% and consequently the deciduous *Larix* lost much of the advantage of its earlier rapid growth rate. Pollard and Wareing (1968) also demonstrated considerable winter growth of gymnosperm seedlings in Wales during the winter months. *Pinus sylvestris* increased in dry weight, primarily in the root and stem. *Pinus radiata* showed an even greater winter increase in dry weight, with most of the increment in the foliage as growth of existing needles and production of new ones occurred.

Several investigators have shown that late season defoliation has an inhibitory effect on the subsequent year's shoot growth by inhibiting bud development. For example, in pines of the northern United States (e.g., *Pinus resinosa, P. strobus*) and in many deciduous trees, shoot growth is a two-season process involving bud formation during the first year and expansion of the bud into a shoot during the next year. In such species the shoots are preformed in the bud and the number of anatomical stem units present in the bud is a prime determinator of ultimate shoot length (Lanner, 1968; van den Berg and Lanner, 1971). A large bud invariably produces a long shoot, and a small bud a short one (Kozlowski, 1971a). Positive correlations have been found between size of the unopened bud and shoot growth of *Picea sitchensis* (Burley, 1966), *Pinus sylvestris* (Szymanski and Szczerbinski, 1955), *Pinus resinosa* (Clements, 1970; Kozlowski *et al.*, 1973), and *Populus tremuloides* (Maini, 1966a,b). Bud size is a good index of the number of cataphylls in *Pinus resinosa* (Lanner, 1968) and number of enclosed needle primordia in *Picea sitchensis* (Burley, 1966). That late season photosynthate has marked effects on the subsequent year's shoot growth is also shown by closer correlation between shoot growth and weather of the year of bud formation than with weather during the year of expansion of the bud into a shoot (Kozlowski, 1964b, 1968a,b, 1971a; Duff and Nolan, 1953; Clements, 1970).

Because of general quantitative correlations between the degree of defoliation by forest pests and amount of growth, there often has been a tendency to attribute growth reduction following defoliation to a deficiency of carbohydrate alone. It is well known that defoliation during the growing season reduces carbohydrate synthesis and leads to a decrease in carbohydrate reserves in plants (Parker and Houston, 1971). Nevertheless, the inhibition of cambial growth following defoliation appears to be considerably more complicated than a mere starvation response (Kozlowski, 1969, 1971a). Accompanying reduction in carbohydrate supplies by defoliation is a reduction in synthesis and basipetal transport of hormonal growth regulators. Evert and Kozlowski (1967) and Evert *et al.* (1972) showed that imposition of phloem blocks by stem girdling rapidly and drastically inhibited production of xylem and phloem in *Populus tremuloides* and *Acer saccharum* trees. The inhibitory effects did not appear to be caused directly by availability of food because axial and ray parenchyma contained abundant starch below the level of the severed phloem. Rather cambial activity appeared to be checked by regulatory influences on food utilization. Such observations place considerable emphasis on the importance of a continuous supply of basipetally translocated hormonal growth regulators for cambial growth. Normal cambial growth appears to be the result of balances among several hormonal growth regulators, including growth promoters and inhibitors, and synergisms between them (Kozlowski, 1971b).

Variations in Response to Defoliation

Removal of approximately the same amount of foliage by various agencies may result in appreciably different amounts of growth inhibition. For example, defoliation by branch pruning of forest trees often inhibits cambial growth less than does removal of an equivalent amount of foliage by certain insects. This is because pruning generally removes only the physiologically inefficient lower branches whereas some insects remove foliage in the upper crown which is much more important in synthesizing carbohydrates and hormonal growth regulators. For example, defoliation by spruce budworm is localized in the upper crown (Kulman, 1971). Furthermore, some insects preferentially defoliate trees that are lacking in vigor and have restricted carbohydrate reserves. In a general way, mortality and growth reduction by defoliation of trees are proportional to the amount of foliage removed (Church, 1949). This has been confirmed by many studies of defoliation by various insects and by artificial defoliation. A few examples will be given.

In a study by McClintock (1955), heavy, medium, and light defoliation of *Abies balsamea* in Quebec by the spruce budworm was followed by 89, 62, and 0% mortality in the 4 years following insect attacks. Kinghorn (1954) reported that heavy defoliation (80–90%) by the hemlock looper (*Lambdina fiscellaria lugubrosa*) in British Columbia resulted in 65–78% mortality of gymnosperms whereas lighter defoliation (50–75%) caused mortality of 10–25%. Removal by clipping 30, 60, and 90% of the needle length of *Pinus palustris* seedlings in Mississippi was followed by a decrease of 24, 34, and 51% in height growth in a study by Bruce (1956). In New England defoliation of 4 species of *Quercus* of 21–40, 41–60, 61–80, or 81–100% by the gypsy moth (*Porthetria dispar*) was followed by decreases in cambial growth of 9, 20, 19, and 30%, respectively, as reported by Baker (1941). Similar amounts of defoliation reduced cambial growth of *Pinus strobus* by 0, 15, 27, and 13%. In Michigan Dils and Day (1950) found that 24 or 100% defoliation of *Populus tremuloides* trees by the forest tent caterpillar (*Malacasoma disstria*) reduced cambial increment by 38 and 67%, respectively.

Marked differences occur among species of trees in response to insect defoliation. Angiosperms and deciduous gymnosperms usually survive a single severe defoliation, probably because their destroyed foliage is rapidly replaced by expansion of dormant buds into shoots (Kozlowski, 1971a). Two severe defoliations in the same year may kill angiosperms but many angiosperms can survive a single annual defoliation repeated for several years. Evergreen gymnosperms often are killed by one defoliation, especially if it occurs late in the growing season and includes removal of new foliage. However, evergreen gymnosperms of the temperate zone generally are able to survive one complete removal of old needles prior to beginning of the growing season (Kulman, 1971).

Growth of both angiosperms and pines is rapidly reduced by insect defoliation which occurs before completion of annual shoot elongation or cambial growth. However, recurrently flushing pines (e.g., *Pinus taeda*, *P. elliottii*) withstand severe defoliation much better than pines which usually produce one annual flush of shoot growth (e.g., *P. resinosa*) (Kozlowski, 1971a). In contrast to responses of pines, growth reduction in spruces (*Picea*) and firs (*Abies*) in response to insect defoliation may be considerably delayed, sometimes for a few years. For example, during a 4-year period of defoliation of *Abies balsamea* by spruce budworm, approximately 50, 80, 40, and 20% of the foliage was removed. In the third year, growth was decreased by 25% and by 37, 45, and 55% in subsequent years, according to Belyea (1952).

The effect of defoliation on shoot growth of gymnosperms may be expected to vary with the time of defoliation and the age class of needles removed. Although shoot growth utilizes both reserve and currently produced carbohydrates it is clear that early season defoliation greatly inhibits current year shoot elongation since reserves in other parts of trees are inadequate. By complete defoliation, girdling of branches, girdling of the main stem above the ground line, and various combinations of these during the dormant season, Kozlowski and Winget (1964) evaluated the importance of stem and branch reserves and of currently produced growth requirements from old needles to shoot growth of 8-year-old *Pinus resinosa* trees. Reserves in tissues other than leaves were relatively unimportant for shoot growth. In contrast, early defoliation inhibited expansion of the terminal leader and weakened apical dominance relations (Table V). The old needles contributed about four-fifths or more to total shoot growth, with the exact contribution depending on whether dry weight of shoots or shoot elongation was used as the measure of growth. It was estimated that phloem-translocated reserves from the branches, main stem, and roots contributed only about 6, 5, and 2%, respectively, to total shoot growth.

To illustrate the variable responses of different species of pines to late season defoliation it is important to recognize that pines vary greatly in seasonal duration of shoot growth (Kozlowski and Ward, 1961; Kozlowski,

TABLE V

EFFECTS OF VARIOUS TREATMENTS ON DRY WEIGHT OF 1963 SHOOTS, DRY WEIGHT AND LENGTH OF 1963 NEEDLES, AND BUD OPENING[a]

Treatment	Contributing parts[b]	Dry wt. of 1963 shoots		Dry wt. of 1963 needles		Average length of 1963 needles		Average number of unopened buds
		In gm	As % of control	In gm	As % of control	In cm	As % of control	
Needles removed	B + S + R	7.26 ± 0.72	12.42	4.99 ± 0.67	12.52	1.15 ± 0.09	34.64	9.75 ± 2.53
Branch girdled, needles removed	B	3.37 ± 2.91	5.76	0.88 ± 0.24	2.21	0.55 ± 0.09	16.57	30.40 ± 3.02
Base girdled, needles removed	B + S	6.20 ± 0.57	10.61	3.52 ± 0.46	8.83	0.75 ± 0.07	22.59	12.85 ± 2.21
Branch girdled	N + B	42.31 ± 4.15	72.43	27.35 ± 2.83	68.62	2.90 ± 0.10	87.35	6.15 ± 1.01
Base girdled	N + B + S	50.46 ± 5.50	86.37	32.31 ± 3.75	81.06	2.52 ± 0.11	75.90	0.0
Control	N + B + S + R	58.42 ± 5.14	100.00	39.86 ± 3.51	100.00	3.32 ± 0.09	100.00	0.20 ± 0.16

[a] From Kozlowski and Winget (1964).
[b] B = Branches, S = stems, R = roots, N = needles.

1964b). In some pines (e.g., *Pinus resinosa, P. strobus, P. lambertiana*) shoot formation involves differentiation of shoot primordia in the bud during one year and expansion of the preformed parts into a shoot during a relatively short part of the frost-free season during the next year. Other pines (e.g., *Pinus taeda, P. echinata, P. palustris, P. radiata*) have a recurrently flushing pattern of shoot growth. Shoot growth of these species involves the expansion of one to several successively formed terminal buds on the same shoot in one growing season. The winter bud expands into a shoot in the initial seasonal growth flush. Then a new bud forms at the apex of the same shoot and shortly thereafter expands into a new shoot. This process of recurrent growth may be repeated several times during the same growing season (Kramer, 1943; Kozlowski, 1964b, 1971a). Most shoots of recurrently flushing species sequentially form 2 or 3 buds which expand in the same growing season, but some shoots may form and expand more. In one year as many as 7 buds formed and expanded on the terminal leader of recurrently flushing pines (Wakeley and Marrero, 1958).

These marked differences in duration of shoot growth of gymnosperms emphasize that a late season defoliation (during or after July) could not affect current year shoot growth of species such as *P. resinosa* since its shoots expand only during the very early part of the frost-free season. It should be pointed out, however, that such late season defoliation of this species would restrict the size of the bud formed and have an inhibitory effect on shoots formed the following year (Kozlowski, 1971a). By comparison, defoliation in July or August will restrict current year shoot elongation of recurrently flushing species which continue to expand their shoots late into the summer. It will also restrict shoot elongation of continuously growing tropical species such as "foxtail" pines (Kozlowski and Greathouse, 1971).

In a species such as *P. resinosa* the effect of defoliation on current year shoot growth will depend on the age class of needles removed and, as mentioned, the time of year when each class is removed. In *P. resinosa* which usually bears 3 age classes of old needles, the 1-year-old needles are the most important contributors to early season shoot growth. For example, Dickmann and Kozlowski (1968) found that 1-year-old needles were supplying more [^{14}C]photosynthate to expanding shoots early in the growing season than were the 2- and 3-year-old needles. The current year needles were not major exporters of carbohydrates until they were fully expanded late in the summer. These variations in seasonal carbohydrate sources emphasize the relative importance at different times of several age classes of foliage of gymnosperms. Obviously defoliation of an age class of foliage at a time at which it is exporting carbohydrates and hormones will be more critical than removal of the same foliage at a time when it is no longer contributing appreciable amounts of substances required to sustain growth.

References

Adams, M. S., and Strain, B. R. (1969). Seasonal photosynthetic rates in stems of *Cercidium floridum* Benth. *Photosynthetica* **3**, 55–62.

Adams, M. S., Strain, B. R., and Ting, I. P. (1967). Photosynthesis in chlorophyllous stem tissue and leaves of *Cercidium floridum*: Accumulation and distribution of ^{14}C from $^{14}CO_2$. *Plant Physiol.* **42**, 1797–1799.

Addicott, F. T. (1954). Abscission and plant regulators. *In* "Plant Regulators in Agriculture" (H. B. Tukey, ed.), Chapter 7, pp. 99–116. Wiley, New York.

Addicott, F. T. (1965). Physiology of abscission. *In* "Handbuch der Pflanzenphysiologie" (W. Ruhland, ed.), Vol. 15, Part 2, pp. 1094–1126. Springer-Verlag, Berlin and New York.

Addicott, F. T. (1970). Abscission. *McGraw-Hill Encycl. Sci. Technol.* **1**, 12–13.

Addicott, F. T., and Lynch, R. S. (1955). Physiology of abscission. *Annu. Rev. Plant Physiol.* **6**, 211–238.

Addicott, F. T., and Lynch, R. S. (1957). Defoliation and desiccation: Harvest aid practices. *Advan. Agron.* **9**, 67–93.

Alvim, P. de T. (1964). Tree growth periodicity in tropical climates. *In* "The Formation of Wood in Forest Trees" (M. H. Zimmermann, ed.), pp. 479–496. Academic Press, New York.

Arend, J. L., and Roe, E. I. (1961). Releasing conifers in the Lake States with chemicals. *U. S., Dept. Agr., Forest Serv., Agr. Handb.* **185**.

Avery, D. J. (1969). Comparisons of fruiting and deblossomed maiden apple trees on a dwarfing and an invigorating rootstock. *New Phytol.* **68**, 323–336.

Baker, F. S. (1950). "Principles of Silviculture." McGraw-Hill, New York.

Baker, W. L. (1941). Effect of gypsy moth defoliation on certain trees. *J. Forest.* **39**, 1017–1022.

Bartlett, J. B., Ruble, W. W., and Thomas, W. P. (1939). The influence of hydrogen peroxide treatments on the exchange capacity of Maryland soils. *Soil Sci.* **44**, 123–138.

Beard, J. S. (1946). The natural vegetation of Trinidad. *Oxford Forest. Mem.* **20**.

Belyea, R. M. (1952). Death and deterioration of balsam fir weakened by spruce budworm defoliation in Ontario. *J. Forest.* **50**, 729–738.

Berkley, E. E. (1931). Marcescent leaves of certain species of *Quercus. Bot. Gaz. (Chicago)* **92**, 85–93.

Bode, H. R. (1959). Über den Zusammenhang zwischen Blattenfaltung und Neubildung der Saugwurzeln bei *Juglans. Ber. Deut. Bot. Ges.* **72**, 93–98.

Bosse, G. (1960). Die Wurzelentwicklung von Apfelkonen und Apfelsämlingen während der ersten drei Standjahre. *ErwObstbl.* **2**, 26–30.

Bower, C. A. (1949). Studies on the form and availability of soil organic phosphorus. *Iowa, Agr. Exp. Sta., Res. Bull.* **362**, 963.

Bray, J. R., and Gorham, E. (1964). Litter production in forests of the world. *Advan. Ecol. Res.* **2**, 101–157.

Broadbent, F. E. (1953). The soil organic fraction. *Advan. Agron.* **5**, 153–183.

Bruce, D. (1956). Effect of defoliation on growth of longleaf pine seedlings. *Forest Sci.* **2**, 31–35.

Burley, J. (1966). Genetic variation in seedling development of Sitka spruce, *Picea sitchensis* (Bong.) Carr. *Forestry* **39**, 68–94.

Chadwick, L. C., and Houston, R. (1948). A preliminary report on the prestorage defoliation of some trees and shrubs. *Proc. Amer. Soc. Hort. Sci.* **51**, 659–667.

Childers, N. F., and White, D. G. (1942). Influence of submersion of the roots on transpiration, apparent photosynthesis, and respiration of young apple trees. *Plant Physiol.* **17**, 603–618.

Church, T. W. (1949). Effects of defoliation on growth of certain conifers. *U. S., Forest Serv., Northeast. Forest Exp. Sta., Pap.* **22**, 1–12.

Clausen, J. J., and Kozlowski, T. T. (1967). Food sources for growth of *Pinus resinosa* shoots. *Advan. Front. Plant Sci.* **18**, 23–32.

Clements, J. R. (1970). Shoot responses of young red pine to watering applied over two seasons. *Can. J. Bot.* **48**, 75–80.

Cook, D. B. (1954). Chemi-peeling hardwood. *Pap. Trade J.* **138**, 54.

Davis, L. D. (1957). Flowering and alternate bearing. *Proc. Amer. Soc. Hort. Sci.* **70**, 545–556.

Del Moral, R., and Muller, C. H. (1970). The allelopathic effects of *Eucalyptus camaldulensis. Amer. Midl. Natur.* **83**, 254–282.

Dickmann, D. I., and Kozlowski, T. T. (1968). Mobilization by *Pinus resinosa* cones and shoots of C^{14}-photosynthate from needles of different ages. *Amer. J. Bot.* **55**, 900–906.

Dickmann, D. I., and Kozlowski, T. T. (1970). Mobilization and incorporation of photoassimilated ^{14}C by growing vegetative and reproductive tissues of adult *Pinus resinosa* Ait. trees. *Plant Physiol.* **45**, 284–288.

Dils, R. E., and Day, M. W. (1950). Effect of defoliation upon the growth of aspen. *Mich., Agr. Exp. Sta., Quart. Bull.* **33**, 111–113.

Dorman, K. W., and Barber, J. C. (1956). Time of flowering and seed ripening in southern pines. *U. S., Forest Serv., Southeast. Forest Exp. Sta., Pap.* **72**.

Duff, G. H., and Nolan, N. J. (1953). Growth and morphogenesis in the Canadian forest species. I. The controls of cambial and apical activity in *Pinus resinosa* Ait. *Can. J. Bot.* **31**, 471–513.

Eames, A. J., and MacDaniels, L. H. (1947). "An Introduction to Plant Anatomy." McGraw-Hill, New York.

Ensminger, L. E., and Pearson, R. W. (1950). Soil nitrogen. *Advan. Agron.* **2**, 81–111.

Erickson, L. C., and Brannaman, B. L. (1960). Abscission of reproductive structures and leaves of orange trees. *Proc. Amer. Soc. Hort. Sci.* **75**, 222–229.

Esau, K. (1965). "Plant Anatomy." Wiley, New York.

Evert, R. F., and Kozlowski, T. T. (1967). Effect of isolation of bark on cambial activity and development of xylem and phloem in trembling aspen. *Amer. J. Bot.* **54**, 1045–1055.

Evert, R. F., Kozlowski, T. T., and Davis, J. (1972). Influence of phloem blockage on cambial growth of *Acer saccharum. Amer. J. Bot.* **59**, 632–641.

Faegri, K., and Iversen, J. (1964). "Textbook of Pollen Analysis." Hafner, New York.

Fahn, A. (1964). Some anatomical adaptations of desert plants. *Phytomorphology* **14**, 93–102.

Fowells, H. A., and Schubert, G. H. (1956). Seed crops of forest trees in the pine region of California. *U. S., Dep. Agr., Tech. Bull.* **1150.**

Funke, G. L. (1943). The influence of *Artemisia absinthium* on neighboring plants. *Blumea* **5,** 281–293.

Galston, A. W. (1967). Changing the environment. Herbicides in Vietnam. II. *Scientist Citizen* p. 123.

Galston, A. W. (1971). Warfare with herbicides in Vietnam. *In* "The Patient Earth" (J. Harte and R. H. Socolow, eds.), pp. 136–150. Holt, New York.

Gordon, J. C., and Larson, P. R. (1970). Redistribution of ¹⁴C-labeled reserve food in young red pines during shoot elongation. *Forest Sci.* **16,** 14–20.

Gosz, J. R., Likens, G. E., and Bormann, F. H. (1972). Nutrient content of litter fall on the Hubbard Brook Experimental Forest, New Hampshire. *Ecology* **53,** 769–784.

Gray, R., and Bonner, J. (1948). An inhibitor of plant growth from the leaves of *Encelia farinosa. Amer. J. Bot.* **35,** 52–57.

Hawley, R. C., and Smith, D. M. (1954). "The Practice of Silviculture." Wiley, New York.

Hayward, F. (1957). Tidal wave of hardwoods. *Amer. Forests* **63,** 29–31 and 50–52.

Holttum, R. E. (1931). On periodic leaf changes and flowering of trees in Singapore *Gard. Bull.* **5,** 173–206.

Holttum, R. E. (1940). On periodic leaf changes and flowering of trees in Singapore. II. *Gard. Bull.* **11,** 119–175.

Hoover, M., and Walhood, V. T. (1969). Chemical harvest aids for cotton. *Univ. Calif., Davis, Ext. Publ.* **AX2-208.**

Hoshaw, R. W., and Guard, A. T. (1949). Abscission of marcescent leaves of *Quercus palustris* and *Q. coccinea. Bot. Gaz. (Chicago)* **110,** 587–593.

Houston, D. R., and Kuntz, J. E. (1964). Pathogens associated with maple blight. *Univ. Wis., Coll. Agr., Res. Bull.* **250,** 59–79.

Huber, B. (1948). Physiologie der Rindenschälung bei Fichte und Eichen. *Forstwiss. Centralbl.* **67,** 129–164.

Jaeger, E. C. (1955). "The California Deserts." Stanford Univ. Press, Stanford, California.

Kinghorn, J. M. (1954). The influence of stand composition on the mortality of various conifers caused by defoliation by the western hemlock looper on Vancouver Island, British Columbia. *Forest Chron.* **30,** 380–400.

Kofranek, A. M., and Leiser, A. T. (1958). Chemical defoliation of *Hydrangea macrophylla,* Ser. *Proc. Amer. Soc. Hort. Sci.* **71,** 555–562.

Kozlowski, T. T. (1949). Light and water in relation to growth and competition of Piedmont forest tree species. *Ecol. Monogr.* **19,** 207–231.

Kozlowski, T. T. (1960). Some problems in the use of herbicides in forestry. *Proc. North Central Weed Contr. Conf.* **17,** 1–10.

Kozlowski, T. T. (1964a). "Water Metabolism in Plants." Harper, New York.

Kozlowski, T. T. (1964b). Shoot growth in woody plants. *Bot. Rev.* **30,** 335–392.

Kozlowski, T. T. (1968a). Water balance in shade trees. *Proc. Int. Shade Tree Conf., 44th, 1968* pp. 29–42.

Kozlowski, T. T. (1968b). Soil water and tree growth. *In* "The Ecology of Southern Forests" (N. E. Linnartz, ed.), pp. 30–57. Louisiana State Univ. Press, Baton Rouge.

Kozlowski, T. T. (1969). Tree physiology and forest pests. *J. Forest.* **69,** 118–122.

Kozlowski, T. T. (1971a). "Growth and Development of Trees," Vol. 1. Academic Press, New York.

Kozlowski, T. T. (1971b). "Growth and Development of Trees," Vol. 2. Academic Press, New York.

Kozlowski, T. T. (1972). Physiology of water stress. In "Wildland Shrubs—Their Biology and Utilization" (C. M. McKell, J. P. Blaisdell, and J. R. Goodin, eds.), pp. 229–244. U. S. Dep. Agr., Forest Serv., Ogden, Utah.

Kozlowski, T. T., and Borger, G. A. (1971). Effect of temperature and light intensity early in ontogeny on growth of Pinus resinosa seedlings. Can. J. Forest Res. 1, 57–65.

Kozlowski, T. T., and Clausen, J. J. (1965). Changes in moisture contents and dry weights of buds and leaves of forest trees. Bot. Gaz. (Chicago) 126, 20–26.

Kozlowski, T. T., and Clausen, J. J. (1966). Shoot growth characteristics of heterophyllous woody plants. Can. J. Bot. 44, 827–843.

Kozlowski, T. T., and Greathouse, T. E. (1971). Shoot growth characteristics of tropical pines. Unasylva 24, 2–9.

Kozlowski, T. T., and Keller, T. (1966). Food relations of woody plants. Bot. Rev. 32, 293–382.

Kozlowski, T. T., and Sasaki, S. (1968). Germination and morphology of red pine seeds and seedlings in contact with EPTC, CDEC, CDAA, 2,4-D, and picloram. Proc. Amer. Soc. Hort. Sci. 93, 655–662.

Kozlowski, T. T., and Sasaki, S. (1970). Effects of herbicides on seed germination and development of young pine seedlings. Proc. Int. Symp. Seed Physiol. Woody Plants, 1968 pp. 19–24.

Kozlowski, T. T., and Ward, R. C. (1961). Shoot elongation characteristics of forest trees. Forest Sci. 7, 357–360.

Kozlowski, T. T., and Winget, C. H. (1964). The role of reserves in leaves, branches, stems, and roots on shoot growth of red pine. Amer. J. Bot. 51, 522–529.

Kozlowski, T. T., Torrie, J. H., and Marshall, P. E. (1973). Predictability of shoot growth from bud size in Pinus resinosa. Can. J. Forest Res. 3, 34–38.

Kramer, P. J. (1943). Amount and duration of growth of various species of tree seedlings. Plant Physiol. 18, 239–251.

Kramer, P. J., and Kozlowski, T. T. (1960). "Physiology of Trees." McGraw-Hill, New York.

Kriedemann, P. E. (1968). ^{14}C translocation patterns in peach and apricot shoots. Aust. J. Agr. Res. 19, 775–780.

Krueger, K. W. (1967). Nitrogen, phosphorous, and carbohydrate in expanding and year-old Douglas-fir shoots. Forest Sci. 13, 352–356.

Kulman, H. M. (1971). Effects of insect defoliation on growth and mortality of trees. Annu. Rev. Entomol. 16, 289–324.

Lahiri, A. N., and Gaur, Y. D. (1969). Germination studies in arid zone plants. V. The nature and role of germination inhibitors present in leaves of Prosopis juliflora. Proc. Nat. Inst. Sci. India, Part B 35, 60–71.

Lanner, R. M. (1968). The pine shoot primary growth system. Ph.D. Thesis, University of Minnesota, St. Paul; Diss. Abstr. 30, 13B (1969).

Licitis-Lindbergs, R. (1956). Branch abscission and disintegration of the female cones of Agathis australis Salisb. Phytomorphology 6, 151–167.

Lott, R. V., and Simons, R. K. (1964). Floral tube and style abscission in the peach and their use as physiological reference points. Proc. Amer. Soc. Hort. Sci. 85, 141–153.

Lott, R. V., and Simons, R. K. (1966). Sequential development of floral-tube and style abscission in the Montmorency cherry (*Prunus cerasus* L.). *Proc. Amer. Soc. Hort. Sci.* **88**, 208–218.

Lott, R. V., and Simons, R. K. (1968a). The developmental morphology and anatomy of floral-tube and style abscission in the Wilson Delicious apricot (*Prunus armeniaca* L.). *Hort. Res.* **8**, 67–73.

Lott, R. V., and Simons, R. K. (1968b). The morphology and anatomy of floral-tube and style abscission and of associated floral organs in the Starking Hardy Giant cherry (*Prunus avium* L.). *Hort. Res.* **8**, 74–82.

McCleary, J. A. (1968). The biology of desert plants. Chapt. V *In* "Desert Biology" (G. W. Brown, Jr., ed.), pp. 141–194. Academic Press, New York.

McClintock, T. F. (1955). How damage to balsam fir develops after a spruce budworm epidemic. *U. S., Forest Serv., Northeast. Forest Exp. Sta., Pap.* **75**, 1–17.

McConkey, T. W. (1958). Helicopter spraying with 2,4,5-T to release young white pines. *U. S., Forest Serv., Northeast. Forest Exp. Sta., Pap.* **101**.

MacGillivray, J. H. (1960). Effects of chemical defoliation of canning tomatoes. *Proc. Amer. Soc. Hort. Sci.* **75**, 625–628.

McIntosh, D. C. (1951). "Effects of Chemical Treatment of Pulpwood Trees," Bull. No. 100. Forest. Br., Forest Prod. Lab., Div. Dep. Forest Prod. Develop., Ottawa, Canada.

Maggs, D. H. (1963). The reduction in growth of apple trees brought about by fruiting. *J. Hort. Sci.* **38**, 119–128.

Maini, J. S. (1966a). Apical growth of *Populus* spp. I. Sequential pattern of internode, bud and branch length of young individuals. *Can. J. Bot.* **44**, 615–622.

Maini, J. S. (1966b). Apical growth of *Populus* spp. II. Relative growth potential of apical and lateral buds. *Can. J. Bot.* **44**, 1581–1590.

Marvin, C. O. (1964). Abscission of marcescent leaves in *Quercus velutina* Lam. M. S. Thesis, University of Wisconsin, Madison.

Mergen, F. (1959). A toxic principle in the leaves of Ailanthus. *Bot. Gaz. (Chicago)* **121**, 32–36.

Millington, W. F. (1963). Shoot tip abortion in *Ulmus americana. Amer. J. Bot.* **50**, 371–378.

Moline, H. E., and Bostrack, J. M. (1972). Abscission of leaves and leaflets in *Acer negundo* and *Fraxinus americana. Amer. J. Bot.* **59**, 83–88.

Möller, C. M., Möller, D., Müller, D., and Nielsen, J. (1954). Loss of branches in European beech. *Forstl. Forsøgsv. Danmark* **21**, 253–271.

Morris, R. F. (1951). The effects of flowering on the foliage production and growth of balsam fir. *Forest. Chron.* **27**, 40–57.

Myers, C. C., and Braga, G. R. (1969). Chemical debarking of Eucalypt. *Papel* **30**, 27–30.

Nye, P. H., and Greenland, D. J. (1960). The soil under shifting cultivation. *Commonw. Bur. Soils, Tech. Commun.* **51**.

Orians, G. H., and Pfeiffer, E. W. (1970). Ecological effects of the war in Vietnam. *Science* **168**, 544–554.

Orlov, A. J. (1960). Rost i vozrastnye izmeneniija sosuscih kornej eli *Picea excelsa* Link. *Bot. Zh. (Leningrad)* **45**, 888–896.

Orshan, G. (1954). Surface reduction and its significance as a hydroecological factor. *J. Ecol.* **42**, 442–444.

Orshan, G. (1963). Seasonal dimorphism of desert and Mediterranean chamaephytes and its significance as a factor in their water economy. *In* "The Water Relations of Plants" (A. J. Rutter and F. W. Whitehead, eds.), pp. 206–222. Blackwell, Oxford.

Orshan, G. (1972). Morphological and physiological plasticity in relation to drought. In "Wildland Shrubs—Their Biology and Utilization" (C. M. McKell, J. P. Blaisdell, and J. R. Goodin, eds.), pp. 245–254. U. S. Dep. Agr., Forest Serv., Ogden, Utah.

Osborne, D. J. (1968). Defoliation and defoliants. Nature (London) 219, 564–567.

Ovington, J. D. (1963). Flower and seed production. A source of error in estimating woodland production, energy flow and mineral cycling. Oikos 14, 148–153.

Parker, J., and Houston, D. R. (1971). Effects of repeated defoliation on root and root collar extractives of sugar maple trees. Forest Sci. 17, 91–95.

Patrick, Z. A. (1971). Phytotoxic substances associated with the decomposition in soil of plant residues. Soil Sci. 111, 13–18.

Phelps, V. H. (1948). White spruce reproduction in Manitoba and Saskatchewan. Can. Dep. Mines Resour., Silvicult. Res. Note 86.

Place, I. C. M. (1955). The influence of seed-bed conditions in the regeneration of spruce and fir. Can. Forest. Bur. Bull. 117.

Pollard, D. F. W., and Wareing, P. F. (1968). Rates of dry matter production in forest tree seedlings. Ann. Bot. (London) [N.S.] 32, 573–591.

Pridham, A. M. S. (1951). Preliminary report on defoliation of nursery stock by chemical means. 5th Annu. Meet., Northeast. States Weed Contr. Conf., Suppl. pp. 127–138.

Pridham, A. M. S. (1952). Preliminary report on defoliation of nursery stock by chemical means. Proc. Amer. Soc. Hort. Sci. 59, 475–478.

Priestley, C. A. (1964). The importance of autumn foliage to carbohydrate status and root growth of apple trees. Annu. Rep., East Malling Res. Sta., Kent pp. 104–106.

Quinlan, J. D. (1969). Mobilization of ^{14}C in the spring following autumn assimilation of $^{14}CO_2$ by an apple rootstock. J. Hort. Sci. 44, 107–110.

Rodin, L. E., and Bazilevich, N. I. (1967). "Production and Mineral Cycling in Terrestrial Vegetation" (English transl. ed. by G. E. Fogg). Oliver & Boyd, Edinburgh.

Rogers, W. S., and Booth, G. A. (1964). Relationship of crop and shoot growth in apple. J. Hort. Sci. 39, 61–65.

Romberger, J. A. (1963). Meristems, growth, and development in woody plants. U. S., Dep. Agr., Tech. Bull. 1293.

Sarup, S. (1952). Plant ecology of Jodhpur and its neighborhood. A contribution to the ecology of northwestern Rajasthan. Bull. Nat. Inst. Sci. India 1, 223–232.

Sarvas, R. (1955a). Investigations into the flowering and seed quality of forest trees. Commun. Inst. Forest. Fenn. 45, 1–37.

Sarvas, R. (1955b). Ein Beitrag zur Fernverbreitung des Blütenstaubes einiger Waldbäume. Z. Forstgenet. 4, 137–142.

Sarvas, R. (1962). Investigations on the flowering and seed crop of Pinus silvestris. Commun. Inst. Forest. Fenn. 53, 1–198.

Sasaki, S., and Kozlowski, T. T. (1968). Effects of herbicides on seed germination and early seedling development of Pinus resinosa. Bot. Gaz. (Chicago) 129, 238–246.

Sasaki, S., and Kozlowski, T. T. (1969). Utilization of seed reserves and currently produced photosynthates of embryonic tissues of pine seedlings. Ann. Bot. (London) [N.S.] 33, 472–482.

Sasaki, S., and Kozlowski, T. T. (1970). Effects of cotyledon and hypocotyl photosynthesis on growth of young pine seedlings. *New Phytol.* **69**, 493–500.

Sax, K. (1962). Aspects of aging in plants. *Annu. Rev. Plant Physiol.* **13**, 489–506.

Schier, G. A. (1970). Seasonal pathways of ^{14}C-photosynthate in red pine labeled in May, July, and October. *Forest Sci.* **16**, 2–13.

Sharp, W. M., and Chisman, H. H. (1961). Flowering and fruiting in the white oaks. I. Staminate flowering through pollen dispersal. *Ecology* **42**, 365–372.

Shaw, J. K. (1934). The lifetime yield of an apple orchard. *Proc. Amer. Soc. Hort. Sci.* **31**, 35–38.

Splittstoesser, W. E., and Meyer, M. M. (1971). Evergreen foliage contributions to the spring growth of *Taxus*. *Physiol. Plant.* **24**, 528–533.

Stuckey, I. H. (1941). Seasonal growth of grass roots. *Amer. J. Bot.* **28**, 486–491.

Sweet, G. B., and Thulin, I. J. (1969). The abortion of conelets in *Pinus radiata*. *N. Z. J. Forest.* **14**, 59–67.

Sweet, G. B., and Wareing, P. F. (1968). A comparison of the seasonal rates of dry matter production of 3 coniferous species with contrasting patterns of growth. *Ann. Bot. (London)* [N.S.] **32**, 721–734.

Sykes, J. M., and Bunce, R. G. H. (1970). Fluctuations in litter fall in a mixed deciduous woodland over a three-year period 1966–68. *Oikos* **21**, 326–329.

Szymanski, S., and Szczerbinski, W. (1955). Paczki jako wskaznik potenczalu xyciowego mlodej sosny. *Rocz. Sekcji Dendrol. Pol. Tow. Bot.* **10**, 275–304.

Tappeiner, J. C. (1969). Effect of cone production on branch, needle, and xylem ring growth of Sierra Nevada Douglas-fir. *Forest Sci.* **15**, 171–194.

Tischler, N., Bates, J. C., and Quimba, G. P. (1950). A new group of defoliant herbicide chemicals. *Proc. Northeast. States Weed Contr. Conf.* pp. 51–84.

Trippi, V. S. (1963). Studies on ontogeny and senility in plants. V. Leaf-fall in plants of different age and the effect of gibberellic acid on *R. pseudoacacia* and *Morus nigra*. *Phyton (Buenos Aires)* **20**, 167–171.

Tschirley, F. H. (1969). Defoliation in Vietnam. *Science* **163**, 779–786.

Tukey, H. B., Jr. (1969). Implications of allelopathy in agricultural plant science. *Bot. Rev.* **35**, 1–16.

van den Berg, D. A., and Lanner, R. M. (1971). Bud development in lodgepole pine. *Forest Sci.* **17**, 479–485.

van der Pijl, L. (1952). Absciss-joints in the stems and leaves of tropical plants. *Proc., Kon. Ned. Akad. Wetensch.* **42**, 574–586.

Voronkov, V. V. (1956). The dying of the feeder root system in the tea plant. *Dokl. Vses. Akad. Sel'skokhoz. Nauk* **21**, 22–24; *Hort. Abstr.* **27**, 2977 (1957).

Wakeley, P. C., and Marrero, J. (1958). Five-year intercept as site index in southern pine plantations. *J. Forest.* **56**, 332–336.

Walhood, V. T., and Addicott, F. T. (1968). Harvest-aid programs: Principles and practices. *In* "Advances in Production and Utilization of Quality Cotton" (F. C. Elliott, M. Hoover, and W. K. Porter, Jr., eds.), Chapter 24, pp. 407–431. Iowa State Univ. Press, Ames.

Walter, H. (1962). "Die Vegetation der Erde in ökologischer Betrachtung." Fischer, Jena.

Weaver, J. E., and Zink, E. (1946). Length of life of roots of ten species of perennial range and pasture grasses. *Plant Physiol.* **21**, 201–217.

Went, F. W. (1952). Fire and biotic factors affecting germination. *Ecology* **33**, 351–364.

Westing, A. H. (1971a). Ecological effects of military defoliation on the forests of South Vietnam. *BioScience* **21**, 893–898.

Westing, A. H. (1971b). Forestry and the war in South Vietnam. *J. Forest.* **69**, 777–783.

Whiteside, T. (1970). "Defoliation." Ballantine, New York.

Whittaker, R. H. (1970). The biochemical ecology of higher plants. *In* "Chemical Ecology" (E. Sondheimer and J. B. Simeone, eds.), Chapter 3, pp. 43–70. Academic Press, New York.

Wilcox, H. (1954a). Some results from the chemical treating of trees to facilitate bark removal. *J. Forest.* **52**, 522–525.

Wilcox, H. (1954b). Preliminary study of the penetration of sodium arsenite and sodium monochloroacetate solutions into trunks of yellow birch and red spruce. *Bot. Gaz. (Chicago)* **116**, 73–81.

Wilcox, H., Czabator, F. J., Girolami, G., Moreland, D. E., and Smith, R. F. (1956). "Chemical Debarking of Some Pulpwood Species," Tech. Publ. No. 77. College of Forestry, Syracuse University, Syracuse, New York.

Williamson, M. J. (1966). Premature abscissions and white oak acorn crops. *Forest Sci.* **12**, 19–21.

Winget, C. H., and Kozlowski, T. T. (1965a). Yellow birch germination and seedling growth. *Forest Sci.* **11**, 386–392.

Winget, C. H., and Kozlowski, T. T. (1965b). Seasonal basal area growth as an expression of competition in northern hardwoods. *Ecology* **46**, 786–793.

Woodfin, R. O., Jr. (1963). Treating and harvesting interval for debarking ponderosa pine pulpwood with sodium arsenite. *Tappi* **46**, 73–75.

Wu, C. C., and Kozlowski, T. T. (1972). Some histological effects of direct contact of *Pinus resinosa* seeds and young seedlings with 2,4,5-T. *Weed Res.* **12**, 229–233.

Wu, C. C., Kozlowski, T. T., Evert, R. F., and Sasaki, S. (1971). Effects of direct contact of *Pinus resinosa* seeds and young seedlings with 2,4-D or picloram on seedling development. *Can. J. Bot.* **49**, 1737–1741.

Zohary, M., and Orshan, G. (1954). Ecological studies in the vegetation of the Near Eastern deserts. V. The zygophylletum dumosi and its hydroecology in the Negev of Israel. *Vegetatio, Haag* **5–6**, 341–350.

. 2 .

Anatomical and Histochemical Changes in Leaf Abscission

Barbara D. Webster

I. Leaf Longevity

Leaves of most flowering plants are determinate in their growth and have limited life spans varying from a few weeks in some desert ephemerals to as long as 30 years in species of *Araucaria*. In the course of development a leaf enlarges and attains its ultimate form. Throughout the growth period of the plant, old leaves abscise and new ones develop; in this way the plant maintains a relatively constant leaf area which is synthetically active.

The length of time that a leaf is retained by a plant varies with species, environment, and age of the plant. Theoretically at least, the retention time of any leaf on a plant could be prolonged indefinitely by pruning competing leaves and flowers. Under normal growing conditions, however, leaves are periodically abscised at predictable intervals. The phenomenon of leaf abscission typically involves formation of an abscission zone through which leaf separation occurs. The zone is most conspicuous in deciduous perennial plants. Deciduous trees and shrubs in temperate regions shed the bulk of their leaves nearly simultaneously during the autumn primarily in response to decreasing day length and decreasing temperatures. Deciduous tropical trees and lianas lose leaves both in season and throughout the year. In the tropical perennial *Plumeria,* shedding is a response to short day conditions; interruptions of a long night preclude foliage drop (Murashige, 1966). In the giant tropical tagalínau (*Gossampinus*) and in rubber (*Hevea*) trees, leaf drop occurs during the dry season (Merrill, 1945). So-called evergreen species in both temperate and tropical zones shed leaves after one or more years of growth and development, but all leaves on an individual plant do not abscise at the same time. In some annual species, sequential, rather than simultaneous, loss of leaves occurs until plant maturity. In other annuals (and monocotyledons as well) leaves do not actually abscise, but shrivel and decay on the plant. Limitations on leaf longevity can be artificially imposed by the application of chemicals which induce defoliation (see Chapter 1).

II. Senescence and Abscission

Leaf abscission in most species of plants is closely associated with aging and senescence; in fact, abscission has been characterized as a senescence phenomenon. The metabolic changes encompassed in the course of aging are part of the normal, morphogenetic program of the leaf. As the changes become degenerative, the leaf becomes senescent—ultimately, irreversibly so—and abscission usually follows (Carr and Paté, 1967). The most conspicuous abscission- and senescence-related change in leaves is that of coloration. The pigmentation changes usually involve loss of chlorophyll and the resultant appearance of carotenoids, as well as increase in anthocyanin synthesis. These pigments subsequently disappear as senescence progresses, although leaves frequently abscise prior to complete loss of color. The visible color changes are accompanied by numerous internal changes associated with senescence, including in some cases a respiratory climacteric (Eberhardt, 1955); translocation of mineral elements from the leaf (Watson and Petrie, 1940); increased production of tannins; deposition of suberin and lignin near the abscission zone (Scott *et al.,* 1948); decreased effectiveness of the photosynthetic apparatus; diminishing auxin supply; and less efficient synthesis of RNA and protein (Meyer, 1918; Leopold, 1961).

It is appropriate to note that the usual symptoms of leaf senescence such as pigment changes do not invariably occur prior to abscission, and that abscission does not always follow this senescence-related characteristic. Pigment changes may not develop in leaves which abscise as a result of disease, insect damage, or extraordinary environmental conditions. On the other hand, discolored leaves may not abscise during the usual shedding period. Marcescent leaves of *Quercus* and *Carpinus,* which become brown and shriveled, remain on the tree through the winter and abscise during the following spring (Tison, 1900; Berkley, 1931; Marvin, 1964), and discolored, withered leaves of *Eupatorium* and *Nicotiana* remain on the plants indefinitely (Lloyd, 1914; Gawadi and Avery, 1950).

The metabolic changes associated with senescence are deleterious to the leaf in terms of longevity, but the role of senescence in terms of the entire plant is distinctly positive, enabling ecological adaptation and providing a means of natural selection. Leaf senescence, according to Leopold (1961), has two primary assets: it permits recovery through retranslocation of the bulk of nutrients from the leaf; and it brings about shedding of ineffective leaves from the plant.

The rate at which foliar senescence proceeds is often correlatively associated with developmental processes occurring in other parts of the plant. In many evergreens, senescence and abscission of the oldest leaves take place after new leaves appear. Nearly mature leaves, buds, and stems also influence the rate of senescence and time of abscission of leaves (Dostál, 1951; Leopold, 1961; Wareing and Seth, 1967; Woolhouse, 1967). During the progressive senescence of foliar appendages on a plant, the first leaf to fall is usually the cotyledon. Its senescence is controlled by the apex, as is readily demonstrated by apical removal at different intervals after germination. In *Phaseolus,* apical excision defers cotyledon senescence from the usual 7–9 days to 1 month. The apical influence also controls the rate of senescence and abscission of other leaves on *Phaseolus* plants. Primary leaves normally senesce approximately 40 days after germination, but apical excision prior to development of 4 or 5 trifoliolate leaves defers yellowing and abscission and results in continued growth of leaves and cotyledons (Leopold, 1961). The nucleic acids of leguminous cotyledons decline with age, and the decline in RNA is matched by a rise in RNA in the growing points of plants. Oota and Takata (1959) have proposed that transport of RNA out of aging organs into meristems is an integral part of the development of senescence.

The intrinsic characteristics of a leaf are important in the control of its own senescence and abscission. This has been demonstrated by experiments in which parts of leaves or petioles were removed and the rate of abscission was regulated. In the Valencia orange, removal of 50% of the leaf blade slightly stimulated abscission, and removal of 75 and 90% markedly stimulated it (Livingston, 1950). Excision of the blade, but not the midrib, of *Coleus* leaves delayed abscission somewhat; excision of all but a small basal portion of the leaf blade more effectively retarded abscission than removal of all but a small tip portion of the blade along the midvein; and a large segment of blade at the midvein tip more effectively retarded abscission than a small amount of tissue there (Myers, 1940). Removal of the entire leaf blade and various lengths of the petioles of *Gossypium* indicated that the shorter the petiole stump, the more rapid its abscission (Addicott and Walhood, 1954).

Decline in total protein content and chlorophyll loss, both symptoms of senescence, were demonstrated during abscission in cells distal to the abscission zone (those of the pulvinus) of *Phaseolus* explants by Osborne (1958). This suggested to Osborne and Moss (1963) that, in addition to the association of abscission of a leaf with senescence of that leaf, separation ultimately depended on senescence in the abscission region itself. Furthermore, they felt that the nature of the agent which brought about senescence was not critical to the actual process of separation, but rather that diverse substances such as auxins, gibberellins, and ethylene, known to accelerate or retard separation at the abscission zone in *Gossypium* and *Phaseolus* plants, caused similar kinds of biochemical changes in abscission zone cells. To the list of substances which exerted control of abscission they added kinetin, which when applied at the abscission zone retarded abscission and when applied away from the zone accelerated abscission. They assumed that kinetin formed local, active sites of accumulation of organic and inorganic constituents, so that applications away from the abscission zone resulted in depletion of metabolites from the zone, thus inducing senescence at the zone and accelerating the time of leaf fall. In support of Osborne and Moss' contentions, Scott and Leopold (1966) noted that during abscission the cells proximal to the region of abscission (those of the petiole) in *Phaseolus* explants increased in dry weight, chlorophyll content, protein, RNA, and total phosphate at the expense of cells distal to the abscission region. They interpreted the mobilization of metabolites as supporting Osborne and Moss' allegation of senescence of abscission zone cells. It is now generally agreed that senescence does occur in cells distal to the abscission zone, that it is accompanied by a concomitant rise in ethylene production in those cells, and that distal senescence is essential to development of abscission

(Abeles *et al.*, 1967; de la Fuente and Leopold, 1968). In the abscission zone itself, however, cell division and cell enlargement precede formation of the separation layer in *Phaseolus*. This favors the assertion that, in addition to senescence of distal cells, separation also involves quite different cellular activities in the abscission region (Carns, 1966; Webster, 1970).

III. Scope of Anatomical Research

Publications on anatomy of leaf abscission fall into three major categories: classic descriptive studies; histochemical investigations of walls of cells in the abscission zone; and discussions of the effects of growth regulators on anatomical changes during abscission.

Comprising the group of descriptive studies are, for example, papers which discuss leaf development in economically important plants, such as *Citrus*, from the time of leaf initiation through the time of abscission (Scott *et al.*, 1948). Also in this group is a spate of descriptive papers with an orientation toward classification. Foremost among them are Tison's (1900) and Lee's (1911) depictions of leaf abscission in 105 and 45 species of mature woody dicotyledons, respectively. In these papers, plant group designations are established based on time of development of various anatomical features of the abscission zone (Tison, 1900) or on structures associated with leaf fall (Lee, 1911). In point of fact, neither systematization scheme is particularly enlightening. The papers are notable to some extent for the extraordinary amount of detailed description which they contain, but more so as reflections of the tenor of the time, during which systems of classification held full sway.

Although descriptions of leaf abscission are in the main noncontroversial, Lloyd (1914) managed to refute nearly every facet of the description of abscission in *Mirabilis* previously reported by Hannig (1913). Lloyd's curiosity was piqued by Hannig's conclusion that separation was the result of solubilization of cells in the abscission zones of *Mirabilis* and *Oxybaphus*, and that, moreover, no chemical changes preceded solubilization. Thus, Hannig claimed this constituted a new type of separation. Lloyd repeated the work on *Mirabilis*, concluding that the mode of separation was not accurately presented by Hannig, that abscission in *Mirabilis* was not unique, and that separation was preceded by chemical alterations in the cells. In fact, the observations of Hannig and Lloyd are so diverse that one wonders whether they actually studied the same plant.

In the second category of papers are those which deal with chemical changes in the walls of cells of the abscission zone prior to leaf detachment. Since the pectic nature of the middle lamella was not established until 1888 (Magnin) and the morphological complexity of the cell wall was not fully appreciated until 1954 (Bailey), many of the papers in this group are regarded as speculative. For example, Molisch (1886) concluded that the breakdown of the middle lamella was a result of "gum ferment." Many other early papers on cell wall changes are rather long on hypothesis and short on procedural data (Wiesner, 1904a,b,c; Sampson, 1918). Major contributions which form the basis of present anatomical concepts of the nature of chemical changes in cell walls have been made by Facey (1950) on *Fraxinus* and, more recently, by Rasmussen (1965) and Morré (1968) on *Phaseolus* (these are further considered in Section VI).

Most of the descriptive studies, as well as the analyses of cell wall changes carried out prior to 1920, have been reviewed by Pfeiffer (1928) in his perceptive treatise "Die pflanzenlichen Trennungswebe." This interesting paper discusses among other matters, mobilization of substances to the abscission area in connection with separation (von Mohl, 1860a), an observation followed up years later by Scott and Leopold (1966); the relationship of size and swelling of buds and abscission, a phenomenon subsequently discussed in connection with the abscission of marcescent leaves of *Quercus* (Marvin, 1964); and the manner in which separation takes place, including changes in cell walls, with which nearly every anatomical and some physiological papers since that time have been concerned. Pfeiffer's survey of the literature is extensive, and particularly valuable for its assessment of research by early foreign workers.

The third group of papers deals with studies of effects of growth regulators, including in particular auxins, ethylene, gibberellins, abscisic acid, and cytokinins on the anatomy of abscission. Some of the investigations have been carried out on intact plants in the field and greenhouse, but more commonly, explants have been used. These consist of small, leafless stem or petiole segments containing an abscission zone. In explants of *Phaseolus* and *Gossypium*, which have been extensively utilized in growth regulator studies, the anatomical changes which occur in the abscission zone following excision appear to be similar to those which develop in intact plants, but in explants the changes are telescoped into a matter of hours rather than spread over a period of weeks (Bornman *et al.*, 1967; Webster, 1970). Explants can be treated easily with various growth regulators by application at the cut surface, or, in the case of ethylene, by enclosing explants in a container and injecting the gas. Jacobs (1968) has warned, however, that contamination must be controlled and other factors associated with excision may be important, so that extrapolation of results from explants to intact plants may not always be valid.

An additional problem in growth regulator studies stems from the fact that although leaf abscission is known to be regulated by plant hormones, no completely acceptable theory has yet been developed to explain the phenomenon, partly because the mode of action of the controlling hormones is not fully understood. Growth regulators can bring about or prevent the structural changes associated with leaf fall, but at present there is no wholly satisfactory interpretation of their effects. Improved techniques for anatomical analyses, particularly autoradiography and electron microscopy, have shown the greatest promise in these studies for providing better understanding of the localization of biosynthetic events and a clearer image of ultrastructural properties of cells involved in abscission. At the present time, although much skepticism surrounds some of the interpretations of anatomical changes associated with growth regulators, it would be unwise to dismiss such studies out of hand.

IV. The Abscission Zone

A. MORPHOLOGY

The abscission of a leaf involves two discrete series of structural changes, the first of which is associated with actual detachment of the leaf from the plant and the second with protection of the surface area exposed on the plant after leaf fall. The distinction between the two series of changes was initially clarified by von Mohl (1860a,b), who delineated the particular characteristics related to each in *Aesculus, Ricinus,* and other dicotyledons. Among contemporary investigations, major attention focuses on separation-related changes; in fact, nothing very significant has been published about protection-related changes since Lee's (1911) discussions of leaf abscission in woody dicotyledons.

Detachment of a leaf is facilitated by development of a distinct anatomical structure, the abscission zone, which differentiates in the petiole. In some plants abscission zone differentiation occurs prior to leaf enlargement, so that the leaf reaches maturity with its abscission zone established. In those herbaceous species which have a differentiated abscission zone, the location of the zone can be readily determined externally. It is seen on the surface of the petiole as a narrow, constricted area which is pale green compared to the adjacent petiolar tissue. In primary leaves of *Phaseolus* the abscission zone is unusually conspicuous, appearing as a narrow, nearly colorless band of tissue between the dark green pulvinus and the paler green petiole (Webster, 1969). The abscission zone of woody species is usually light brown and typically denoted by a distinct furrow at the flared or swollen petiolar base (Lee, 1911; van der Pijl, 1952).

The number and position of abscission zones of an individual leaf are constant for a given species. Many simple leaves, including those of *Coleus* and *Poinsettia*, develop one abscission zone at the base of the petiole (Sampson, 1918; Myers, 1940; Gawadi and Avery, 1950). The primary leaves of *Phaseolus* develop two zones: one at the base of the lamina, subtending the pulvinus, and another near the stem at the base of the petiole and adjacent to the pulvinus. The locations of the two zones are denoted by arrows in Fig. 1. Compound leaves ordinarily have one abscission zone at the base of each leaflet lamina, one at the base of each petiolule, and one at each petiolar base. The bipinnately compound leaf of *Acacia decurrens*, which is approximately 6 inches (15 cm) long, has about 4000 abscission zones.

FIG. 1. Fourteen-day-old plant of *Phaseolus vulgaris* L. with fully expanded primary leaves, showing location of laminar and petiolar abscission zones (arrows) at juncture of the dark pulvinus and the lighter colored petiole. ×0.5.

B. ANATOMY

Within the petiole, the abscission zone is often distinguishable from the adjacent petiolar tissue by characteristics of the cortical parenchyma cells within it. These are typically thin-walled, densely protoplasmic, closely packed, and uniformly smaller than other cortical cells of the petiole.

The degree of internal elaboration and time of development of the abscission zone vary widely from one species to another. Woody plants generally have more specialized and conspicuous zones. In *Fraxinus* leaves, the abscission zone is clearly recognizable at the time of leaf development during the spring by its characteristically small and compactly arranged cortical cells, and it remains prominent through the growing season (Facey, 1950). At the other extreme are the primary leaves of *Phaseolus*, which reach full expansion without development of a structurally distinct abscission zone. Figure 2 shows the juncture of the pulvinus and petiole of a mature *Phaseolus* leaf and the complete absence of a differentiated abscission zone. The general locale of the prospective

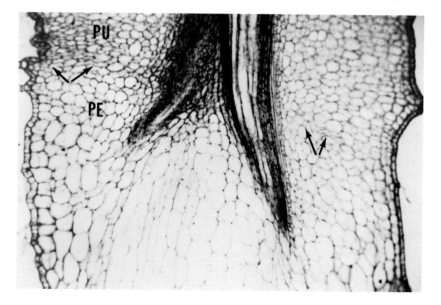

FIG. 2. Photomicrograph of a longitudinal section of the region of abscission in *Phaseolus* showing differences in sizes of petiole (PE) and pulvinal (PU) cells and the vascular configuration in the pulvinus and the petiole regions. Note the lack of a differentiated abscission zone. Arrows denote future region of abscission in part of the cortex. ×40. (From Brown and Addicott, 1950.)

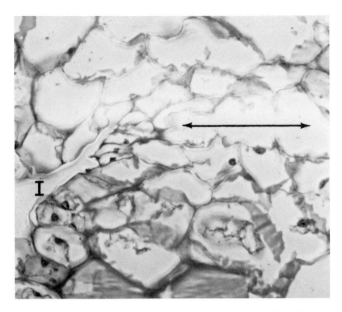

FIG. 3. Photomicrograph of a longitudinal section at the abscission zone of *Phaseolus* at the time of cortical cell separation showing disruption of the vertical walls of a single layer of cells, typically seen following treatment of an explant with ethylene, and invagination (I) of tissue at the abscission zone. Arrows denote separation cavity. ×380.

abscission zone in *Phaseolus* leaves is denoted by an invagination of tissue on the adaxial surface at the base of the pulvinus (Fig. 3) and by a change in vascular configuration at the juncture of the pulvinus and the petiole (Figs. 4, 5). However, the abscission zone does not differentiate until the final 2 weeks of the 5-week life span of the leaf (Webster, 1970). Leaves of *Nicotiana, Eupatorium,* and *Parthenium* never do develop structurally distinct abscission zones; they also never abscise (Gawadi and Avery, 1950). But apparently anatomical differentiation is not necessary for abscission to occur. *Citrus* leaves abscise in the nodal position without prior formation of a distinct abscission zone. Interestingly, abscission at the base of the lamina of *Citrus* is preceded by development of a discrete abscission zone (Scott *et al.,* 1948).

The vascular configuration in the abscission zone sometimes differs from that in other parts of the petiole, and certain structural components, such as xylem and phloem fibers, may be unusually small or absent. The zone is commonly characterized as a region of weakness, but break-strength measurements indicate that in some plants it is as strong as adjacent areas of the petiole. As a matter of interest, in many plants

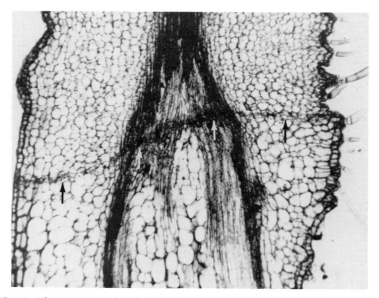

FIG. 4. Photomicrograph of a longitudinal section of the abscission zone of *Phaseolus* after cell divisions have occurred at the juncture of the pulvinus and petiole (arrows denote abscission zone). ×40. (From Brown and Addicott, 1950.)

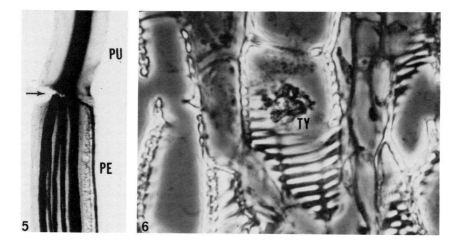

FIG. 5. Explant of *Phaseolus* cleared in chloral hydrate and stained to show vascular patterns of the pulvinus (PU) and petiole (PE) and the region of separation (arrow). ×6.

FIG. 6. Tylosis (TY) in a short, broad vessel element of the abscission zone of *Phaseolus*. ×570.

the abscission zone develops in petiolar regions which have the especially strong arrangement of vascular bundles around the periphery of the petiole (van der Pijl, 1952). In *Phaseolus* and *Nicotiana* plants, in which the future region of separation is initially as strong as adjacent pulvinal and petiolar tissue, decreased strength at the abscission zone gradually becomes apparent with development of cellular changes leading to separation (de la Fuente and Leopold, 1968; Morré, 1968; Valdovinos *et al.*, 1972).

The laminar abscission zone of the primary leaf of *Phaseolus* develops where the individual vascular strands which traverse the petiole coalesce to form a single, amphicribral bundle of the pulvinus (Figs. 4, 5). The tracheary elements in the abscission zone are short and broad, as illustrated in Fig. 6, and the pith cells are unsclerified (unlike those in the adjacent pulvinus). On the basis of such features, perhaps the abscission zone of *Phaseolus* could be characterized as structurally weak, but considering the break-strength tests, it may be more accurately depicted as a region of abrupt structural transition (Webster, 1968).

In many plants the appearance of the abscission zone is followed by a static period during which the abscission zone maintains its full strength. This is the time during which auxin inhibition of abscission can be demonstrated in *Phaseolus* explants and the zone is insensitive to applied ethylene (Rubinstein and Leopold, 1963; Abeles and Rubinstein, 1964). The static period usually comprises a matter of hours in explants, and may extend for weeks in intact plants.

1. Cell Inclusions

Prior to leaf abscission, increasing numbers of tyloses commonly materialize in the vascular elements at the abscission zone (Fig. 6), and in *Phaseolus* tanniniferous compounds may appear in sieve elements (Fig. 7). The relation between formation of tyloses and progress of abscission is not completely clear, but in *Phaseolus* explants tylosis formation is promoted by treatment with ethylene, which accelerates abscission, and inhibited by treatment with indoleacetic acid, which retards abscission. Scott *et al.* (1964, 1967) suggested that the increased numbers of tyloses associated with abscission in *Phaseolus* explants resulted in water stress in distal (pulvinar) tissue and that callose dissolution in sieve elements, which accompanied formation of tyloses, expedited the mobilization of materials from the pulvinus to the petiole. In cotton (*Gossypium*) explants, the more rapidly abscission is induced by various growth regulators, the fewer tyloses are formed. Bornman (1967a) concluded that in cotton, although a correlation existed between formation of tyloses and rate of abscission, the relation was not causal.

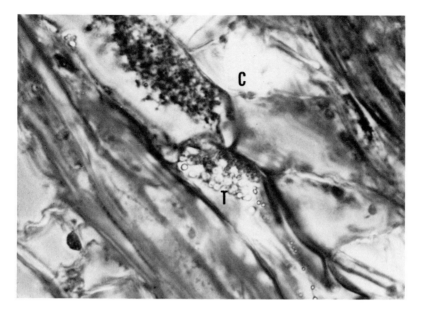

FIG. 7. Tanniniferous compounds (T) and calcium oxalate crystal (C) in phloem cells at the abscission zone of *Phaseolus*. ×900.

In *Phaseolus, Coleus,* and *Citrus,* starch is present in freshly cut sections of the abscission zone in cells of the endodermis, pith, and cortex. Localized starch accumulation in pith cells at the future abscission zone is particularly conspicuous in *Phaseolus* explants following excision (Fig. 8) (Sampson, 1918; Scott *et al.*, 1948; Brown and Addicott, 1950; Webster, 1970). Starch accumulates in the stalk and abscission zone of *Phaseolus* explants treated distally with sucrose, indicating basipetal carbohydrate movement and sugar to starch conversion (Brown and Addicott, 1950). In *Coleus* leaves, reducing sugars gradually increase in aging leaves and stems, but the increase is least pronounced in the separation layer (Sampson, 1918). This suggests that as abscission progresses sugars are utilized in starch synthesis in the abscission zone. Starch in cortical abscission cells of *Gossypium* explants increases initially during abscission, and this is correlated with an increase in the number of cells in the region (Bornman *et al.*, 1966). During the first 24 hr following excision of *Phaseolus* explants, starch increases in cortical cells and disappears from pith cells (Webster, 1968, 1970). According to Lee (1911) there is no starch in the abscission zone of *Salix,* but in most other trees starch is abundant (Lee, 1911; Scott *et al.*, 1948; Ramsdell, 1954).

Distributed ubiquitously in cells of the petiole, and prominent in the abscission zone, are crystals of calcium oxalate (Fig. 7). These may

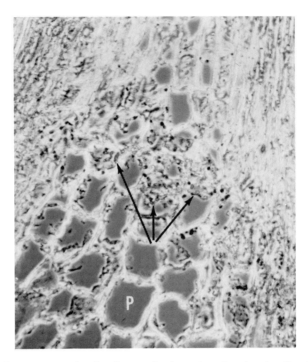

FIG. 8. Photomicrograph of a longitudinal section showing localization of starch grains (arrows) in cells of the pith (P) at the region of abscission in the *Phaseolus* explant. ×600.

gradually enlarge, until in some parenchyma cells of the abscission zone they occupy most of the cell lumen. A calcium oxalate–starch "balance" exists in some parts of *Citrus* leaves, with calcium oxalate replacing starch as leaves mature, but in the abscission zone both calcium oxalate and starch are retained and are present at the time of separation (Scott *et al.,* 1948). Calcium oxalate crystals are also abundant in cortical cells of the abscission zone of other woody plants and are the most conspicuous intracellular inclusions of cells of the protective zone (Lee, 1911).

A particularly interesting characteristic which should be mentioned at this point is the unusually large size of nuclei and nucleoli in certain cells in the abscission region of *Phaseolus* explants and intact plants. Nuclei and nucleoli in cells at and proximal to the abscission zone are larger and show a greater affinity for nuclear stains than do those in cells distal to the zone. Figure 9 shows large nuclei with prominent nucleoli in cells of the abscission zone of *Phaseolus* prior to separation. A histochemical and autoradiographic study of distribution of nuclear

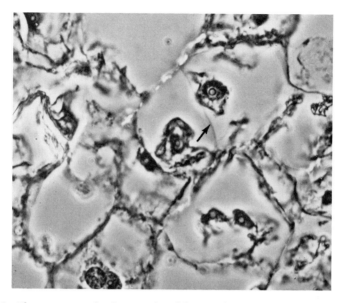

FIG. 9. Photomicrograph showing breakdown of intercellular walls (arrow) of cells of the abscission zone of *Phaseolus,* and the appearance of binucleate cells with large nuclei and conspicuous nucleoli. ×1600.

and nucleolar RNA in *Phaseolus* explants showed that initially there was little difference in RNA distribution between cortical and vascular cells proximal and distal to the region of future separation. After the explants aged for 24 hr, there was an increase in localization of nuclear and nucleolar RNA in cells at the abscission zone (Fig. 10). The localization pattern was more pronounced following treatment with ethylene for 4 or 8 hr. There was also an ethylene-enhanced decrease in protein localization in cortical cells distal to the abscission zone and an increase in cells at and proximal to the separation region (Fig. 11). The increased localization of protein and RNA was restricted to 2–6 layers of cells at the abscission zone (Webster, 1968). Increased protein and RNA localization in connection with treatment with ethylene, which is widely regarded as a potent accelerator of abscission, is of unusual interest. Ethylene has a relatively direct effect on the transcription and translation process (Holm *et al.,* 1968), and the effects of ethylene on abscission may be mediated through this mechanism (Abeles, 1968; Burg, 1968).

2. Cell Wall Constituents

In addition to cellulose, hemicellulose, and pectic compounds, suberin is deposited consistently in abscission zones of higher plants. Suberin

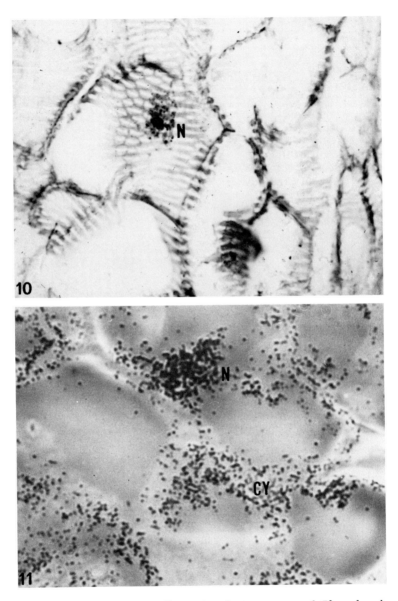

FIG. 10. Radioautograph of cells at the abscission zone of *Phaseolus* showing silver grains denoting incorporation of [³H]uridine and dark color denoting positive reaction to pyronin Y stain (specific for RNA) in the nucleus (N) and nucleolus of vessel elements. ×1300.

FIG. 11. Radioautograph of cells at the abscission zone of *Phaseolus* showing silver grains denoting incorporation of [³H]L-leucine and localization of protein in the nucleus (N) and cytoplasm (CY) of cortical cells at the abscission zone. ×1100.

occurs in cells of the cortical, pith, xylem, and phloem parenchyma as a fine film lining the inner surface of the cell walls. Prior to separation at both the nodal and laminar abscission zones of *Citrus*, suberization increases markedly within the cells as a film and as a well-defined pellicle. Suberin may be deposited in intercellular spaces also, in *Citrus* to such an extent that after separation it forms a blocking surface on the leaf scar (Scott *et al.*, 1948).

Lignin is characteristically absent or present only in small amounts in walls of cells of the abscission zone, although it occurs abundantly in adjacent cells in the petiole and stem. In woody plants lignin deposits are conspicuous in the protective zone; there the cell walls may become completely lignified either prior to or after leaf fall (Lee, 1911). Tison (1900) described a lignin-positive staining reaction with phloroglucinol–hydrochloric acid in cortical cells distal to the region of separation in the abscission zone of *Tilia*, and noted that the effect of such "lignification" was to make the separation layer markedly weaker in comparison to adjacent tissue. However, it is questionable whether parenchymatous cortical cells which reacted positively to phloroglucinol–hydrochloric acid actually were lignified. The stain is not specific for lignin, but also indicates the presence of uronic acids, pentoses, and polysaccharides (Robinson, 1963).

In some woody plants, extensive deposits of cutin are associated with the general region of abscission. In the species of *Salix* examined by Lee (1911) a single layer of cells within the protective zone reacted positively to a test for cutin. The outer walls of the cells thickened so extensively as cutin was deposited that the cell lumen became obliterated. This internal cuticle was usually continuous with the external cuticle of the stem, and was chemically identical to it. As additional cells were subsequently formed by cambial activity in the protective zone, supplementary layers of cuticle were produced. The various species of *Salix* differed in both the time and extent of deposition of cutin.

V. Structural Changes in the Abscission Zone

A. Cell Division

The first conspicuous structural change in the abscission zone involves meristematic activity of cells. Cell divisions are usually confined to 2 or 3 vertical rows, and they occur in cells of the pith, cortex, epidermis, and vascular tissue. The newly formed cells tend not to be strictly and neatly oriented in one direction, as new cell walls may be anticlinal, periclinal, or transverse.

The abscission zone often has been erroneously depicted as being comprised of many uniform layers of cortical cells, neatly oriented perpendicular to the length of the petiole. This is the result of confusion regarding those cell divisions which are related to detachment of a leaf from the plant and those which are related to formation of a protective zone of cells. As a general rule, in the latter, cell divisions are more extensive and cells are more uniform in size and shape.

The cell divisions associated with abscission may take place within a preformed, readily recognizable abscission zone, such as that of *Fraxinus,* in which case they usually are confined to cells in the distal sector of the zone. Alternatively, they may take place through a structurally undistinguished region of the petiole, in which case they give rise to a readily recognizable abscission zone. *Phaseolus* is a prime example of a plant in which the abscission zone is undifferentiated prior to cell division (Fig. 2) and is comprised entirely of recently divided cells after mitoses take place (Fig. 4). The future location of the abscission zone (between the pulvinus and the petiole) can be generally determined anatomically by the vascular configuration at that locale by an adaxial invagination of epidermal and cortical cells and by a size difference between the smaller cortical cells of the pulvinus and the larger ones of the petiole, indicating the boundary between them (Fig. 2). When mitosis is completed, the abscission zone of *Phaseolus* is comprised of recently divided cortical, epidermal, and some vascular cells, and some nondividing vascular cells (Fig. 4).

In many woody dicots and also in *Phaseolus,* cytokinesis following mitosis is incomplete (Lee, 1911; Webster, 1968). Recently divided nuclei remain within the mother cell walls and are separated from each other by newly formed, tenuous walls. In *Phaseolus,* successive nuclear divisions, followed by formation of new transverse walls within the mother cell, give rise to conspicuous nests of cells [which Torrey (1966) has called "meristemoids"] in the cortex (Fig. 12). Histochemical tests of the cell walls of meristemoids indicate increased pectin localization in the abscission zone in comparison to that of surrounding regions. This may relate to electron microscope analyses of pectinase activity which indicate that the Golgi-derived vesicles which fuse to form the cell plates are sites of pectin synthesis (Albersheim, 1965; Mühlethaler, 1967).

Meristematic activity is frequently, but not invariably, associated with leaf abscission, and von Mohl (1860a) regarded mitosis as a prerequisite for leaf separation. Tison (1900) supported this view, adding that in many woody species separation frequently took place through newly formed cells. In other woody dicots, according to Lee (1911), separation

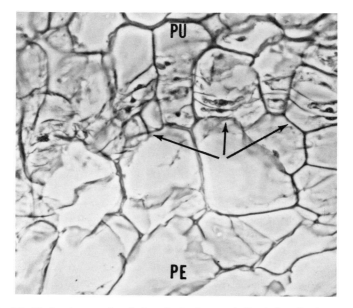

FIG. 12. Photomicrograph of the single layer of divided cells (arrows) at the juncture of the pulvinus (PU) and the petiole (PE) of *Phaseolus* which gives rise to the abscission zone. Note that recently divided nuclei are separated by newly formed transverse walls but are retained within the mother cell walls. ×500.

occurred through cells of the abscission zone which had not divided. In an effort to determine whether cell divisions were causally related to abscission, Gawadi and Avery (1950) treated several different plants with ethylene chlorohydrin, an abscission stimulant, and reported that since it promoted abscission but precluded meristematic activity, cell divisions were not a necessary part of separation. However, ethylene chlorohydrin is a very toxic chemical and not equivalent to ethylene (Pratt and Goeschl, 1969), and any effects which it might produce could readily be attributed to the injury it causes. A causal relationship between cell divisions and abscission might exist in plants which normally abscise through the layers of divided cells; for example, in *Gossypium* and *Phaseolus* separation takes place through the recently divided cells. In *Impatiens,* according to Gawadi and Avery, separation occurred without prior mitotic activity, and thus no causal relationship would appear to exist.

Growth Regulators and Cell Division

In explants of the primary leaf of *Phaseolus,* cell divisions are initiated in the outer cortical cells 6–8 hr after excision. Meristematic activity continues in 1 or 2 rows of cells until abscission occurs. Divisions are commonly anticlinal, primarily in the transverse plane, but occasionally longitudinal. The time when cell divisions commence (6 hr after excision) coincides with a decrease in the initially high endogenous level of ethylene in *Phaseolus* explants (Abeles, 1967), and it is possible that the ethylene decrease triggers the onset of cell divisions. In explants which are air-exposed, cell divisions continue in the abscission zone for about 40 hr and abscission takes place soon after. In intact *Phaseolus* plants, cell divisions occur on an extended time scale, beginning after 19 days. Abscission takes place 35–40 days after germination. Ethylene treatment of explants interrupts the cell division sequence and accelerates abscission. These observations suggest that the control of cell division in explants of *Phaseolus,* as well as the subsequent changes which culminate in separation, may depend on the level of ethylene in the tissue, whether endogenous or applied (Webster, 1970).

Mitoses similar to those observed in *Phaseolus, Gossypium,* and many woody plants in which meristemoids are formed have been observed in tissue cultures to which various growth regulators were added. When an auxin, such as indoleacetic acid (IAA), and a cytokinin, such as kinetin, are included in the culture medium for *Nicotiana* pith tissue, nests of cells develop within the walls of a mother cell (Skoog and Miller, 1957; Murashige, 1964). The similarity between tissue culture nuclear divisions and those in the abscission region suggests an auxin and cytokinin involvement in cell divisions in the abscission zone of some plants.

Treatment with indoleacetic acid results in increased cell divisions, which are apparent 72 hr after excision in cotyledonary petioles of *Gossypium.* Meristematic activity in *Gossypium* explants is also increased by applications of gibberellic acid. Treatment with optimal gibberellic acid concentration (0.01 gm) results in transverse cell divisions; treatment with supraoptimal concentrations (100 gm) results in transverse, oblique, and longitudinal divisions and production of a massive callus tissue. Cell divisions are evident within 24 hr after gibberellin treatment. Fewer cell divisions take place following treatment with abscisic acid (Bornman *et al.,* 1967, 1968).

B. Cell Enlargement

Cell enlargement in the abscission region takes place both prior to and after leaf fall. Before leaf detachment, active growth of cells takes place in distal tissue, and this can prevent the progress of abscission (Jacobs *et al.,* 1964). Differential enlargement of cells prior to abscission also takes place in other tissues, and may bring about the development of shear forces across the cell walls of the abscission zone (Leopold, 1967). Cells exposed on the stem surface of the plant after leaf separation frequently exhibit a marked potential for enlargement and may puff out in a manner suggesting that the leaf was being forced off the plant by the expansion of certain layers of cells. In the abscission region of *Citrus*, the expanded surface cells become suberized (Scott *et al.*, 1948).

The expansion of *Citrus* cells involves those which lie immediately proximal to the separation region. Growth of the cells is longitudinal and as the cells balloon, the exposed stem surface becomes papillate (Scott *et al.*, 1948). Both differential growth and cell elongation take place in abscission zones of *Quercus* and *Aristolochia*, and according to Pfeiffer (1924, 1928), the increase in cell size ultimately results in tissue break. In many trees and in *Gossypium*, intact cells of the separation layer enlarge during leaf severance (Tison, 1900; Bornman *et al.*, 1967). In *Acer* leaves the enlargement could be correlated with the development of stress forces which bring about epidermal rupture prior to leaflet fall (Moline and Bostrack, 1972). Enlargement of parenchyma cells in the abscission zone of *Capsicum* results in the rupture of walls of vessel elements as abscission proceeds (Gawadi and Avery, 1950).

A consequence of cell growth, according to Tison (1900), is the development of "reciprocal pressures" which force apart cells of the abscission zone. The growth pattern of the "papillate" cortical cells of *Citrus*, which expand toward heavily suberized tissue (Scott *et al.*, 1948), might constitute one example of this phenomenon. Another might be the development in woody species of suberized tissue in connection with formation of a protective zone of cells. However, Lee (1911), although repeatedly describing cell elongation in the abscission zone of woody dicotyledons, did not consider the idea of reciprocal pressures credible. It seems reasonable, though, that when cell growth occurs within an abscission zone, cell expansion may constitute a mechanical factor in tissue break.

Growth Regulators and Cell Enlargement

Enlargement of recently divided cells in the abscission zone of *Phaseolus* is primarily in a transverse, rather than longitudinal direction. The lateral expansion of cells is particularly marked after treatment with ethylene (Fig. 13) (Webster, 1970). Lateral expansion of cells following ethylene application involves reorientation of microfibrils of the cell walls in stems of *Pisum* (Apelbaum and Burg, 1971). If cell walls in the abscission zone were affected in a similar manner, the reorientation could influence the development of stress forces across the cell walls.

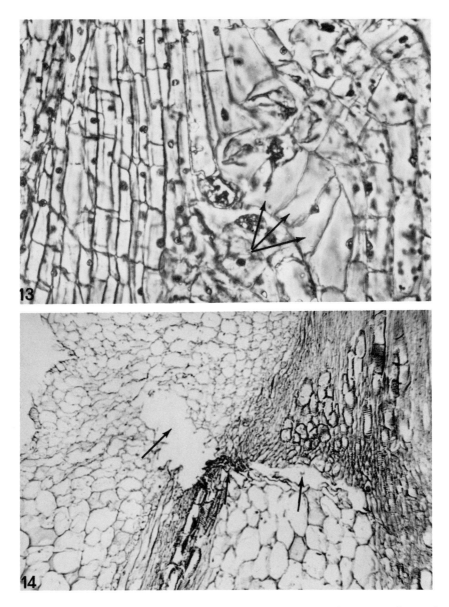

FIG. 13. Photomicrograph showing the pronounced lateral enlargement of recently divided cells of the abscission zone of ethylene-treated of *Phaseolus* explants (arrows). The abscission zone runs from the midbottom to upper right of photograph. ×600.

FIG. 14. Photomicrograph of a longitudinal section of the abscission region of *Phaseolus* showing internal separation and cellular disruption in pith, cortical, and vascular cells of the abscission zone (arrows). ×500.

In explants of the cotyledonary node of *Gossypium*, cell enlargement takes place at two different times. Following the application of IAA, epinasty of petioles is evident, and after 48 hr petioles increase 50 to 100% in length. Cell enlargement also occurs in connection with gibberellic acid treatment, and it precedes cell division in the petiolar abscission region. The enlargement of cells by stretching of periclinal walls results in bulging of the epidermis and underlying cortical parenchyma in a region distal to the abscission zone. Continued cell growth leads to rupture of the cell wall (Bornman *et al.*, 1967).

C. CELL SEPARATION

According to von Mohl (1860a), two sets of phenomena are brought into operation in connection with leaf detachment. The first is the formation of a discrete separation layer, through which abscission subsequently takes place, and the second is the actual separation of adjacent cells as a result of chemical alterations in the cell walls.

The separation layer typically comprises one or more rows of cells in the distal sector (toward the blade) of the abscission zone. Its exact location can sometimes be deduced prior to any histochemically detectable wall changes in its cells by a staining affinity of the parenchymatous cells lying immediately distal to it. When the abscission zone and contiguous tissue are treated with phloroglucinol-hydrochloric acid, the walls of parenchymatous cells distal to the zone are stained red. Separation subsequently takes place through unstained cells within the abscission zone which lie adjacent to stained cells.

This staining reaction has actually been known for some time, but until recently it has not been clearly evaluated or correlated with the sequence of changes which takes place during abscission. In point of fact, the basis for the reaction is unclear. In woody species the walls of parenchymatous cells immediately distal to the future separation layer invariably stain red or red-violet with phloroglucinol-hydrochloric acid, and Tison (1900) interpreted this as evidence of lignification. In *Gossypium*, Bornman (1965) noted a similar red-violet reaction in cells distal to the abscission zone and speculated that it might be indicative of the presence of polysaccharides.

Major interest in this staining procedure stems from the fact that cell walls distal to the separation layer show the reaction; those proximal to it do not, and interestingly, the intensity of the red stain can be demonstrated to parallel the abscission process. *Gossypium* explants show no evidence of the reaction when stained immediately after excision, but several hours later the response is positive. In *Gossypium* and *Phaseolus* the development of the staining response can be inhibited or delayed by auxinlike regulators. Carns (1966) has suggested that the phloroglucinol and hydrochloric acid reaction is the first observable indication of the initiation of abscission, and that the stain reaction provides visual evidence of a response mechanism which completes the initial processes required for abscission to subsequently occur.

1. *Initiation of Separation*

Separation of cells can apparently commence in any tissue of the separation layer. The separation break in *Coleus* leaves begins on the lower (abaxial) side of the petiole and gradually extends through the epidermal cells and cortex until the petiole is supported only by the upper part of the cortex and the xylem elements (Myers, 1940). In pepper (*Capsicum*) plants, separation starts in the epidermis and hypodermis, and according to Gawadi and Avery (1950), separation need not be preceded by formation of an abscission layer. Separation also commences in epidermal cells in leaves of *Impatiens* and neither cell division nor abscission layer differentiation precedes the actual cell separation (Gawadi and Avery, 1950). In *Phaseolus* plants and explants, pith cells are normally the first to separate (Webster, 1970) and separation subsequently continues through vascular and cortical cells (Fig. 14). When the pulvini of *Phaseolus* leaves are supplied with large amounts of water or the plant is treated with ethylene, separation commences internally. But if the water content of the pulvinus is low, separation begins externally and progresses across the abscission layer (Brown and Addicott, 1950).

The concave appearance of the pulvinus after separation in *Phaseolus* leaves suggests that tension and compression of cells exist in the abscission zone at the time of separation. Cortical pulvinar cells protrude beyond the cells of the vascular cylinder, accounting for the central depression region of the dislodged pulvinus. The turgor of pulvinar cortical cells appears to regulate tensions and compressions within pulvini of both *Phaseolus* and *Citrus,* and turgor pressures facilitate separation in the latter (Livingston, 1948; Brown and Addicott, 1950). Mechanical resistances (presumably in thick-walled vascular elements) influence the plane of separation in *Mirabilis.* Cells of the inner cortex separate first, and this is followed by progressive separation inward to the pith and outward toward the epidermis (Lloyd, 1914).

Treatment with various growth regulators influences not only the time but also the position of initial separation. In *Gossypium* explants, separation commences adaxially in untreated plants and in those treated with auxin and either adaxially or abaxially after application of abscisic acid (Bornman *et al.*, 1967).

2. Arrested Separation

Although differentiation of an abscission zone and mitotic activity of cells within the zone take place in most deciduous trees, further changes in the zone leading to separation may be temporarily or permanently arrested so that leaves are not shed during the usual leaf-fall season. Temporary arresting of leaf or leaflet abscission is common in certain species of *Quercus, Carpinus, Fagus,* and *Acer.* A significant number of marcescent leaves are retained on these trees during the winter following leaf emergence, and leaves are abscised the following spring (Tison, 1900; Berkley, 1931; Hoshaw and Guard, 1949; Marvin, 1964). In the abscission zones of *Quercus palustris, Q. coccinea,* and *Q. velutina,* some cell divisions take place during the autumn months, but further progress toward separation ceases until the following spring. When the axial buds begin to swell, the separation which ensues commences adaxially, suggesting that to some extent the leaf is "pushed" off the plant as the buds increase in size (Hoshaw and Guard, 1949; Marvin, 1964).

Leaves of some lower plants, some monocotyledons and some annuals remain attached indefinitely, although they become discolored and withered. In *Nicotiana* (tobacco) plants, in which retention of aged leaves commonly occurs, some cell divisions take place in the petiole, but no organized abscission zone ever develops (Gawadi and Avery, 1950).

3. Development of the Protective Region

On the stem side of the abscission zone a number of changes take place in several layers of cells, resulting in formation of a resistant protective tissue which ultimately becomes the leaf scar. The cellular changes related to development of the protective region may be initiated prior to or after leaf fall. The major cell modifications include alterations in cell inclusions and cell wall composition and in meristematic activity of cells.

As development of the protective region proceeds, the walls of 3 or 4 layers of cells in the proximal sector of the abscission zone typically become lignified and suberized. Other substances, including cutin and tannins, are also deposited. Tyloses commonly appear in the vascular tissue, and protoplasts of the parenchyma cells gradually disappear. After some or all of these changes have occurred, the cells constitute a protective layer which lies over the exposed surface of the stem after leaf fall.

Beneath the protective layer, meristematic activity takes place in several rows of cells and a cambiumlike zone (the periderm) forms. As the periderm develops more extensively it becomes continuous with the periderm of the stem. In cambial fashion, it cuts off cells toward the stem surface; these gradually undergo wall changes and become incorporated into the surface–protective layer. The periderm also produces cells toward the inner part of the stem; these comprise layers of phelloderm. Inner cells retain their protoplasts and undergo no wall changes.

The protective region is extensively developed in many woody plants. Both Tison (1900) and Lee (1911) described it at length, noting that beneath the leaf scar of many woody plants several meristematic areas developed and functioned over a period of 3 or 4 years. Protective cells at and close to the surface were gradually sloughed off, and the layer was replenished from below as new cells were produced by activity of the periderm.

VI. Histochemical Studies of Cell Walls

Wall changes leading to separation of cells involve (1) hydrolysis or dissolution of the middle lamella, which results in loss of cementing effectiveness between adjacent cell walls; (2) dissolution of the lamella plus breakdown of all or parts of the cellulosic cell wall; and (3) mechanical breakage of nonliving elements. These three types of separation are not mutually exclusive. In *Phaseolus*, for example, separation at both the laminar and nodal abscission zones involves all three types (Figs. 15–17) (Webster, 1970). It is the separation aspect of abscission which has commanded most of the attention of both anatomists and physiologists, for both histochemical and enzymatic changes in the cells and cell walls appear to be involved. The provinces of anatomists and physiologists have overlapped, with both groups utilizing light and electron microscopy, histochemical staining, and extraction procedures for cell wall analyses. These techniques have provided a somewhat clearer conception of the nature of leaf abscission.

A. Changes in the Middle Lamella

The major thrust of studies of cell wall modifications has focused on the changes which take place in the middle lamella, ultimately resulting in separation of adjacent intact cells. Research efforts which preceded Magnin's (1888) discovery of the pectic nature of the middle lamella

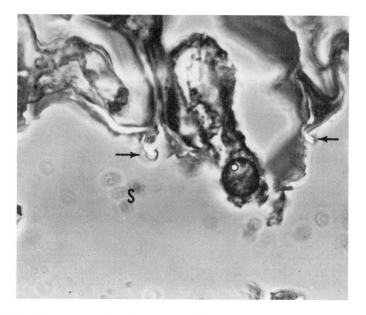

FIG. 15. Photomicrograph of a cortical cell adjacent to the separation cavity (S) at the abscission zone of *Phaseolus* showing extrusion of cell contents following disruption of the primary wall (arrows). ×900.

were somewhat like tilting at windmills; witness, for example, Wiesner's (1871) conclusion that lamellar dissolution resulted from the action of organic acids contained therein and Molisch's (1886) decision that separation resulted from accumulation of "gum ferment" between adjacent cells. Contemporary studies have come a long way since then, but still harken back frequently to some of the basic observations on the middle lamella made in 1911 by Lee and in 1950 by Facey.

Lee's (1911) concern was the structural alterations in the middle lamella during abscission in woody dicotyledons. He noted that the lamella frequently swelled and that it invariably acquired a mucilaginous consistency. It also reacted positively to "pectic stains." Unfortunately, Lee gives no clue to the nature of the stains nor any indication of the histochemical procedures employed to discern lignin, suberin, and cellulose changes, all of which he described in conjunction with lamellar modifications. In addition to lamellar swelling, Lee noted a change in size of the primary wall, but no further primary wall changes. The lamella gradually deteriorated and separation was effected as contiguous, intact cells parted from one another. Such a separation procedure is not confined only to woody plants which Lee studied, but occurs, for example, in some

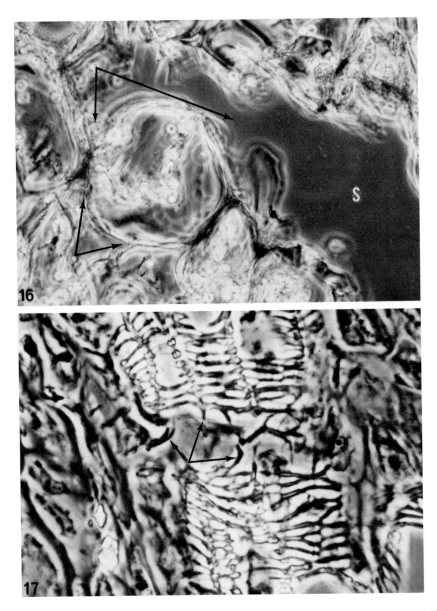

FIG. 16. Photomicrograph of a cortical cell adjacent to the separation cavity (S) at the abscission zone of *Phaseolus* showing areas of middle lamella dissolution (arrows) and loosening of the cell from those adjacent to it, with the primary wall intact. ×700.

FIG. 17. Photomicrograph of vessel elements at the abscission region of *Phaseolus* showing disrupted thick, secondary walls (arrows). ×700.

cortical cells of *Phaseolus* (Fig. 16). Both Lee and von Mohl (1860b) concluded that final separation of leaves in woody plants was brought about by mechanical breakage of xylem elements. This appears to be the case also in *Phaseolus* and in *Gossypium* (Bornman *et al.*, 1968).

Facey's (1950) investigation centered on the pattern of pectic changes analyzed by staining and extraction techniques in leaves of *Fraxinus*. Her techniques are carefully delineated and her results are assessed and evaluated in the light of similar studies which preceded hers. Despite some question about the specificity of ruthenium red which she used as a stain for pectins (see below) her observations on *Fraxinus* have stood the test of time and were recently corroborated by Morré's (1968) and Rasmussen's (1965) more elaborate studies employing more refined techniques on *Phaseolus*.

Basically, Facey (1950) found two successive transformations in the nature of pectic compounds as *Fraxinus* leaves approached abscission: a change from insoluble pectates, among which calcium pectate predominated, to pectic acid; and methylation of pectic acid to form pectin. The analytical technique involved staining fresh-cut sections from the abscission zone with ruthenium red; results were positive. This was followed by water-soaking the cells for 1 hr; the stain was retained. The tissue was then treated with 2% sodium hydroxide, and very little staining was evident. The change in stain intensity indicated a change from calcium pectate to pectinic acid (which is soluble in sodium hydroxide). The tissue was then resoaked in water, and subsequent staining was negative. Facey attributed this to methylation of pectinic acid to water-soluble pectin.

The only problem in interpreting these results arises from the fact that pectic substances stain with ruthenium red only when the pectin concentration is high and when no interfering substances are present (Jensen, 1962). On the face of it, presence of interfering substances would appear to be a distinct possibility, as Facey (1950) noted that "lignification" was occasionally evident in cells of the separation layer [but see Carns (1966) and Section V,C] and that just prior to abscission, tests for calcium pectate were unsatisfactory without preliminary treatment with sodium hydroxide. Nine years after Facey's results appeared, Reeve (1959) published a specific hydroxylamine–ferric chloride reaction for histochemical localization of pectins, and recently Luft (see Webster, 1970, 1973) communicated a technique for purification of ruthenium red which enhances its specificity for pectins. Both of these latter techniques have been widely employed by contemporary anatomists and physiologists studying abscission. Their observations in the main substantiate those of Facey.

Before concluding the discussion of Facey's (1950) studies on pectin changes in *Fraxinus* leaves, it is relevant to note that she also commented on the condition of the primary wall, noting that cellulose was present at the time of abscission. Cellulose seemed to have a mucilaginous consistency and the walls had a very slight resistance to breakage. The primary walls swelled as abscission progressed and subsequently became distorted and stretched. Wall attenuation gave the impression that cellulose had disappeared.

One particularly useful addition to studies of middle lamellar changes during abscission is Morré's (1968) analysis of cell wall dissolution and enzyme secretion during leaf abscission in *Phaseolus*. Recalling Carns' (1966) contention that the anatomical changes attending leaf separation are often difficult to recognize in early stages of development of the abscission zone, Morré employed break-strength measurements across the zone to determine quantitatively the time of formation of the separation layer. He then related changes in break strength to changes in composition of cell walls. In *Phaseolus* he reported that the general solubilization of pectin fractions which ensued as break strength declined probably involved synthesis of polygalacturonase. Pectinase activity preceded separation. The cell wall changes which he observed by electron microscopy were consistent with physiological tests, indicating pectin dissolution through depolymerization and removal of divalent ions. Morré's results confirmed those of Facey (1950) and other recent histochemical studies, all of which have indicated increases in soluble pectins during leaf abscission (Rasmussen, 1965; Bornman, 1967b).

Morré (1968) noted that even though dissolution of pectic substances might be necessary and sufficient for formation of the separation layer, pectin dissolution alone was probably insufficient to insure actual separation of the leaf from the plant. As evidence he cited his observation that break-strength measurements reached a minimum between 48 and 72 hr after excision and deblading in explants of *Phaseolus,* yet separation did not occur until later (ca. 100 hr). At that time the thick, cellulosic walls of the vascular elements were severed. While the separation of intact cells following lamellar dissolution is not the most common mode of separation in *Phaseolus* (Webster, 1970), nevertheless, pectin dissolution followed by loosening of intact cells is sufficient under certain circumstances for separation in other species, including *Coleus, Gossypium,* and *Psoralea* (Myers, 1940; Becker, 1968; Bornman *et al.,* 1969).

The appearance of calcium in oxalate crystals or as free ions in the region of the abscission zone suggests that dissolution of pectic substances of the middle lamella leads to release of cations that formerly bound the pectins to a cell wall cement (Scott *et al.*, 1948; Rasmussen and Bukovac, 1969; Stösser *et al.*, 1969). In abscising pedicels, ultra-structural evidence suggests that at least some of the enzymes involved in wall degradation may be released from lysozymelike bodies associated with cells of the abscission zone (Jensen and Valdovinos, 1968). The fact that certain abscission zones can be forced to weaken and then to restrengthen after chemical treatment indicates that the wall degradation process may be in a state of equilibrium with wall-strengthening processes (Ben-Yehoshua and Biggs, 1970).

One final observation of pectin dissolution prior to separation relates to the newly formed cell walls (those laid down during mitosis) in *Phaseolus* explants treated with ethylene. These broke down readily as cells enlarged. The walls were primarily pectinaceous, and their dissolution resulted in appearance of bi- and multinucleate cells in the abscission zone (Fig. 9) (Webster, 1968).

B. CHANGES IN THE PRIMARY WALL

Although pectin dissolution is commonly associated with formation of the separation layer, histological studies frequently show evidence of some degree of hydrolysis of other cell wall substances, particularly cellulose and hemicelluloses. All levels of dissolution of pectic and other wall constituents have been reported, including complete cytolysis.

In *Phaseolus* and many other plants, separation of cortical cells involves changes in both the middle lamella and the primary walls of the abscission layer. Rasmussen (1965), observing the changes in intact, debladed *Phaseolus* plants, described swelling of the lamella and subsequent breakdown of noncellulosic polysaccharides and cellulose. The changes were reflected in a fading of distinct cell outlines, a loss of cell continuity, and ultimately, in cell collapse. The actual breakdown of cells points up the fact that wall components other than pectin were altered or degraded in the process of separation. Facey (1950) concluded that in leaves of *Fraxinus* cellulose in the abscission zone was altered by hydration but was not broken down. Recent reports by Horton and Osborne (1967) and Lewis and Varner (1970), however, have indicated the synthesis of cellulases capable of acting on cellulosic walls associated with the abscission zone of *Phaseolus* and the release of cellulases from some bound form into the solution bathing the cell wall (Abeles and Leather, 1971). These physiological observations have been substantiated by polarizing microscopy. Primary cell wall breakdown was frequently observed in *Phaseolus* explants, and it appeared to be under enzymatic control. In addition to facilitating separation of cortical cells, the enzymes (cellulases) might also be involved in the rupture of nonlignified vascular elements (Osborne, 1968).

The chronology of changes which preceded separation layer formation in *Phaseolus,* according to Rasmussen (1965) and Morré (1968), involved induction, increase in divalent cations with activation of pectin esterase, pectinase activity, decrease in divalent ions, and solubilization and swelling of pectins. The changes which took place concomitant with separation included cell wall collapse, inactivation of enzymes, and loss of membrane permeability. According to Morré, these latter characteristics might arise as senescence phenomena which would precede death of the separating leaf. The relation between senescence and the stimulation and secretion of lytic enzymes such as pectinase is still not clear; however, endogenous ethylene is regarded as a potential enzyme inducer (Hall, 1958; Rubinstein and Leopold, 1964; Abeles, 1967; Lewis and Varner, 1970).

Extending the observations of histochemical and physiological changes in cells of the abscission zone in *Phaseolus,* Rasmussen and Bukovac (1969) reported a decrease in calcium in the abscission zone prior to formation of the separation layer and an increase in calcium in the petiole. The first visual change in abscission zone cells was a swelling of pectin compounds in the cell walls, and this was followed by breakdown of other cell wall components. In a sequential extraction procedure, Rasmussen and Bukovac noted darker-staining cells distal to the abscission zone, but they obtained negative results when these cells were stained with phloroglucinol-hydrochloric acid. They concluded that the dark-staining material was not lignin, but probably cellulose. The swelling of pectic materials was correlated with loss of calcium from the abscission layer, and delay in calcium loss was correlated with application of auxin. Recently, Poovaiah and Leopold (1972) reported retardation of abscission when calcium solutions were applied to *Phaseolus* petiole explants. Their results indicated hardening of cell walls and maintenance of cell membranes, as well as deferral of pulvinar senescence. They suggested that the onset of abscission in explants might be related to deprivation of calcium.

Varying degrees of hydrolysis of cell walls, in addition to pectin dissolution have been reported in *Gossypium* and *Coleus* (Myers, 1940; Leinweber and Hall, 1959; Bornman, 1967b). Loss of both cellular and cell wall integrity in *Coleus* approached the condition of complete cytolysis, described by Pfeiffer (1928) and Eames and MacDaniels (1947). Such complete cytolysis involved plasmolysis, loss of membrane integrity, and swelling and gelatinization of cell walls (Brown and Addicott, 1950; Sacher, 1957).

C. Mechanical Disruption of Secondary Walls

Mechanical disruption of the secondary walls of vascular elements commonly follows lamellar and primary wall changes and may ultimately effect detachment of the leaf from the plant. Mechanical breaking is usually ascribed to vagaries of the environment; however, it readily occurs under controlled conditions in a growth chamber. In certain species of *Quercus,* separation by mechanical disruption of vascular elements is the rule, rather than the exception, since the separation layer may be non-functional or undeveloped. In guayule (*Parthenium*), the abscission zone was characterized as a weak region, and the separation which took place after the demise of the leaf was mechanical (Addicott, 1945).

D. Growth Regulators and Cell Wall Changes

In addition to its effects on RNA and protein localization in the abscission zone, ethylene treatment results in disruption of cell walls and accelerated separation of cells in *Phaseolus* explants. Studies at the electron microscope level indicate that separation does not involve extensive cell wall lysis. Wall breaks in cells of the abscission zone may be confined to a single locale on the vertical wall, and the breaks may not be preceded by dissolution of the middle lamella (Figs. 3, 18–19). In nonethylene-treated material, lamellar dissolution precedes wall breakdown (Fig. 20). Ethylene treatment also affects cell contents causing configurational changes in mitochondria, breakdown of chloroplasts, and loss of the selective permeability characteristics of the plasma membrane. The changes occur rapidly, suggesting that ethylene interrupts the fairly lengthy sequence of anatomical changes and accelerates a series of events related to cell separation (Webster, 1973).

Abscisic acid, phosphon, and gibberellic acid also accelerate abscission, and auxin retards it in explants of the cotyledonary node of *Gossypium.* The degree of retardation is enhanced by gibberellic acid and inhibited by abscisic acid. Treatment with growth regulators also affects the manner of separation. In nontreated plants, rupture of parts of the cell wall occurs, followed by disruption of cell contents. In explants treated with abscisic acid, no well-defined separation layer develops, and separation takes place by breakdown of cells. When gibberellic acid is

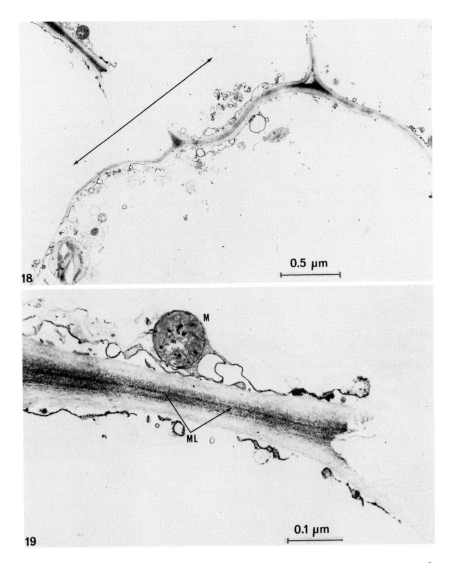

FIG. 18. Electron micrograph of a longitudinal view of the abscission zone of an ethylene-treated *Phaseolus* explant showing the separation cavity (arrows) and disruption at a single location on the vertical wall. Note degradation of cell inclusions.

FIG. 19. Electron micrograph of a partial longitudinal wall at the abscission zone of an ethylene-treated *Phaseolus* explant showing appearance of the primary wall and presence of the middle lamella (ML) at the time of wall breakage. Note the plasmolyzed condition of the cell and the mitochondrion (M) with a conspicuous DNA-like region and lack of cristae.

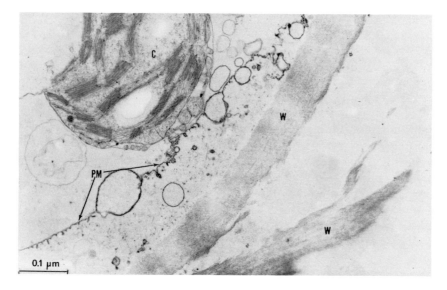

FIG. 20. Electron micrograph of a portion of a cell wall in the abscission zone of a *Phaseolus* explant showing appearance of the primary wall (W) and lack of a middle lamella at the time of wall breakage. Note the plasmolyzed condition of the cell and general degradation of cell contents. C, chloroplast; PM, plasma membrane.

applied, a separation layer forms, and the effects of gibberellic acid are primarily on the middle lamella (Bornman *et al.*, 1967).

References

Abeles, F. B. (1967). Mechanism of action of abscission accelerators. *Physiol. Plant.* **20**, 442–454.

Abeles, F. B. (1968). Role of RNA and protein synthesis in abscission. *Plant Physiol.* **43**, 1577–1586.

Abeles, F. B., and Leather, G. R. (1971). Abscission: Control of cellulase secretion by ethylene. *Planta* **97**, 87–91.

Abeles, F. B., and Rubinstein, B. (1964). Regulation of ethylene evolution and leaf abscission by auxin. *Plant Physiol.* **39**, 963–969.

Abeles, F. B., Holm, R. E., and Gahagan, H. E. (1967). Abscission: The role of aging. *Plant Physiol.* **42**, 1351–1356.

Addicott, F. T. (1945). The anatomy of leaf abscission and experimental defoliation in guayule. *Amer. J. Bot.* **32**, 250–256.

Addicott, F. T., and Walhood, V. T. (1954). Leaf abscission as affected by blade, petiole and defoliant. *Proc. Cotton Defoliation-Physiol. Conf., 8th, 1954* pp. 38–41.

Albersheim, P. (1965). The substructure and function of the cell wall. *In* "Plant

Biochemistry" (J. Bonner and J. E. Varner, eds.), 2nd ed., pp. 151–188. Academic Press, New York.

Apelbaum, A., and Burg, S. P. (1971). Altered cell microfibrillar orientation in ethylene-treated *Pisum sativum* stems. *Plant Physiol.* **48**, 648–652.

Bailey, I. W. (1954). "Contributions to Plant Anatomy." Chronica Botanica, Waltham, Massachusetts.

Becker, D. A. (1968). Stem abscission in the tumbleweed, *Psoralea. Amer. J. Bot.* **55**, 753–756.

Ben-Yehoshua, S., and Biggs, R. H. (1970). Effects of iron and copper ions in promotion of selective abscission and ethylene production by citrus fruit. *Plant Physiol.* **45**, 604–607.

Berkley, E. E. (1931). Marcescent leaves of certain species of *Quercus. Bot. Gaz.* (*Chicago*) **92**, 85–93.

Bornman, C. H. (1965). Histological and histochemical effects of gibberellin and auxin on abscission. Ph.D. Thesis, University of California, Davis.

Bornman, C. H. (1967a). The relation between tylosis and abscission in cotton (*Gossypium hirsutum* L.) explants. *S. Afr. J. Agr. Sci.* **10**, 143–154.

Bornman, C. H. (1967b). Some ultrastructural aspects of abscission in *Coleus* and *Gossypium. S. Afr. J. Sci.* **63**, 325–331.

Bornman, C. H., Addicott, F. T., and Spurr, A. R. (1966). Auxin and gibberellin effects on cell growth and starch during abscission in cotton. *Plant Physiol.* **41**, 871–876.

Bornman, C. H., Spurr, A. R., and Addicott, F. T. (1967). Abscisin, auxin, and gibberellin effects on the developmental aspects of abscission in cotton (*Gossypium hirsutum*). *Amer. J. Bot.* **54**, 125–135.

Bornman, C. H., Addicott, F. T., Lyon, J. L., and Smith, O. E. (1968). Anatomy of gibberellin-induced stem abscission in cotton. *Amer. J. Bot.* **55**, 369–375.

Bornman, C. H., Spurr, A. R., and Addicott, F. T. (1969). Histochemical localization by electron microscopy of pectic substances in abscising tissue. *J. S. Afr. Bot.* **35**, 253–263.

Brown, H. S., and Addicott, F. T. (1950). The anatomy of experimental leaflet abscission in *Phaseolus vulgaris. Amer. J. Bot.* **37**, 650–656.

Burg, S. P. (1968). Ethylene, plant senescence and abscission. *Plant Physiol.* **43**, 1503–1511.

Carns, H. R. (1966). Abscission and its control. *Annu. Rev. Plant Physiol.* **17**, 295–314.

Carr, D. J., and Paté, J. S. (1967). Aging in the whole plant. *Symp. Soc. Exp. Biol.* **21**, 559–599.

de la Fuente, R. K., and Leopold, A. C. (1968). Senescence processes in leaf abscission. *Plant Physiol.* **43**, 1496–1502.

Dostál, R. (1951). Odlučovani listů jake jev korelační. (Leaf abscission as a correlative phenomenon.) [Czech with English summary.] *Acta Acad. Sci. Natur. Moravosilesiacae* **23**, 67–106.

Eames, A. J., and MacDaniels, L. H. (1947). "An Introduction to Plant Anatomy," 2nd ed. McGraw-Hill, New York.

Eberhardt, F. (1955). Der Atmungsverlauf alternder Blätter und reifender Früchte. *Planta* **45**, 57–67.

Facey, V. (1950). Abscission of leaves in *Fraxinus americana* L. *New Phytol.* **49**, 103–116.

Gawadi, A. G., and Avery, G. S., Jr. (1950). Leaf abscission and the so-called "abscission layer." *Amer. J. Bot.* **37**, 172–180.

Hall, W. C. (1958). Physiology and biochemistry of abscission in the cotton plant. *Tex. Agr. Exp. Sta., Misc. Publ.* **285**, 1–23.

Hannig, E. (1913). Untersuchungen über das Abstossen von Blüten u.s.w. *Z. Bot.* **5**, 417–469.

Holm, R. E., O'Brien, T. J., Cherry, J. H., and Key, J. L. (1968). Effect of 2,4-dichlorophenoxyacetic acid (2,4-D) and ethylene on soybean chromatin. *Plant Physiol.* **43**, 5–19.

Horton, R. F., and Osborne, D. J. (1967). Senescence, abscission and cellulase activity in *Phaseolus vulgaris*. *Nature* (*London*) **214**, 1086–1088.

Hoshaw, R. W., and Guard, A. T. (1949). Abscission of marcescent leaves of *Quercus palustris* and *Quercus coccinea*. *Bot. Gaz.* (*Chicago*) **110**, 587–593.

Jacobs, W. P. (1968). Hormonal regulation of leaf abscission. *Plant Physiol.* **43**, 1480–1495.

Jacobs, W. P., Kaushik, M. P., and Rochmis, P. G. (1964). Does auxin inhibit the abscission of *Coleus* leaves by acting as a growth hormone? *Amer. J. Bot.* **51**, 893–897.

Jensen, T. E., and Valdovinos, J. G. (1968). Fine structure of abscission zones. III. Cytoplasmic changes in abscising pedicels of tobacco and tomato flowers. *Planta* **83**, 303–313.

Jensen, W. A. (1962). "Botanical Histochemistry." Freeman, San Francisco, California.

Lee, E. (1911). The morphology of leaf-fall. *Ann. Bot.* (*London*) **25**, 51–107.

Leinweber, C. L., and Hall, W. C. (1959). Foliar abscission in cotton. I. Effect of age and defoliants on the respiratory rate of blade, petiole, and tissues of the abscission zone. *Bot. Gaz.* (*Chicago*) **120**, 144–151.

Leopold, A. C. (1961). Senescence in plant development. *Science* **134**, 1727–1732.

Leopold, A. C. (1967). The mechanism of foliar abscission. *Symp. Soc. Exp. Biol.* **21**, 507–516.

Lewis, L. N., and Varner, J. E. (1970). Synthesis of cellulase during abscission of *Phaseolus vulgaris* leaf explants. *Plant Physiol.* **46**, 194–199.

Livingston, G. A. (1948). Contributions to the knowledge of leaf abscission in the Valencia orange. Ph.D. Thesis, University of California, Los Angeles.

Livingston, G. A. (1950). *In vitro* tests of abscission agents. *Plant Physiol.* **25**, 711–721.

Lloyd, F. E. (1914). Abscission. *Ottawa Natur.* **28**, 41–52 and 61–75.

Magnin, L. (1888). Sur la constitution de la membrane des végétaux. *C. R. Acad. Sci.* **107**, 144–146.

Marvin, C. O. (1964). Abscission of marcescent leaves in *Quercus velutina* Lam. M.S. Thesis, University of Wisconsin, Madison.

Merrill, E. D. (1945). "Plant Life of the Pacific World." Macmillan, New York.

Meyer, A. (1918). Eiweissstoffwechsel und Vergilben der Laubblätter von *Tropaeolum majus*. *Flora* (*Jena*) **11–12**, 85–127.

Moline, H. E., and Bostrack, J. M. (1972). Abscission of leaves and leaflets in *Acer negundo* and *Fraxinus americana*. *Amer. J. Bot.* **59**, 83–88.

Molisch, H. (1886). Untersuchungen über Laubfall. *Bot. Zentralbl.* **25**, 393–394.

Morré, D. J. (1968). Cell wall dissolution and enzyme secretion during leaf abscission. *Plant Physiol.* **43**, 1543–1559.

Mühlethaler, K. (1967). Ultrastructure and formation of plant cell walls. *Annu. Rev. Plant Physiol.* **18**, 1–24.

Murashige, T. (1964). Analysis of the inhibition of organ formation in tobacco tissue culture by gibberellin. *Physiol. Plant.* **17**, 636–643.

Murashige, T. (1966). The deciduous behavior of a tropical plant, *Plumeria acuminata*. *Physiol. Plant.* **19**, 348–355.

Myers, R. M. (1940). Effect of growth substances on the absciss layer in leaves of *Coleus*. *Bot. Gaz. (Chicago)* **102**, 323–328.

Oota, Y., and Takata, K. (1959). Changes in microsomal ribonucleoproteins in the time course of the germination stage as revealed by electrophoresis. *Physiol. Plant.* **12**, 518–525.

Osborne, D. J. (1958). Changes in the distribution of pectin methylesterase across leaf abscission zones of *Phaseolus vulgaris*. *J. Exp. Bot.* **9**, 446–457.

Osborne, D. J. (1968). Hormonal mechanisms regulating senescence and abscission. In "Biochemistry and Physiology of Plant Growth Substances" (F. Wightman and G. Setterfield, eds.), pp. 815–840. Runge Press, Ottawa.

Osborne, D. J., and Moss, S. E. (1963). Effect of kinetin on senescence and abscission in explants of *Phaseolus vulgaris*. *Nature (London)* **200**, 1299–1301.

Pfeiffer, H. (1924). Neue Untersuchungen über die Trennungsgewebe zur Ablösung pflanzlicher Organe. I. Die Entwicklungsgeschichte der schlauchförmigen Trennungs-Zellen bei *Coleus hybridus* Hort und *Camellia japonica* L. *Mikrokosmos* **18**, 28–30.

Pfeiffer, H. (1928). Die pflanzlichen Trennungsgewebe. In "Handbuch der Pflanzenanatomie" (K. Linsbauer, ed.), Vol. 5, No. 1, Part 2, p. 22. Borntraeger, Berlin.

Poovaiah, B. W., and Leopold, A. C. (1972). Retardation of abscission by calcium. *Plant Physiol.* **49**, 17.

Pratt, H. K., and Goeschl, J. D. (1969). Physiological roles of ethylene in plants. *Annu. Rev. Plant Physiol.* **20**, 541–584.

Ramsdell, V. H. (1954). The anatomy of cotton leaf abscission. M.A. Thesis, University of California, Los Angeles.

Rasmussen, H. P. (1965). Chemical and physiological changes associated with abscission layer formation in the bean (*Phaseolus vulgaris* L. cv. Contender). Ph.D. Thesis, Michigan State University, East Lansing.

Rasmussen, H. P., and Bukovac, M. J. (1969). A histochemical study of abscission layer formation in the bean. *Amer. J. Bot.* **56**, 69–76.

Reeve, R. M. (1959). A specific hydroxylamine-ferric chloride reaction for histochemical localization of pectin. *Stain Technol.* **34**, 209–211.

Robinson, T. (1963). "The Organic Constituents of Higher Plants." Burgess, Minneapolis, Minnesota.

Rubinstein, B., and Leopold, A. C. (1963). Analysis of the auxin control of bean leaf abscission. *Plant Physiol.* **38**, 262–267.

Rubinstein, B., and Leopold, A. C. (1964). The nature of leaf abscission. *Quart. Rev. Biol.* **39**, 356–372.

Sacher, J. A. (1957). Relationship between auxin and membrane-integrity in tissue senescence and abscission. *Science* **125**, 1199–1200.

Sampson, H. C. (1918). Chemical changes accompanying abscission in *Coleus blumei*. *Bot. Gaz. (Chicago)* **66**, 32–53.

Scott, F. M., Schroeder, M. R., and Turrell, F. M. (1948). Development, cell shape, suberization of internal surface, and abscission in the leaf of the Valencia orange, *Citrus sinensis. Bot. Gaz. (Chicago)* **109**, 381–411.

Scott, P. C., and Leopold, A. C. (1966). Abscission as a mobilization phenomenon. *Plant Physiol.* **41**, 826–830.

Scott, P. C., Webster, B. D., and Leopold, A. C. (1964). Formation of tyloses during bean leaf abscission. *Plant Physiol.* **39**, Suppl., XIV.

Scott, P. C., Miller, L. W., Webster, B. D., and Leopold, A. C. (1967). Structural changes during bean leaf abscission. *Amer. J. Bot.* **54**, 730–734.

Skoog, F., and Miller, C. O. (1957). Chemical regulation of growth and organ formation in plant tissues cultured *in vitro. Symp. Soc. Exp. Biol.* **11**, 118–131.

Stösser, R., Rasmussen, H. P., and Bukovac, M. J. (1969). Histochemical changes in the developing abscission layer in fruits of *Prunus cerasus. Planta* **68**, 151–164.

Tison, A. (1900). Recherches sur la chute des feuilles chez les Dicotylédoneés. *Mem. Soc. Linn. Normandie* **20**, 121–132.

Torrey, J. G. (1966). The initiation of organized development in plants. *Advan. Morphog.* **5**, 39–91.

Valdovinos, J. G., Jensen, T. E., and Sicko, L. M. (1972). Fine structure of abscission zones. IV. Effect of ethylene on the ultrastructure of abscission cells of tobacco flower pedicels. *Planta* **102**, 324–333.

van der Pijl, L. (1952). Absciss-joints in the stems and leaves of tropical plants. *Proc., Kon. Ned. Akad. Wetensch., Ser. C* **55**, 574–586.

von Mohl, H. (1860a). Ueber die anatomischen Veränderungen des Blattgelenkes, welche das Abfallen der Blätter herbeiführen. *Bot. Ztg.* **18**, 1–17.

von Mohl, H. (1860b). Ueber den Ablösungsprozess saftiger Pflanzenorgane. *Bot. Ztg.* **18**, 273–274.

Wareing, P. F., and Seth, A. K. (1967). Ageing and senescence in the whole plant. *Symp. Soc. Exp. Biol.* **21**, 543–558.

Watson, R., and Petrie, A. H. K. (1940). Physiological ontogeny in the tobacco plant. *Aust. J. Exp. Biol. Med. Sci.* **18**, 313–339.

Webster, B. D. (1968). Anatomical aspects of abscission. *Plant Physiol.* **43**, 1512–1544.

Webster, B. D. (1969). Abscission. *In* "McGraw-Hill Yearbook of Science and Technology," pp. 85–88. McGraw-Hill, New York.

Webster, B. D. (1970). A morphogenetic study of leaf abscission in *Phaseolus. Amer. J. Bot.* **57**, 443–451.

Webster, B. D. (1973). Ultrastructural studies of abscission in *Phaseolus:* Ethylene effects on cell walls. *Amer. J. Bot.* **60**, 436–447.

Wiesner, J. (1871). Untersuchungen über die herbstliche Entlaubung der Holzgewächse. *Sitzungsber. Akad. Wiss. Wien, Math.-Naturwiss. Kl.* **64**, 465–509.

Wiesner, J. (1904a). Über Laubfall infolge Sinkens des absoluten Lichtgenusses. (Sommerlaubfall.) *Ber. Deut. Bot. Ges.* **22**, 64–72.

Wiesner, J. (1904b). Über Trieblaubfall. *Ber. Deut. Bot. Ges.* **22**, 316–323.

Wiesner, J. (1904c). Über Hitzelaubfall. *Ber. Deut. Bot. Ges.* **22**, 501–505.

Woolhouse, H. W. (1967). The nature of senescence in plants. *Symp. Soc. Exp. Biol.* **21**, 179–211.

. 3 .

Physiological
Ecology of Abscission

Fredrick T. Addicott and Jessye Lorene Lyon

I. Introduction

This chapter extends information on the physiological ecology of abscission, treated less comprehensively in earlier reviews (Addicott, 1965, 1968). Our continuing interest in the subject stems from the conviction that the physiology of abscission can not be appreciated fully until we understand the manner in which each of the ecological stimuli elicits its characteristic response. We have attempted to give a broad view of the variety of ecological factors affecting abscission and to indicate the probable physiological effects of these factors on abscission. Limitations of time and space have permitted us to refer to only a small fraction of the literature, but we have cited one or more examples of each significant aspect of our subject. Comments for which sources are not indicated represent our own observations or interpretations.

A. Natural History of Abscission

1. *General*

Abscission is one mechanism of the shedding of plant parts. It is a complex physiological process influenced by many external and internal factors. The process of abscission encompasses a group of biochemical steps that culminate in enzymatic dissolution of cell walls in an abscission zone. Another mechanism of shedding involves the nonphysiological separation of plant parts, such as the falling of a dead branch from a tree. It would be difficult, if not impossible, to draw a sharp line of distinction between the two kinds of shedding, either morphologically or physiologically. Within the plant kingdom there are wide and continuous variations in the pattern of shedding: morphologically, from organs that have a well-defined abscission zone to those that have none; and physiologically, from organs that can be abscised with great rapidity to those apparently lacking any physiological mechanism to assist shedding. Except where noted, this chapter will be concerned with the common types of physiological abscission displayed in leaves, flowers, and fruits.

The ability to abscise parts appears to be a primitive character as ancient as the ability of two recently divided cells to separate. Among the lower plants the ability to fragment the plant body, or to shed or release portions of the plant body, is widespread. Within two large taxonomic groups, bacteria and yeasts, the separation of recently divided cells is a common occurrence.

With the evolution of the multicellular plant body the ability to separate cells was retained, but the ability is usually localized in such morphologically distinct organs or tissues as ripening fleshy fruits and zones of abscission or dehiscence.

In a vast array of situations the process of abscission is sensitive to and influenced by environmental factors and our knowledge of these factors and influences is called physiological ecology of abscission. However, abscission is also sensitive to the various factors of internal plant physiology (as discussed in Chapter 4). It is largely through influences on the internal factors that the environmental factors exert their effects.

Broadly viewed, the internal physiological factors in abscission express themselves in phenomena of correlation and competition. Correlation phenomena (Dostál, 1967) are those in which events in one part of the organism are correlated in a causal way with events in another part. For example, changes in levels of carbohydrates, nitrogenous substances, or hormones in the leaf blade influence the cells of the abscission zone at the base of the leaf and can affect both time of onset and rate of the abscission process. Similarly, competition among organs of a plant is a reflection of differences in nutritional, hormonal, metabolic, and other internal factors. When competition becomes sufficiently severe, many species react by abscising their weaker organs. The onset and intensity of expression of the correlative and competitive phenomena are strongly influenced by such environmental factors as light, temperature, water, and mineral nutrients. The main body of this chapter will be concerned with ecological factors that are known to influence abscission and with the effects of those factors on the portions of the internal physiology that appear to be involved with abscission.

2. Leaf Abscission Habits of Plants

a. AUTUMNAL ABSCISSION. In the Northern Hemisphere the most conspicuous leaf abscission habit is that of the autumnal leaf fall displayed by many species of trees and shrubs. The phenomenon has been celebrated in the literature, legends, and songs of the western world since ancient times. And, indeed, it was recognized as a scientific phenomenon by Theophrastus (285 B.C.) who in his lectures included comments on the shedding of leaves as well as on the influence of ecological factors on the time of leaf shedding (also see Wiesner, 1871).

Autumnal leaf abscission is a striking phenomenon. Although the species involved will shed an occasional leaf from late spring onward, an individual tree or shrub sheds most of its leaves within a few days or weeks. Thus, the appearance of a typical deciduous forest changes dramatically during a short period in the autumn. The green of the trees in full foliage rapidly gives way to the reds and yellows of the senescent leaves as day length decreases and temperature drops. Soon thereafter the main flush of leaf abscission commences and all the leaves are shed exposing the bare trunks and branches. Sometimes, when weather is unusually conducive, the majority of the leaves of a tree can be shed within an hour or two.

A variety of ecological factors can influence the onset of autumnal leaf fall, but among these cold temperatures (Rutland, 1888) and decreasing day length (Olmsted, 1951; Wareing, 1956) appear to be the most important. In various experiments absence of one or the other of these stimuli has delayed leaf abscission indefinitely. Other factors can have important modifying effects so that under favorable conditions leaves tend to be retained somewhat longer than under unfavorable conditions. For example, Theophrastus (285 B.C.) commented that trees growing in moist places tend to keep their leaves late, whereas the same species growing in dry places, and in general where the soil is light, shed their leaves earlier.

b. SUMMER (DROUGHT) ABSCISSION. There are many plant species in which the onset of abscission is correlated with moisture stress. These species fall into two groups: those that are *summer deciduous* and those that shed only a portion of their leaves after a period of moisture stress.

Many of the summer deciduous species are tropical and for these the term "summer" is used to indicate the dry period rather than a particular part of the calendar year. The dry period occurs at quite different times in various parts of the tropics and its duration varies greatly. Many species characteristically abscise their leaves early in the dry period and do not leaf out again until after the advent of rain. For example, in Malaysia *Hevea* is usually leafless for 2 months out of the year (Chua, 1970), while in the more arid tropics trees may remain leafless much longer. Where the dry period is brief or does not occur every year as in the tropics of the South Pacific, a number of species have been recognized as *facultatively deciduous*, retaining their leaves in the wetter seasons and shedding them in the drier seasons (Merrill, 1945). Bews (1925) described a similar tendency shown by a few species of trees in subtropical Africa. He also noted that in the thorn-veld of South Africa most tree species are facultatively deciduous but some are regularly deciduous (e.g., *Erythrina* spp., *Brachystegia* spp.).

The summer deciduous habit is also a conspicuous adaptation to the dry season of Mediterranean climates (Mooney and Dunn, 1970). This habit is shown by *Encelia californica*, *Salvia mellifera*, *S. leucophylla*, and *Artemsia californica* and several other species of the coastal sage community of California; a comparable group of species occurs in the climatically similar region of Chile (Mooney *et al.*, 1970; Harrison *et al.*, 1971). In the Negev of Israel there are a number of species that abscise their winter leaves in early summer and replace them by much smaller summer leaves, e.g., *Poterium spinosum*, *Ononis natrix*, *Artemesia monosperma* (Orshan, 1954). A similar adaptation is shown by *Zygophyllum dumosum* from which the pair of leaflets is abscised from each leaf in early summer, leaving the succulent petiole to function during the dry season (Zohary and Orshan, 1954).

Another example in the wide range of summer deciduous habits is exhibited by the shrub *Fouquieria splendens* of the deserts of southwest North America. This spiny plant is leafless most of the year and puts out short leafy branchlets (short shoots) only after a rain (Henrickson, 1972) (Fig. 1). The leaves seldom persist for long; they are abscised soon after the return of hot, dry weather.

The second group of species showing summer leaf fall are those that abscise only a portion of their leaves. The group is represented by broad-leaved evergreens such as *Eucalyptus* spp., *Citrus* spp., and presumably many other tropical and subtropical evergreen species. Under unstressed

FIG. 1. *Fouquieria splendens* (ca. natural size). (A) Portion of a stem showing leafy short shoots that develop within a few days following rain. (B) Portion of a shoot during dry period. Leafy short shoots have abscised. The spines are persistent abaxial tissues of the petioles of the primary, long shoot leaves. Dormant buds at the nodes will again give rise to leafy short shoots after the next rain.

conditions these species abscise an occasional leaf at any time during the year. The abscised leaves are the disadvantaged ones that, because of shading or limited supplies of nutrients, lose out in the competition with younger, more vigorous leaves. After a period of hot, dry weather these species will have a flush of leaf abscission, shedding a portion of their older, less vigorous leaves. The amount of foliage abscised is a direct reflection of the intensity of stress to which the tree was exposed. In Mediterranean climates which have a relatively long dry summer, these species will commonly show several flushes of leaf abscission.

c. VERNAL ABSCISSION. Vernal (spring) abscission is characteristic of a large number of broad-leaved evergreen species. Typically all the leaves of the previous year's foliage are shed at a time that is correlated with flowering and/or appearance of new leaves. Species that show vernal abscission include *Magnolia grandiflora, Cinnamomum camphora, Persea americana,* and the live oaks such as *Quercus suber* and *Q. agrifolia.* Vernal leaf fall usually extends over a period of a few weeks, beginning shortly after the expansion of new foliage; thus, the trees always have an appreciable amount of foliage. However, in some individuals and in some varieties leaf fall precedes the emergence of new foliage; consequently for a very brief period such individuals may resemble a deciduous tree. While trees having vernal abscission are properly classified as evergreen by virtue of having some foliage at all times of the year, most of them barely qualify, retaining each year's foliage by only a little more than 12 months. Thus, trees which show vernal leaf abscission are similar to autumnal deciduous species in having a complete change of foliage every year.

The distinctive quality of vernal abscission is that it comes at the *beginning* of the season of growth, in contrast to autumnal (and summer) abscission which occurs at the *end* of a season of growth. The ecological factors associated with these two habits of leaf abscission, vernal and autumnal, are essentially opposite. Vernal abscission is correlated with rising temperature and increasing day length, while autumnal abscission is correlated with falling temperature and decreasing day length. If we assume that the mechanisms of cell separation are identical in vernal and autumnal abscission, how can we account for two such divergent sets of ecological factors bringing about identical biochemical results? The answer appears to lie in the complexity and variety of intermediate hormonal and metabolic control systems that receive and interpret ecological stimuli (discussed in more detail later in this chapter) and in the variety of internal factors that can affect abscission (discussed in Chapter 4).

The ecological changes of late summer and early autumn contribute to substantial changes in the hormonal and metabolic balance of the leaves and help to bring on leaf senescence. The internal changes combine to initiate and perhaps foster the ultimate biochemical steps of separation.

In vernal leaf abscission the leaves that are about to be shed show fewer signs of senescence than preabscission autumn leaves. What *is* apparent is rapid growth and development of young buds and flowers brought on by increasing temperature and lengthening photoperiod. Growing buds are rich sources of auxins (Went and Thimann, 1937; Aldén, 1971) and other growth hormones, some of which move basipetally to the older portions of twigs and branches. Auxins and gibberellins (GA) approaching an abscission zone from a proximal direction disturb the existing hormonal balance. A change in auxin gradient across the abscission zone, with increased amounts on the proximal side, is well recognized as an important promoting factor in abscission (Addicott, 1970). Changes in hormone gradients, together with the development of new sinks in the growing buds, appear to be important intermediate factors in the chain of events leading to vernal leaf abscission.

d. ABSCISSION OF MARCESCENT LEAVES. *i. Physiological separation.* In genera that are commonly deciduous in the autumn a few species retain an appreciable number of leaves through the winter and abscise them in the spring. The blade and most of the petiole die in the autumn, but the tissues at the very base of the petiole, including the abscission zone, remain alive and hold the marcescent leaf on the tree through winter. Tison (1900) described the anatomy of such abscission in *Carpinus betulus, Fagus sylvatica, Quercus hispanica,* and *Q. pedunculata.* Abscission of marcescent leaves of the American species *Quercus coccinea, Q. velutina, Q. marilandica, Q. rubra, Fagus grandifolia, Ostrya virginiana,* and *Acer saccharum* has been studied by Berkley (1931) and that of *Quercus palustris* and *Q. coccinea* by Hoshaw and Guard (1949). Examination of the abscission zones showed that in some cases abscission commenced in the autumn and then was arrested, presumably by the advent of cold weather. In any event, the abscission processes of marcescent leaves follow an anatomical pattern essentially identical with that of leaves that are abscised in the autumn (Berkley, 1931).

The abscission of marcescent leaves can most reasonably be interpreted as a kind of delayed autumnal abscission, delayed because the species concerned are rather slow to initiate autumnal abscission and cold weather slows metabolic activities. With the higher temperature of spring the process of abscission is resumed and goes to completion. It is possible that hormones from the developing buds may contribute to abscission of marcescent leaves as they appear to do in the case of vernal leaf abscission. Hoshaw and Guard (1949) observed in *Q. coccinea* that although the majority of the leaves had fallen by the time the buds began to swell, some abscission did not occur until later.

ii. Nonphysiological separation. At the end of its life the typical marcescent leaf withers, dies, and remains attached to the parent plant indefinitely. This habit is common in ferns, grasses, and palms, and in herbaceous annual plants generally. In woody species the marcescent leaves may eventually break off, but the separation is due to mechanical forces and does not involve physiological (biochemical) weakening of cell walls in the abscission zone. For example, on the desert shrub, *Parthenium argentatum,* marcescent leaves do not wither appreciably but become dry and brittle. In due course the dry leaves are broken off by such mechanical factors as wind. Commonly the leaf breaks from the stem at the abscission zone; this is the weakest part of the leaf because of generally thin cell walls and absence of fibers (Addicott, 1945). Some marcescent leaves of *Quercus* spp. may also be broken off during winter.

In this case the relatively brittle petioles snap at their narrowest point and the petiole stumps are abscised later by physiological action in the abscission zone (Hoshaw and Guard, 1949). In the case of many tree ferns, such as the New Zealand species *Cyathea medullaris*, and *C. dealbata*, the weight of the leaves causes the senescent petioles to buckle a few centimeters away from the trunk. The leaves then hang until the continued disintegration of petiolar tissues and the action of wind enable them to fall. The petiole stumps and especially the stumps of the vascular bundles may persist on the trunk indefinitely. However, after several years the stumps of vascular bundles on *Cyathea medullaris* usually are polished away by the abrasion of hanging dead leaves or adjacent shrubbery. Thus, a few feet below the crown of the tree the leaf scars are smooth and give the false impression that abscission was the result of active physiology. Other tree ferns and palms (e.g., *Dicksonia squarrosa*, *Washingtonia* spp.) hold their dead, pendant leaves tenaciously and these can accumulate as massive skirts around the trunks.

e. Origin and Evolution of Leaf Abscission Habits. The ability to separate cells quite likely appeared early in the evolution of multicellular organisms. Once this ability was established, the next likely steps were development and localization of control mechanisms so that individual cells, tissues, or organs could be abscised when conditions were appropriate. The value of such control mechanisms to survival and reproduction of a plant is readily apparent, and it is not surprising that every group of plants has developed abscission mechanisms of one kind or another.

The earliest fossil record of deciduous leaves that has come to our attention is in the Glossopteridae of the Southern Hemisphere during the Carboniferous period some 300,000,000 years ago. This group included deciduous woody plants of arborescent habit that shed their leaves in clusters on short shoots, apparently in much the same manner as the needle clusters of present day *Pinus* are shed (Plumstead, 1958). The deciduous habit of angiosperm trees of the Northern Hemisphere developed during the early Cretaceous period, over 100,000,000 years ago (Axelrod, 1966). Figure 2 shows a few fossil leaves representative of the several hundred species that occur in the Dakota formation of the Lower Cretaceous (Lesquereux, 1891). These leaves and many others in the formation can be assigned to present-day genera that are deciduous in habit. It is noteworthy that in both the Southern Hemisphere Carboniferous and the Northern Hemisphere Cretaceous the deciduous habit evolved in conjunction with establishment of a strongly seasonal climate, indicating that ecological factors had a decisive influence on the evolution of the habit.

From the limited information available we can surmise something of

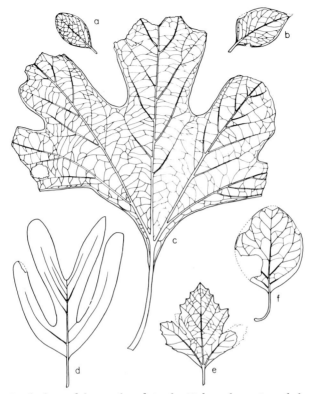

FIG. 2. Fossil abscised leaves found in the Dakota formation of the Lower Cretaceous (ca. one-half natural size; after Lesquereux, 1891). The present-day species of the genera represented are all autumn deciduous. (a) *Diospyros rotundifolia* Lesq. (b) *Nyssa snowiana* Lesq. (c) *Sassafras dissectum* Lesq. (d) *Liriodendron wellingtonii* Lesq. (e) *Crataegus tenuinervis* Lesq. (f) *Populus hyperborea* Heer.

the evolution of leaf abscission habits in the higher plants. The primitive condition appears to be one in which leaves fall singly in response to ecological stress and internal competition. The advanced condition is the deciduous habit in which leaves are abscised simultaneously in response to such ecological factors as drought, cold, or photoperiod. Present-day higher plants that are unable to abscise leaves appear to represent a specialized, advanced condition in which the ability to abscise has been lost. This condition is common in annual plants and in other species showing special adaptation to extended periods of dry weather in which retention of dead leaves on the plant is of little consequence to the survival of the species. It may be noted in passing that both the deciduous habit and loss of leaf abscission ability appear to have evolved independently in many of the families and genera in which they are found.

3. *Abscission of Other Organs*

While shedding of leaves is at times the most conspicuous manifestation of abscission in plants, abscission of other organs also has a significant role in the growth, development, and survival of many species. Apparently almost all discrete organs of higher plants are capable of being abscised in one species or another (see Addicott, 1965). For many of these organs little is known of the physiology or ecology of their abscission responses, but the abscission of terminal buds, twigs, branches, and bark has received some attention (Chapters 5 and 6) and that of flowers and fruits is becoming the object of systematic investigations (Chapters 9–12). This section will include a few comments on ecological aspects of some of the better known examples.

A number of woody species terminate a flush of shoot growth by abortion and abscission of shoot tips rather than by formation of a terminal bud enclosed in bud scales (Romberger, 1963). In various experiments, abortion and abscission of shoot tips have occurred earlier when water or nutrients were deficient, day length decreased, or temperature lowered. Conversely, ample water and nutrients, long days, and warm temperature tended to delay shoot tip abortion and abscission (see Romberger, 1963; and Chapter 5 of this volume). Internal competition from nearby buds and young leaves also influences the process; removal of nearby buds and leaves usually has favored continued growth of the shoot tip. From the limited observations so far made it is not yet possible to put together a simple picture of the control mechanism of shoot tip abortion and abscission. However, since the process can occur in different species at different times of the year under rather different conditions, it would appear that any of several factors may be the initiating one.

Twig and branch abscission brought about by physiological processes is uncommon, being restricted to relatively few genera, and the physiology and ecology have as yet received little attention (see Chapter 5). Many gymnosperms shed branchlets with leaves still attached. *Sequoia sempervirens* annually sheds some of its branchlets, mostly toward the end of the dry summer. In contrast *Taxodium distichum*, *Metasequoia glyptostroboides*, and *Larix* spp. are completely deciduous, abscising their branchlets in response to autumnal stimuli. Many species of *Quercus* and *Populus* have a swollen abscission zone at the base of their twigs and small branches (Eames and MacDaniels, 1947). The effectiveness of the process of separation appears to vary from species to species. In some, the twigs will fall by their own weight usually after most or all of the leaves have abscised; in others (e.g., *Quercus suber*), a moderate wind may be necessary to complete the separation. The pattern of branch abscission varies with species and leads to some distinctive growth habits such as those of the tree and shrub Euphorbias (Leach, 1970a,b). Some podocarp species can abscise branches up to 5 cm in diameter. Many of these species are tropical or subtropical and little information is available on the ecology of their branch abscission. However, *Agathis australis*, a forest tree, abscises its lower branches in response to the crowding and shading of nearby trees so that at maturity the lower 40 to 60 ft of the main stem is free of branches, a characteristic that contributes substantially to the high quality of the timber.

Bark abscission is a conspicuous feature of a number of woody species (see Chapter 6). There is no evidence that shedding of bark is accomplished by active physiological processes; thus it is more aptly termed "exfoliation" (Eames and MacDaniels, 1947). The pattern of bark exfoliation varies greatly from species to species and is determined by the location of layers of thin-walled cells that fracture mechanically to permit shedding. Shedding of bark is often seasonal, occurring toward the end of a period of growth and after hot weather. Shrinkage induced by desiccation of the outer bark facilitates separation from the trunk.

The abscission of flower buds, flowers, flower parts, fruits, and seeds is a normal part of the growth and development of many species. However, adverse conditions such as drought, flooding, frost, heat, mineral deficiencies, and insect infestation usually promote abscission of reproductive organs more strongly than vegetative organs (Chandler, 1951; Wittwer, 1954). Specific instances of the effects of these conditions on abscission of reproductive parts will be discussed later in this chapter.

B. Ecological Factors Affecting Abscission

An actively growing plant is a highly sensitive organism, responsive to very small changes in the environment. This is especially evident in relation to abscission for which plants have evolved a set of physiological mechanisms that can come into action rapidly and help to maintain a functional homeostasis in the face of adverse conditions. The plant is able to sense a variety of environmental changes, often to a fine degree, integrate them, and respond when indicated with the abscission of one or more organs. Conversely, in the presence of unusually favorable conditions, the plant will delay abscission appropriately (Table I).

The climatic factors to which abscission processes are sensitive include: (1) light, in relation to both photosynthesis and photoperiod; (2) temperature, particularly the extremes of heat and cold; (3) rainfall (especially its periodicity) and humidity; and (4) wind.

TABLE I

ABSCISSION EFFECTS OF SOME ECOLOGICAL FACTORS[a]

Ecological factors	Abscission[b]
Light	
Photosynthesis	
Normal	+
Deficient	− −
Excessive	− −
Photoperiod	
Long day	− − −
Short day	+ + +
Temperature	
Moderate	+
Light frost or brief heat	+ + +
Extreme cold or heat	− − −
Water	
Drought	+ + +
Flooding	+ +
High humidity	+ +
Wind	+ +
Soil	
Nitrogen	− − −
Deficiencies of	
N, P, K, S, Ca, Mg, Fe, Zn, B	+ +
Toxic levels of	
B, Fe, Zn, Cu, Mn, Cl, I	+ + +
Salinity/alkalinity	
Moderate	−
Severe	+ +
Atmosphere	
Oxygen (above 20%)	+ + +
Carbon dioxide	− − −
Pollutants	
ETH,[c] CO, SO_2, H_2S, O_3, PAN,[d]	+ + +
NO, NH_3, Hg, Cl_2, fluorides	
Biotic factors	
Fungal and bacterial pathogens	
e.g., *Verticillium, Omphalia,*	+ +
Cronartium	
Insects and mites	
e.g., *Lygus, Anthonomus,*	+ +
Tetranychus	

[a] See text for references and discussion.
[b] Abscission: "+" promotion; "−" retardation.
[c] ETH = Ethylene.
[d] PAN = Peroxyacetyl nitrates.

Edaphic factors that can influence abscission include: (1) soil moisture and aeration; (2) deficiencies and toxicities of mineral constituents; (3) salinity and alkalinity; and (4) soil-borne diseases.

Other factors that can influence abscission include: (1) atmospheric pollutants; (2) oxygen and carbon dioxide; (3) fire; (4) gravity; and (5) biotic factors such as insects and pathogenic microorganisms.

II. Ecological Factors and Their Physiological Implications for Abscission

A. INTRODUCTION

The variety of ecological factors that can influence abscission suggests that there must also be a considerable variety of internal "receptor mechanisms" that are sensitive to external changes and whose responses alter the internal physiology sufficiently to affect the course of abscission. This view is supported by a growing body of experimental evidence such as that presented in Chapter 4 (also see Addicott, 1965). The discussions in this section will be introduced by a summary of the major physiological factors involved with abscission (Table II).

The critical and final step in the abscission process is dissolution of cell walls in the abscission layer (Chapter 2; Addicott, 1965). Dissolution is brought about through the action of hydrolytic enzymes such as cellulases and pectinases (Chapter 4) apparently synthesized *de novo* in the abscission layer (Lewis and Varner, 1970). Enzyme synthesis requires: (1) a supply of amino acids for construction of the protein moieties of enzymes and (2) a supply of energy (presumably adenylate). The need for energy explains why respiration is essential to abscission, and why the abscission process requires oxygen and a modest supply of respiratory substrates (Carns *et al.*, 1951; Addicott, 1965). Further, certain amino acids are essential precursors in the biosynthesis of three of the plant hormones, indoleacetic acid (IAA), cytokinin (CK), and ethylene (ETH), and of course photosynthate and adenylate are necessary for synthesis of all of the hormones. Some of the principal relationships of physiological factors in abscission are shown in Fig. 3. The scheme is designed to direct attention to some of the more important nutritional and respirational aspects of abscission [the roles of hormones in abscission were recently reviewed (Addicott, 1970) and are further considered in Chapter 4]. The supply of products of photosynthesis, particularly sugars, is central to the metabolism of cell walls. If sugars are in ample supply,

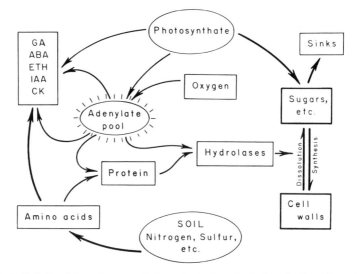

FIG. 3. Relationships of some of the principal ecological and physiological factors in abscission. Abbreviations: GA, gibberellins; ABA, abscisic acid; ETH, ethylene; IAA, indoleacetic acid and related auxins; CK, cytokinins. See text for explanation.

cell wall deposition will be heavier and the walls more difficult to hydrolyze. Lower levels of sugars lead to weaker cell walls and easier abscission. When sugars are in limited supply, they tend to be mobilized to the more vigorous parts of the plant (sinks) further weakening the less vigorous parts and encouraging hydrolysis of cell walls. Supplies of photosynthate are usually adequate to meet the energy requirements of abscission, but under some experimental conditions external supplies of sugar have been required for a reasonable rate of abscission (Biggs and Leopold, 1957). The mineral nutrients from the soil that are essential to normal metabolism can influence abscission in several ways. One of the most interesting relationships is that between nitrogen nutrition and hormone levels. The scheme in Fig. 3 is based on the fact that auxin levels are correlated with levels of available nitrogen (via the amino acid, tryptophan) (Avery *et al.*, 1937; H. R. Carns, unpublished). A similar relationship undoubtedly exists for CK; its molecule contains 5 nitrogen atoms derived from glycine, aspartic acid, and glutamine. Ethylene is released from the sulfur-containing amino acid, methionine (Yang, 1968). The scheme shown in Fig. 3 indicates that energy for the various reactions comes from an "adenylate pool" (e.g., ATP) which is maintained by oxidative respiration of photosynthate. The amount of energy needed for abscission is small relative to the requirements of the entire plant, but the energy must be made available within the spatial limits of the abscission layer and during the period of abscission. The foregoing explains the absolute requirement of oxygen for abscission (Carns *et al.*, 1951) and the need for at least small amounts of photosynthate (Biggs and Leopold, 1957). The actual interrelationships of the factors in abscission are far more intricate than could be displayed in such a scheme as that of Fig. 3; for example, the several known sites of hormonal action are not indicated (but see Addicott, 1970).

B. Climatic Factors

1. *Light*

Photosynthesis, through its products, influences abscission in several ways. The carbohydrates of photosynthesis are utilized directly for deposition of cell walls. When the environmental factors combine to favor photosynthesis but restrict overall growth, cell walls are heavier and abscission is more difficult. Conversely, if photosynthesis is minimal relative to growth requirements, then cell walls are thinner and abscission is easier. The equilibrium between sugars and cell walls appears to be fairly fluid; application of sucrose to explants substantially delayed abscission (Livingston, 1950; Martin, 1954). However, in explants from seedlings deficient in photosynthate, additional sugar was required for abscission (Biggs and Leopold, 1957). Presumably in such cases insufficient carbohydrate is available to meet the energy requirements for enzyme synthesis.

Abscission effects of photosynthesis may be localized, e.g., shaded leaves are abscised more readily than unshaded leaves (Meyers, 1940; Schaffalitzky de Muckadell, 1961). In deciduous fruit trees the abscission of young fruit is greater when nearby twigs are deficient in carbohydrate or have a lower than normal number of leaves (Chandler, 1951).

In 1923 Garner and Allard suggested that autumnal leaf fall is a photoperiodic phenomenon initiated by decreasing day length. This view is supported by a growing body of evidence. Street lights can delay the autumnal leaf fall of trees near the lights (Matzke, 1936; Schroeder, 1945). Young sugar maple (*Acer saccharum*) trees maintained in a constant day length of 16 hr retained their leaves up to 5 months longer than control trees in the natural day length (Olmsted, 1951). The normal autumnal leaf fall of *Plumeria* was completely prevented by interrupting the long nights of autumn and winter with artificial light (Murashige, 1966).

Although the close correlation between changing photoperiod and both autumnal and vernal leaf abscission is one of the most conspicuous as-

TABLE II

SOME PHYSIOLOGICAL RESPONSES TO ECOLOGICAL FACTORS
THAT INFLUENCE ABSCISSION[a]

Ecological factor	Physiological response[b]	
Light		
Photosynthesis	Carbohydrates	++
	Cell walls	++
Long days	Auxins; gibberellins	+++
	Abscisic acid, etc.	++
Short days	Auxins; gibberellins	− −
	Abscisic acid, etc.	++[c]
Temperature		
Moderate	Respiration; metabolism	+
Extreme	Injury	+++
Water		
Drought	Abscisic acid	+++
Flooding	Mineral uptake	− − −
High humidity	Cell wall solubilization	++
Soil		
Nitrogen	Amino acids; auxins; cytokinins; ethylene	+++
Zinc	Auxin (IAA)	++
Calcium	Calcium pectate	+++
Boron	Pollen tube growth	+++
Sulfur	Methionine/ethylene	+++
Excessive Fe, Cu, B, Cl, etc.	Toxic injuries	+++
Atmosphere		
Oxygen	Respiration	+++
Carbon dioxide	Respiration	− −
SO₂, NH₃, H₂S, etc.	Toxic injuries	++
Ethylene	Respiration; cellulase	+++
Fluorides	Binding of calcium	+++
Biotic factors		
Omphalia	IAA-oxidase	+++
Verticillium	Auxin	++
	Abscisic acid	+++
	Ethylene	+++
Penicillium, Alternaria, etc.	Pectinases; ethylene	+++
Lygus	Polygalacturonase	+++
Tetranychus	Ethylene	++

[a] See text for references, further examples, and discussion.

[b] Physiological response: "+" increase; "−" decrease.

[c] Some investigators have found no differences in abscisic acid levels between long and short day treatments.

pects of the abscission phenomenon, there is only fragmentary information on the physiology of the responses (Hendricks and Borthwick, 1965). In particular, the involvement of phytochrome in abscission remains obscure, but probably in time a role will emerge in the translation of photoperiodic stimuli into biochemical messages that can affect abscission. There are indications that long days produce high levels of auxins and gibberellins and low levels of inhibitors, and conversely, that short days produce low levels of auxins and gibberellins and high levels of inhibitors (Phillips and Wareing, 1958; Nitsch, 1963; but see Zeevaart, 1971a,b; Lenton *et al.*, 1972). [The inhibitor most frequently encountered in such investigations has been the hormone abscisic acid (ABA) (Eagles and Wareing, 1964; Rudnicki *et al.*, 1968; Bornman, 1969; Seeley, 1971).] Experiments with assays for hormones and application of hormones indicate that the hormonal changes in leaves correlated with decreasing day length could well be major factors in initiation of autumal leaf abscission. With regard to vernal leaf abscission, the increasing day length leads to an upsurge of hormone synthesis, especially in buds. Consequently, buds and developing shoots become strong sinks, and the auxin moving downward from the buds shifts the auxin gradient across the abscission zones of the old leaves. Both changes would serve to promote vernal leaf abscission.

The possibility that phytochrome might serve as the connecting link between photoperiodic stimuli and hormonal responses has been an attractive hypothesis. Both phytochrome and the hormones affect many of the same developmental processes of plants (e.g., see Goeschl *et al.*, 1967; Reid *et al.*, 1968; Russell and Galston, 1969), but evidence of a direct link between phytochrome and the levels or activity of hormones is still limited. Most of the present evidence suggests that the action sites and action mechanisms of phytochrome and plant hormones are different, and that the interactions between them take place indirectly in the course of lengthy pathways leading to morphogenetic expressions (Furuya, 1968).

2. *Temperature*

In excised abscission zones (explants) the abscission response to temperature shows a curve typical of complex physiological processes with a maximum near 30°C for beans (*Phaseolus vulgaris*) (Yamaguchi, 1954) and near 35°C for cotton (*Gossypium hirsutum*) (M. L. Chase, unpublished). Such data support the view that respiration-dependent synthesis of hydrolytic enzymes is a limiting factor in the process of abscission.

In addition to direct effects on metabolism in the abscission zone, temperature can have indirect effects through changes in composition and transport of materials from the subtending organ. Exposure to sublethal levels of cold has a variety of effects on physiological and biochemical processes in leaves. Cold-induced changes include: reduced photosynthesis and profound changes in chloroplast structure (Taylor and Rowley, 1971; Taylor and Craig, 1971), reduced protein synthesis and accumulation of amino acids (Spencer and Titus, 1972; Taylor *et al.*, 1972), and a wide variety of enzymatic changes (Spencer and Titus, 1972). Significant changes in hormonal metabolism accompany the biochemical changes. Young and Meredith (1971) reported that after exposure to subfreezing temperature the rate of ETH release by *Citrus* leaves increased approximately 100-fold. In another case, pear (*Pyrus communis*) fruits on trees kept under cool conditions (7°–18°C) for a 30-day preharvest period accumulated 5 times as much ABA as did fruit kept under warm conditions (16°–24°C) (Wang *et al.*, 1972).

The above observations and comments provide some insight into the leaf abscission behavior of cold-sensitive plants (e.g., *Citrus* and *Gossypium*) in which a mild freeze can induce leaf abscission while a severe freeze inhibits the process. The mild freeze brings about changes in leaf blade metabolism that induce abscission with but little injury to the abscission zone. Thus, abscission is initiated and can proceed. The severe freeze injures all tissues including the abscission zone so drastically that the tissues die before they can complete the process of abscission.

The time of advent of cold weather appears to influence the pattern of leaf abscission from the species of *Quercus* and *Fagus* that retain most of their leaves in marcescent condition through winter. Such species normally do abscise a small portion of their leaves during autumn. If it is assumed that the metabolic changes of abscission proceed very slowly in these species, then the advent of cold weather could be sufficient to arrest the process and delay its completion until spring. If this explanation is correct we would expect to find more leaves abscising in years when the autumn months are mild than in years when cold weather sets in early. A minor but still interesting effect of cold comes from its ability to increase the brittleness of petioles and thus facilitate breakage of the petioles by the winds of winter storms (Hoshaw and Guard, 1949).

Attention has recently been drawn to a phenomenon in which apparently sound branches fall from mature trees (Harris, 1972). This has been observed only on hot, still summer afternoons. Such limb breakage has been reported for *Platanus acerifolia*, *Cedrus deodara*, *Ficus retusa*, and several species of elm, eucalyptus, oak, and pine from widely distributed inland areas of California. The fallen branches ranged in diameter at the break from 10 to 50 cm. As yet there is no explanation to account for these occurrences.

Excessive heat contributes to abscission in many situations (e.g., see Wiesner, 1904). Abscission of flower buds, flowers, and young fruits is frequently correlated with periods of hot weather (see Wittwer, 1954). Petal abscission of several genera has been accelerated by brief exposure of flowers to temperatures of 33° to 40°C (Fitting, 1911). Leaf abscission of mature cotton plants can be induced by brief exposure to air heated to 150°–300°C (Batchelder *et al.*, 1971). However, only rarely is it possible to separate the effects of heat *per se* from the effects of moisture stress that usually accompanies heat. In a few instances at least, such as in abscission of young fruits of irrigated cotton during periods of hot weather, heat would seem to be the damaging factor as ample supplies of water are available to the plants. Various metabolic processes are adversely affected by excessive heat (Treshow, 1970); such adverse effects probably underly the heat-induced failure of pollen development, pollination, fertilization, and young fruit development. In most plants such failures inevitably result in abscission of aborted ovaries and other flower parts.

3. *Water*

The availability of water is a major factor controlling abscission. Regions of the world having relatively uniform rainfall are characterized by evergreen vegetation, but regions having periods of drought often have many species that are drought (summer) deciduous or are capable of abscising an appreciable number of their leaves when stressed for moisture. Moisture stress initiates the abscission of buds, flowers, or young fruits as well as leaves of many crop species.

The physiological effects of drought on leaves have been studied for many years (see Laude, 1971; Crafts, 1968; Vaadia *et al.*, 1961); some of these effects can be linked to control of abscission. Moisture stress induces numerous biochemical changes such as decreased activity of phenylalanine ammonialyase and nitrate reductase (Bardzik *et al.*, 1971), loss of RNA (Gates and Bonner, 1959; Bourque and Naylor, 1971), and increased activity of IAA-oxidase (Darbyshire, 1971a,b). Such changes can serve to assist the plant in adapting to moisture stress (Darbyshire, 1971b) and quite conceivably some of the changes could tend to promote abscission. Several hormonal responses to moisture stress have now been reported, including: (1) decrease in diffusible auxin (Hartung and Witt, 1968); (2) decrease in CK activity (Itai and Vaadia, 1965, 1971); (3) increase in ETH release (McMichael *et al.*, 1972); and (4) increase in ABA (Wright and Hiron, 1969, 1972; Zeevaart, 1971b; Mizrahi *et al.*, 1971). Abscisic acid induces stomatal closure and thus reduces transpiration (Mittleheuser and Van Steveninck, 1969); this mechanism functions as an early line of defense against the injury of moisture stress. If the ABA levels remain high for more than a brief period they would tend to promote abscission as would reduced auxin and increased ETH levels (Addicott, 1970).

Flooding of soil can lead to a number of deleterious effects on plants (e.g., see Kramer, 1951), including the abortion and subsequent abscission of flower buds (Lloyd, 1920). A well-known consequence of flooding is impaired mineral uptake, especially nitrogen uptake; flooded plants soon develop nitrogen deficiency symptoms. Nitrogen deficiency not only hampers metabolism generally, it specifically leads to lower levels of auxin (Avery *et al.*, 1937) and presumably to lower levels of CK and methionine from which ETH is released. Reid and Crozier (1971) observed that flooding of tomato roots reduced the levels of GA in the roots, shoots, and bleeding sap. Such hormonal changes are likely to render the buds less effective as sinks and hence less able to prevent their own abscission.

Rutland (1888) reported an interesting influence of running water on leaf abscission. He observed that trees of weeping willow (*Salix babylonica*) growing on the bank of a cool stream with many of their roots in the water absciced their leaves earlier in the autumn than did nearby trees growing in soil that was warmer than the stream. Rutland was inclined to attribute the earlier leaf fall of the trees on the stream bank to the colder temperature of their roots. However, there is another possible explanation. Trees with an abundant water supply would be able to utilize fully the available photosynthate and mineral nutrients for development of a sizable plant body. Trees with a less abundant water supply would develop less extensive canopies and would tend to accumulate photosynthate, some of it being deposited as heavier cell walls. These trees would also accumulate larger amounts of nitrogenous substances. The higher levels of photosynthate and nitrogenous substances would act to retard abscission as discussed in Sections II,A and II,C,2,a.

Moisture is involved in abscission in other ways. For example, as abscission proceeds there is often a rapid loss of water from tissues distal to the abscission zone; the consequent shrinkage of the distal tissues facilitates their separation from the proximal tissues. Separation is also facilitated by tissue tensions within the abscission zone (Brown and Addicott, 1950), tensions that result from turgor differences among the tissues. When citrus and almond trees under irrigation in arid climates become sufficiently injured from moisture stress to induce leaf abscission, the leaves do not separate from the tree as long as the irrigation water is withheld. However, the leaves fall very promptly after the reapplication of water. Also, during periods of hot, dry weather, the petals of

Magnolia grandiflora are retained on the flower in a semimarcescent condition for 2 or 3 days longer than when the weather is cooler and more humid (Griesel, 1954). A related phenomenon is observed occasionally in autumn deciduous plants in mid or late autumn when leaf fall has been relatively light. After a cool night of high humidity or fog, a massive leaf fall can occur as the temperature rises during the early daylight hours (8 to 10 AM). In these cases of moisture-induced abscission it appears that most of the steps in abscission had been completed well in advance of actual separation, but that the supply of water had been insufficient to complete the process. Whether the water is needed merely to solubilize the end products of cell wall dissolution or to participate in some earlier chemical reaction is not yet known.

Short-term weather conditions can have a variety of important effects on abscission and shedding. For example, a heavy rain falling on cotton flowers during the late morning hours (the period of pollination) can wash away the pollen, with the result that the flowers that were at anthesis abort and abscise. A similar rain falling at other hours during the day has no ill effects on pollination and subsequent fruit development (Lloyd, 1920). Of course a heavy downpour of rain or hail with stones of sufficient size can mechanically separate leaves, flowers, and fruit, sometimes to a disastrous degree (see Treshow, 1970).

4. Wind

The mechanical force of wind is an appreciable factor contributing to the final separation of leaves and other organs. Wind assists in removal of marcescent leaves, breaking the blades away from petioles that have become brittle from desiccation or cold. Also, we have observed that almost all of the small leafless twigs that are abscised by *Quercus suber* each winter fall during periods of windy weather.

Almost 100 years ago Darwin (1877) recorded his observations of very rapid petal abscission that results from a slight jarring of the flowers of *Verbascum* spp. Fitting (1911) extended these experiments and noted that petal abscission of *Verbascum* usually occurred 1–3 min after shaking, and that petals of flowers of *Veronica* spp. and *Cistus* spp. were also abscised rapidly after shaking.

Although the physiological effects of wind have received some attention (Satoo, 1961; Whitehead, 1957), the influence of these effects on abscission is essentially unstudied. We might expect, however, that dry winds would accentuate and perhaps accelerate the abscission responses to drought. Also, it is well known that when plants are manipulated regularly or subjected to moderate shaking their growth is reduced (e.g., see Neel and Harris, 1971). It has been suggested that this growth reduction could be the result of increased release of ETH (Turgeon and Webb, 1971). Also, mechanical irritation of *Bryonia* shoots has decreased auxin concentrations (Boyer, 1967). Both hormonal changes could be promotive of abscission.

C. EDAPHIC FACTORS

1. *Soil Moisture and Aeration*

Some of the information on soil moisture that is pertinent to abscission has been dealt with in Section II,B,3. Here we will add a few comments on intermediate factors that could be involved with abscission induced by soil flooding. Flooded soil rapidly becomes anaerobic; in consequence root metabolism and plant development generally are greatly modified (Kramer, 1951; Treshow, 1970; Reid and Crozier, 1971). The impaired mineral uptake under anaerobic conditions can have a variety of deleterious effects on photosynthesis and respiration in the shoots; these tend to retard the development of organs such as flower buds and thereby make them more susceptible to abscission.

Several hormonal responses to flooding are known. In *Helianthus,* flooding induced a threefold increase in auxin levels that persisted for about 2 weeks (Phillips, 1964). The symptoms of epinasty and adventitious root formation described by Kramer (1951) suggest that auxin also increases for a time in tomato and a few other plants. In yellow poplar (*Liriodendron tulipifera*) such symptoms were slight (Kramer, 1951), and Lloyd (1920) reported none from his many observations of flooding effects on cotton. Presumably plants such as cotton contain few nitrogenous reserves from which to synthesize auxin after the nitrogen supply from soil is cut off. This view is supported by the readiness with which cotton sheds its buds and young fruits after flooding (Lloyd, 1920). High amounts of soil moisture, short of flooding, can also decrease auxin levels. Hartung and Witt (1968) found that soil moisture levels above 60% of field capacity reduced appreciably the amount of diffusible auxin in stems of *Helianthus;* and in the leaves of *Anastatica* the diffusible auxin was reduced when soil moisture was above 60% of field capacity and reached zero at 90% of field capacity. Levels of other hormones are also sensitive to flooding. In the exudate from flooded *Helianthus* roots the levels of CK fell steadily after 12 hr and sharply after 48 hr as reported by Burrows and Carr (1969). Similarly levels of GA in tomato plants fell rapidly after the roots were flooded (Reid and Crozier, 1971). Although these investigations are only indirectly related to abscission, they indicate one important mechanism, hormone metabolism, whereby high soil moisture can influence abscission.

2. Mineral Elements

a. DEFICIENCIES. Abscission of leaves, buds, flowers, and fruits is a common response to mineral deficiency. Elements whose deficiency has led to abscission include N, P, K, S, Ca, Mg, Zn, B, and Fe (Sprague, 1964; Cook *et al.*, 1960; Seeley, 1950; Smith and Reuther, 1949; Treshow, 1970; Oberly and Boynton, 1966). Each of these elements is essential to the physiology and biochemistry of the plant in one or more ways, so it is not surprising that a deficiency will seriously affect developmental processes. What at first may seem unusual is that deficiency of so many elements can affect abscission. However, this fact emphasizes the complexity of the channels of abscission control. And when one considers how important the process of abscission is to successful maintenance of a homeostatic balance among competing organs, it is not surprising that plants have developed mechanisms capable of sensing environmental changes and coordinating appropriate abscission responses (see Section I,A,1,e). What is known of some of these mechanisms is discussed in the following paragraphs.

The involvement of nitrogenous compounds in abscission has already been mentioned in a preliminary way (Section II,A; and see Addicott and Lynch, 1955). The amino acids serve as a pool of intermediate products from which many of the more specialized compounds are synthesized. Among these compounds are some of the major plant hormones. The biosynthesis of IAA proceeds directly from the amino acid, tryptophan (Gruen, 1959; Muir and Lantican, 1968), and the levels of IAA are correlated with availability of nitrogen in the soil (Avery *et al.*, 1937). Since IAA is a powerful inhibitor of abscission (see Chapter 4) the consequent reduced levels of IAA from nitrogen deficiency strongly favor abscission and defoliation (Table III). Presumably the levels of CK in the plant are also very sensitive to the supply of nitrogen, although we are unaware of data bearing directly on this point. The cytokinins can delay abscission by increasing the ability of an organ to serve as a sink. Ethylene is released by breakdown of the amino acid, methionine, and tissues that are higher in proteins and amino acids are richer sources of ETH (Hall *et al.*, 1957). There appear to be no investigations of ETH release in relation to soil nitrogen, but as far as ETH for abscission is concerned the availability of external nitrogen may be relatively unimportant. Ordinarily there are enough protein and amino acid reserves in the cells of the abscission zone to enable the release of considerable ETH. With

TABLE III

RELATION OF SOIL NITROGEN, PETIOLE AUXIN, AND SUSCEPTIBILITY TO
CHEMICAL DEFOLIATION[a]

Soil nitrogen[b]	Petiole auxin[c]	Defoliation[d] (%)
0	0.1	95
75	0.4	80
105	0.8	70
150	1.4	50

[a] Data of H. R. Carns (unpublished).

[b] Pounds N per acre, applied as NO_3^-.

[c] IAA equivalent in micrograms per kilogram fresh weight, measured 60 hr after defoliant application.

[d] Percent leaf abscission following application of 20 lb/acre of calcium cyanamide.

respect to leaf abscission, we can summarize the responses to nitrogen deficiency by indicating that reduced availability of nitrogen will reduce the levels of IAA and CK in the blade, thus reducing the ability of the blade to inhibit abscission. At the same time the abscission zone tends to conserve nitrogenous substances and its ability to release ETH remains essentially unaltered.

Zinc is a micronutrient whose deficiency symptoms include leaf abscission. Its presence in the plant is essential to synthesis of auxin (IAA) (Skoog, 1940; also see Takaki and Kushizaki, 1970). Thus, auxin deficiency appears to be the route by which Zn deficiency induces abscission. Skoog (1940) also observed that plants deficient in Cu and Mn became low in auxin, but the deficiency symptoms had to be severe before auxin levels decreased. This may account for the fact that deficiencies of Cu and Mn do not induce abscission; at least we have found no such reports in the literature.

Calcium deficiency leads to considerable abscission as well as to several other responses. Calcium functions as a constituent of calcium pectate in the cell wall and particularly in the middle lamella; Ca ions act as bridges between the carboxyl groups of the branching chains of pectic substances. When Ca is deficient there are fewer bridges, the pectic substances are weaker, and abscission is facilitated.

Boron deficiency also leads to abscission in susceptible species. The biochemical role of boron in plants is still incompletely known, but boron is very much involved in carbohydrate physiology. Its deficiency favors the pentose shunt pathway for degradation of hexoses and leads to accumulation of phenolic acids (Epstein, 1972). For some years boron has been implicated in auxin metabolism. Under some conditions IAA can alleviate symptoms of B deficiency (Eaton, 1940), although under other conditions it cannot (MacVicar and Tottingham, 1947). Still other experiments showed that B deficiency can lead to an excess of auxin, possibly by decreasing activity of IAA-oxidase (Coke and Whittington, 1968). This view is supported by the fact that phenolic acids (in particular the dihydroxyphenols) mentioned above can inhibit the action of IAA-oxidase. As yet a clear relationship between B deficiency and auxin physiology has not emerged, but the results to date indicate that B-induced changes in auxin levels and gradients can have appreciable effects on abscission.

Boron is essential to germination of pollen and is present in stigmatic fluid in rather high concentrations (see Linskens, 1964). Thus, B deficiency could be expected to lead to flower abortion and abscission.

In a study of 7 forest community types in north central Florida Monk (1966) found a preponderance of deciduous species at the sites having the greater rainfall and the more fertile soils. In contrast, evergreen species were more numerous at the drier and less fertile sites. Monk suggested that segregation of evergreen species to the dry, less fertile sites may be related to the more gradual return of nutrients to the soil by year-round leaf fall of the evergreens. This would permit a more tightly closed mineral cycle and enable the evergreens to survive where deciduous species could not. In a subsequent investigation the stems of deciduous and evergreen species were innoculated with ^{45}Ca. In the course of a year 85% of the Ca was lost from the deciduous trees with their abscised leaves, while only 10 to 25% of the Ca was lost from the evergreen species with their abscised leaves (Monk, 1971).

b. Toxicities. A number of mineral substances, particularly the metallic ions, can be quite toxic to plants (Epstein, 1972; Levitt, 1972; Childers, 1966), but the literature contains only a few comments on abscission responses to toxic ions. High levels of B, from either excessive applications of borax or excessive levels of B in irrigation water, has produced a number of toxicity symptoms including reduced fruit set and premature leaf abscission (Richards, 1954; Oberly and Boynton, 1966; Ballinger et al., 1966). High levels of mineral ions applied to foliage can be sufficiently toxic to induce leaf abscission, and some of the ions have been considered for use as agricultural defoliants. The more effective ones include Fe^{3+}, Zn^{2+}, Cu^{2+}, Mn^{2+}, Cl^-, and I^- (Addicott and Lynch, 1957; Herrett et al., 1962; Biggs, 1971). The mechanism of action of the ions in promoting abscission is unknown but a reduction of IAA levels in the leaf blade by I^- has been suggested (Herrett et al., 1962). Increased release of ETH accompanied abscission acceleration by Fe^{3+} and Cu^{2+} (Ben-Yehoshua and Biggs, 1970). However no increase in ETH release could be detected when abscission was accelerated by Mn^{2+} (Biggs, 1971). A number of other changes are likely as a result of the stress of toxicity, such as higher levels of ABA, lower levels of GA and CK, and greatly modified metabolism, generally.

3. Salinity and Alkalinity

Soils that are saline (high in neutral salts) or alkaline (high in exchangeable Na) occur in arid regions, where crop production is often attempted if irrigation water is available. If salinity and alkalinity are not too severe, crops can be produced by good management (Richards, 1954). However, symptoms of toxicity are common, especially among the less tolerant crops (see Richards, 1954), and these symptoms can include considerable leaf abscission (Hayward and Wadleigh, 1949; Treshow, 1970).

One of the main effects of soil salinity is to limit availability of water to plants. This effect of salinity serves to promote abscission in the same way as does moisture stress (Section II,B,3). Also saline soils frequently accumulate toxic levels of Ca^{2+}, Mg^{2+}, K^+, Cl^-, HCO_3^-, or NO_3^-; but as indicated in the previous section, there is little in the literature on the abscission physiology of these ions.

Alkaline soils have a dispersed structure that leads to poor water infiltration and poor aeration. Lack of aeration can be sufficiently severe to prevent adequate absorption of both water and nitrogen, so that the symptoms and abscission responses of plants growing in alkaline soils can resemble those of plants under flooded conditions (Section II,C,1).

4. Soil-borne Diseases

A number of soil-borne pathogenic organisms, particularly *Verticillium* and *Fusarium* spp., induce vascular diseases with symptoms including wilting and leaf abscission (Talboys, 1968). Little attention has been given to the mechanism of abscission induction as other symptoms have been more important to understanding the disease. However, many observations have been made of hormonal and biochemical responses, some of which could well be related to abscission.

Auxin (IAA) is readily synthesized by a large number of soil microorganisms, including pathogens. Higher than normal auxin level is a common (although not universal) symptom of diseased tissues (Gruen, 1959). Also levels of scopoletin and other growth inhibitors may rise (Sequeira and Kelman, 1962; Wood, 1967). A study by Wiese and DeVay (1970) showed that stems and leaves of cotton infected with *Verticillium albo-atrum* contained higher amounts IAA than controls and this is correlated with reduced ability of the infected tissues to decarboxylate IAA. The latter property was ascribed to higher levels of caffeic acid and other IAA-oxidase inhibitors in the affected tissues. Levels of ABA in the affected leaves were double the levels in the controls, and ETH released by the infected plants was 2 to 5 times that by the control plants. In addition, different isolates of *V. albo-atrum* varied in their ability to induce the hormonal responses. Measurements of hormone changes correlated with these diseases have not yet been made in enough tissues to enable a suggestion as to the manner in which the hormones might be inducing abscission. However, the increased ABA and ETH levels are consistent with the usual acceleration of abscission by those hormones.

Pectolytic enzymes such as pectin methylesterase and polygalacturonase are produced by most of the major vascular pathogens *in situ,* and cellulases are produced at least *in vitro* (Talboys, 1968). If released in or near the abscission zone these enzymes could be major factors facilitating abscission. Probably there are still other pathways through which plant diseases facilitate abscission.

D. ATMOSPHERIC FACTORS

1. *Oxygen and Carbon Dioxide*

The importance of oxygen to abscission has already been indicated (Section II,A); it is absolutely essential. Although there have been few investigations of the effects of O_2 and CO_2 on abscission in intact plants, there have been several such investigations with abscission in explants. With bean explants the rate of abscission increased with increasing oxygen concentration from 0 to 40%; from 40 to 100% O_2 the abscission rate remained maximal (Carns *et al.*, 1951; Yamaguchi, 1954). Working in the presence of half-maximal levels of ETH (0.1 ppm), Abeles and Gahagan (1968) found the rate of abscission rose steeply as O_2 was increased up to about 10% and was maximal when O_2 was 20% or higher. With explants of *Coleus* Rosen and Siegel (1963) found that lowering the O_2 in the atmosphere to 5% inhibited abscission. That low level of O_2 also greatly reduced the acceleration of abscission by ETH or GA. The rate of abscission from treatments combining 5% O_2 with either ETH or GA was less than half of the control rate. Siegel and Gerschman (1959) observed a variety of toxic effects following exposure of plants to O_2 under elevated pressure for a few hours. Exposure for 15 hr at 6 atm was severely damaging to most species and the symptoms included leaf abscission, particularly in *Begonia* and *Euphorbia*.

Carbon dioxide is a retardant of abscission in explants; a few percent of CO_2 in air reduced the rate of abscission appreciably (Abeles and Gahagan, 1968). However, in mixtures of CO_2 and O_2, more than 15% CO_2 was required to reduce the rate of abscission to half the rate in pure O_2 (Yamaguchi, 1954). The investigations cited here also included experiments with interactions of O_2 and CO_2 with ETH; O_2 tends to augment the acceleration by ETH whereas CO_2 is antagonistic to ETH. However, the mechanisms of these interactions are not yet clear. In contrast to the foregoing, exposure of intact flowers to CO_2 as well as to many other gases can induce petal abscission (Fitting, 1911; and see the following section) but this abscission appears to be a response to injury to the petals rather than to a direct effect of the gas on the abscission zone.

2. Pollutants

Toxic effects of air pollutants from both natural and industrial sources have been known for many centuries (see Weinstein and McCune, 1971; Middleton, 1961). Pollutants that have produced serious injury to plants include: illuminating gas, ETH, CO, SO_2, H_2S, fluorides, O_3, peroxyacetyl nitrates (PAN), nitrogen oxides, NH_3, Cl_2, and mercury vapor; also there are likely to be many others whose effects are not yet recognized. The agricultural, physiological, and biochemical effects of pollutants have been intensively studied (Treshow, 1970, 1971; Weinstein and McCune, 1971; Middleton, 1961; Dugger and Ting, 1970; Crocker, 1948; Rich, 1964), and most of the pollutants have been reported to induce abscission (Kendall, 1918; Crocker, 1948; Treshow, 1970; Weinstein and McCune, 1971). With a few exceptions, the action of the pollutants in accelerating abscission has not been investigated. But in each case enough is known of the effects of the pollutant to support some speculation as to the mode of action in abscission via pathways of carbohydrate, nitrogen, or hormone physiology. Several such speculations have been advanced in previous sections of this chapter. We will however discuss briefly some of the interesting and/or useful aspects of a few of the pollutants.

Ethylene, long recognized as the most toxic constituent of illuminating gas, is also a very common product of combustion and is considered responsible for much of the crop damage near metropolitan areas (Treshow, 1970). However ETH is unique among the pollutants in being also a plant product and in functioning as a plant hormone in some kinds of growth, fruit ripening, and abscission. The role of ETH in abscission is discussed in Chapter 4. It is of interest to note that larger than normal amounts of ETH are released by plant tissues in response to wounding and other kinds of stress and that ETH promotes adaptive responses that help to overcome the adverse effects of stress (Burg, 1968; Pratt and

Goeschl, 1969). Thus, ETH the *pollutant* (in high concentrations) is a cause of stress and injury to the plant, while ETH the *hormone* (in lower concentrations) is an important physiological agent in the recovery of the plant from injury or stress.

Ozone is a major constituent of photochemical smog and appears to be the principal cause of the chlorotic decline that has seriously damaged *Pinus ponderosa* in the San Bernardino Mountains of southern California. Symptoms of ozone injury include chlorotic mottling of needles, abscission

FIG. 4. Effect of ozone on *Pinus ponderosa*. Right: branch fumigated with 0.5 ppm ozone, 9 hr per day for 18 days. Note increased abscission of older needle clusters. Left: control branch, unfumigated. (Photograph of P. R. Miller, U. S. Forest Service.)

of needle clusters, and terminal dieback (Fig. 4) (Miller *et al.*, 1963).

Fluorides produce a variety of toxic symptoms in plants, but the only abscission responses yet in the literature are two reports of increased abscission of young fruits (Treshow, 1971). This response appears to be the result of reduced pollen germination and tube growth that leads to reduced fertilization and consequent increased abortion and shedding of young fruit. It has been suggested that this response to fluorides results from their binding of Ca; the latter is essential for normal pollen germination and tube development (Treshow, 1971).

Sulfur dioxide in low concentrations is not especially toxic to plants; indeed it enters readily into the S metabolism of the plant. Much of the S of SO_2 can be incorporated into the S-containing amino acids and some accumulates as sulfates (Treshow, 1970). It is at the higher concentrations and in susceptible species that symptoms appear after toxic ions accumulate. There are some potentially valuable applications of SO_2 in agriculture. Preliminary experiments have demonstrated that treatment with SO_2 speeds conversion of freshly harvested plants into silage and that SO_2 can be used to defoliate cotton plants prior to harvest.

Air pollution by NH_3 results from accidents in connection with its widespread industrial and agricultural use. As a result of such accidents it was discovered that brief exposure to NH_3 can induce leaf abscission of cotton plants. Since NH_3 is relatively inexpensive, some attention has been given to development of equipment for the application of NH_3 to field cotton prior to harvest (Elliott, 1964; Walhood and McMeans, 1965).

E. Other Factors

1. *Fire*

Fire is such a drastic agent that injuries from it are seldom sufficiently mild to permit the abscission zone to function. Obviously fire is a very potent defoliating agent and is sometimes so used in forestry (Hough, 1968). Further, fire can have a significant pruning effect on branches. In forest regions that are subject to periodic understory fires, such as the yellow pine (*Pinus ponderosa*) forest of western America, trees whose bark is resistant can survive an understory fire with only the loss of their lower branches (see Daubenmire, 1967). *Sequoiadendron gigantea*, the big tree, is the prime example of a tree whose mature habit is determined to a considerable degree by "fire pruning."

Fire is also an important factor in shedding of seeds of serotinous conifers (Daubenmire, 1967). For example, cones of *Pinus attenuata* can remain closed for many years and seldom open to release seeds unless they have been heated by a forest fire. Thus, a burned forest of *P. attenuata* promptly reseeds itself (Jepson, 1925).

2. Gravity

Gravity usually provides a mechanical *coup de grâce* to abscission of an organ by effecting final separation after cell wall dissolution has advanced sufficiently. The physiological influence of gravity on abscission of debladed petioles of *Coleus* has been studied in some detail (Terpstra, 1956; Vendrig, 1960). When plants were placed in a horizontal position abscission of petioles was accelerated, and the response was accompanied by hypertrophy and root development on the lower side of the stem. Auxin physiology, particularly transport, was greatly modified in the horizontal plants, but the manner in which abscission was affected is not yet clear. When horizontal plants were rotated about their axes at various speeds from 0.01 to 1.0 rpm, the slower the speed of rotation, the more rapid was the rate of petiole abscission (Vendrig, 1960).

3. Biotic Factors

a. PATHOGENIC MICROORGANISMS. Leaf abscission is a common response to infection by pathogenic microorganisms; it can serve as a valuable defense mechanism. Almost complete defoliation can occur in a relatively short period, as in diseases of almond (*Prunus amygdalus*) caused by *Coryneum beijerinckii* and *Cladosporium carpophilum* (Ogawa *et al.,* 1955). Another interesting example occurs in *Ribes roezli* which is one of the intermediate hosts of white pine blister rust (*Cronartium ribicola*). Kimmey (1945) observed that in *R. roezli* normal leaf abscission began with the dry weather of summer and continued throughout the season, but rust infection caused considerable premature leaf abscission. Premature leaf abscission was greater from plants growing in the open where moisture stress was higher and in seasons when there were no summer rains. The first light frosts of autumn promoted the fall of infected leaves more than that of uninfected leaves (Kimmey, 1945).

As indicated in Section II,C,4 very little is yet known of the mechanisms whereby pathogenic microorganisms induce abscission; but many metabolic, hormonal, and enzymatic changes brought about by pathogens (Beckman, 1964; Wood, 1967) can easily be implicated. In addition to those changes in levels of auxin already mentioned, the rapid leaf abscission induced by infections of *Omphalia flavida* in *Coffea* was correlated with large amounts of IAA-oxidase produced by the fungus (Sequeira and Steeves, 1954). Also, increased release of ETH is a common response to infection (Williamson and Dimock, 1953). Before we can fully understand the mechanism of pathogen-induced abscission, many more investigations involving the simultaneous examination of several hormonal and physiological factors are needed (similar to the investigation of Wiese and DeVay, 1970).

In most situations plant diseases appear to affect shedding by modifying the physiological factors that normally influence or control abscission. However, this is not always the case, and a few of the more unusual examples of shedding responses to plant pathogens are worthy of note. One example is that of leaf shedding induced by plane tree anthracnose caused by *Gnomonium venata*. This disease leads to considerable necrosis of young leaves and twigs, particularly at the base of petioles near the abscission zone, and sometimes in the tender new stem tissues above a node. Thus, individual leaves, or short twigs bearing leaves, are shed as a result of petiole or stem necrosis.

An unusual kind of abscission occurs in a shot-hole disease of almond (*Prunus amygdalus*) caused by *Clasterosporium carpophilum*. In this disease an adventitious abscission layer forms in a ring around each small area of infection and the infected tissue shrinks from loss of moisture and falls away. The "proximal" portion of the abscission layer develops into a corky, protective layer around the shot-hole (Samuel, 1927). The physiology of this response has not been investigated, but it probably involves the release of a wound substance, possibly ETH. A wound to the leaf blade of cherry-laurel (*Prunus laurocerasus*) with a hot needle can induce a similar shot-hole abscission response (Samuel, 1927).

Finally, there is at least one example of a disease in which abscission is inhibited. The corky bark disease of grape (*Vitis* spp.) can rapidly and completely kill the branches of a vine. Apparently the cells of the abscission zones are killed before they have time to abscise the leaves and the dead leaves remain in place on the vines indefinitely (Beukman and Goheen, 1970).

b. INSECTS. In the millions of years that insects and plants have been interacting, a variety of abscission relationships (or lack of them) have evolved. Perhaps the simplest example is that of the "defoliator" insects whose feeding consumes the blades of leaves (Kulman, 1971). Usually after such feeding the main veins and petioles remain attached to the plant. Although the leaf has been destroyed as an effective photosynthetic organ, the remaining tissues appear capable of producing sufficient hormones to at least delay abscission.

In other cases the feeding of insects can have more subtle and far-reaching effects. Even partial consumption of a flower bud or young fruit will often be followed by abortion and abscission of the affected organ. Loss of seeds can be critical. Developing seeds are a major source of growth hormones, essential to fruit development; if seeds are eaten, fruits are usually abscised.

The saliva of some insects contains toxic substances that produce many symptoms including necrosis (Addicott and Romney, 1950), abortion of flowers and young fruits, and abscission (Carter, 1939). In the case of *Lygus hesperus* the injuries have been shown to be due to salivary polygalacturonase (Strong, 1970). Another example is the abscission of flower buds of cotton caused by feeding of larvae of the boll weevil (*Anthonomus grandis*). Abscission is induced after only slight physical damage and appears to be due to secretion of a protein by the larvae. Larval extracts containing the protein induced abscission of flower buds that was preceded by epinasty of the floral bracts and hypertrophy in the abscission zone (King and Lane, 1969). Whether the abscission activity is due to the protein itself or to some *in vivo* breakdown product has not yet been determined.

Some insects and the mite, *Tetranychus atlanticus,* also can induce considerable abscission while seemingly having few other effects on the plant. As few as 4 *Antestia* bugs feeding on a *Coffea* plant can cause economically damaging abscission of green berries (Le Pelley, 1942). In some cases ETH may be the active agent. Rose leaves infested with the mite, *Tetranychus bimaculatus,* released much more ethylene than did healthy leaves (Williamson and Dimock, 1953).

In closing this section we should draw attention to the role of insect pollination in the prevention of abortion and abscission of flowers. For many species, pollination by insects is essential to fertilization and normal fruit development, and absence of the appropriate insect inevitably results in abortion, typically followed by abscission.

Acknowledgments

We gratefully acknowledge the assistance of Alice B. Addicott in preparation of the manuscript and figures, and helpful comments from Marilyn L. Chase.

References

Abeles, F. B., and Gahagan, H. E., III. (1968). Abscission: The role of ethylene, ethylene analogues, carbon dioxide, and oxygen. *Plant Physiol.* **43**, 1255–1258.

Addicott, F. T. (1945). The anatomy of leaf abscission and experimental defoliation in guayule. *Amer. J. Bot.* **32**, 250–256.

Addicott, F. T. (1965). Physiology of abscission. *In* "Handbuch der Pflanzenphysiologie" (W. Ruhland, ed.), Vol. 15, Part 2, pp. 1094–1126. Springer-Verlag, Berlin and New York.

Addicott, F. T. (1968). Environmental factors in the physiology of abscission. *Plant Physiol.* **43**, 1471–1479.

Addicott, F. T. (1970). Plant hormones in the control of abscission. *Biol. Rev. Cambridge Phil. Soc.* **45**, 485–524.

Addicott, F. T., and Lynch, R. S. (1955). Physiology of abscission. *Annu. Rev. Plant Physiol.* **6**, 211–238.

Addicott, F. T., and Lynch, R. S. (1957). Defoliation and desiccation: Harvest-aid practices. *Advan. Agron.* **9**, 67–93.

Addicott, F. T., and Romney, V. E. (1950). Anatomical effects of *Lygus* injury to guayule. *Bot. Gaz.* (*Chicago*) **112**, 133–134.

Aldén, T. (1971). Seasonal variations in the occurrence of indole-3-acetic acid in buds of *Pinus sylvestris*. *Physiol. Plant.* **25**, 54–57.

Avery, G. S., Jr., Burkholder, J. R., and Creighton, H. B. (1937). Nutrient deficiencies and growth hormone concentration in *Helianthus* and *Nicotiana*. *Amer. J. Bot.* **24**, 553–557.

Axelrod, D. I. (1966). Origin of deciduous and evergreen habits in temperate forests. *Evolution* **20**, 1–15.

Ballinger, W. E., Bell, H. K., and Childers, N. F. (1966). Peach nutrition. *In* "Temperate to Tropical Fruit Nutrition" (N. F. Childers, ed.), pp. 276–390. Somerset Press, Somerville, New Jersey.

Bardzik, J. M., Marsh, H. V., Jr., and Havis, J. R. (1971). Effects of water stress on the activities of three enzymes in maize seedlings. *Plant Physiol.* **47**, 828–831.

Batchelder, D. G., Porterfield, J. G., and McLaughlin, G. (1971). Thermal defoliation of cotton. *Proc. Cotton Defoliation-Physiol. Conf., 25th, 1971* p. 36.

Beckman, C. H. (1964). Host responses to vascular infection. *Annu. Rev. Phytopathol.* **2**, 231–252.

Ben-Yehoshua, S., and Biggs, R. H. (1970). Effects of iron and copper ions in promotion of selective abscission and ethylene production by citrus fruit and the inactivation of indoleacetic acid. *Plant Physiol.* **45**, 604–607.

Berkley, E. E. (1931). Marcescent leaves of certain species of *Quercus*. *Bot. Gaz.* (*Chicago*) **92**, 85–93.

Beukman, E. F., and Goheen, A. C. (1970). Grape corky bark. *In* "Virus Diseases of Small Fruits and Grapevines" (N. W. Frazier, ed.), pp. 207–209. Univ. of California Press, Div. Agr. Sci., Berkeley.

Bews, J. W. (1925). "Plant Forms and Their Evolution in South Africa." Longmans, Green, New York.

Biggs, R. H. (1971). Citrus abscission. *HortScience* **6**, 388–392.

Biggs, R. H., and Leopold, A. C. (1957). Factors influencing abscission. *Plant Physiol.* **32**, 626–632.

Bornman, C. H. (1969). Laminal abscission in *Streptocarpus*. *Proc. Int. Bot. Congr., 11th, 1969* Abstracts, p. 18.

Bourque, D. P., and Naylor, A. W. (1971). Large effects of small water deficits on chlorophyll accumulation and ribonucleic acid synthesis in etiolated leaves of jack bean (*Canavalia ensiformis* [L.] DC.). *Plant Physiol.* **47**, 591–594.

Boyer, N. (1967). Modification de la croissance de la tige de Bryone (*Bryonia dioica*) a la suite d'irritations tactiles. *C. R. Acad. Sci., Ser. D* **264**, 2114–2117.

Brown, H. S., and Addicott, F. T. (1950). The anatomy of experimental leaflet abscission in *Phaseolus vulgaris*. *Amer. J. Bot.* **37**, 650–656.

Burg, S. P. (1968). Ethylene, plant senescence and abscission. *Plant Physiol.* **43**, 1503–1511.

Burrows, W. J., and Carr, D. J. (1969). Effects of flooding the root system of sunflower plants on the cytokinin content in the xylem sap. *Physiol. Plant.* **22**, 1105–1112.

Carns, H. R., Addicott, F. T., and Lynch, R. S. (1951). Some effects of water and oxygen on abscission *in vitro*. *Plant Physiol.* **26**, 629–630.

Carter, W. (1939). Injuries to plants caused by insect toxins. *Bot. Rev.* **5**, 273–326.

Chandler, W. H. (1951). "Deciduous Orchards." Lea & Febiger, Philadelphia, Pennsylvania.

Childers, N. F., ed. (1966). "Temperate to Tropical Fruit Nutrition." Somerset Press, Somerville, New Jersey.

Chua, S. E. (1970). Physiology of Foliar Senescence and Abscission in *Hevea brasiliensis* Muell. Arg. *Res. Arch. Rubber Res. Inst. Malaya, Docu.* **63**.

Coke, L., and Whittington, J. (1968). The role of boron in plant growth. IV. Interrelationships between boron and indol-3-yl-acetic acid in the metabolism of bean radicles. *J. Exp. Bot.* **19**, 295–308.

Cook, J. A., Lynn, C. D., and Kissler, J. J. (1960). Boron deficiency in California vineyards. *Amer. J. Enol. Viticult.* **11**, 185–194.

Crafts, A. S. (1968). Water deficits and physiological processes. *In* "Water Deficits and Plant Growth" (T. T. Kozlowski, ed.), Vol. 2, pp. 85–133. Academic Press, New York.

Crocker, W. (1948). "Growth of Plants." Van Nostrand-Reinhold, Princeton, New Jersey.

Darbyshire, B. (1971a). The effect of water stress on indoleacetic acid oxidase in pea plants. *Plant Physiol.* **47**, 65–67.

Darbyshire, B. (1971b). Changes in indoleacetic acid oxidase activity associated with plant water potential. *Physiol. Plant.* **25**, 80–84.

Darwin, C. R. (1877). "The Different Forms of Flowers on Plants of the Same Species." Murray, London.

Daubenmire, R. F. (1967). "Plants and Environment." Wiley, New York.

Dostál, R. (1967). "On Integration in Plants." Harvard Univ. Press, Cambridge, Massachusetts.

Dugger, W. M., and Ting, I. P. (1970). Air pollution oxidants—their effects on metabolic processes in plants. *Annu. Rev. Plant Physiol.* **21**, 215–234.

Eagles, C. F., and Wareing, P. F. (1964). The role of growth substances in the regulation of bud dormancy. *Physiol. Plant.* **17**, 697–709.

Eames, A. J., and MacDaniels, L. H. (1947). "Introduction to Plant Anatomy." McGraw-Hill, New York.

Eaton, F. M. (1940). Interrelations in the effects of boron and indole-acetic acid on plant growth. *Bot. Gaz. (Chicago)* **101**, 700–705.

Elliott, F. C. (1964). Defoliation with anhydrous ammonia and desiccation with arsenic acid in Texas. *Proc. Cotton Defoliation-Physiol. Conf., 18th, 1964* pp. 8–10.

Epstein, E. (1972). "Mineral Nutrition of Plants: Principles and Perspectives." Wiley, New York.

Fitting, H. (1911). Untersuchungen über die vorzeitige Entblätterung von Blüten. *Jahrb. Wiss. Bot.* **49**, 187–263.

Furuya, M. (1968). Biochemistry and physiology of phytochrome. *Progr. Phytochem.* **1**, 347–405.

Garner, W. W., and Allard, H. A. (1923). Further studies in photoperiodism, the response of the plant to relative length of day and night. *J. Agr. Res.* **23**, 871–920.

Gates, C. T., and Bonner, J. (1959). The response of the young tomato plant to a brief period of water shortage. IV. Effects of water stress on the ribonucleic acid metabolism of tomato leaves. *Plant Physiol.* **34**, 49–55.

Goeschl, J. D., Pratt, H. K., and Bonner, B. A. (1967). An effect of light on the production of ethylene and the growth of the plumular portion of etiolated pea seedlings. *Plant Physiol.* **42**, 1077–1080.

Griesel, W. O. (1954). Cytological changes accompanying abscission of perianth segments of *Magnolia grandiflora*. *Phytomorphology* **4**, 123–132.

Gruen, H. E. (1959). Auxins and fungi. *Annu. Rev. Plant Physiol.* **10**, 405–440.

Hall, W. C., Truchelut, G. B., Leinweber, C. L., and Herrero, F. A. (1957). Ethylene production by the cotton plant and its effects under experimental and field conditions. *Physiol. Plant.* **10**, 306–317.

Harris, R. W. (1972). High-temperature limb breakage. *Arborist's News* **48**, 133–134.

Harrison, A. T., Small, E., and Mooney, H. A. (1971). Drought relationships and distribution of two Mediterranean-climate California plant communities. *Ecology* **52**, 869–875.

Hartung, W., and Witt, J. (1968). Über den Einfluss der Bodenfeuchtigkeit auf den Wuchsstoffgehalt von *Anastatica hierochuntica* und *Helianthus annuus*. *Flora (Jena), Abt. B.* **157**, 603–614.

Hayward, H. E., and Wadleigh, C. H. (1949). Plant growth on saline and alkali soils. *Advan. Agron.* **1**, 1–38.

Hendricks, S. B., and Borthwick, H. A. (1965). The physiological functions of phytochrome. *In* "Chemistry and Biochemistry of Plant Pigments" (T. W. Goodwin, ed.), pp. 405–436. Academic Press, New York.

Henrickson, J. (1972). A taxonomic revision of the Fouquieriacea. *Aliso* **7**, 439–537.

Herrett, R. H., Hatfield, H. H., Crosby, D. G., and Vlitos, A. J. (1962). Leaf abscission induced by the iodide ion. *Plant Physiol.* **37**, 358–363.

Hoshaw, R. W., and Guard, A. T. (1949). Abscission of marcescent leaves of *Quercus palustris* and *Q. coccinea*. *Bot. Gaz. (Chicago)* **110**, 587–593.

Hough, W. A. (1968). Carbohydrate reserves of saw-palmetto: Seasonal variation and effects of burning. *Forest Sci.* **14**, 399–405.

Itai, C., and Vaadia, Y. (1965). Kinetin-like activity in root exudate of water-stressed sunflower plants. *Physiol. Plant.* **18**, 941–944.

Itai, C., and Vaadia, Y. (1971). Cytokinin activity in water-stressed shoots. *Plant Physiol.* **47**, 87–90.

Jepson, W. L. (1925). "A Manual of the Flowering Plants of California." Associated Students Store, Berkeley, California.

Kendall, J. N. (1918). Abscission of flowers and fruits in the Solanaceae, with special reference to *Nicotiana*. *Univ. Calif., Berkeley, Publ. Bot.* **5**, 347–428.

Kimmey, J. W. (1945). The seasonal development and the defoliating effect of *Cronartium ribicola* on naturally infected *Ribes roezli* and *R. nevadense*. *Phytopathology* **35**, 406–416.

King, E. E., and Lane, H. C. (1969). Abscission of cotton flower buds and petioles caused by protein from boll weevil larvae. *Plant Physiol.* **44**, 903–906.

Kramer, P. J. (1951). Causes of injury to plants resulting from flooding of the soil. *Plant Physiol.* **26**, 722–736.

Kulman, H. M. (1971). Effects of insect defoliation on growth and mortality of trees. *Annu. Rev. Entomol.* **16**, 289–324.

Laude, H. M. (1971). Drought influence on physiological processes and subsequent growth. *Crop Sci. Soc. Publ.* **2**, 45–56.

Leach, L. C. (1970a). *Euphorbia* species from the flora zambesiaca area. IX. *J. S. Afr. Bot.* **36**, 13–52.

Leach, L. C. (1970b). Euphorbiae succulentae angolenses. III. *Bol. Soc. Broteriana* **44**, 185–206.

Lenton, J. R., Perry, V. M., and Saunders, P. F. (1972). Endogenous abscisic acid in relation to photoperiodically induced bud dormancy. *Planta* **106**, 13–22.

Le Pelley, R. H. (1942). The food and feeding habits of *Antestia* in Kenya. *Bull. Entomol. Res.* **33**, 71–89.

Lesquereux, L. (1891). The flora of the Dakota Group. *U. S. Geol. Surv., Monogr.* **17**, 1–400.

Levitt, J. (1972). "Responses of Plants to Environmental Stresses." Academic Press, New York.

Lewis, L. N., and Varner, J. E. (1970). Synthesis of cellulase during abscission of *Phaseolus vulgaris* leaf explants. *Plant Physiol.* **46**, 194–199.

Linskens, H. F. (1964). Pollen physiology. *Annu. Rev. Plant Physiol.* **15**, 255–270.

Livingston, G. A. (1950). *In vitro* tests of abscission agents. *Plant Physiol.* **25**, 711–721.

Lloyd, F. E. (1920). Environmental changes and their effect upon boll-shedding in cotton (*Gossypium herbaceum*). *Ann. N. Y. Acad. Sci.* **29**, 1–131.

McMichael, B. L., Jordan, W. R., and Powell, R. D. (1972). An effect of water stress on ethylene production by intact cotton petioles. *Plant Physiol.* **49**, 658–660.

MacVicar, R., and Tottingham, W. E. (1947). A further investigation of the replacement of boron by indoleacetic acid. *Plant Physiol.* **22**, 598–602.

Martin, L. B. (1954). Abscission and starch distribution following application of sucrose and indoleacetic acid to excised abscission zones of *Phaseolus vulgaris*. Ph.D. Thesis, University of California, Los Angeles.

Matzke, E. B. (1936). The effect of street lights in delaying leaf-fall in certain trees. *Amer. J. Bot.* **23**, 446–452.

Merrill, E. D. (1945). "Plant Life of the Pacific World." Macmillan, New York.

Meyers, R. M. (1940). Effect of growth substances on the absciss layer of *Coleus*. *Bot. Gaz.* (*Chicago*) **102**, 323–338.

Middleton, J. T. (1961). Photochemical air pollution damage to plants. *Annu. Rev. Plant Physiol.* **12**, 431–448.

Miller, P. R., Parmeter, J. R., Jr., Taylor, O. C., and Cardiff, E. A. (1963). Ozone injury to the foliage of *Pinus ponderosa*. *Phytopathology* **53**, 1072–1076.

Mittelheuser, C. J., and Van Steveninck, R. F. M. (1969). Stomatal closure and inhibition of transpiration induced by (RS)-abscisic acid. *Nature* (*London*) **221**, 281–282.

Mizrahi, Y., Blumenfeld, S., Bittner, S., and Richmond, A. E. (1971). Abscisic acid and cytokinin contents of leaves in relation to salinity and relative humidity. *Plant Physiol.* **48**, 752–755.

Monk, C. D. (1966). An ecological significance of evergreenness. *Ecology* **47**, 504–505.

Monk, C. D. (1971). Leaf decomposition and loss of ^{45}Ca from deciduous and evergreen trees. *Amer. Midl. Natur.* **86**, 379–384.

Mooney, H. A., and Dunn, E. L. (1970). Convergent evolution of Mediterranean-climate evergreen sclerophyll shrubs. *Evolution* **24**, 292–303.

Mooney, H. A., Dunn, E. L., Shropshire, F., and Song, L. (1970). Vegetation comparisons between the Mediterranean climatic areas of California and Chile. *Flora (Jena)* **159**, 480–496.

Muir, R. M., and Lantican, B. P. (1968). Purification and properties of the enzyme system forming indoleacetic acid. *In* "Biochemistry and Physiology of Plant Growth Substances" (F. Wightman and G. Setterfield, eds.), pp. 259–272. Runge Press, Ottawa.

Murashige, T. (1966). The deciduous behavior of a tropical plant. *Physiol. Plant.* **19**, 348–355.

Neel, P. L., and Harris, R. W. (1971). Motion-induced inhibition of elongation and induction of dormancy in *Liquidambar*. *Science* **173**, 58–59.

Nitsch, J. P. (1963). The mediation of climatic effects through endogenous regulating substances. *In* "Environmental Control of Plant Growth" (L. T. Evans, ed.), pp. 175–193. Academic Press, New York.

Oberly, G. H., and Boynton, D. (1966). Apple nutrition. *In* "Temperate to Tropical Fruit Nutrition" (N. F. Childers, ed.), pp. 1–50. Somerset Press, Somerville, New Jersey.

Ogawa, J. M., Nichols, C. W., and English, H. (1955). Almond scab. *Calif. Dep. Agr., Bull.* **44**, 59–62.

Olmsted, C. E. (1951). Experiments on photoperiodism, dormancy, and leaf age and abscission in sugar maple. *Bot. Gaz. (Chicago)* **112**, 365–393.

Orshan, G. (1954). Surface reduction and its significance as a hydroecological factor. *J. Ecol.* **42**, 442–444.

Phillips, I. D. J. (1964). Root-shoot hormone relations. II. Changes in endogenous auxin concentration produced by flooding of the root system in *Helianthus annuus*. *Ann. Bot. (London)* [N.S.] **28**, 37–45.

Phillips, I. D. J., and Wareing, P. F. (1958). Effect of photoperiodic conditions on the levels of growth inhibitors in *Acer pseudoplatanus*. *Naturwissenschaften* **45**, 26.

Plumstead, E. P. (1958). The habit of growth of Glossopteridae. *Trans. Geol. Soc. S. Afr.* **61**, 81–94.

Pratt, H. K., and Goeschl, J. D. (1969). Physiological role of ethylene in plants. *Annu. Rev. Plant Physiol.* **20**, 541–584.

Reid, D. M., and Crozier, A. (1971). Effects of waterlogging on the gibberellin content and growth of tomato plants. *J. Exp. Bot.* **22**, 39–48.

Reid, D. M., Clements, J. B., and Carr, D. J. (1968). Red light induction of gibberellin synthesis in leaves. *Nature (London)* **217**, 580–582.

Rich, S. (1964). Ozone damage to plants. *Annu. Rev. Phytopathol.* **2**, 253–266.

Richards, L. A., ed. (1954). "Diagnosis and Improvement of Saline and Alkaline Soils," Agr. Handb. No. 60. U. S. Dep. Agr., Washington, D. C.

Romberger, J. A. (1963). Meristems, growth, and development in woody plants. *U. S. Dep. Agr., Forest Serv., Tech. Bull.* 1293.

Rosen, L. A., and Siegel, S. M. (1963). Effect of oxygen tension on the course of ethylene- and gibberellin-induced foliar abscission. *Plant Physiol.* **38**, 189–191.

Rudnicki, R., Pieniążek, J., and Pieniążek, N. (1968). Abscisin II in strawberry plants at two different stages of growth. *Bull. Acad. Pol. Sci., Ser. V* **16**, 127–130.

Russell, D. W., and Galston, A. W. (1969). Blockage by gibberellic acid of phytochrome effects on growth, auxin responses, and flavonoid synthesis in etiolated pea internodes. *Plant Physiol.* **44**, 1211–1216.

Rutland, J. (1888). The fall of the leaf. *Trans. N. Z. Inst.* **21**, 110–120.

Samuel, G. (1927). On the shot-hole disease caused by *Clasterosporium carpophilum* and on the "shot-hole" effect. *Ann. Bot. (London)* **41**, 375–404.

Satoo, T. (1961). Wind, transpiration, and tree growth. *In* "Tree Growth" (T. T. Kozlowski, ed.), pp. 299–310. Ronald Press, New York.

Schaffalitzky de Muckadell, M. (1961). Environmental factors in development stages of trees. *In* "Tree Growth" (T. T. Kozlowski, ed.), pp. 289–297. Ronald Press, New York.

Schroeder, C. A. (1945). Tree foliation affected by street lights. *Arborist's News* 10, 1–3.

Seeley, J. G. (1950). Potassium deficiency of greenhouse roses. *Proc. Amer. Soc. Hort. Sci.* 56, 466–470.

Seeley, S. D. (1971). Electron capture gas chromatography of plant hormones with special reference to abscisic acid in apple bud dormancy. Ph.D. Thesis, Cornell University, Ithaca, New York.

Sequeira, L., and Kelman, A. (1962). The accumulation of growth substances in plants infected by *Pseudomonas solanacearum*. *Phytopathology* 52, 439–448.

Sequeira, L., and Steeves, T. A. (1954). Auxin inactivation and its relation to leaf drop caused by the fungus *Omphalia flavida*. *Plant Physiol.* 29, 11–16.

Siegel, S. M., and Gershman, R. (1959). A study of the toxic effects of elevated oxygen tension on plants. *Physiol. Plant.* 12, 314–323.

Skoog, F. (1940). Relationships between zinc and auxin in the growth of higher plants. *Amer. J. Bot.* 27, 939–951.

Smith, P. F., and Reuther, W. (1949). Observations on boron deficiency in citrus. *Proc. Fla. State Hort. Soc.* 62, 31–38.

Spencer, P. W., and Titus, J. S. (1972). Biochemical and enzymatic changes in apple leaf tissue during autumnal senescence. *Plant Physiol.* 49, 746–750.

Sprague, H. B., ed. (1964). "Hunger Signs in Crops," 3rd ed. McKay, New York.

Strong, F. E. (1970). Physiology of injury caused by *Lygus hesperus. J. Econ. Entomol.* 63, 808–814.

Takaki, H., and Kushizaki, M. (1970). Accumulation of free tryptophan and tryptamine in zinc deficient maize seedlings. *Plant Cell Physiol.* 11, 793–804.

Talboys, P. W. (1968). Water deficits in vascular disease. *In* "Water Deficits and Plant Growth" (T. T. Kozlowski, ed.), Vol. 2, pp. 255–311. Academic Press, New York.

Taylor, A. O., and Craig, A. S. (1971). Plants under climatic stress. II. Low temperature, high light effects on chloroplast ultrastructure. *Plant Physiol.* 47, 719–725.

Taylor, A. O., and Rowley, J. A. (1971). Plants under climatic stress. I. Low temperature, high light effects on photosynthesis. *Plant Physiol.* 47, 713–718.

Taylor, A. O., Jepson, N. M., and Christeller, J. T. (1972). Plants under climatic stress. III. Low temperature, high light effects on photosynthetic products. *Plant Physiol.* 49, 798–802.

Terpstra, W. (1956). Some factors influencing the abscission of debladed leaf petioles. *Acta Bot. Neer.* 5, 157–170.

Theophrastus. (285 B.C.). "Enquiry into Plants" (English translation by A. Hort, Vol. I. Putnam, New York, 1916).

Tison, A. (1900). Recherches sur la chute des feuilles chez les dicotyledonées. *Mem. Soc. Linn. Normandie* 20, 121–137.

Treshow, M. (1970). "Environment and Plant Response." McGraw-Hill, New York.

Treshow, M. (1971). Fluorides as air pollutants affecting plants. *Annu. Rev. Phytopathol.* 9, 21–44.

Turgeon, R., and Webb, J. A. (1971). Growth inhibition and mechanical stress. *Science* 174, 961–962.

Vaadia, Y., Raney, F. C., and Hagan, R. M. (1961). Plant water deficits and physiological processes. *Annu. Rev. Plant Physiol.* **12**, 265–292.

Vendrig, J. C. (1960). On the abscission of debladed petioles in *Coleus rhenaltianus* especially in relation to the effect of gravity. *Wentia* **3**, 1–96.

Walhood, V. T., and McMeans, J. L. (1965). The action of common liquid fertilizers as desiccants in the cotton harvest-aid program. *Proc. Cotton Defoliation-Physiol. Conf., 19th, 1965* pp. 88–91.

Wang, C. Y., Wang, S. Y., and Mellenthin, W. M. (1972). Identification of abscisic acid in Bartlett pears and its relationship to premature ripening. *Agr. Food Chem.* **20**, 451–453.

Wareing, P. F. (1956). Photoperiodism in woody plants. *Annu. Rev. Plant Physiol.* **7**, 191–214.

Weinstein, L. H., and McCune, D. C. (1971). Effects of fluoride on agriculture. *J. Air Pollut. Contr. Ass.* **21**, 410–413.

Went, F. W., and Thimann, K. V. (1937). "Phytohormones." Macmillan, New York.

Whitehead, F. H. (1957). Wind as a factor in plant growth. *In* "Control of the Plant Environment" (J. P. Hudson, ed.), pp. 84–95. Butterworth, London.

Wiese, M. V., and DeVay, J. E. (1970). Growth regulator changes in cotton associated with defoliation caused by *Verticillium albo-atrum. Plant Physiol.* **45**, 304–309.

Wiesner, J. (1871). Untersuchungen über die herbstliche Entlaubung der Holzgewäche. *Sitzungsber. Akad. Wiss. Wien, Math.-Naturwiss. Kl.* **64**, 465–509.

Wiesner, J. (1904). Über den Hitzlaubfall. *Ber. Deut. Bot. Ges.* **22**, 501–505.

Williamson, C. E., and Dimock, A. W. (1953). Ethylene from diseased plants. *Yearb. Agr. (U. S. Dep. Agr.)* pp. 881–886.

Wittwer, S. H. (1954). Control of flowering and fruit setting by plant regulators. *In* "Plant Regulators in Agriculture" (H. B. Tukey, ed.), pp. 62–80. Wiley, New York.

Wood, R. K. S. (1967). "Physiological Plant Pathology." Blackwell, Oxford.

Wright, S. T. C., and Hiron, R. W. P. (1969). (+)-Abscisic acid, the growth inhibitor induced in detached wheat leaves by a period of wilting. *Nature (London)* **224**, 719–720.

Wright, S. T. C., and Hiron, R. W. P. (1972). The accumulation of ABA in plants during wilting and under other stress conditions. *In* "Plant Growth Substances, 1970" (D. J. Carr, ed.), pp. 291–298. Springer-Verlag, Berlin and New York.

Yamaguchi, S. (1954). Some interrelations of oxygen, carbon dioxide, sucrose and ethylene in abscission. Ph.D. Thesis, University of California, Los Angeles.

Yang, S. F. (1968). Biosynthesis of ethylene. *In* "Biochemistry and Physiology of Plant Growth Substances" (F. Wightman and G. Setterfield, eds.), pp. 1217–1228. Runge Press, Ottawa.

Young, R., and Merideth, F. (1971). Effect of exposure to subfreezing temperatures on ethylene evolution and leaf abscission in *Citrus. Plant Physiol.* **48**, 724–727.

Zeevaart, J. A. D. (1971a). Effects of photoperiod on growth rate and endogenous gibberellins in the long-day rosette plant spinach. *Plant Physiol.* **47**, 821–827.

Zeevaart, J. A. D. (1971b). (+)-Abscisic acid content of spinach in relation to photoperiod and water stress. *Plant Physiol.* **48**, 86–90.

Zohary, M., and Orshan, G. (1954). Ecological studies in the vegetation of the near eastern deserts V. The Zygophylletum dumosi and its hydroecology in the Negev of Israel. *Vegetatio, Haag* **5–6**, 340–350.

. 4 .

Internal Factors
Regulating Abscission

Daphne J. Osborne

I. Introduction

Abscission of plant parts is an unmistakable biological phenomenon if the organ being shed drops to the ground. Separation processes occurring in tissues such as the bark of trees and in the periderm of roots are slower and much less dramatic. For this reason they are less amenable to experimental study and we therefore have less information on the factors that regulate them. For an understanding of how abscission processes are controlled the plant physiologist has directed most of his studies to the abscission of leaves or fruits and it remains to be shown how far the information acquired from these systems is applicable to separation processes as a whole. Rather than attempt to survey in this chapter all

125

facets of the endogenous control of leaf and fruit abscission, I have assembled evidence from experiments with a number of plants which, taken together, suggest a model for regulation of the abscission process. The model will be developed in three stages: stimulus, signal, and response. The stimulus sets in motion a progressive development which at a certain stage culminates in a signal. Cell separation and shedding is the biochemical response to the signal.

II. The Signal

The first question one may ask is, "Is there any condition that is recognizable as the forerunner of abscission?" The answer is a clear affirmative. Senescence of the leaf or ripening of a fruit (a special form of senescence) appears always to precede the shedding process and is therefore considered to be the signal. In general, senescence can be recognized by the color change of the organ from green to yellow and at the ultrastructural level by the progressive disruption first of chloroplast organization (Shaw and Manocha, 1965) and then of other organelles (Butler and Simon, 1971). Occasionally, however, following a stress shock, leaves and fruits may separate while still green, and this can result from dehydration by drought or from excessive heat or cold. In such circumstances biochemical studies show (Gates and Bonner, 1959; Mothes, 1964) that leaf blades exhibit some symptoms before leaf fall that are similar to those encountered in senescence and it is likely, though not demonstrated, that the abscission region itself may react prematurely in a way closely parallel to normal abscission. So whereas senescence naturally progresses from the distal to the proximal regions of an organ with senescence occurring last in the cells that abut the actual point of separation, under stress conditions the sequence may be altered such that the cells close above an abscission zone become senescent first and the leaf is then shed green.

An important aspect of the signal is therefore not so much whether the organ itself is senescent, but whether the cells immediately above the abscission zone are senescent. Because senescence is normally contagious from cell to cell from the apex to the base of an organ, the cells just above the separation point are normally the last to attain this state. At this point the senescence wave is halted and the zone forms a distinct yellow/green junction comprising senescent cells from the organ next to nonsenescent cells of the body of the plant. It is at this junction of the two physiologically dissimilar tissues that the processes of abscission take place.

III. What Is Senescence?

Although in general physiological terms senescence has been defined as changes which result in a decreased survival capacity (Strehler, 1962), we can be somewhat more precise in this present context. In leaves, yellowing and loss of chlorophyll is associated with a fall in total protein, RNA, and DNA (Wollgiehn, 1961; Osborne, 1962; Parthier, 1964), a decline in the synthesis of RNA and protein (Osborne, 1962), a decrease in enzymes such as nitrate reductase (McKee, 1962), and pectin methylesterase (Osborne, 1958a), and an increase in hydrolytic enzymes such as proteases (Anderson and Rowan, 1966) and nucleases (Srivastava and Ware, 1965). Some of these changes, e.g., loss of chlorophyll, occur when a plant is transferred to darkness, but this is etiolation not senescence, for on returning to the light the processes are reversed and the leaves regreen. Senescence occurs *in the light*, and unequivocal demonstrations of reversal or rejuvenation of leaves are few. Many of the instances reported may well be more properly attributed to nutrient deficiencies, for senescence can be only temporarily arrested by additions of metal ions, phosphate, nitrogen, or carbohydrates (Wood *et al.*, 1944; Vickery *et al.*, 1937, 1946). However, the addition of one of the plant hormones, auxin, gibberellin, or a cytokinin, will effectively halt the progress of senescence for considerable periods of time, and as a result, the time at which abscission occurs can be correspondingly postponed.

The marked retardation of senescence afforded by supplying the correct hormone is a clue to the answer to the next question: What is the *stimulus* that causes the leaf or the fruit to become senescent? Because so much more is understood for leaves, the question will be discussed primarily in relation to leaf senescence with analogies to fruits only when this seems appropriate.

IV. The Stimulus for Senescence

A. Deciduous Species

The timing of the galaxy of golden colors that heralds the synchronous autumnal fall of leaves can largely be attributed to the shortening day length that follows the summer solstice. Many people will have noticed that where trees are exposed to extra light from street lamps the leaves closest to the lights remain green longest and are shed last. One of the best experiments illustrating this was carried out by F. W. Went in the controlled conditioned laboratories of the first Phytotron at Pasadena. He exposed trees of *Prunus persica* to extended day length (16-hr photoperiod) during the autumn, keeping the temperature and other conditions similar to those of trees in natural day length. The trees with extra light retained their leaves and the shoots continued to grow until the following year, whereas on trees in a 13-hr photoperiod or less, the leaves became senescent and subsequently abscinded (Went, 1957).

How are changes in the environment translated into biochemical responses within the leaf? Although the processes of photosynthesis are clearly under light control the evidence quoted indicates that it is not the major metabolites within a leaf that control senescence, but rather substances present in microgram/kilogram quantities such as hormones. It has been shown that both auxin (Avery *et al.*, 1937) and gibberellin (Radley, 1963) contents of leaves are increased by light, so that during the long days of spring and summer leaves might well be supposed to contain sufficient auxin and gibberellin (Nitsch, 1963) to maintain a nonsenescing state. Further, in *Nicotiana*, the transfer of plants to darkness causes a decrease in auxin content of the young leaves within 18 hr, but this increases again if the plant is returned to light (Avery *et al.*, 1937). Leaf age is also important for in *Phaseolus* (Shoji *et al.*, 1951) and *Coleus* (Wetmore and Jacobs, 1953), the older the leaf, the lower is its content of extractable and diffusible auxin. These facts suggest that by late summer, the leaves of a deciduous tree could be of an age when auxin levels are falling and the decreasing day length and light intensity limit auxin synthesis.

If this interpretation is correct and hormone levels *do* control senescence, then the application of either auxin or gibberellin to leaves that are still green in autumn should retard their senescence and retain them in the green state for many days longer than untreated controls. This is indeed the case; Brian *et al.* (1959) found that autumn coloration of the leaves of many species of deciduous English trees was delayed by spraying them with gibberellin A_3 in the early autumn. Similar results were obtained with auxins by Osborne (1959) and Osborne and Hallaway (1960) on the cherry, *Prunus senriko serrulata*. They showed that not only would auxin maintain the green color of the leaves but also the auxin-treated part remained photosynthetically active and, in contrast to control leaves, protein levels did not decline for many days. With a second application of auxin, leaf senescence was postponed even longer.

In a through-the-year study of auxin control of leaf senescence in cherry, Osborne and Hallaway (1964) found that senescence of detached but young leaves taken from April to July was not immediately retarded by the synthetic auxin, 2,4-dichlorophenoxyacetic acid (2,4-D). Some loss of protein always occurred before the added auxin prevented further decline, indicating that at the time of year of long photoperiod and high light intensity, auxin was not a limiting substance regulating senescence. However, as the season progressed, less protein was degraded in these leaves before the 2,4-D was effective in preventing further loss and by early October, when protein levels in the leaves were already declining naturally, an application of 2,4-D immediately arrested any further loss. Wherever the blades of these autumn cherry leaves were treated with the auxin, incorporation of radioactive precursors into protein and RNA was maintained at the initial level, whereas untreated parts of the blade showed progressive loss of synthetic activity typical of normal senescent leaves. It is significant that auxins [indoleacetic acid (IAA), its methyl ester or nitrile, or the synthetic auxin 2,4-D] or gibberellin were the hormones found to retard senescence of deciduous species; the cytokinins were ineffective.

It seems, therefore, that the low light intensity and shortening photoperiod of autumn days is an environmental stimulus for leaf senescence. In certain deciduous species so far studied, the lowered levels of auxin or gibberellins then set in motion senescence changes in the leaf blade that lead to the eventual development of the senescent/nonsenescent junction at the petiole/stem abscission zone, and to the final shedding of the leaf.

B. Herbaceous Species

When herbaceous species shed their leaves, they do so sequentially in the order of their development, the oldest leaves at the base of the plant becoming senescent and abscinding first. At any one time therefore, the plant may bear senescent leaves at the base and expanding leaves at the apex. Although the shedding of leaves of herbaceous species is so clearly sequential during the lifetime of the plant, even deciduous synchronous shedders show some evidence of a sequential order. The leaves that are shed first (a few days before the others) are usually those that emerged and expanded first in the spring. Younger leaves, therefore, appear always to have some advantage over older leaves in terms of nearness to senescence.

What is the stimulus to leaf senescence in herbaceous plants? Some simple experiments with the dwarf bean *Phaseolus vulgaris* have revealed much of the answer. *Phaseolus vulgaris* is a useful plant to work with for it has a well-defined abscission zone at the junction of the petiole and

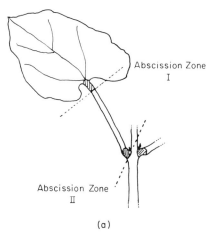

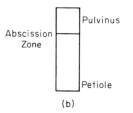

FIG. 1. (a) Primary leaf of bean *Phaseolus vulgaris* showing distal (I) and proximal (II) abscission zones. (b) Explant (1 cm) cut to include distal abscission zone.

pulvinus just below the leaf blade of the primary leaf (Fig. 1a). If this plant is reduced to its primary leaves by removing the apex and all the axillary buds, then the pair of primary leaves will remain green and functional for years, slowly becoming thicker and larger. Senescence does not occur as long as new shoot growth is prevented, and, therefore, there is no abscission at the petiole/pulvinar junction.

If leaves are removed and their petioles rooted, their lifespan will be extended many times beyond that of the life expectancy of one attached to the plant. In fact, their death is usually from accident rather than from senescence. In contrast, an excised unrooted leaf becomes senescent much sooner than it would if it remained attached, and, therefore, abscission too is advanced.

Clearly something is provided from the roots that is necessary to retain the leaf blade in a nonsenescent condition. Chibnall (1939) and Chibnall and Wiltshire (1954) called this a "root factor" and in the early days it seemed that this might be auxin. Now we know that auxin is involved primarily in inducing initiation of roots (Mullins, 1972) and that it is a product of the roots themselves, other than auxin, which is responsible for the retardation of senescence.

The experiments of Richmond and Lang (1957) with *Xanthium pennsylvanicum* were the first to demonstrate that leaf senescence of herbaceous plants could be retarded by a cytokinin in the absence of roots. Since then, the number of species in which a cytokinin will retard the loss of chlorophyll, RNA, DNA, and protein is legion, although we still have little idea how the cytokinin brings this about. One attractive suggestion is that it functions as an inhibitor of specific tRNA nucleases (Cherry and Anderson, 1972), but so far there is no evidence to support the idea.

There is, however, a clear association of cytokinins with the roots. Extraction and assay of root tips of *Helianthus* (Weiss and Vaadia, 1965) and of root exudates of *Vitis* (Skene and Kerridge, 1967) indicate that the root apex is the site of natural cytokinin activity and that the hormone is transported to the leaves through the vascular system. Certainly, removal of root tips from single rooted leaves starts senescence changes in the leaf, and heat, drought, or salt stress to the roots leads to a lowering of cytokinin levels in the leaves (Itai and Vaadia, 1971). It is highly suggestive, therefore, that cytokinins from the roots play a major role in retarding leaf senescence. The question to be answered is why the older leaves should senesce and abscind first.

Considerable work has shown that applied hormones or substrates move preferentially to parts of plants that are metabolic sinks, namely, those parts of greatest metabolic activity (Davies and Wareing, 1965; Wareing and Seth, 1967). Therefore, in competition for cytokinins from the roots, young leaves will always be more favorable sinks than old leaves, and old leaves will tend to become deficient in cytokinin unless the younger, more vigorously growing tissues are removed.

Retardation of leaf senescence in herbaceous leaves is not, however, specific to cytokinins and other hormones play their part. Both gibberellins and auxin may be involved. In dandelion, *Taraxacum officinale*, Fletcher and Osborne (1966) found gibberellin A_3 to be considerably more effective than kinetin, and Fletcher *et al.* (1969) showed this was true for leaves produced in the low light and short photoperiod of autumn and winter, but that kinetin was as effective as gibberellin in leaves grown in the long days of spring and summer. They related the different senescence retarding effects at various times of the year to higher endogenous levels of gibberellins in the long photoperiod.

Last, although senescence in most herbaceous leaf blades so far studied can be retarded by a gibberellin or a cytokinin it is rare for auxin to be effective in this way. In fact, in *Xanthium* addition of an auxin will actually enhance senescence (Osborne, 1967). Petioles, however, behave quite differently. From all the early classic work on abscission, we know that a debladed leaf stump can be retained green and healthy without abscinding by the addition of an auxin to the cut end of the petiole stump (La Rue, 1936; Addicott and Lynch, 1955). In the more modern work, carried out with excised abscission zones, called explants (see later section), auxin is rarely without a senescence retarding and abscission retarding effect on the tissue (Rubinstein and Leopold, 1964; Addicott, 1970; Jackson and Osborne, 1972). Cytokinins too will retard senescence and abscission in explant systems (Osborne and Moss, 1963), but only when they are applied directly to the zone. This is probably due to either relatively poor transport or rapid metabolism of the cytokinins for the retardation of senescence is localized to the treated region.

All the foregoing details indicate that the causes of senescence may be attributed to hormonal imbalance caused either by an environmental stimulus or by endogenous competition. If in the resulting distribution one of the hormones falls below a critical level, senescence is then initiated and becomes the signal that precedes abscission.

C. Evergreen Species

The leaves of most temperate evergreens are shed in an essentially synchronous manner but because the life span of each leaf extends for more than one year and only the leaves of the oldest year's growth are shed at each season, the tree always bears foliage.

For example, leaves of *Euonymus japonica* are shed in their third year of growth, and shedding generally follows the flush of growth and new leaf production at the shoot apex that occurs each spring. All the third year leaves than yellow and are shed together, though those that emerge from the bud first are slightly more advanced than the later leaves of that year's growth. The stimulus for shedding is not day length, for removal of the growing buds prevents it. It seems, therefore, that evergreens of this type behave like herbaceous plants and it is competition from the new leaf growth that is the stimulus to senescence of the oldest leaves.

This would appear to hold for tropical evergreens, too, in which growth of the shoot apex is more or less continuous. In these species, leaves are sequentially lost as new ones are produced.

The hormonal relationships controlling senescence have been worked on very little in evergreens, and this rewarding field of research awaits further study.

V. The Processes of Abscission

A. BIOCHEMISTRY

The questions to consider now are what is the nature of the processes that occur in an abscission zone when abscission is initiated and what is the property of senescent tissue that initiates the changes.

Probably more is known about the separation processes in the distal abscission zone (see Section IV, Fig. 1a) of the primary leaf of *Phaseolus* than of any other species, and most of the information has been obtained from work with "explants." These are 1-cm segments cut from just below the leaf blade and include the whole of the distal pulvinus, the zone, and part of the petiole (Fig. 1b).

As soon as the explant is excised senescence symptoms appear in the small cells of the pulvinus. Not only do they lose chlorophyll but their capacity to incorporate radioactive precursors into RNA and protein falls. The somewhat larger cells on the petiole side of the zone do not, however, become senescent at this stage. Instead, 2 or 3 rows of cells (approximately 200–300 μm) immediately adjacent to the senescing pulvinus start to enlarge. Autoradiography has shown that when the appropriate radioactive precursors are supplied to thin sections of the explant, the synthesis of both protein and RNA in these petiole cells is enhanced above the original level at excision (Osborne, 1968). If senescence of the pulvinus is postponed by supplying auxin to the cut end, neither enlargement of the petiole cells nor their rise in protein or RNA synthesis occurs until such time as abscission is initiated. In other words, high synthetic activity and growth of cells extending for some 300 μm back from the zone occurs only when senescent tissue is present distal to the zone. Eventually by a combination of cell enlargement on the petiole side and senescence and dehydration of the cells on the pulvinus side, the cell walls between petiole and pulvinus are sheared and a complete separation of the two parts is achieved. The process takes about 48 hr from the time of excision, if the explants are kept in darkness at 25°C. According to Caldwell and Mercer (quoted by Osborne, 1968), the increased synthesis and activity of cells proximal to the zone is associated, in *Coleus*, with an increase in the number of ribosomes, rough endoplasmic reticulum, dictyosomes, and dictyosome vesicles on the sides of the cells nearest to the zone. Such changes suggest considerable metabolic activity and cytoplasmic differentiation and it is not surprising, therefore, that the activities of several enzymes have been found to increase in this proximal tissue prior to abscission.

Because there is an actual separation between the cells on either side of an abscission zone, the possible involvement of enzymes concerned in cell wall dissolution or cell wall plasticity has been a focus of interest. Hydrolysis of the middle lamella pectins alone could theoretically account for abscission. Morré (1968) and Mussel and Morré (1969) found evidence for the synthesis and secretion of a polygalacturonase. Decreases in the levels of pectin methylesterase in the distal senescent tissue of *Phaseolus* pulvini have been shown to occur prior to abscission (Osborne, 1958) and Valdovinos and Muir (1965) showed that in *Coleus* this was accompanied by an increase in extent of methylation of the pectin. Such changes could account for an increase in the level of soluble pectins in the middle lamellae and walls of the cells of the pulvinus that abut onto abscission zones and can be correlated with their ease of separation from the adjacent petiole. Indeed, measurable increases in total soluble pectins have been reported for separating abscission zones (Facey, 1950; Rasmussen, 1965; Morré, 1968). In addition, decrease in calcium content at the zone which could be inversely related to the degree of methylation of pectins was shown long ago by Sampson (1918) and confirmed by Rasmussen and Bukovac (1965). However, since considerable expansion growth of the proximal cells of a zone is also involved in separation, other wall enzymes are clearly involved as well (Leopold, 1967).

The most spectacular enzyme is probably cellulase, β-1,4-glucanase first reported by Horton and Osborne (1967) in *Phaseolus* and later demonstrated for *Coleus* and *Gossypium* (M. G. Mullins and D. J. Osborne, unpublished), citrus leaves (Ratner *et al.*, 1969), and in fruit abscission zones of the squirting cucumber (*Ecballium elaterium*) (Jackson *et al.*, 1972). In *Phaseolus*, by far the highest increase in cellulase activity occurs in those cells that are closely associated with and proximal to the zone and, furthermore, the rise occurs only at zones that are about to separate. When abscission is prevented by auxin, no rise in cellulase occurs. Evidence suggests that the rise in cellulase results from new synthesis (Lewis and Varner, 1970) of enzymes and not from an activation of existing enzyme proteins. In addition, by separating the 4 major cellulase isoenzymes in *Phaseolus* abscission zones, Lewis *et al.* (1972) showed that only one of them increased prior to and during abscission. Such results indicate the specific nature of the abscission cellulase. Although the rise takes place in the cells proximal to the zone (where the rise in protein and RNA synthesis also occurs), increased levels of cellulase are found in the senescent cells of the pulvinus that abut on the active petiolar cells. Since these pulvinar cells synthesize little protein, and since cellulases are extracellular enzymes, it seems reasonable to conclude that activity in the pulvinar tissue results from secretion of enzymes from the petiolar cells. These cellulases could have 2 functions. In the senescing pulvinus where we suspect the cellulose to be associated with a more highly methylated pectin, the enzyme could be involved in a cellulose or hemicellulose degradation which is *not* associated with new wall synthesis. In contrast, on the petiole side of the zone, the enzyme could be involved with increases in cell wall plasticity and synthesis of the walls of the growing cells (Fig. 2).

Although the breakage of cell to cell contact between living tissues at the abscission zone is clearly a biochemical process, the final snapping of vascular connections at abscission may well be a purely mechanical one. Seen in this context, the special juxtaposition of senescent cells and expanding cells at the yellow/green junction of the zone and the special and distinct biochemical events that occur in these physiologically dis-

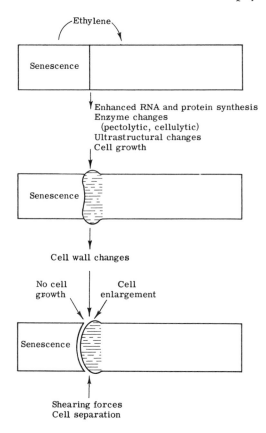

FIG. 2. Sequence of events during abscission of the distal zone of the bean *Phaseolus vulgaris.*

similar tissues set the stage for the eventual mechanical rejection of the senescent organ by the shearing force of the actively growing cells from below.

B. The Initiation of Abscission

What is the property of senescent tissue distal to an abscission zone that initiates the biochemical changes in the cells below the zone?

We now believe that the substance that initiates abscission is ethylene. In all the species where ethylene production has been monitored so far, the tissues distal to an abscission zone have been found to produce relatively large amounts of the gas during senescence of the tissue and prior to abscission of the organ (Jackson and Osborne, 1970). In retrospect, the history of the association of ethylene with abscission is of considerable interest. For over a century various air pollutants have been known to accelerate leaf fall, illuminating gas and smokes from organic materials were particularly effective. The enclosure of plants with ripe fruits would

TABLE I

ETHYLENE PRODUCTION BY GREEN OR SENESCING (YELLOWING) LEAF BLADES AND PETIOLES OF DIFFERENT PLANT SPECIES[a]

	Ethylene production nl/gm/hr	
	Green	Senescing
Leaf blades		
Phaseolus vulgaris	0.11 (0.02)	0.12 (0.03)
Xanthium pennsylvanicum	0.08 (0.02)	0.09 (0.02)
Ecballium elaterium	0.4	4.5
Prunus serrulata senriko	1.91 (0.58)	8.30 (1.6)
Stranvaesia davidiana	0.22	2.39
Hevea brasiliensis	0.8	5.0
Euonymus japonica	0.02	7.3
	(2nd year)	(3rd year)
Leaf petioles		
Phaseolus vulgaris (pulvini)	0.40	2.10 (0.38)
Xanthium pennsylvanicum	0.13	0.35
Ecballium elaterium	0.2	5.5
Prunus serrulata senriko	0.28	2.87
Stranvaesia davidiana	0.43 (0.3)	3.2
Parthenocissus quinquefolia	0.06 (0.02)	0.82
Hevea brasiliensis	0.07	0.70
Euonymus japonica	0.04	12.3
	(2nd year)	(3rd year)

[a] Variations from the mean (in parentheses) are shown only where they exceed 10%.

also induce premature abscission (Crocker, 1948). We now know that ethylene is a common contaminant in these volatile mixtures and that ethylene can be produced at rates as high as 20–30 nl/gm/hr by ripening fruits. But the demonstration that high levels of ethylene could be a normal pattern for senescing vegetative tissues was not provided until relatively recently (Jackson and Osborne, 1970). Although all deciduous and evergreen species tested produce relatively large amounts of ethylene from senescing blades and petioles compared with those that are green (Table I), not all herbaceous species do so. The senescing blades of *Phaseolus* and *Xanthium*, for example, produce no more than those of green leaves. However, in these species, the tissue of pulvinus or petiole that is distal to and adjoining the abscission zone does produce relatively large amounts of ethylene at senescence and it is to this ethylene production that the initiation of abscission in *Phaseolus* is attributed.

Other evidence suggests that applied ethylene (Abeles *et al.*, 1971) or the large amounts of ethylene produced when leaf blades of certain evergreens are treated with an auxin (Hallaway and Osborne, 1969) will actually accelerate blade and petiole senescence. This, in turn, leads to an earlier rise in ethylene production.

C. The Role of Ethylene

Abscission can be initiated by concentrations of ethylene in the air as low as 0.1 μl/liter, but the leaves of different ages vary greatly in their sensitivity to the gas. From any plant exposed to an abscission-accelerating concentration of ethylene in the air, the oldest leaves are shed first in sequential order with the youngest leaves last. The sensitivity to ethylene is, therefore, greatest in the oldest leaf. If, however, the auxin content of a leaf blade of the evergreen *Euonymus japonica* is increased, for example, by painting the whole leaf surface with an auxin solution, the abscission accelerating effect of applied ethylene can be postponed for many days (Hallaway and Osborne, 1969). Similar auxin treatments to leaf blades of *Phaseolus* or application of drops of auxin solutions to the cut end of the pulvinus of bean explants will delay or negate the abscission-accelerating effects of applied ethylene. In all these instances, auxin appears to (a) prevent senescence of the tissue distal to the abscission zone and (b) prevent the cells below the zone from responding to ethylene by undergoing the normal changes observed during abscission. However, it is quite clear that once the abscission process has been initiated and the increase in synthetic activity and cellulase activity has commenced in petiolar cells, an addition of auxin is without effect for abscission cannot then be retarded (Barlow, 1950; Jackson and Osborne, 1972). The addition of more ethylene will, however, cause an already initiated abscission zone to proceed faster to complete separation (de la Fuente and Leopold, 1968). This indicates that normal abscission probably never occurs at the maximum rate and that ethylene can be a limiting factor.

Abscission is one of the fields of plant physiology in which most detailed work has been carried out on somewhat abnormal systems of excised abscission zones. The petiole–pulvinus explants of *Phaseolus* or the petiole–stem explants of *Coleus* and *Gossypium* have been the most favored. These excised pieces of tissue are usually enclosed in small containers at high humidity in darkness. It is only in the last few years, since the demonstration of the very high, though temporary productions of ethylene resulting from cutting, bruising (Jackson and Osborne, 1970), or even handling (Osborne, 1972) plant tissues, that the importance of this ethylene has been fully appreciated in abscission experiments. Now proper precautions are usually taken to prevent an accumulation of the "wound ethylene" inside experimental containers by including dishes of an ethylene absorbent such as a solution of mercuric perchlorate.

Wound ethylene is not the only source of ethylene to accumulate within containers. It was suspected by Zimmerman and Wilcoxon in 1935 and clearly demonstrated by Morgan and Hall (1962, 1964) that when plant tissues are treated with an auxin, the endogenous rate of ethylene production is enhanced. It has been clearly demonstrated that regions of the plant that are rich in auxin (the young expanding parts) (Burg and Burg, 1968) are also regions of high ethylene production. Ethylene production from senescing tissue is the exception to the rule and this aspect of abscission is discussed in a later section.

The buildup of ethylene in enclosed containers means that some of the paradoxical data for abscission of explants should be considered anew with this point in mind. The anomalous situation in which the youngest leaves abscind last on the whole plant, even when it is treated with ethylene, while explants of the youngest leaves abscind first in experimental dishes (Biggs and Leopold, 1957; Osborne, 1958) is a good example. Another is the apparent stimulation of abscission observed when auxin is added to an explant in which abscission is already initiated. This latter situation could be caused by the extra ethylene that is then produced.

It is clear from much of the early work that when an explant is cut from the plant it has an initial insensitivity to ethylene. In *Phaseolus*, this insensitivity lasts for some 18–24 hr depending on the age of the leaf from which the explant is cut. Yamaguchi (1954) and Rubinstein and Leopold (1964) referred to this period as phase 1 or stage 1. By using explant systems where ethylene accumulation in the surrounding air was reduced to a minimum (i.e., by including mercuric perchlorate), certain of the kinetics of abscission have been established for *Phaseolus* (Jackson and Osborne, 1972).

During the period of insensitivity to ethylene, neither wound ethylene production (1–2 nl/gm/hr) nor applied ethylene (2 μl/liter) initiates abscission. If explants are then maintained in an ethylene-free environment, they do not become senescent or abscind for some 100–145 hr. However, at any time from 24–90 hr, an exposure to 2 μl/liter ethylene for 24 hr is sufficient to trigger and complete separation during that 24-hr period. In addition, if 2 μl of IAA (10^{-3} M) is applied to the pulvinar end at or just before the start of the 24 hr in ethylene, abscission can be completely prevented. It seems, therefore, that for explants maintained in ethylene-free air, the control of abscission and initiation of the changes in pulvinus and petiole can be attributed to an auxin–ethylene interaction at the zone.

There is a wealth of literature concerned with the effects of other hormones on the control of abscission in explants, and clearly, cytokinins and gibberellins can also play a part. But it may be that they do so by modulating the basic auxin–ethylene control system, perhaps by modifying the endogenous levels of auxin or ethylene or by extending or contracting the period of insensitivity to ethylene. The length of time a polar auxin transport system is maintained within the explant may also be of considerable importance (Osborne *et al.*, 1968; Beyer and Morgan, 1971). Readers who wish to learn more of these aspects of abscission control are referred to the excellent review by Addicott (1970).

The general picture of leaf shedding that we have so far is of an environmental or endogenous *stimulus* to senescence which is translated into a rise in ethylene production when senescence reaches the cells distal to the abscission zone. Because the period of insensitivity to ethylene has already passed at the time senescent distal cells produce their ethylene, abscission is initiated as soon as their ethylene production reaches the critical level (approximately 1 nl/gm/hr). This still leaves a major question, namely, what controls the production of ethylene in senescing cells distal to an abscission zone?

VI. The Regulation of Ethylene Production

Reference has already been made to auxin enhancement of ethylene production (Morgan and Hall, 1964; Burg and Burg, 1968). Generally, senescing tissues or ripening fruits are low in auxin although they produce relatively large amounts of the gas. In senescence, therefore, some factor other than auxin would appear to regulate the level of ethylene production. The discovery of a nonvolatile substance in leaves that both accelerates abscission and stimulates ethylene production may be the last major link in the chain for our understanding of the control of abscission in plants. The existence of such a substance is now established (Osborne *et al.,* 1972) and sets in context the abscission accelerator called senescence factor (SF) which was first demonstrated in aqueous diffusates from senescent but not green leaves of deciduous, evergreen, and herbaceous plants (Osborne, 1955).

A. ABSCISIC ACID

In the search for this naturally occurring nonvolatile accelerator of abscission, an active substance from mature cotton bolls was isolated, purified, and characterized (Okhuma *et al.,* 1965). It was the same component as the dormancy-inducing substance that increased in the leaves of deciduous trees exposed to short days (Cornforth *et al.,* 1965). This chemical material, later called abscisic acid, was for a long time thought to be the naturally occurring abscission accelerator and probably the original senescence factor, but although nearness to abscission could in many instances be correlated with the extractable content of abscisic acid (Davis, 1968), there seemed to be too many exceptions for this to hold true (Wareing and Saunders, 1971). Nor could abscisic acid contents be correlated with the onset of senescence for the degree of water stress in a leaf plays an overriding role in determining abscisic acid levels (Wright and Hiron, 1969). Whereas abscisic acid content does not correlate well with abscission, there is considerable evidence to show that applications of abscisic acid will enhance the rate of senescence, particularly in old leaf tissue where the levels of other hormones are already low (see Addicott, 1970). Not only do the color changes of senescence occur sooner, but RNA and protein synthesis are reduced. In young tissue, reductions in RNA and protein synthesis also occur (Osborne, 1968; Mullins and Osborne, 1970), but because the metabolism of abscisic acid itself is more rapid in the younger tissue, the effects are not sufficiently long lasting for senescence to ensue. This may be the reason why single applications of abscisic acid to whole plants have for most species been signally unsuccessful in causing defoliation, whereas repeated treatments were more effective (El-Antably *et al.,* 1967; Smith *et al.,* 1968).

Abscisic acid is, however, very active in accelerating abscission in *explants* of many plant species (Smith *et al.*, 1968; Addicott, 1970) and here it is clear that the acceleration is related to an enhanced senescence of the tissue distal to the zone. The enhanced growth of the petiole tissue and the increase in cellulase activity occur (Craker and Abeles, 1969) despite the fact that abscisic acid depresses total protein synthesis. By monitoring the amounts of ethylene produced following the treatment of explants with abscisic acid (Jackson and Osborne, 1972; Osborne *et al.*, 1972) showed that abscisic acid does not directly increase the production, but does so indirectly by accelerating senescence of the pulvinus. In experiments where explants were kept in ethylene-free air (except when their ethylene production was being monitored), the onset of the senescence-associated production of ethylene was brought forward from 80 hr to 40 hr. Abscission was accelerated accordingly, for this ethylene production was occurring when explants had passed their ethylene insensitive period.

B. SENESCENCE FACTOR

In the original work with the senescence factor, SF, diffusion techniques into agar or water were used and activity was found only in the diffusates from senescent tissue (Osborne, 1955, 1958b). The lack of a clear difference in abscisic acid contents in young and senescent leaves and the relatively low activity compared to SF led to a renewal of the search for the abscission accelerator with activity comparable with SF. It was evident that an active fraction which was not abscisic acid could be separated from aqueous diffusates and alcoholic extracts of senescent leaves of *Phaseolus* and the evergreen *Euonymus japonica* (Osborne *et al.*, 1972). Equally active fractions could however be obtained by alcoholic extraction of green leaves indicating that although SF was always present in the leaf it could be obtained by aqueous diffusion only when the leaf was senescent, or, as other experiments showed, by the tissue becoming damaged in some way such as water-logging. Following exhaustive thin-layer and column chromatography, considerably purified fractions of SF were obtained. These were quite distinct from either IAA or abscisic acid in chemical and physiological properties. For whereas, auxin stimulates ethylene production and retards abscission, abscisic acid enhances senescence and accelerates abscission but causes no immediate stimulation of ethylene production, SF both accelerates abscission and has an immediate *and* stimulatory effect on ethylene production, increasing the level by as much as eightfold.

It has been proposed (Osborne *et al.,* 1972) that SF is a regulator of ethylene biosynthesis and that in young nonsenescent tissue, SF could be contained within membrane-bound compartments of the cell. A release of SF through an alteration in membrane integrity would result in a stimulation of ethylene biosynthesis. Senescence, applications of auxin, or wounding could all modify or disorganize the containing membrane in such a way that SF is released. Subsequent repair (reassembly or resynthesis) of the membrane could account for the temporary nature of the wound ethylene rise and the eventual recovery from an auxin treatment, but there would be no recovery from the membrane breakdown and loss of permeability associated with senescence. The large rise in ethylene production associated with yellowing of leaves and ripening of fruits can, therefore, be a reflection of the loss of organization heralding the eventual death of the tissue.

VII. The Model for Abscission

In this chapter, the course of leaf abscission has been outlined together with the series of biochemical events that take place before the organ is finally shed. These can be set out in a model of stimulus, signal, and response (Fig. 3).

The stimulus for abscission results from unfavorable modifications of the external environment or internal competitions between different parts of the plant which are translated within the cells of the organ into altered levels of hormonal substances. Thus displacements of the homeostatic hormone balance by factors such as drought stress may be only temporary, and in such circumstances homeostasis may be restored. With factors such as autumnal short days or the competition of old leaves with young leaves, the stimulus is continuous and the changes progressive and cumulative, so an inevitable sequence of events follows, culminating in senescence.

Because the cells below the separation junction of an abscission zone are physiologically distinct and act as a barrier to the further basipetal progress of senescence, the junction is eventually a meeting point of adjacent senescent and nonsenescent cells. A large rise in ethylene is produced by senescent cells if the normally membrane-enclosed factor (SF) is released to the cytoplasm and there enhances the ethylene biosynthesis in a way we do not yet understand. Loss of membrane integrity is, therefore, an important feature of the senescence signal. Ethylene triggers the nonsenescent cells below the zone to produce en-

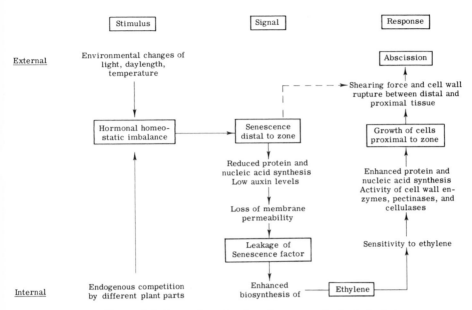

FIG. 3. The model of stimulus, signal, and response for leaf abscission.

zymes involved in their own cell growth, and cell wall degradation in the senescent cells above. This is the response, and as a result, the two tissues separate. It should be pointed out, however, that not all leaves abscind, although they all become senescent and die back to a yellow/green stem junction, and even though they may produce ethylene distal to this junction [e.g., leaves of *Ecballium elaterium* (Jackson *et al.*, 1972)], cellulase activity may not develop. It may be, therefore, that certain leaf–stem junctions at which leaves do not abscind are lacking the ethylene-responsive cells found in abscinding species.

Considerably less work has been done with fruits, though the involvement of ethylene in ripening and shedding does not seem in doubt. In *Ecballium* (Jackson *et al.*, 1972), where the fruit separates from the peduncle with explosive force, ejecting seeds for many feet, the ethylene produced by the ripening fruit does trigger cell growth and the production of cellulases by the peduncle. But in this case we do not know the stimulus that causes the onset of the senescence signal that evokes the biochemical response.

In order to present a model for the internal factors regulating abscission, it is inevitable that much relevant work has not been included. This model should, therefore, be considered as a framework for interpretation of the very many aspects of shedding that are discussed elsewhere in this volume.

References

Abeles, F. B., Craker, L. E., and Leather, G. R. (1971). Abscission: The phyto-gerontological effects of ethylene. *Plant Physiol.* **47,** 7–9.

Addicott, F. T. (1970). Plant hormones in the control of abscission. *Biol. Rev. Cambridge Phil. Soc.* **45,** 485–524.

Addicott, F. T., and Lynch, R. S. (1955). Physiology of abscission. *Annu. Rev. Plant Physiol.* **6,** 211–238.

Anderson, J. W., and Rowan, K. (1966). The effect of 6-furfuryl-aminopurine on senescence in tobacco leaf tissue after harvest. *Biochem. J.* **98,** 401–404.

Avery G. S., Burkholder, P. R., and Creighton, H. B. (1937). Growth hormones in terminal shoots of *Nicotiana* in relation to light. *Aust. J. Bot.* **24,** 666–673.

Barlow, H. W. B. (1950). Studies in abscission. I. Factors affecting the inhibition of abscission by synthetic growth-substances. *J. Exp. Bot.* **1,** 264–281.

Beyer, E. M., and Morgan, P. W. (1971). The role of ethylene modification of auxin transport. *Plant Physiol.* **48,** 208–212.

Biggs, R. H., and Leopold, A. C. (1957). Factors influencing abscission. *Plant Physiol.* **32,** 626–632.

Brian, P. W., Petty, J. H. P., and Richmond, P. T. (1959). Effects of gibberellic acid on development of autumn colour and leaf-fall of deciduous woody plants. *Nature (London)* **183,** 58–59.

Burg, S. P., and Burg, E. A. (1968). Auxin stimulated ethylene formation. Its relationship to auxin inhibited growth, root geotropism and other plant processes. *In* "Biochemistry and Physiology of Plant Growth Substances" (F. Wightman and G. Setterfield, eds.), pp. 1275–1294. Runge Press, Ottawa.

Butler, R. D., and Simon, E. W. (1971). Ultrastructural aspects of senescence in plants. *Advan. Gerontol. Res.* **3,** 73–129.

Cherry, J. H., and Anderson, M. B. (1972). Cytokinin-induced changes in transfer RNA species. *In* "Plant Growth Substances 1970" (D. J. Carr, ed.), pp. 181–189. Springer-Verlag, Berlin and New York.

Chibnall, A. C. (1939). "Protein Metabolism in the Plant." Yale Univ. Press, New Haven, Connecticut.

Chibnall, A. C., and Wiltshire, G. H. (1954). Protein metabolism in rooted runner bean leaves. *New Phytol.* **53,** 38–43.

Cornforth, J. W., Milborrow, B. V., Ryback, G., and Wareing, P. F. (1965). Identity of Sycamore "dormin" with abscisin II. *Nature (London)* **205,** 1264–1270.

Cornforth, J. W., Milborrow, B. V., Ryback, G., and Wareing, P. F. (1966). Isolation of Sycamore dormin and its identity with abscisin II. *Tetrahedron, Suppl.* **8,** 603–610.

Craker, L. E., and Abeles, F. B. (1969). Abscission: Role of abscisic acid. *Plant Physiol.* **44,** 1144–1149.

Crocker, W. (1948). "Growth of Plants." Van Nostrand-Reinhold, Princeton, New Jersey.

Davies, C. R., and Wareing, P. F. (1965). Auxin induced transport of radiophosphorous in stems. *Planta* **65,** 139–156.

Davis, L. A. (1968). Gas chromatographic identification and measurement of abscisic acid and other plant hormones in the developing cotton fruit. Ph.D. Dissertation, University of California, Davis.

de la Fuente, R. K., and Leopold, A. C. (1968). Senescence processes in leaf abscission. *Plant Physiol.* **43**, 1496–1501.

El-Antably, H. M. M., Wareing, P. F., and Hillman, J. (1967). Some physiological responses to D,L-abscisin (dormin). *Planta* **73**, 74–90.

Facey, V. (1950). Abscission in leaves of *Fraxinus americana* L. *New Phytol.* **49**, 103–116.

Fletcher, R. A., and Osborne, D. J. (1966). Gibberellin as a regulator of protein and ribonucleic acid synthesis during senescence in leaf cells of *Taraxacum officinale*. *Can. J. Bot.* **44**, 739–745.

Fletcher, R. A., Oegema, T., and Horton, R. F. (1969). Endogenous gibberellin levels and senescence in *Taraxacum officinale*. *Planta* **86**, 98–102.

Gates, C., and Bonner, J. (1959). The response of the young tomato plant to a brief period of water shortage. 4. Effects of water stress on the RNA metabolism of tomato leaves. *Plant Physiol.* **34**, 49–55.

Hallaway, H. M., and Osborne, D. J. (1969). Ethylene: A factor in defoliation by auxins. *Science* **163**, 1067–1068.

Horton, R. F., and Osborne, D. J. (1967). Senescence, abscission and cellulase activity in *Phaseolus vulgaris*. *Nature (London)* **214**, 1086–1088.

Itai, C., and Vaadia, Y. (1971). Cytokinin activity in water-stressed shoots. *Plant Physiol.* **47**, 87–90.

Jackson, M. B., and Osborne, D. J. (1970). Ethylene, the natural regulator of leaf abscission. *Nature (London)* **225**, 1019–1022.

Jackson, M. B., and Osborne, D. J. (1972). Abscisic acid, auxin and ethylene in explant abscission. *J. Exp. Bot.* **23**, 849–862.

Jackson, M. B., Morrow, I. B., and Osborne, D. J. (1972). Abscission and dehiscence in the squirting cucumber, *Ecballium elaterium*. Regulation by ethylene. *Can. J. Bot.* **50**, 1465–1471.

La Rue, C. D. (1936). The effect of auxin on the abscission of petioles. *Proc. Nat. Acad. Sci. U. S.* **22**, 254–259.

Leopold, A. C. (1967). The mechanism of foliar abscission. *Symp. Soc. Exp. Biol.* **21**, 507–516.

Lewis, L. N., and Varner, J. E. (1970). Synthesis of cellulase during abscission of *Phaseolus vulgaris* leaf explants. *Plant Physiol.* **46**, 194–199.

Lewis, L. N., Lew, F. T., Reid, P. D., and Barnes, J. E. (1972). Isozymes of cellulase in the abscission zone of *Phaseolus vulgaris*. In "Plant Growth Substances 1970" (D. J. Carr, ed.), pp. 234–239. Springer-Verlag, Berlin and New York.

McKee, H. S. (1962). "Nitrogen Metabolism in Plants." Oxford Univ. Press, London and New York.

Morgan, P. W., and Hall, W. C. (1962). The effect of 2,4-D on the production of ethylene in cotton and sorghum. *Physiol. Plant.* **15**, 420–427.

Morgan, P. W., and Hall, W. C. (1964). Accelerated release of ethylene by cotton following applications of indolyl-3-acetic acid. *Nature (London)* **207**, 99.

Morré, D. J. (1968). Cell wall dissolution and enzyme secretion during leaf abscission. *Plant Physiol.* **43**, 1545–1550.

Mothes, K. (1964). The role of kinetin in plant regulation. In "Régulateurs naturels de la croissance végétale" (J. P. Nitsch, ed.), pp. 131–140. CNRS, Paris.

Mullins, M. G. (1972). Auxin and ethylene in adventitious root formation in *Phaseolus aureus* (Roxb). In "Plant Growth Substances 1970" (D. J. Carr, ed.), pp. 526–533. Springer-Verlag, Berlin and New York.

Mullins, M. G., and Osborne, D. J. (1970). Effect of abscisic acid on growth correlation in *Vitis vinifera* L. *Aust. J. Biol. Sci.* **23**, 479–483.

Mussel, H. W., and Morré, D. J. (1969). A quantitative bioassay specific for polygalacturonases. *Anal. Biochem.* **28**, 353–360.

Nitsch, J. P. (1963). The mediation of climatic effects through endogenous regulating substances. *In* "Environmental Control of Plant Growth" (L. T. Evans, ed.), pp. 175–193. Academic Press, New York.

Ohkuma, K., Addicott, F. T., Smith, O. E., and Thiessen, W. E. (1965). The structure of abscisin II. *Tetrahedron Lett.* pp. 2529–2535.

Osborne, D. J. (1955). Acceleration of abscission by a factor produced in senescent leaves. *Nature (London)* **176**, 1161–1163.

Osborne, D. J. (1958a). Changes in the distribution of pectin methylesterase across leaf abscission zones of *Phaseolus vulgaris. J. Exp. Bot.* **9**, 446–457.

Osborne, D. J. (1958b). The role of 2,4,5-T butylester in the control of leaf abscission in some tropical woody species. *Trop. Agric., Trin.* **35**, 145–158.

Osborne, D. J. (1959). Control of leaf senescence by auxins. *Nature (London)* **183**, 1459.

Osborne, D. J. (1962). The effect of kinetin on protein and nucleic metabolism in *Xanthium* leaves during senescence. *Plant Physiol.* **37**, 595–602.

Osborne, D. J. (1967). Hormonal regulation of leaf senescence. *Symp. Soc. Exp. Biol.* **21**, 305–321.

Osborne, D. J. (1968). Hormonal mechanisms regulating senescence and abscission. *In* "Biochemistry and Physiology of Plant Growth Substances" (F. Wightman and G. Setterfield, eds.), pp. 815–840. Runge Press, Ottawa.

Osborne, D. J. (1972). Hormonal mediation of plant responses to the environment. *In* "Crop Processes in Controlled Environments" (A. R. Rees *et al.*, eds.), pp. 251–264. Academic Press, New York.

Osborne, D. J., and Hallaway, H. M. (1960). Auxin control of protein levels in detached autumn leaves. *Nature (London)* **188**, 240–241.

Osborne, D. J., and Hallaway, H. M. (1964). The auxin, 2,4-dichlorophenoxyacetic acid, as a regulator of protein synthesis and senescence in detached leaves of *Prunus. New Phytol.* **63**, 334–347.

Osborne, D. J., and Moss, S. E. (1963). Effect of kinetin on senescence and abscission in explants of *Phaseolus vulgaris. Nature (London)* **200**, 1299–1301.

Osborne, D. J., Horton, R. F., and Black, M. K. (1968). Senescence in excised petiolar segments. The relevance to auxin and kinetin transport. *In* "The Transport of Plant Hormones" (Y. Vardar, ed.), pp. 79–96. North-Holland Publ., Amsterdam.

Osborne, D. J., Jackson, M. B., and Milborrow, B. V. (1972). Physiological properties of abscission accelerator from senescent leaves. *Nature (London)* **240**, 98–101.

Parthier, B. (1964). Protein synthese in grünen Blättern. II. *Flora (Jena)* **154**, 230–244.

Radley, M. (1963). Gibberellin content of spinach in relation to photoperiod. *Ann. Bot. (London)* [N.S.] **27**, 373–377.

Rasmussen, H. P. (1965). Chemical and physiological changes associated with abscission layer formation in the bean (*Phaseolus vulgaris* L. cv. Contender). Doctoral Dissertation, Michigan State University, East Lansing.

Rasmussen, H. P., and Bukovac, M. J. (1965). Some chemical and physiological changes associated with abscission layer formation in *Phaseolus vulgaris. Plant Physiol.* **40**, XXVI.

Ratner, A., Goren, R., and Monselise, S. P. (1969). Activity of pectin esterase and cellulase in the abscission zone of *Citrus* leaf explants. *Plant Physiol.* **44**, 1717–1723.

Richmond, A. E., and Lang, A. (1957). Effect of kinetin on protein content and survival of detached *Xanthium* leaves. *Science* **125**, 650–651.

Rubinstein, B., and Leopold, A. C. (1964). The nature of leaf abscission. *Quart. Rev. Biol.* **39**, 356–372.

Sampson, H. C. (1918). Chemical changes accompanying abscission in *Coleus blumei*. *Bot. Gaz. (Chicago)* **66**, 32–53.

Shaw, M., and Manocha, M. S. (1965). Fine structure in detached senescing wheat leaves. *Can. J. Bot.* **43**, 747–755.

Shoji, K., Addicott, F. T., and Swets, W. A. (1951). Auxin in relation to leaf blade abscission. *Plant Physiol.* **26**, 189–191.

Skene, K. G. M., and Kerridge, G. H. (1967). The effect of root temperature on cytokinin activity in root exudate of *Vitis vinifera* L. *Plant Physiol.* **42**, 1131–1139.

Smith, O. E., Lyon, J. L., Addicott, F. T., and Johnson, R. E. (1968). Abscission physiology of abscisic acid. *In* "Biochemistry and Physiology of Plant Growth Substances" (F. Wightman and G. Setterfield, eds.), pp. 1547–1560. Runge Press, Ottawa.

Strehler, B. L. (1962). "Time, Cells, and Aging." Academic Press, New York.

Srivastava, B. I. S., and Ware, G. (1965). The effect of kinetin on nucleic acids and nucleases of excised barley leaves. *Plant Physiol.* **40**, 62–64.

Valdovinos, J. G., and Muir, R. M. (1965). Effects of D and L amino acids on foliar abscission. *Plant Physiol.* **40**, 335–340.

Vickery, H. B., Pucher, G. W., Wakeman, A. J., and Leavenworth, C. S. (1937). *Conn., Agr. Exp. Sta., Bull.* **399**.

Vickery, H. B., Pucher, G. W., Wakeman, A. J., and Leavenworth, C. S. (1946). *Conn., Agr. Exp. Sta., Bull.* **496**.

Wareing, P. F., and Saunders, P. F. (1971). Hormones and dormancy. *Annu. Rev. Plant Physiol.* **22**, 261–288.

Wareing, P. F., and Seth, A. K. (1967). Ageing and senescence in the whole plant. *Symp. Soc. Exp. Biol.* **21**, 543–558.

Weiss, C., and Vaadia, Y. (1965). Kinetin-like activity in root apices of sunflower plants. *Life Sci.* **4**, 1323–1326.

Went, F. W. (1957). Experimental control of plant growth. *Chron. Bot.* **17**, 169.

Wetmore, R. H., and Jacobs, W. P. (1953). Studies on abscission: The inhibiting effect of auxin. *Amer. J. Bot.* **40**, 272–276.

Wollgiehn, R. (1961). Untersuchungen über den Einfluss des Kinetins auf den Nucleinsäure und Proteinstoffwechsel isolierter Blätter. *Flora (Jena)* **151**, 411–437.

Wood, J. S., Mercer, F. V., and Pedlow, C. (1944). The metabolism of starving leaves. 4. Respiration rate and metabolism of leaves of Kikuyu grass during air–nitrogen transfers. *Aust. J. Exp. Biol. Med. Sci.* **22**, 137–143.

Wright, S. T. C., and Hiron, R. W. P. (1969). (±)-Abscisic acid, the growth inhibitor induced in detached wheat leaves by a period of wilting. *Nature (London)* **224**, 719–720.

Yamaguchi, S. (1954). Some interrelations of oxygen, carbon dioxide, sucrose and ethylene in abscission. Ph.D. Thesis, University of California, Los Angeles.

Zimmerman, P. W., and Wilcoxon, F. (1935). Several chemical growth substances which cause initiation of roots or other responses in plants. *Contrib. Boyce Thompson Inst.* **7**, 209–229.

. 5 .

Shedding of
Shoots and Branches

W. F. Millington and W. R. Chaney

I. Introduction

Abscission of leaves, flowers, and fruits is a familiar phenomenon. Less well known is the regular shedding of shoots and branches of a number of woody perennials. Abortion of the tip of the growing shoot with its apical meristem and several young leaves, usually in the spring as new shoots expand, seems puzzling from the standpoint of efficiency when the presence of more leaves would increase plant growth. Shedding of branches also has its paradoxical aspect, since in many cases relatively large branches are self-pruned from the tree. These phenomena have received rather limited research attention in recent years, yet they are of

149

academic interest from the standpoint of mechanism and evolution. Furthermore, the shedding of shoots and the shedding of branches are of practical significance in forestry and horticulture because both influence growth pattern and form. This chapter will discuss shedding of shoots and lateral branches with emphasis on developmental anatomy, physiological mechanisms, environmental relations, and significant implications.

II. Abortion of Shoot Tips

A. OCCURRENCE

In recent years several investigators have called attention to the common occurrence of shoot tip abortion in many woody plants (Romberger, 1963; Kozlowski, 1964, 1971a; Zimmermann and Brown, 1971). As pointed out by Garrison and Wetmore (1961), abscission of shoot tips occurs in broad-leaved trees and shrubs and among species in a wide array of taxonomic groups in diverse environments. Shedding of the shoot tip has been reported in species of *Ailanthus, Albizzia, Betula, Catalpa, Carpinus, Castanea, Celtis, Cercidiphyllum, Citrus, Corylus, Diospyros, Fagus, Gleditsia, Gymnocladus, Maclura, Platanus, Rhamnus, Robinia, Salix, Sambucus, Staphylea, Syringa, Tilia, Ulmus, Vaccinium,* and *Viburnum* (Barker and Collins, 1963; Kozlowski, 1964, 1971a; Lyashenko, 1967; Millington, 1963; Romberger, 1963; Smith, 1963; Zimmermann and Brown, 1971). In addition, *Bumelia, Cercis, Morus, Paulownia,* and *Sapindus* are recorded in a taxonomic key (Harlow, 1946) as having "pseudoterminal buds," with an axillary bud terminating the season's growth. The phenomenon usually involves a paling or yellowing of the shoot tip and ultimately its abscission (Fig. 1). Shoot tip abortion, by inducing continuing development from one or more subjacent axillary buds, results in sympodial branching in contrast to monopodial branching. In the latter type a persistent terminal bud continues growth from year to year.

While initially presumed to be a result of frost injury, shoot tip abortion was recognized early as nonpathological in nature (Romberger, 1963). Klebs (1917) referred to the seasonal phenomenon as a "normal disease." Death and abscission of the shoot tip in May and sympodial assumption of development by an axillary pseudoterminal bud was reported in *Tilia* as early as 1822 by Vaucher (Romberger, 1963). The phenomenon was studied experimentally by Berthold in 1904 (Garrison and Wetmore, 1961) and was recounted in Goebel's widely distributed "Organography of Plants" (1905) and in Büsgen and Münch's (1931) "The Structure and

FIG. 1. Elm shoot tip at the time of abortion showing yellowing terminal region. Arrow notes demarcation at abscission zone. (Millington, 1963.)

Life of Forest Trees." It is somewhat surprising that interest in shoot tip abortion was not revived until the 1960's, probably in association with interest in effects of photoperiod on plant growth.

While this section is limited to termination of growth through shedding or abortion of the shoot tip, other examples of cessation of function in apical meristems undoubtedly are related to this phenomenon. Parenphymatization of cells of the shoot apex occurs in *Theobroma* and *Alstoma* (Lyashenko, 1967), resulting in sympodial growth, and in the short shoots of *Pinus* which normally are determinate (Sacher, 1955). An expression of senescence of the apical meristem through differentiation of its cells occurs in thorns such as those of *Rhamnus cathartica* (Wiesner cited by Klebs, 1917), *Gleditsia* (Neville, 1969), and *Ulex* (Bieniek and Millington, 1967) and in tendrils as in *Parthenocissus* (Millington, 1966).

Abortion of the shoot tip in the spring as subjacent parts of the shoot continue growth seems anomalous behavior. Although the terminal meristem can no longer function, meristems in lower axillary buds continue development, forming bud scales then leaf primordia, which will expand the ensuing spring. In these persistent axillary buds, and in the terminals of monopodial perennials which do not exhibit shoot tip abortion, the shoot apex thus constructs an overwintering bud. Several investigators have commented on the apparent inability of the terminal shoot apex of sympodial plants to make cataphylls, in contrast to the apices of axillaries on the same shoot. Neville (1969, 1970) in discussing this disparity of behavior among buds of the same plant called attention also to the observation of Teuch and Roux that the shoot apex of orthotropic shoots of *Paliuris australis* makes bud scales, whereas the apex of plagiotropic shoots, which undergo abortion, does not. Codacionni (1963) found that even excised terminal shoot tips of *Castanea* seedlings, cultured *in vitro* and stimulated to make successive sets of foliage leaves by destruction of axillary buds or by explanting to new media, failed to produce bud scales between the sets of leaves. Even though the apex could be induced to resume organogenesis, it made only foliage leaves.

A modification of the usual pattern of shoot tip abortion in which the terminal aborts early is seen in *Parthenocissus vitacea* (W. F. Millington, unpublished). In this vine, trailing or climbing long shoots, a meter and more in length and bearing several leaves, differ markedly from lateral short shoots of a few centimeters which produce 1 or 2 leaves a year. The shoot apex of the long shoot functions throughout the summer producing leaves and tendrils, finally aborting as late as September in Wisconsin. In contrast, the shoot tip of the short shoot, which developed from an axillary bud of the previous season, aborts with its young leaves in 4 or 5 weeks after expansion of usually only one basal leaf (Fig. 2). The effect of the early abortion of the shoot tip over several years is to make a short shoot of several segments, each representing 1 year's growth (Fig. 3). Thus while the apex of the long shoot grows through summer, that of the short shoot aborts early in the spring (May, June), and growth the next year is carried on from the distal axillary bud. If the single leaf of a short shoot is removed in May before shoot tip abortion, a new leaf expands (in our study, 30 of 30 shoots) and abortion is delayed for more than 30 days. Removing the successively formed leaves stimulated production of a new leaf with each excision, although internodes remained short. Short shoots can be induced to escape early abortion and continue development as long shoots by pruning back the parent long shoot to the short shoot node. A similar response occurs in *Cercidiphyllum japonicum*, in which decapitation of long shoots induces short shoot growth (Titman and Wetmore, 1955). Thus timing of shoot tip abortion is influenced by correlative interactions between long and short shoots, shown to involve auxin by Titman and Wetmore. In *Cercidiphyllum* the short shoot is usually terminated by a flower or, if not, the apex becomes quiescent and sympodial growth from an axillary bud is induced, whereas long shoots develop 2 months and then abort, as in *Parthenocissus*. Unlike the situa-

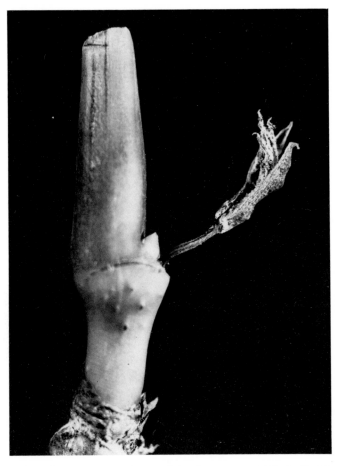

FIG. 2. Aborted tip of short shoot of *Parthenocissus* (right). Petiole of single expanded leaf at left with pseudoterminal bud at base.

tion in *Parthenocissus* and *Cercidiphyllum,* in *Maclura pomifera,* the short shoot apex persists from year to year making alternate sets of bud scales and leaves, whereas the long shoot tip aborts annually (Neville, 1970).

The timing of shoot tip abortion varies with age, vigor, and environment. In many genera, such as *Ulmus, Fagus, Tilia, Syringa,* and *Vaccinium,* the shoot tip aborts with decline in growth as the shoot and its leaves complete expansion in the spring. In *Ulmus americana* (Millington, 1963), shoot tip abortion occurs at the same time that fruits are being shed, but in young trees which have not yet borne fruits, abortion occurs concurrently so fruit abscission is not essential. There is a delay in

FIG. 3. Short shoot of *Parthenocissus* showing 5 years growth. Shoot tip of current season (arrow) in front of petiole base of single leaf, ready to abort. Axillary bud not yet developed. Supernumerary latent axillary buds evident at leaf axils of previous seasons' growth.

abortion, however, in young saplings and seedlings and in vigorous shoots, such as suckers or water sprouts (Klebs, 1917; Garrison and Wetmore, 1961; Millington, 1963; Smith, 1963). *Tilia* seedlings, for example, are reported sometimes to retain shoot tips over winter in southwestern Michigan (Ashby, 1962). The date of shoot abortion in adult *Ulmus* trees varies somewhat in successive years with climatic factors (Millington, 1963). Timing of abortion in some genera (e.g., *Salix nigra, Albizzia julibrissin*) is delayed until late in the growing season (Zimmermann and Brown, 1971), suggesting different triggering events. Other environmental factors are also involved as shown by abortion of shoot tips in tropical species (Koriba, 1968; cited by Romberger, 1963).

B. ANATOMY OF SHOOT TIP ABORTION

The earliest external indication of abortion of the shoot tip is a yellowing of the tissue of the stem at the site where abscission will occur, just above the node bearing the axillary bud which will serve as the pseudoterminal bud (Fig. 1, arrow) (Garrison and Wetmore, 1961; Millington, 1963; Neville, 1969). Yellowing rapidly progresses acropetally from the oldest leaf involved until the entire tip is yellow, in *Ulmus* within 48 hr. In *Syringa* (Garrison and Wetmore, 1961) and *Gleditsia* (Neville, 1969) sections of the shoot tip sampled as shoot elongation ceased and before yellowing was evident showed no visible evidence of cellular breakdown. When the tip has yellowed, the cytoplasm of cells throughout the tip stains less intensely. In *Ulmus* the mesophyll cells of stipules are first to show cellular breakdown, and indeed the stipules may be the first to show yellowing externally (Millington, 1963). Necrosis of older leaves at the apex follows next in *Ulmus*, evident in sections as a distinct color response in the protein stain, Fast Green. Necrosis progresses distally into young leaves, and then cells of the meristem lose their cytoplasmic density as in *Syringa* and *Gleditsia*. In *Maclura pomifera* there is a similar decrease in density of cytoplasm, initially in the metrameristem (Smith, 1963). Prior to this, with decline in shoot growth, the files of rib meristem cells of *Maclura* become fewer in number as cells differentiate below without replacement from the meristem. In the final stages of abortion, the necrotic cells of the *Syringa* apex stain intensely and disintegrate, but abscission of the tip usually occurs before this happens in *Ulmus*. Following abscission in *Ulmus* a phellogen develops across the stem in the abscission zone, whereas in *Syringa* and *Gleditsia* it precedes abscission. The nuclei on either side of this abscission zone in *Gleditsia* are notably different in appearance. An additional periderm develops below the abscission zone in *Gleditsia*. Anatomical studies in *Ulmus* showed no occlusion of xylem or phloem below the site of abscission, which might indicate a cause for tip abortion (Millington, 1963).

C. PHYSIOLOGICAL ASPECTS: EXTERNAL ENVIRONMENT

Klebs (1917) concluded that abortion of shoot tips was induced by external factors. He noted that Wiesner (1889), seemingly with inadequate evidence, attributed death of the shoot tip to deprivation of water through transpiration. Wiesner had reasoned that rapidly elongating shoots which aborted tips ahead of slowly expanding shoots lost water faster. Prolonged rains delayed abortion. Codacionni (1962) noted that *Castanea* seedlings undergoing water stress aborted early. That water deficiency caused abortion was regarded as unlikely by Garrison and Wetmore (1961), since abortion occurred in *Syringa* tips cultured *in vitro* in high humidity, and Romberger (1963) found *Tilia* tips collected immediately after abortion to have water contents of 75–80%. Millington (1963) found no acceleration of abortion in flaccid tips and Neville (1969) pointed out that, with continued leaf expansion in *long* days, competition for water increased but *Gleditsia* seedlings aborted in *short* days. Culturing experiments with seedlings convinced Klebs that in addition to water supply, soil nutrients, temperature, and light led to unknown internal changes which could prolong growth of the shoot tip and were thus involved in control of shoot tip abortion. He questioned that abortion could be a hereditary, autonomous phenomenon, since environmental factors were involved. Delaying abortion of shoot tips by adding nutrients has been noted by others. Abortion of seedlings of *Ulmus* was delayed by addition of organic fertilizer and accelerated in soil deficient in nitrogen (Millington, 1963), just as Knop's solution delayed abortion in *Castanea* seedlings growing in vermiculite (Codacionni, 1962). Temperature affects the rate of abortion slightly. Phenological observations of *Ulmus* correlated early shoot tip abortion 1 year with a prolonged period of warm weather (Millington, 1963), and Nitsch (1957) reported delay of abscission in low temperatures. Senescence in *Pisum* shoots (maturation of the apex as a flower) is delayed at low temperatures (Lockhart and Gottschall, 1961).

Accelerated abortion in short days and delay in long days have been well documented. Young *Ulmus* plants 3–4 years old in short days had lost 90% of the shoot tips in 61 days, whereas in long days, or in daily cycles in which a 17-hr dark period was interrupted by ½ hr of light, 40–50% had aborted (Millington, 1963). The effect of photoperiod on abortion is influenced by plant age and vigor. Vigorous 1-year-old *Ulmus* seedlings in short days showed 60% abortion in 83 days and 27% in long days, compared to almost 100% abortion in weak seedlings in both photoperiods. Although experiments with mature trees were not performed, Millington noted that branches of mature *Ulmus* trees adjacent to street lights aborted shoot tips at the same time as nonilluminated branches.

Abortion of shoot tips is intimately associated with arrest of growth and onset of dormancy, phenomena well known to be responsive to photoperiod. Neville (1968, 1969) studied the effect of photoperiod on abortion as it related to rest and dormancy in seedlings of *Gleditsia triacanthos*.

In an 8-hr-day, abortion occurred in 35–42 days, whereas in 16-hr days growth continued as long as a year. Transfer to long days of seedlings, whose growth was inhibited in short days and which lacked terminal buds, permitted latent axillary buds to resume growth, if not maintained too long in short days. Although growth had been inhibited, the plants were not dormant, as shown by renewed growth of axillaries, but they became progressively more dormant with increased short days. Neville argued that senescence was not a consequence of dormancy, but rather that the two were expressions of the same cause. Neville further emphasized that not all buds on a tree were arrested equally. Some of the buds expanding on a tree in the spring are dominant and become long shoots, while others are more or less inhibited and become arrested while the dominant buds continue to develop. Thus, growth inhibition depends on internal relations among buds as well as external factors. Neville, and Garrison and Wetmore (1961) emphasized that senescence of the terminal bud occurs when the subjacent stem and its leaves are in active elongation and the axillary buds are producing the scale and leaf primordia for next season.

D. Internal Factors

Differences in timing of shoot tip abortion in seedlings and adult trees in response to environmental stimuli and among buds on the same tree indicate involvement of internal regulating factors. Illustrative of this is the observation in several species that the shoot tip persists longer in seedlings and in young trees than in adult plants in the same environment and in suckers or water shoots on a given tree (Garrison and Wetmore, 1961; Barker and Collins, 1963; Millington, 1963; Romberger, 1963; Smith, 1963). As noted above, abortion of the shoot tip is associated with termination of seasonal growth and inception of dormancy in woody perennials, and it is no doubt related to senescence of the entire aboveground parts of herbaceous perennials and the complete demise of annuals. Shoot tip abortion can be considered an aspect of senescence. Hence studies of senescence may contribute to understanding the mechanism of shoot tip abortion. In studies with *Pisum* Lockhart and Gottschall (1961) were able to induce apical senescence in which the terminal shoot apex differentiated with the formation of a terminal flower. Noting that removal of flowers or fruits has long been known to delay senescence, they showed prolonged stem growth with an increase in number of leaves in response to deflowering or application of gibberellic acid (GA) to the

shoot tip. Removal of fruits or seeds was also effective in delaying senescence. In all treatments, however, stem growth ultimately ceased and
the shoot apex differentiated with flower formation. That the cause of
senescence originated in the shoot tip and not in older parts was indicated in graft experiments. Young shoots grafted on old ones grew as
long as untreated young controls, whereas old shoots grafted to young
shoots senesced at the same time as intact old plants. No evidence for
transmission of aging or juvenile factors emerged in other grafting experiments. Accompanying apical senescence in *Pisum* is a decline in the
shoot tip in total RNA, nitrogen, protein, and inorganic phosphorus in
comparison to deflowered controls (Ecklund and Moore, 1968, 1969),
similar to the decline in these components in cells of other senescing
plant systems. Specific activity of ribonuclease increased with shoot tip
senescence, indicating enzymatic degradation of RNA.

Correlation, the interaction among plant organs, was evident in early
studies of shoot tip abortion. In 1899 Wiesner noted that removal of lateral
buds of *Rhamnus* seedlings delayed shoot tip abortion (cited in Klebs,
1917). Klebs referred to the hypothesis of DeCandolle and Goebel and
noted that competition for nutrients by the leaves deprived shoot tips
and led to abortion. He was critical of Mogk, who attributed termination
of tip growth to mutual correlative influences between main and lateral
branches along with constant changes in the capacity of the meristem to
utilize nutrients. Klebs regarded Mogk's conclusions as unsupported and
sought explanations for abortion in terms of limitations of water, temperature, and light. Support for the view that limitations in growth
requirements to the apex contribute to tip abortion came from experiments with *Betula papyrifera* reported by Kozlowski and Clausen (1966).
Abortion of the distal portion of the shoot was induced when early leaves
or all leaves of a shoot were covered with foil.

Considerable evidence indicates involvement of growth hormones in
abortion. Defoliation delayed shoot tip abortion in *Gleditsia* seedlings
placed in an aborting photoperiod (Neville, 1969). Removing additional
leaves as they formed further delayed shoot tip senescence. Young leaves
thus contributed to tip abortion induced in short days. Neville raised
the question of whether the apex in short days became more sensitive to
inhibitors from young leaves, or whether the inhibitory effect of young
leaves was increased.

Removal of axillary buds and leaves of long shoots of *Cercidiphyllum*
promoted continued development at the shoot apex, with production of
up to 12 additional leaves before the shoot tip aborted (Titman and
Wetmore, 1955). Removing only leaves did not promote continued
growth, indicating that axillary buds were primarily involved in inducing
abortion. In studying long and short shoot growth, Titman and Wetmore

(1955) found that the auxin level increased in long shoots until arrest of growth and abscission, when it gradually declined, but auxin supply seemed not to be causal. The auxin assays revealed possible production of an auxin inhibitor in the axillary buds which accelerated abortion. In shoots with apices and leaves removed but axillary buds intact, auxin levels were about half those in intact shoots. However, if axillary buds were also removed, auxin levels were somewhat higher, indicating inhibition by buds.

In *Phaseolus* plants, excision of the terminal bud causes abscission of the tip at the first expanded internode below the bud. This can be duplicated by application of inhibitors of auxin transport such as TIBA, triiodobenzoic acid, suggesting auxin regulation. Ringing the stem with lanolin–TIBA in the second internode or spraying the plants caused abscission of the tip, usually at the fourth internode (Whiting and Murray, 1948). Concurrent application of indoleacetic acid (IAA) with TIBA to a leaf prevented abscission of the tip at the subjacent internode, which occurred if TIBA alone was applied (Weintraub *et al.*, 1952). This inhibiting influence of auxin on abscission of the shoot tip was contrary to results obtained in experimental spraying of *Vaccinium* plants in the field (Barker and Collins, 1963). Spraying with IAA 3 times weekly for 3 weeks gave slight promotion of abortion (60% aborted) over water-sprayed controls (51% aborted). Correlation in decline of endogenous auxin level with shoot tip abortion of *Syringa* was reported by Garrison and Wetmore (1961). Expanding buds in spring showed an auxin increase in the whole bud during bud enlargement as well as only in the distal region with 3–4 leaf primordia. The high auxin level continued in expanding shoots bearing an enlarging leaf pair, but assays at this time of only the distal aborting portion with its 3–4 leaf primordia showed a decline to 0. Although a decrease in auxin level at the tip occurred as elongation ceased, Garrison and Wetmore noted that it could not be shown that reduced auxin production in the tip caused abortion, nor could they find evidence for an auxin inhibitor in aborting tips. They called attention to the fact that death of the apex did not initiate growth inhibition; the apex had potential for further growth when axillary buds were removed. This was demonstrated by destruction of the uppermost pair of axillary buds on several centimeter-long shoots of *Syringa* shrubs in the greenhouse, a treatment which induced "compensatory growth" in 8 of 10 shoots. At the time of treatment, the shoot tip was near abortion and green or yellow in color. Tips of 5 shoots resumed activity and produced several bractlike structures, while 2 produced only a pair of small leaves before aborting. Outdoor shrubs showed no added growth on debudding. Evidence for a role of auxin in regulation of abortion is thus unconvincing.

A trophic effect of distal axillary buds on the terminal apex was also reported in *Castanea* seedlings (Codacionni, 1962). Suppressing a sub-distal axillary bud delayed abortion, with the terminal shoot apex producing 3–5 additional leaves. The response depended on the stage of the terminal meristem. At earlier stages resumption of terminal growth was assured by suppressing an axillary bud. In the stage just before abscission, suppressing 2 axillary buds induced resumption of growth, but suppressing 1 bud was rarely effective. The effect of bud removal was evident cytologically in the apex. Seven days after axillary bud excision, microscopic observation of sections showed recovery of the meristematic state in cells of the terminal shoot apex.

The paradox of demise of the shoot tip when subjacent tissues were flourishing led to attempts to excise and culture the shoot tip *in vitro*. Garrison and Wetmore (1961) cultured *Syringa* shoot tips from dormant and actively growing buds. Shoot tips from quiescent buds after the dormant period unfolded 1–5 pairs of leaves followed by death of the tip. Tips from vigorously growing shrubs in the greenhouse unfolded the outermost leaf pair but usually no additional leaves unfolded. Explanted tips of shoots growing outdoors turned brown at varying times without production of additional leaves. Excised tips thus seemed programmed to die even when removed from subjacent tissues.

Codaccione (1963) attempted to prevent abortion of shoot tips of seedlings of *Castanea sativa* by *in vitro* culture. Intact seedlings germinated on simple nutrients (Knop's solution/2 and 2% glucose) developed 5–6 leaves and then a small bud of 2–3 leaves which abscised at 4 months. On agar without nutrients the bud aborted a month earlier. Transferring the plant before bud abortion to filter paper in liquid medium induced production of 3–4 supplemental leaves. Growth could be renewed several times by such transfer, but survival of the meristem was not indefinite: an axillary bud assumed growth in spring. Excising and explanting the shoot tip a month before abortion also stimulated resumption of leaf production following formation of basal callus, but required IAA (10^{-7}). Successive explanting of the tip permitted production of as many as 30 leaves. Tip explants made less than a week before abscission failed to form callus and produce leaves, although the shoot apex appeared viable. If the apical meristem *alone* was excised from shoots shortly before bud abortion, or even several hours after, it could be induced to resume growth when rooting occurred in response to auxin. Three to five leaves were produced. By subculturing the apical meristem in this way Codaccione was able to maintain a functioning meristem for 2 years. She noted, however, that although subculturing the meristem alone permitted indefinite survival, a winter bud with bud scales was never produced. Contrary to the *in vitro* studies with *Syringa*, these results indicate an influence from older tissues which promotes abortion.

The extensive literature dealing with the role of inhibitors and growth substances in arrest of growth, dormancy, and breaking of dormancy is particularly relevant for an understanding of shoot tip abortion, which seems to be an expression of the onset of dormancy. The factors that cause arrest and dormancy bring about death and abscission in the distal part of the shoot. The identification of inhibitors in woody plants and demonstration of their seasonal fluctuations, triggered by photoperiod, which correlated with growth and dormancy, have thus provided the best leads for explaining shoot tip abortion. Isolation of "dormin," from dormant leaves and buds (Wareing *et al.*, 1964) and its characterization as being identical with the abscission promoter, Abscisin-II (Cornforth *et al.*, 1965), now called abscisic acid (ABA), have enabled demonstration of a hormone which induced dormancy (Wareing, 1954; Wareing and Ryback, 1970). Although ABA also induces leaf abscission (Addicott *et al.*, 1964), and probably tip abscission, we have not yet seen a report of the specific induction of shoot tip abortion on application of ABA although it induces resting buds in nonaborting species. That control of arrest of growth, dormancy, and senescence involves an interaction between inhibitors and growth-promoting hormones is evident in the stimulation of growth in dormant buds with gibberellic acid (GA) and the increase in endogenous gibberellin with breaking of bud dormancy. It has been suggested also that cytokinins might be one of the factors regulating shoot senescence, since cytokinins can protect leaves from the senescing influences of ABA (Wareing and Ryback, 1970), and since the cytokinin level in root exudate decreases when plants approach senescence (Sitton *et al.*, 1967). Although auxins seem to have no effect on dormancy (Phillips, 1971), they may in some cases have an effect on abortion, as indicated in earlier studies already described. Ethylene, shown to regulate leaf abscission, may also play a role in natural abscission of shoot tips. Zimmermann and Brown (1971) suggest that some type of inhibitor is formed in axillary buds and/or leaves and translocated to the shoot tip where abortion is induced. Short days could enhance inhibitor production while long days might promote higher GA synthesis, countering inhibition under this scheme.

Shoot tip abortion occurs in different patterns on branches of different species. Most commonly the terminal shoot tip aborts and axillaries continue growth sympodially the next year, at which time the tips of these shoots in turn abort. In some species with long and short shoots the terminal shoot tips abort, but apices of lateral short shoots function from year to year. In vines like *Parthenocissus*, shoot tips of laterals abort early, whereas the apices of main shoots function until the end of summer.

Localization of inhibitors like ABA, with a decline in gibberellins or cyto-kinins, may regulate abortion of shoot tips, but the pattern in which the process occurs in the plant seems to involve some genetic regulation characteristic of the species. A mutant phenotype in cotton, "abortive terminal," has been described (Quisenberry and Kohel, 1971). Although the description suggests more a failure of the shoot apex to develop than its death, the phenomenon seems related to tip abortion and shows genetic control. Quisenberry and Kohel (1971) believe "abortive terminal" to be cytoplasmically inherited. As Neville (1970) noted, the different fates of apical meristems (in *Maclura* long shoot tips only can abort, while short shoot tips may or may not) represent differences in repression and de-repression of genes. Such genetic differences dictate whether a shoot apex can make a set of bud scales and then switch to leaf production, or whether it senesces and is lost with abortion of the shoot tip.

It is difficult to find the significance of shoot tip abortion in growth of woody plants. If one speculates that its advantage is induction of sym-podial branching and possible proliferation of branches and ultimately increased photosynthetic surface, it must also be recognized that species without tip abortion compete just as favorably. Even in nonaborting species, lateral buds develop into branches. The sympodial branching pattern in which the entire aerial portion dies back, as in herbaceous perennials with bulbs, corms, droppers, runners, and rhizomes, is a much more drastic senescence apparently related to winter survival. Codaccione (1963) has pointed out that aborting shoot tips have meristems which seem unable to make bud scales. One might speculate that the abortion pattern evolved in response to loss of this ability. If one looks to the environment for selective pressure, perhaps injury to growing tips in adverse environmental situations, as when tips of *Platanus* trees are winter killed, led to association of sealing off the injured part by a pro-tective zone with inhibition of growth in shoots. Wound healing had no doubt been established early. Internal biochemical changes involved in cellular responses that yielded a protective layer now became associated with the triggering of cessation of growth and elimination of the entire shoot tip. The first cytological appearance of necrosis back of the meri-stem rather than in the meristem itself supports such a view. Shoot tip abortion remains a paradoxical developmental event.

III. Shedding of Lateral Branches

Shedding of lateral twigs and well-developed branches is a common phenomenon inherent to a host of woody plants (Kozlowski, 1971a).

Many families of gymnosperms and angiosperms contain species which shed lateral branches. The paucity of studies on this phenomenon is astounding in view of its influence on form of woody plants and on value of lumber obtained. Available data indicate that branch shedding occurs by 2 distinct mechanisms; one involving physiological processes and the second involving an interaction of biotic and mechanical agents. The physiological mechanism periodically results in abscission of branches from woody plants by a process similar to that involved in leaf abscission. Separation of branches from the parent plant occurs through well-defined cleavage zones and is preceded by tissue weakening and periderm formation (Barnard, 1926; Licitis-Lindbergs, 1956; van der Pijl, 1952, 1953; Schaffner and Tyler, 1901). This process, called cladoptosis (Brown *et al.*, 1949; Büsgen and Münch, 1931), is usually limited to twigs and small branches in the crown of trees, but large branches also can be shed from the bole of some species.

Natural pruning and self-cleaning are terms used for branch shedding which involves biotic and mechanical agents. Branches low on the bole often die, especially in dense stands, from shading and competition. The dead branches are attacked by saprophytic fungi and insects which eventually render them decayed, weakened, and susceptible to breakage by animals, wind, snow, or other elements of the environment (Baker, 1950; Smith, 1962; Köstler, 1956; Toumey and Korstian, 1947).

A. Cladoptosis in Gymnosperms

Among the 4 orders of living Gymnospermae, only Coniferales and Gnetales consist of species reported to demonstrate cladoptosis. The phenomenon is found in 5 families of the Coniferales.

Some of the most remarkable examples of lateral branch abscission occur in species of Taxodiaceae. Of the 15 living species which are found in 9 genera, 5 of which are monotypic (Harlow and Harrar, 1968), 11 are reported to exhibit some form of lateral branch abscission.

Most of the lateral branchlets of *Taxodium distichum* which bear leaves are shed in the autumn (Fig. 4). Terminal shoots, in contrast, are persistent and individual leaves abscise from them (Everett, 1968; Harlow and Harrar, 1968; Vines, 1960). *Taxodium mucronatum*, an evergreen species also abscises lateral branchlets with attached leaves, but these are retained on the tree until their second season (Bailey, 1933).

Metasequoia glytostroboides, like *Taxodium distichum*, sheds lateral branchlets with attached leaves in the autumn (Fig. 5). This intriguing tree is native to only a small area of China and was described and named from fossil records before it was known to exist in present world flora

FIG. 4. Lateral branchlet with attached leaves and branch scars on *Taxodium distichum.*

FIG. 5. Abscised branchlets with attached leaves of *Metasequoia glyptostroboides.*

(Everett, 1968). Another deciduous form indigenous to Asia which annually sheds lateral branchlets with attached foliage is *Glyptostrobus pensilis* (Barnard, 1926).

Other genera of Taxodiaceae reported to exhibit branch abscission consist of evergreen species. *Sequoia gigantea* and *Sequoia sempervirens* shed branches with several years' shoot growth and attached leaves each fall. The branches shed from *Sequoia sempervirens* exhibit 2, 3, 4, or occasionally 5 seasons' growth. The majority, however, show 3 years' growth (Foster and Gifford, 1959; Stark 1876). In *Cryptomeria japonica* and the 3 species of *Athrotaxis*, individual leaves as well as branches with attached leaves are ultimately deciduous (Barnard, 1926; Everett, 1968). In *Sciadopitys verticillata*, 2 long flat needles fused at the margins are borne on a spur shoot and a whorl of these spur shoots with foliage leaves is clustered at the tip of long shoots (Bailey, 1933; Chamberlain, 1934). The spur shoots are ultimately deciduous as in *Pinus* (Barnard, 1926).

Only 2 of the 9 genera of Pinaceae, are reported to exhibit cladoptosis. Hickman (1903) briefly mentioned shedding of twigs and branches of *Tsuga canadensis*. The twigs and branches observed were dead when abscised, but apparently were cast after development of a cleavage zone. The best documented occurrence of cladoptosis in Pinaceae is in *Pinus*. The shoot system of *Pinus* is characterized by long and short shoots as in *Larix*. In contrast to the condition in *Larix*, however, the photosynthetic leaves of *Pinus* are restricted to short shoots (except in seedlings) which ultimately abscise from trees with leaves attached. A short shoot develops from a bud in the axil of a primary scale leaf occurring on a long shoot. When approximately 250 μm long, apical growth ceases. Secondary leaves arise from the short shoot and number from only 1 in *Pinus monophylla* to generally 6 but sometimes 7 or 8 in *Pinus durangensis*. The scales of the short shoot bud elongate into a basal sheath that may be persistent or deciduous. A diagram of a "needle fascicle" and its relation to a long shoot appears in Fig. 6. Persistence of short shoots with attached leaves is extremely variable among species. *Pinus strobus* and *Pinus palustris* short shoots normally abscise at the end of the second growing season, whereas short shoots of other species persist for 3 to 5 years and up to 12 to 14 years in *Pinus balfouriana* and *Pinus aristata* (Bold, 1957; Foster and Gifford, 1959; Mirov, 1967). Other genera of Pinaceae, including *Cedrus* and *Larix* which have prominent short shoots, have leaves which are individually deciduous (Foster and Gifford, 1959).

Several examples of cladoptosis are evident in the Cupressaceae. The phenomenon is apparently restricted to mature trees since no juvenile trees are purported to shed branches. *Thuja occidentalis*, for example,

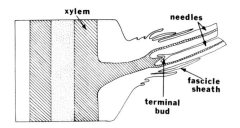

FIG. 6. Diagram of a longitudinal section of a long shoot of *Pinus* with lateral short shoot and attached needles. (Redrawn from Cooperrider, 1938.)

generally sheds 3-year-old branchlets from mature trees in the autumn (Fig. 7). The scale leaves are firmly attached to the twigs and in general do not separate from them. *Libocedrus decurrens, Chamaecyparis (Cupressus) lawsoniana,* and *Chamaecyparis nootkatensis (Cupressus nutkoensis)* also annually shed small branchlets of undetermined age in the fall. Cleavage planes occur in basal joints of *Juniperus* and 3-year-old branchlets are noted to abscise in great numbers from *Juniperus virginiana* and *Juniperus communis* (Büsgen and Münch, 1931; Schaffner, 1902b; Stark, 1876).

Of the 6 genera of Podocarpaceae, 2 are reported to show branch abscission. *Podocarpus imbricata,* for example, sheds some small secondary branchlets up to 20 cm long and older branches as large as 7 mm in diameter. Specimens of *Podocarpus vitiensis* exhibit cladoptosis to a much

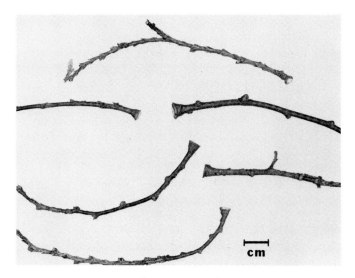

FIG. 7. Abscised branches of *Thuja occidentalis.*

greater degree. Branchlets about 25 cm long are shed in large numbers and some older branches up to 80 cm long and 13 mm in diameter also are abscised. Van der Pijl (1953) noted that this phenomenon was recorded for *Podocarpus* as early as 1878.

In the aberrant *Phyllocladus,* the short shoots become flattened and function as leaves while the very deciduous real leaves are shed (Chamberlain, 1934). Barnard (1926) implied that these cladodes ultimately abscised from this evergreen genus.

Araucariaceae consists of two genera, *Agathis* and *Araucaria,* both of which contain species reported to exhibit cladoptosis. As early as 1877 Hooker described a specimen of *Agathis* (*Dammara*) *robusta* growing under adverse conditions in a large public building in London. The tree shed its leaves and afterward some branches 2 to 3 ft in length. Under favorable conditions the leaves of *Agathis* normally are persistent on foliar branches that are ultimately deciduous (Barnard, 1926). An exception is *Agathis macrophylla* which sheds its leaves before the short foliar branches abscise (Whitmore, 1966). In addition to shedding young foliar branches, older laterals are shed by young trees and, consequently, the wood is exceptionally free of knots and highly prized for veneer (Barnard, 1926; Shattock, 1888; Streets, 1962). This phenomenon has been thoroughly investigated in *Agathis australis* by Licitis-Lindbergs (1956) who examined abscised branches. Van der Pijl (1953) reported an abscised branch of *Agathis palmerstoni* 18 mm in diameter which still bore green leaves and a young living cone. He also observed that the side branches of *Agathis alba* often bend down at the junction of the branch with the stem during periods of drought, and it is at the point of bending that the cleavage zone occurs when branches are shed. Second and third order branches about a centimeter in diameter and less than a meter long are abscised most commonly, but occasionally branches of the first order also are shed (van der Pijl, 1952, 1953).

Araucaria also exhibits the peculiar habit of shedding entire foliar branches rather than individual leaves (Barnard, 1926; Corner, 1951; van der Pijl, 1953) and, like *Agathis,* some species also shed first order lateral branches. For example, lateral branches of the monkey puzzle tree, *Araucaria araucana,* are deciduous after several years, with age at abscission markedly affected by site as discussed later. *Araucaria columnaris* has a bizarre appearance because the lower branches of trees are shed and replaced by short shoots, producing the effect of a dense green column which suddenly widens at the top (Dallimore and Jackson, 1967; Streets, 1962). Branch abscission in this genus was apparently first described by Muhldorf (1925).

Species of Ephedraceae and Gnetaceae (order Gnetales) exhibit lateral branch abscission. The approximately 40 species of *Ephedra* are xerophytic plants characterized mostly by shrubby or trailing forms. Seven species occur in the arid regions of the southwestern United States where they often are important range plants and are known as "joint fir." In older portions of stems which have become woody, the young branches are fasciculate, and many of them are shed during droughts by abscission layers which extend across the pith and wood (Bold, 1957). Many branches also are shed at the end of the growing season (Chamberlain, 1934).

Excellent examples of branch abscission are shown by species of *Gnetum*. Abscission is especially striking in climbing lianas of *Gnetum* which may shed thick and woody branches. For example, van der Pijl (1952) reported a branch abscission scar with a diameter of 64 mm on *Gnetum gnemonoides*. Large quantities of abscised branches which resemble bones often are observable under species of *Gnetum* (van der Pijl, 1952).

B. Cladoptosis in Angiosperms

A large number of woody and herbaceous Angiospermae also are reported to abscise lateral branches. Species which exhibit the phenomenon occur in a number of families including Ulmaceae, Salicaceae, Fagaceae, Juglandaceae, Oleaceae, Aceraceae, Rosaceae, Geraniaceae, Euphorbiaceae, Piperaceae, Moraceae, Annonaceae, Leguminosae, Casuarinaceae, Lauraceae, Chloranthaceae, Sonneratiaceae, Begoniaceae, Nyctaginaceae, Galsaminaceae, Vitaceae, Loranthaceae, Tamaricaceae, Ericaceae, Rhamnaceae, Celastraceae, Thymelaeaceae, Saxifragaceae, Crassulaceae, and Elaeagnaceae.

Among woody angiosperms, cladoptosis has been most extensively documented in *Populus*. In *Populus alba*, leafy twigs and branches with well-formed winter buds routinely are shed each autumn. Abscised branches may be as large as 2 cm in diameter and may vary from 1 to 20 years old (Eames and MacDaniels, 1947; Schaffner and Tyler, 1901). Branches abscised from *Populus deltoides* (Fig. 8) purportedly vary from 1 to 8 years old and are shed from the time leaves appear in spring until they abscise in autumn, with most branch shedding occurring in the autumn. As in *Populus alba*, the branches apparently are vigorous and have well-formed terminal buds (Anonymous, 1930; Bessey, 1900; Hickman, 1903; Schaffner, 1902b; Trelease, 1884). Other species which exhibit a similar shedding habit are *Populus grandidentata*, *P. tremuloides*, *P. pyramidalis* (*dilatata*), *P. balsamifera*, *P. tremula*, *P. serotina*, *P. canadensis*, and

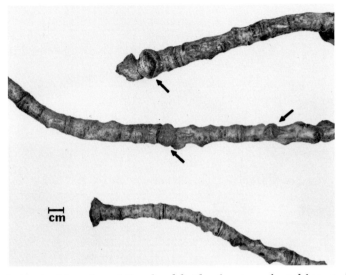

FIG. 8. Abscised branches of *Populus deltoides* showing enlarged base and branch scars (indicated by arrows).

P. canadensis var. *eugenii* (Broadhurst, 1917; Büsgen and Münch, 1931; Schaffner, 1902b; Schaffner and Tyler, 1901; Shattock, 1888; Thomas, 1933).

Species of *Salix* also shed twigs and branches. Two types of separation zones are recognized, a true cleavage zone and a brittle zone. The true cleavage zone reportedly occurs in *Salix interior, S. fragilis,* and *S. amygdaloides.* A brittle zone allows for shedding of some branches one to several years old from the above species as well as from *Salix babylonica, S. nigra,* and *S. alba* var. *vitellina* (Eames and MacDaniels, 1947; Schaffner, 1902a; Schaffner and Tyler, 1901; Shattock, 1888; Trelease, 1884). Branches also are shed from the tropical *Salix tetrasperma* by an undetermined type of separation zone. A large number of branchlets under 20 cm in length and less than 5 mm in diameter are shed as well as a considerable number from 40 to 60 cm long and 7 mm thick. The former are usually 1-year-old shoots and the latter mainly 3-year-old branches (van der Pijl, 1953).

Tamarix gallica sheds twigs each autumn as part of the annual process of defoliation (Bessey, 1900; Everett, 1968; Schaffner and Tyler, 1901).

In addition to annually shedding shoot tips (Section II) *Ulmus* often sheds twigs in the autumn. *Ulmus americana,* for example, may shed twigs 2 to 4 inches long and 1 to 8 years old (occasionally up to 12). On abscission the twigs usually break into segments because cleavage zones occur not only at the base of the twigs but at the annual nodes as well

(Broadhurst, 1917; Büsgen and Münch, 1931; Eames and MacDaniels, 1947; Schaffner, 1902a; Schaffner and Tyler, 1901; Trelease, 1884). A similar habit is exhibited by *Ulmus alata* and *Ulmus thomasii* (Schaffner, 1906).

Species of *Acer* are somewhat unique in that branches apparently are shed predominantly in the spring and early summer. *Acer saccharinum* may shed thousands of twigs from 1 to 10 years old during the period of May to July. One- to 10-year-old branches of *Acer rubrum* also are abscised in spring. Likewise, *Acer pseudoplatanus* and *A. platanoides* are reported to exhibit cladoptosis, whereas *A. saccharum* and *A. negundo* do not (Büsgen and Münch, 1931; Schaffner, 1902b, 1906).

Several other genera of temperate zone trees and shrubs shed branches. These include *Juglans, Fraxinus, Prunus,* and *Quercus.* Abscised branches of *Prunus serotina* have well-formed terminal buds and vary from 1 to 6 years old. Branches are shed in the autumn from several species of *Quercus* including *Q. alba* (Fig. 9), *Q. macrocarpa, Q. acuminata, Q. velutina, Q. imbricaria, Q. platanoids, Q. alexanderi, Q. primus, Q. robur,* and *Q. sessiliflora.* Branches cast from *Quercus alba* vary from 1 to 4 years old whereas those shed from *Q. acuminata* may be up to 7 years old. *Quercus velutina* and *Q. imbricaria* shed small twigs, but not very abundantly. Branches up to 2½ ft long may abscise from *Quercus sessiliflora* (Büsgen and Münch, 1931; Schaffner, 1902a,b; Schaffner and Tyler, 1901; Shattock, 1888; Trelease, 1884). *Euonymus atropurpureus* and *E. europaeus* shed branches from 1 to 4 and 1 to 8 years old, respectively.

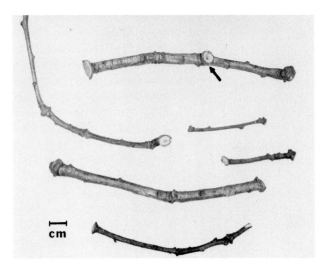

FIG. 9. Twigs abscised from a mature *Quercus alba* showing enlarged bases and a branch scar (indicated by arrow). (From Chaney and Leopold, 1972.)

Small twigs of undetermined age are shed from *Shepherdia* (*Lepargyrae*) *canadensis, Vaccinium vacillans, V. stamineum,* and some species of *Rhamnus* and *Dirca* (Eames and MacDaniels, 1947; Schaffner, 1902a,b, 1906). *Ribes* bushes also occasionally shed shoots. Swarbrick (1928) observed that "blowing out" of shoots of black currant during July and August was spuriously regarded as wind damage. Examination of shoots revealed a cleavage zone which formed a perfect ball and socket joint upon separation.

Van der Pijl (1952, 1953) reported cladoptosis in a number of tropical trees. In *Albizzia procera,* for example, a majority of the previous year's branches were shed at the beginning of the growing season. Many 2-year-old branchlets also were shed, leaving smooth abscission scars. Most abscised branches measured 8 to 10 mm in diameter and 30 to 70 cm long. Older branches (20 mm in diameter) did not routinely abscise but were easily broken from trees, indicating retention of preformed cleavage zones.

Another interesting example of branch abscission in the Leguminosae is shown by the prairie perennial, *Psoralea argophylla.* Shoots of this species arise from the upper portion of a woody taproot which is 5 to 15 cm below the soil surface. Prominent cleavage zones are located at the base of stems, and abscission of the entire aerial portion of the plant occurs during August or September (Becker, 1968).

A number of Euphorbiaceae varying from trees to herbaceous plants shed lateral branches. Scale leaves occur on the main stem and branches of these plants, whereas foliage leaves are restricted to lateral branches which may be up to 40 inches long. Leaves generally fall individually from the branches which subsequently abscise. This habit is found in tropical trees and shrubs of the genera *Phyllanthus, Breynia, Cicca, Emblica,* and *Glochidion* and in the herbaceous *Sauropus* (Corner, 1951; Lynch, 1878; van der Pijl, 1952, 1953).

Several other tropical trees are known to abscise branches. *Sonnerata acida* sheds numerous branches up to 11 mm in diameter. The branches are similar to those of *Gnetum* and *Ulmus* in that cleavage zones occur in every node and the branches frequently break into several parts upon abscission. A similar habit is exhibited by *Chloranthus.* In *Persea gratissima,* 1-year-old branchlets are shed and occasionally, older branches with laterals also are abscised (van der Pijl, 1953).

Two species of Annonaceae, *Xylopia curtisii* and *Canangia* (*Canangium*) *odoratum,* readily shed lateral branches and produce straight, clean stems. Abscised branches often litter the ground under these trees and have been observed with diameters as large as 21 mm in *Xylopia* and 60 mm in *Canangia* (van der Pijl, 1952, 1953).

The scale leaves of *Casuarina* occur at the nodes of slender deciduous branchlets which function as leaves (Everett, 1968). Van der Pijl (1953) observed a specimen of *Casuarina sumatrana* which shed 99% of its small branchlets. Older branches commonly abscise also. *Casuarina equiseti-folia*, for example, may shed branches as large as 9 mm in diameter. This thickness is considerable compared to the primary branchlet which is approximately ⅓ mm in diameter.

Cladoptosis is well known in the Central American rubber tree, *Castilloa elastica* (Fig. 10). Permanent branches are not produced until the third or fourth year. Temporary branches seldom attain an inch in diameter, but may be 10 or 12 ft long. They are regularly shed from the bole beginning with the oldest or lowest. Mature trees produce few deciduous branches. The two branch forms are distinguishable by the angle of insertion at the main stem. Permanent branches are directed upward at an angle of about 45°, whereas temporary branches are almost or quite horizontal (Cook, 1903; Lynch, 1878). Another member of the Moraceae, the upas tree, (*Antiaris toxicaria*) sheds lower lateral branches much like *Castilloa*, and produces a straight bole with the crown restricted to the upper third of the tree (Corner, 1951; Lynch, 1878; Shattock, 1888).

FIG. 10. *Castilloa elastica* with abscised branch. (From Cook, 1903.)

Several lianas abscise branches. About the time that the leaves of *Ampelopsis cordata* (Vitaceae) are shed, nearly all the slender green branches break apart at the nodes. During the winter the plant resembles an artificially pruned vine. *Parthenocissus quinquefolia* shows a similar habit, but to a lesser degree (Schaffner and Tyler, 1901). A number of species of *Vitis* also shed twigs through cleavage zones at the leaf nodes (Schaffner, 1902a). Similarly, in climbing species of *Piper*, some branches are shed either as a whole or at successive nodes (van der Pijl, 1952).

Among herbaceous angiosperms, shoots and internodes are reportedly shed by species of *Impatiens, Begonia, Mirabilis, Viscum, Phoradendron, Crassula*, and *Bryophyllum* (van der Pijl, 1952; Schaffner, 1902a; Shattock, 1888).

C. Cladoptosis in Fossil Plants

Good evidence exists that branch abscission occurred among the Lycopsida of the Paleozoic era. Paleobotanists have described several branch specimens which are characterized by large scars. Many theories have been promulgated through the years to account for these scars, but the preponderance of evidence indicates that the scars result from branch abscission (Seward, 1959, 1963; Scott, 1962).

Fossil shoots which exhibit large and more or less circular cup-shaped scars are known as *Ulodendron*. This form genus is applied to certain shoots of plants which may belong to the genera *Lepidodendron, Bothrodendron*, and possibly *Sigillaria* (Seward, 1963).

At the center of abscission scars on *Ulodendron* shoots of *Lepidodendron veltheimianum* is a slightly projecting boss or umbilicus which represents the vascular tissue. The sides of the scars are usually marked with radiating ridges (Seward, 1963). It is generally accepted that these scars indicate the point of attachment of branches, but the nature of the attachment is still in question. Watson (1914) maintained that branches were attached to the whole area of the scar, the umbilicus corresponding to the central vascular cylinder and radial marks on the scar representing leaf traces of the branch. Reiner, in contrast, concluded that branches were attached to the umbilicus only and that the sides of the scar resulted from pressure of the branch base against the expanding stem (Lindsey, 1915).

Ulodendron shoots of *Bothrodendron punctatum* are distinguished from those of *Lepidodendron* by the eccentric position of the umbilicus in the bottom of the cup-shaped scars. These scars may be quite large. A *Bothrodendron* specimen from the Coal-Measures in Belgium had scars

9 cm in diameter. Another specimen from the Lancashire Coal-Measures had 2 rows of scars 11 to 12 cm in diameter on a stem 112 cm in girth and 233 cm long (Seward, 1963).

Two specimens of *Bothrodendron minutefolium* were described by Lindsey (1915). Although these were not *Ulodendron* shoots, they were interesting because one was an abscised branch and the other indicated the attachment of lateral branches. The base of the branches was trumpet-shaped, about 2½ times larger than the major portion of the branch, and the cortex of the branches was continuous with that of the main stem. An enlarged base also is characteristic of many branches which abscise from present-day flora.

Kidston (1885) described the *Ulodendron* forms of two species of *Sigillaria*. The specimens were distinguished from those of *Bothrodendron* by the almost central position of the umbilicus in the scar. *Lepidodendron* scars also have a centrally positioned umbilicus, and the distinction between *Sigillaria* and *Lepidodendron* is difficult to make (Seward, 1963). For these reasons Seward argued that conclusive evidence for *Ulodendron* shoots of *Sigillaria* was lacking.

Another shoot form that exhibits scars which probably represent branch abscission sites is *Halonia*. The *Halonia* shoot is usually smaller than *Ulodendron*, but still often several inches in diameter and is characterized by a number of spirally arranged protuberances. The surface of specimens is generally poorly preserved, having been decorticated before fossil-ization, accounting for the protuberances. Well-preserved *Halonia* shoots occasionally are found and have been shown to belong to the *Lepido-phloios* genus and possibly the *Lepidodendron*. On specimens having the outer cortex intact, the protuberances appear as scars. These scars are distinct from those of *Ulodendron* because of the absence in *Halonia* of the cup-shaped depressions (Arnold, 1947; Scott, 1962).

D. NATURAL PRUNING

Lower branches of trees, particularly of trees under forest conditions, often have little physiological importance. The photosynthetic surface of these branches is usually reduced and the leaves present are heavily shaded and at a definite competitive disadvantage. Such branches con-tribute little or nothing to the growth of trees (Kramer, 1958). In *Pinus taeda*, for example, Labyak and Schumacher (1954) found that the upper branches contributed most of the food for stem growth and the lower branches made such a small contribution that they could be removed to 60% of tree height without reducing stem diameter growth. Natural pruning of these useless branches proceeds slowly during the life of a tree. The initial step is death of branches followed by gradual weakening and decay and eventual mechanical shedding. Trees vary tremendously in the rate at which lower branches die, decay, and ultimately break away (Table I). As a general rule, dead branches in angiosperm trees usually decay rapidly, whereas in gymnosperms the resinous branches persist for a long time (Spaulding and Hansbrough, 1944). Exceptions occur in both groups, however. For example, dead branches of both *Quercus palustris* and *Q. coccinea* are retained long after they die, whereas dead branches on pines of the southern United States prune quite well (Smith, 1962). Figure 11 illustrates the well-pruned boles of a stand of *Pinus palustris*. For contrast, the persistent branches of *Pinus virginiana* which is a notoriously poor natural pruner are shown in Fig. 12.

An unusual shedding of living, solid, nondiseased branches recently has been reported in *Platanus acerifolia, Cedrus deodara, Ficus retusa* and in species of *Ulmus, Eucalyptus, Quercus,* and *Pinus* (Harris, 1972). Branches 4 to 20 inches in diameter fell from trees late on hot, still, summer afternoons. No plausible explanation was given for the cause of this limb breakage which apparently involved temperature, moisture, and wood strength.

E. ANATOMY OF CLADOPTOSIS

Information on anatomical changes associated with branch and twig abscission is meager. Observations have been made, however, for *Taxodium distichum, Populus grandidenta, Populus tremula,* and *Agathis australis.* The anatomy of branch abscission is apparently quite similar to that of leaves (Esau, 1960).

Barnard (1926) reported that early in the season cells of the cortex of *Taxodium distichum* branchlets were continuous with those in the main stem, and no abscission zone was evident. The separation zone was discernible only after changes occurred which preceded branch fall. In the latter part of the growing season, a band of cells at the base of the branch became active and the cells divided. This activity commenced at the periphery of branches and extended toward the vascular tissue. A layer of 7 to 12 cells with high protoplasmic contents and large nuclei was finally produced. After the separation zone was differentiated, a layer of cells situated further down the branch became active and formed a phellogen, which subsequently produced a band of cork about 4 cells wide. Between the cork and separation layers, a zone about 5 cells wide was present. These cells were practically devoid of protoplasm and the

TABLE I

NATURAL PRUNING CHARACTERISTICS OF VARIOUS TREE SPECIES

Species	Pruning characteristics	Reference
Abies balsamea	Prunes poorly. Dead branches persistent	Paille, 1966; Vezina and Paille, 1967
Abies concolor	Prunes slowly, especially in young trees	Maul, 1958
Abies lasiocarpa	Prunes well in dense stands. May be free of branches on 20 to 40 ft or more of bole. Branches extend to ground when open grown	Alexander, 1958b
Abies procera	Generally prunes well in closed stands	Staebler, 1958
Acer macrophyllum	Dense side shade results in long boles 100 ft or more and free of branches for $\frac{1}{2}$ to $\frac{2}{3}$ of length	Ruth and Muerle, 1958
Aesculus glabra	Prunes moderately well under side competition and shade	Merz, 1957a
Aesculus octandra	Has a long clean bole under forest conditions	Carmean, 1958
Betula alleghaniensis	Prunes moderately well in dense stands	Gilbert, 1960
Betula lenta	Prunes well in dense stands. Low, large branches produced on open-grown trees	Leak, 1958
Carya cordiformis	Prunes more readily than other species of *Carya*	Nelson, 1965
Carya illinoensis	Prunes well in well-stocked stands	Nelson, 1965
Carya laciniosa	Bole may be clean of branches for half its length	Merz, 1957b
Celtis laevigata	Does not prune well in natural stands	Putnam *et al.*, 1960
Chamaecyparis thyoides	In closed stands lower branches die at early age but persist for many years	Little, 1959a
Fagus grandifolia	Prunes well under forest conditions; open-grown trees limby	Rushmore, 1961
Fraxinus americana	Prunes well in closed stands. Small branches shed within 1 to 2 years and larger ones within 4 to 5 years after death	Wright, 1959b
Fraxinus pennsylvanica	Prunes moderately well but not as well as *F. americana*	Wright, 1959a

TABLE I (*Continued*)

Species	Pruning characteristics	Reference
Gleditsia triacanthos	Lower limbs of forest grown trees do not survive in deep shade, but dead limbs persist for several years	Funk, 1957
Juglans nigra	Does not prune well. Light shade of walnut in plantations is insufficient to cause early pruning; open-grown trees limby	Brinkman, 1957b
Juniperus occidentalis	Prunes poorly. Rarely has as much as 4 to 8 ft of clear bole	Sowder and Mowat, 1958
Juniperus scopulorum	Lower branches on trees in dense stand or in deep shade die, but persist	Herman, 1958
Juniperus virginiana	Branches are persistent	Williamson, 1957
Larix laricina	Excellent pruner. Bole often clear of branches for $\frac{1}{2}$ to $\frac{2}{3}$ of length at 25 to 30 years of age	Roe, 1957
Larix occidentalis	Small dead branches persist throughout lower bole of 50-yr-old trees	Boe, 1958
Libocedrus decurrens	Does not prune well even in dense stands	Schubert, 1957b
Liquidambar styraciflua	Prunes readily under forest conditions	Martindale, 1958
Liriodendron tulipifera	Prunes well except in very sparsely stocked stands	Olson, 1969
Lithocarpus densiflorus	Prunes well in closed stands	Roy, 1957
Magnolia grandiflora	Forest grown trees frequently clear of branches for 40 ft above ground	Bennett, 1961
Picea engelmannii	Does not prune well	Alexander, 1958a
Picea mariana	Prunes well in dense stands; open-grown trees retain branches nearly to ground	Heinselman, 1957
Picea rubens	Prunes well in dense stands	Hart, 1959
Picea sitchensis	Lower branches die early in dense stands but branch stubs persist many years	Ruth, 1958
Pinus banksiana	Prunes moderately well in dense stands, but where stocking is less prunes poorly	Rudolf, 1958
Pinus contorta	Does not prune well; branches retained nearly to ground in	Tackle, 1959

(*Continued*)

TABLE I (*Continued*)

Species	Pruning characteristics	Reference
	open-grown stands. Partial pruning of 10–25% of bole common in dense stands	
Pinus coulteri	Prunes poorly; lower branches remain alive and functional	Baker, 1950
Pinus echinta	Prunes more slowly than *P. taeda* or *P. palustris;* branches on lower 20 ft of bole remain alive an average of 8 years; dead branches may persist for 12 years	Paul and Smith, 1950
Pinus edulis	Shaded lower branches die but are persistent	Fowells, 1965
Pinus elliottii	Prunes well, particularly in dense stands; dead branches may persist for about 6 years	Cooper, 1957; Paul and Smith, 1950
Pinus lambertiana	Does not prune well, even in dense stands; branches die early but persist	Fowells and Schubert, 1956
Pinus monticola	Does not prune well, even in dense stands	Rapraeger, 1939; Wellner, 1962
Pinus palustris	In dense stands on good sites natural pruning begins at 10 years of age; usually branches on lower 20 ft of bole remain alive for 7 years and dead branches may persist for 6 years	Allen, 1955; Paul and Smith, 1950
Pinus ponderosa	Young trees typically limby and natural pruning develops slowly; old-growth trees have clean bole; on the average, bole clear for 17 ft at 78 years and 33 ft at 103 years	Curtis and Lynch, 1957; Kotok, 1951
Pinus radiata	Is not self-pruning; branches persist indefinitely	Fowells, 1965
Pinus resinosa	Prunes very well beginning at the age of about 25 years in dense stands; in more open stands pruning is delayed to a greater age; on poor sites, lower branches pruned even when crown closure does not occur	Rudolf, 1957
Pinus rigida	In dense stands prunes about as well as *P. echinata*, but in	Little, 1959b

TABLE I (*Continued*)

Species	Pruning characteristics	Reference
	understocked stands it tends to produce larger and more persistent branches than *P. echinata*	
Pinus strobus	Trees in pure second-growth stands noted for branchiness; branches live for about 15 years and persist on the bole for more than 25 years after they die; on the average, 60 limbs occur in the first 16 ft of the bole	Mollenhauer, 1938; Wilson and McQuilkin, 1963
Pinus taeda	Prunes readily at younger ages, before branches develop heartwood; branches on lower 20 ft of bole remain alive for 7 years; in 15-year-old trees, average of 8.6 years needed for complete shedding of dead branches	Paul, 1933; Reynolds, 1967; Wenger, 1958
Pinus virginiana	Prunes poorly; retains dead branches for an extremely long period	Fenton and Bond, 1964
Platanus occidentalis	Under forest conditions produces a bole that may be clear of branches for 70 or 80 ft	Merz, 1958
Populus balsamifera	Prunes well; trees commonly clear of branches for 30 to 50 ft	Roe, 1958a
Populus deltoides var. *occidentalis*	Mature trees 80 to 90 ft tall have clear boles for 30 ft or more	Read, 1958
Populus tremuloides	Prunes extremely well when side shade present	Strothmann and Zasada, 1957
Populus trichocarpa	Prunes well; in closed stands lower $\frac{2}{3}$ of bole clear of branches; in open stands, lower $\frac{1}{2}$ of bole clear	Roe, 1958b
Prunus serotina	Lower branches prune well when shaded	Hough, 1960
Pseudotsuga menziesii	Lower branches die early in closed stands but persist for many years; on the average, bole clean for 17 ft at 77 years and for 33 ft at 107 years	Isaac and Dimock, 1958; Smith, 1956

(*Continued*)

TABLE I (*Continued*)

Species	Pruning characteristics	Reference
Quercus alba	Prunes well in moderate to heavily stocked stands; branches persist along bole if strong side light present	Minckler, 1957b
Quercus bicolor	Prunes well under forest conditions; lower brancher persistent when open grown	Clark, 1958
Quercus coccinea	Prunes poorly; branch stubs very persistent	Campbell, 1957
Quercus falcata var. *pagodaefolia*	Usually produces a branch-free lower bole	Lotti, 1957
Quercus laurifolia	Prunes poorly and large limbs persist on the bole many years, even under a dense canopy	Fowells, 1965
Quercus lyrata	In mixed stands with other hardwoods on a good site can develop a long trunk clear of branches for 40 ft or more	Fowells, 1965
Quercus nigra	Slow to prune itself; small dead limbs or their stubs remain on the tree for a considerable time	Putnam and Bull, 1932
Quercus palustris	In closed stands, profuse lower branches die at an early age but dead branches persist for many years	Minckler, 1957a
Quercus phellos	Pruning is poor, especially on poor sites; numerous small live and dead branches are long persistent on the trunk	Putnam and Bull, 1932
Quercus prinus	Prunes moderately well in forest stands	Fowells, 1965
Quercus rubra	Prunes well under forest conditions	Sander, 1957
Quercus velutina	Prunes moderately well in dense stands where branches die before larger than 1 to $1\frac{1}{2}$ inches in diameter; in open stands branches persist for many years	Brinkman, 1957a
Robinia pseudoacacia	When forest grown produces clean, straight bole; in the open bole is limby	Roach, 1958

TABLE I (*Continued*)

Species	Pruning characteristics	Reference
Salix nigra	Prunes well in dense stands; open-grown trees generally limby	Fowells, 1965
Sassafras albidum	Pruning good in well-stocked stands	Fowells, 1965
Sequoia gigantea	Pruning from trees less than 100 years old is poor; bole of mature trees usually free of branches to a height of 100 to 150 ft	Schubert, 1957a
Sequoia sempervirens	Pruning in young stands is poor; dead branches may persist for over 50 years; trees 85 years old in dense stands may have trunks clear of branches for 75 to 100 ft	Fowells, 1965
Simaruba amara	Exhibits no pruning even in dense stands	Schulz and Vink, 1966
Taxodium distichum	Prunes well in dense stands; limby above buttswell in poorly stocked stands	Langdon, 1958; Mattoon, 1915
Tilia americana	In closed stands, bole clear of branches for 50% of its length	Scholz, 1958
Tilia heterophylla	Bole typically free of lower branches	Renshaw, 1961
Tsuga canadensis	Branches remain alive for many years and when dead prune poorly even in dense stands	Harlow and Harrar, 1968
Tsuga heterophylla	Prunes early in dense even-aged stands; branches persistent when open grown	Berntsen, 1958
Tsuga mertensiana	Prunes poorly even in crowded stands with branch stubs persistent to ground	Dahms, 1958
	Good pruning in even-aged stands in British Columbia	Fowells, 1965
Ulmus americana	In dense stands, develops a clear bole for 50 to 60 ft in length	Guilkey, 1957
Ulmus thomasii	May have bole clear of branches to a height of 50 to 60 ft in well-stocked stands	Scholz, 1957
Umbellularia californica	Prunes well even when open grown	Stein, 1958

FIG. 11. Stand of *Pinus palustris* showing good natural pruning. (Photo courtesy of U. S. Forest Service.)

walls became highly lignified. As autumn approached, the middle lamellae of cells of the separation zone dissolved and the cells separated. Actual separation of a branchlet generally occurred along the inner side of the separation layer, with most of this layer falling with the branch. The vascular tissue was broken by gravity or other mechanical action. After branch fall, cells between the remaining separation zone and phellogen became strongly lignified and formed a protective layer. The phellogen produced more cork cells until a thick periderm formed. At about the time of abscission and later, a large amount of a "brownish gum" was produced as globules in cells of the separation layer and in adjoining cells, particularly in the phloem. When larger branches of *Taxodium distichum* were shed, the anatomical changes which occurred were similar to those in the smaller foliar branchlets. Distinct separation and cork layers formed.

In *Populus grandidentata* small branches are swollen at the base where they are attached to the trunk or larger branches (as in *Populus deltoides*, Fig. 8). A parenchymatous cortex is largely responsible for the swelling.

FIG. 12. Persistent branches of *Pinus virginiana*. (From Fenton and Bond, 1964.)

Stone cells are present in this region, but not fibers. Xylem vessels are few in number and considerably modified, with abundant scalariform or reticulate pits instead of the normal circular pits. Parenchyma cells are abundantly scattered among the vessels. The separation layer is formed through the living cells of the xylem and cortex. Nonliving elements of the xylem rupture easily because of transverse scalariform pitting. As in *Taxodium distichum*, a periderm layer is formed just below the face of the scar after fall of the branch or possibly before its separation (Eames and MacDaniels, 1947).

In *Populus tremula* the base of branches also is enlarged due to an increase of cortical parenchyma and branching type sclereids. The xylem is thin at the bases of branches and is further weakened by clefts much as in *Agathis australis* (Fig. 13). Parenchyma cells in the zone of weak xylem multiply by transverse divisions producing a layer of cork. The thin xylem is subsequently broken under the weight of the branch and separation occurs through the corky layer (Shattock, 1888).

Licitis-Lindbergs (1956) found several modifications in the zone of branch abscission of *Agathis australis*. The bases of branches were enlarged because of dilated cortex and pith (Fig. 14). The cortex contained

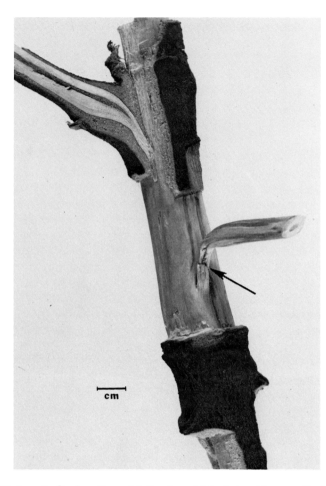

FIG. 13. Longitudinal section of foliar branchlet of *Agathis australis* showing enlarged cortex and narrowed xylem at branch attachment (upper). Foliar branchlet with cortex removed showing grooves in xylem and point of breakage (arrow) (lower). (From Licitis-Lindbergs, 1956.)

a large number of sclerenchymatous cells scattered among diffuse and irregularly arranged parenchyma. The xylem was modified and reduced. Parenchyma cells within the xylem divided and eventually separated the vascular tissue into 4 parts. In addition to being weakened by parenchyma in the zone of abscission, the xylem tissue tapered from branches to the point of attachment with the main stem (Fig. 13). Weakening of xylem in this zone was shown by removal of bark and cortex. After a few days, branches bent downward and broke easily (Fig. 13). The breaking point

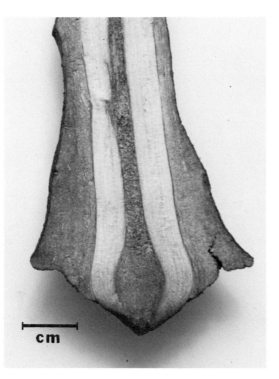

FIG. 14. Longitudinal section of abscised branch of *Agathis australis* cut perpendicular to the axis of the main bole. Note dilated pith and cortex. (From Licitis-Lindbergs, 1956.)

of the xylem occurred slightly proximal to the place where the branch cortex met that of the trunk, but above the point where the xylem arose. A periderm formed on the scar before separation of the branch. The proximal end of abscised branches was smooth but a periderm was lacking. Lack of protective periderm permitted desiccation and shrinkage so that, after abscission, branch bases did not fit the scars.

F. BRANCH SCARS

Scars on branches and trunks of trees can arise from two mechanisms, branch abscission or markings or distortions caused by branches before they are naturally pruned. Abscission scars will be considered first.

Abscission of branches often reveals a ball and socket arrangement. The base of an abscised branch has a convex surface and the scar left by a fallen branch is concave. The surface of a scar is usually smooth because the separation extends chiefly through living cells of the cortex and vascular tissue. Typical smooth, concave branch scars occur on *Populus deltoides* (Fig. 8), *Quercus alba* (Fig. 9), and *Populus gradidentata* (Eames and MacDaniels, 1947). In *Populus*, a periderm layer which is continuous with periderm of the main branch is formed just below the face of the scar. Periderm formation in *Quercus* is similar to that in *Populus* but occurs deeper in the scar (Eames and MacDaniels, 1947).

A fully developed periderm (phellem, phellogen, and phelloderm) is present on the scar resulting from abscission of *Agathis australis* branches. When a branch is shed, the reduced xylem breaks leaving a small hole in the middle of the scar. Periderm extends over the hole later. The scars are circular and somewhat concave and may be as large as 8 cm in diameter (Licitis-Lindbergs, 1956). Dallimore and Jackson (1967) reported circular scars on stems of *Agathis* which were ½ to ¾ inches deep. A scar on an *Araucaria cunninghamii* stem measured 15 cm in diameter.

Not all branch scars are prominently concave. Many are somewhat flat, circular, and smooth surfaces as illustrated by scars resulting from abscission of short shoots of *Pinus* and branchlets of *Taxodium distichum* (Fig. 4) and *Thuja occidentalis*.

Markings which are caused by lateral branches that do not abscise often occur on stems of trees, particularly those with smooth bark. When a living branch is present on a tree of the temperate zone, a new layer of wood is added each year to the main bole as well as to the branch. The additional wood forces the bark outward. When this occurs in the angle between the upper side of a branch and the bole, the bark is of necessity buckled because the surface to accommodate the bark actually is reduced. This also occurs on the lower side of a branch, but to a lesser degree. When a branch dies a similar effect gives a much accentuated result. As the bole of a tree continues to grow in girth, the bark is forced out, whereas on the dead branch no corresponding thickening takes place and the bark remains closely attached to the old wood surface. Consequently, there is considerable folding of the bark on the trunk around the branch stub insertion.

The folds which develop in the bark remain visible for many years, and indicate the former existence of a branch on the stem (Fig. 15). The wrinkles in the bark are perpetuated because the phellogen adheres to the bark when it is forced into folds, and the living cells following this contour continue to add new cork and maintain the same pattern of folds year after year. As the trunk increases in girth, the pattern of folds tends to retain the same relative size in proportion to the circumference of the tree. Consequently, markings formed early in the life of a tree may become very wide. For example, Priestley and Scott (1932) mea-

FIG. 15. Folds in the bark of *Malus* indicating the former existence of branches.

sured a 41-inch-wide scar on a stem of *Fagus.* Smaller but equally
prominent scars were observed on *Betula, Platanus,* and *Pseudotsuga.*

The area of bark affected by branch-induced folding is related to the
angle at which a branch leaves the trunk. If a branch forms an acute
angle with the bole of a tree, the area affected by folding is deep. The
bark area influenced by folding is wider but not as deep when a branch
grows out more nearly at right angles to the axis (Priestley and Scott,
1932).

Occlusion of gaps which result from the eventual breaking away of
dead and decayed branches often produces aberrations in the bark.
Proliferation of callus is an important part of the healing process and is
considered below.

G. HEALING AND OCCLUSION OF BRANCH SCARS AND STUBS

One of the earliest observations on healing of branch scars was made
by Shattock (1888) on *Agathis (Dammara) robusta.* Beneath the corky
layer on the surface of scars of this tree, parenchyma cells are present

which develop into a secondary wood-forming meristem. A layer of wood soon forms under the scar. The wood produced over the scar area and that formed in the rest of the tree bole are not a uniform continuous sheath. In longitudinal sections of old scars, wood over branch scars is distinct and appears as a cylindrical core with little connection to the vascular tissue around it. Whether the pattern of healing of branch scars of *Agathis robusta* is unique or is characteristic of healing of all branch scars is unknown. This interesting process appears to have been overlooked by anatomists.

Healing of wounds created by the mechanical breaking of branches is different from that of branch scars. Neither a periderm nor cambium initially forms in these wounds, but rather gradually extends from the edges toward the center. The callus formed around branch wounds initially is uniform in size and is concentric. As callus continuous to grow in successive seasons, a vertical polarity is established, presumably under the influence of vertical transport of materials in the phloem. Consequently, callus on the sides of wounds grows more rapidly than that at the top and bottom resulting in formation of a spindle-shaped callus tissue (Neely, 1970; Roth, 1948; Zimmermann and Brown, 1971).

Formation of callus in the healing of branch wounds results from abundant proliferation of cambial cells with an associated production of masses of parenchyma. The outer cells of this tissue either become suberized themselves or periderm develops within them. Beneath the protective layer a cambium forms and produces new vascular tissue. Around a branch wound the cambium is at an angle to the face of the wound at the point of intersection with it. In this position, new vascular tissue extends the growing layer over the wound surface until the 2 opposite sides meet. The cambium layers eventually unite and the branch stub is completely occluded (Eames and MacDaniels, 1947).

The rate of healing of branch wounds usually is correlated with tree vigor (Smith, 1962; Neely, 1970; Roth, 1948). Wounds heal more readily on rapidly growing trees than on slow growing ones. The position of wounds on the trunks of trees also affects the rate of healing since the rate of growth varies along the bole. The vertical distribution of radial trunk growth decreases from the base of the crown downward in suppressed trees. In dominant trees a similar pattern exists except at the base where increased radial growth occurs (Kozlowski, 1971b). Consequently, branch wounds in the upper stem heal more rapidly than those low on the bole (Neely, 1970; Rapraeger, 1939). In vigorous or open-grown trees with deep crowns there is little difference in radial trunk growth in relation to height and also little difference in the rate of wound healing.

H. Factors Affecting Cladoptosis

Abscission of twigs and branches occurs in response to a number of physiological and environmental factors. Low tree vigor, water supply, site factors, and aging influence the number of branches shed. However, the intricacies of the relationships are little understood and apparently have been almost entirely ignored by plant physiologists.

Many of the branches that abscise from trees are weak and nonvigorous. For example, insect or disease infestations which reduce the vigor of twigs can elicit abundant abscission (Büsgen and Münch, 1931). In *Populus serotina* Thomas (1933) found that branches which arise from small buds and make poor growth are usually shed in the fall. Branches which produce a preponderance of flower buds year after year produce progressively shorter internodes. Such reproductive branches produce little current photosynthate, eventually become depleted of carbohydrate reserves, and abscise. In *Quercus alba*, twigs that were retained on the tree had longer terminal shoots than abscised twigs, indicating that the less vigorous twigs were shed (Chaney and Leopold, 1972).

The capacity for branch abscission is apparently associated with maturation of woody plants. For example, twig abscission does not occur in juvenile *Quercus alba* but is common in mature trees (Chaney and Leopold, 1972). The retention of brown and withered leaves through the winter on juvenile *Q. alba* is possibly related to retention of branches. It is interesting to note that marcescent leaves also are characteristic of *Quercus imbricaria* which abscises only limited numbers of branches (Schaffner, 1902b). Stark (1876) observed that branches were shed from several species of Cupressaceae but only from mature trees. In *Castilloa elastica* a different response to aging occurs. Young trees readily shed lateral branches, whereas mature trees develop branches which do not abscise (Cook, 1903).

Water deficits apparently stimulate branch abscission. Schaffner (1902a), for example, observed branch abscission in a number of angiosperm trees in Ohio before July 15 during a dry summer. These trees usually shed branches in autumn. Species of *Ephedra* exhibit branch abscission as a drought-resistance mechanism and it is probable that the abscission is stimulated by water stress (Bold, 1957; Chamberlain, 1934). Water supply also influences branch abscission in *Araucaria araucana*. On well-drained but moist loamy soil this species produces a full crown and retains its lower branches for 50 or more years. On dry, shallow, or gravelly soils, lower branches are shed early in life (Dallimore and Jackson, 1967).

The persistence of short shoots in *Pinus* varies widely among species as indicated previously and undoubtedly is genetically controlled. However, the period of persistence is markedly influenced by environmental factors. Weidman (1939) concluded that persistence of short shoots of *Pinus ponderosa* was not inherited. In parent trees short shoots persisted for 4 to 5 years on a moderate site, 6 years on a more severe site, and 8 years on a very rigorous site. When progeny of these trees were planted on a moderate site, the short shoots uniformly persisted for 3 to 4 years regardless of origin. Similar results were obtained for *Pinus sylvestris* in Switzerland (Weidman, 1939).

I. Factors Affecting Natural Pruning

The seemingly simple phenomenon of natural pruning is actually the product of a number of interacting physiological and ecological factors. Light intensity, fungi, insects, soil factors, and weather conditions are all inextricably involved. Management measures which influence any of these factors will necessarily also influence natural pruning and can be used to advantage.

1. Physiological Factors

Initial steps in natural pruning include physiological senescence and eventual death of a branch. Undoubtedly the moribund condition of a branch is influenced by several factors, but light intensity appears to be the most important.

Foresters indirectly classify trees on the basis of their light requirements. Shade-tolerant trees such as *Tsuga canadensis* and *Fagus sylvatica* can grow and reproduce in dense shade, whereas most pines, for example, are shade intolerant and require higher light intensity. The rapidity with which lower branches are naturally pruned is considered a good indicator of the degree of tolerance. Lower branches generally die earlier on intolerant than on tolerant trees. Exceptions to this general rule occur, however, and *Pinus coulteri* is an example. Its lower branches remain alive and functional, whereas lower branches of the equally shade intolerant *Pinus palustris* are short-lived (Baker, 1950; Toumey and Korstian, 1947).

Trees are remarkable reflectors and absorbers of short wave radiation and little visible light penetrates the upper canopy to the lower branches. For example, on a clear midsummer day only about 10% of the short wave radiation which impinges on a closed canopy generally will penetrate to the forest floor (Reifsnyder and Lull, 1965). Angiosperm trees in full leaf exclude as much light as gymnosperms do. Ovington and Madgwick (1955) reported that light under broad-leaved trees was 2 to 6% of that in an open field, and under gymnosperms was 0.5 to 6.7%. Geiger (1965) summarized the findings of several investigators and reported much broader ranges for light transmission. Light intensity under angiosperms was 2 to 60% of that in the open, whereas under gymnosperms it was 2 to 20%. The broad ranges occur because the amount of light which penetrates a canopy varies with species composition of stands. According to Kittredge (1948), less light is transmitted by stands of shade tolerant species than by intolerant ones.

Decrease in light intensity is very rapid through the canopy. For example, Trapp, as cited by Geiger (1965), showed that in 120- to 150-year-old *Fagus sylvatica* stands, 90% of the light on a sunny day was caught in the crown space and 80% of it in the upper two-thirds. The resultant shading of leaves on lower branches reduces their photosynthetic rate (Kramer, 1958). The effects of shading on *Pinus taeda* are illustrative. Under full sunlight (10,000 ft-c) the photosynthetic rate was 14.3 mg $CO_2/dm^2/hr$, whereas it was only 2 mg $CO_2/dm^2/hr$ under 2000 ft-c (Spector, 1956). The compensation point of leaves on lower branches of many trees may not even be attained (Heath, 1969). Branches with leaves with low photosynthetic rates or with leaves which are nonself-supporting become weak and predisposed to attack by disease and insects, and eventually die.

When a branch dies it is usually sealed off from the living bole by materials deposited in the base of the dying or dead branch. These deposits are usually composed of resins in gymnosperms and tyloses or gums in angiosperms. The extent of deposit is variable among species. Dead branches of highly resinous gymnosperms, such as pines, are more effectively sealed from the rest of the tree than those of less resinous species such as spruces. Although the deposits prevent spread of decay fungi into the bole of the tree, they sometimes have the unfortunate effect of hampering growth of fungi in the branches and thus retard natural pruning (Mayer-Wegelin, as cited by Smith, 1962).

That natural pruning is influenced at least somewhat by physiological factors is indicated by the fact that one of the criteria for selecting superior specimens for breeding in tree improvement programs is often a tree's self-pruning ability (Anderson and Smith, 1970). Vezina and Paille (1967) observed that variation in natural pruning of *Abies balsamea* was determined largely by genetic factors. For *Picea mariana* × *P. rubens* hybrids, in contrast, most variation in natural pruning resulted from differences in age and environment.

2. Ecological Factors

Early death of a branch does not necessarily ensure its early shedding. Dead branches do not fall until weakened by fungi, insects, or other agencies. The activity of these organisms is influenced by species and environmental factors.

Activity of fungi, which is influenced by temperature and moisture, is the most important factor determining the rate of pruning of dead branches. Pines of the southern United States (e.g., *Pinus taeda, P. echinata*) occur in a warm, humid climate and their dead branches fall off in half the time required by branches of *Pinus resinosa* and *P. strobus* in the cooler and drier climate of north central United States (Paul, 1938). Small branches which dry out readily sometimes persist for unusually long periods. Conditions that favor retention of moisture in dead branches hasten decay and eventual shedding of branches. Loose, flaky bark of branches can retard desiccation sufficiently to account for faster decay than of branches with smooth bark. *Pinus strobus*, with smooth bark, retains branches almost indefinitely (Smith, 1962). Good natural pruning of pines in Sweden occurs in areas along the coast which have high rainfall and high humidity during the period when decay organisms are most active (Åkebrand, 1957).

Many saprophytic fungi attack dead branches. Toole (1961) identified 25 decay fungi associated with dead branches of several angiosperms of the southern United States. Four of these, however, caused 52% of the decay. These were *Stereum gausapatum, S. subpileatum, Poria andersonii,* and *Pleurotus corticatus*.

Parasitic fungi may attack branches that are weakened by shade, drought, or other factors and cause their eventual death. For example, Long (1924) found *Cenangium abietis* on lower branches of *Pinus ponderosa*. Live branches up to 2 inches in diameter were infected. The fungus began growing on branches near the ground where moisture supply was greatest and spread to a height of 30 ft. Lack of moisture apparently limited further spread of the fungus which girdled and killed branches. Long suggested that in *P. ponderosa* shading may be a secondary cause of natural pruning in that it provides conditions conductive to growth of fungi which kill branches.

Bark beetles and ambrosia beetles are responsible for tremendous damage to standing trees and cut timber, but some may be beneficial since they breed only in dead limbs and hasten natural pruning. Good examples of beetles which never attack living parts of trees are various species of *Hypothenemus* and *Stephanoderes*, certain species of *Micracis*, and *Lymantor decipiens*. Other species such as *Pityophthorous nudus* and

P. granulatus breed in dead branches but also may become injurious to
the rest of the tree (Blackman, 1922).

Site can influence natural pruning. For example, *Pseudotsuga menziesii*
on a poor site produced smaller branches than on a good site. The small
branches decayed and were shed more rapidly than large ones (Paul,
1947).

Dead branches which are weakened by fungi and insects ultimately
break off the tree under their own weight or sooner, depending on
weather and activity of animals. Strong winds, ice, and heavy snow
undoubtedly hasten shedding of dead and decaying branches.

3. Silvicultural Factors

Natural pruning can be enhanced by manipulation of stand density and
composition and possibly by judicious use of fire. Of these silvicultural
alternatives, control of stand density is the most important.

Proper density is achieved by encouraging regeneration which ensures
establishment of at least several thousand trees per acre rapidly enough to
prevent much variation in height of a young stand. Superdominant trees
develop large branches and prune slowly, whereas suppressed trees are
likely to succumb before any advantage can be taken of their good
natural pruning (Smith, 1962). It is desirable to have stands dense enough
that the lower branches do not grow larger than 1 to 1½ inches in diameter. Such branches usually decay and are shed rapidly, producing small
knots. Initial stand density is critical in some species. If a stand of *Pinus
strobus*, for example, is understocked when young, it is rare that any
amount of crowding later on will be effective either in reducing the size
of branches or enhancing natural pruning (Tarbox and Reed, 1924).
Some investigators have suggested that the best stand density is that
which maintains a crown to height ratio of 40% on the majority of trees
(Chapman, 1953; Holsoe, 1950). Such a ratio allows for commercially
acceptable radial growth on well-pruned boles. When the crowns fall
below 40% of tree height, thinning of stands is recommended.

Mixed stands often exhibit good natural pruning. Pine stands with a
rather uniform though subordinate broad-leaved component often shed
lower branches more quickly because of the increased side competition,
shade, and possibly humidity which promotes more rapid decay. Gymnosperms under an angiosperm overstory also can promote pruning. For
example, in some angiosperm stands, maintenance of a *Tsuga* understory
beneath less tolerant species forces the development of long, straight,
well-pruned boles in trees of the overstory (Smith, 1962; Wahlenberg,
1960).

Mild ground fire which kills the foliage on lower limbs and results in their early death also promotes natural pruning. Rapid pruning of lower branches of *Pinus taeda* has been successfully attained in this manner (Wahlenberg, 1960).

J. CHEMICAL PRUNING

Both natural pruning and cladoptosis have been enhanced by application of chemicals. The use of herbicides to control broad-leaved species in gymnosperm stands indicated that these chemicals not only were effective in killing entire trees but also could kill lower branches without reducing tree vigor (MacConnell and Bond, 1961, 1962). MacConnell and Kenerson (1964) tested 2,4-D and 2,4,5-T for chemical pruning potential on several angiosperm species of the northern United States. The lower boles of trees, mainly *Quercus* species, were sprayed with 2,4,5-T in kerosene or an invert emulsion of 2,4,5-T and 2,4-D at rates of 0.5 to 1.0 lb acid equivalent per acre. All sprays killed branches up to 16 to 20 ft from the ground with no damage to the trees when applied from late July to early September. Branches up to 5 cm in diameter fell off in 4 to 7 years after spraying. Similar results were obtained by Splettstosser (1957) in Germany. The lower boles of *Quercus* and *Fagus* were sprayed with 0.2 or 0.4% Tormona 80 (2,4,5-T preparation), respectively. Lower branches died and were shed approximately 3 years later. Petschke (1962) showed that the best method for pruning *Fagus sylvatica* trees was application of a 1% aqueous solution of Select (2,4-D + 2,4,5-T) to branches in full leaf. Branches up to 2 cm in diameter were killed and were shed in about 4 years. Lower branches of *Populus nigra* also were killed when sprayed with 0.15, 1, or 2% Tormona 100 (2,4-D). A protective layer formed at the base of dead branches which became heavily infected with fungus (Liese, 1957).

In contrast to the killing of lower branches and their subsequent decay,

TABLE II

PERCENTAGE OF TWIGS ABSCISED FROM A MATURE *Quercus alba*[a]

Treatment	Abscission (%); date of treatment			
	Aug. 30	Sept. 11	Sept. 28	Oct. 11
Control	28	36	36	16
1% Ethephon	44	56	48	4
2% Ethephon	52	80	48	8

[a] From Chaney and Leopold (1972).

ethephon (2-chloroethylphosphonic acid) was shown to enhance cladoptosis of living twigs from a mature *Quercus alba* tree (Chaney and Leopold, 1972). One or 2% ethephon in lanolin was applied to the base of young twigs 4 times during late summer and early autumn. The greatest stimulation resulted from treatment on September 11 (Table II) when 2% ethephon enhanced twig abscission approximately 122%. However, similar treatments on mature *Juglans nigra* did not promote cladoptosis (W. R. Chaney, unpublished data).

K. Significance of Branch Shedding

Cladoptosis and natural pruning have tremendous influence on tree form and quality of lumber. The branch system which develops and supports the photosynthetic surface of trees is radically different and much reduced from the branch system which potentially could develop. As a consequence of competition among twigs and branches for light, water, and nutrients, there develops instead of the potential branch system, another which could be called the physiological branch arrangement. If all twigs and branches produced were retained on trees, a 100-year-old *Quercus* tree would have 99 orders of branches instead of the 5 or 6 which actually exist. Many of the branches are partially or entirely shed, however. The pattern of branch shedding varies widely among trees and results in a complex array of often identifiable tree crown shapes (Büsgen and Münch, 1931).

In selection of trees and shrubs for ornamental use, branch abscission and pruning characteristics often are unconsciously a factor. The most desirable ornamentals often have dense foliage and limbs which extend to the ground. *Pinus nigra,* for example, is highly prized for landscaping, partially because of its dense foliage. The short shoots with attached needles of this species generally are retained through the fourth season. In contrast short shoots of *Pinus strobus* are shed at the end of the second season, resulting in an open and, to some people, unattractive crown (Smith, 1962). Another popular ornamental is the monkey puzzle tree, *Araucaria araucana,* which develops a large crown of pendulous branches that are retained low on the trunk when the tree is open grown (Everett, 1968).

It is not coincident that the most valuable commercial trees are generally intolerant species which usually are the best natural pruners of lower branches. Knots and grain distortion which result from persistent branches are two of the most serious defects in lumber. Early and rapid shedding of branches is desirable to reduce the size of knots and proportion of wood affected. Knots may be tight or loose. As long as a branch is alive its cambium is continuous with that of the main bole and the knot produced is connected to the surrounding wood. When a branch dies the cambium no longer produces a sheath of wood which is continuous with that of the bole. Wood is produced around the branch stub and eventually buries it, producing a loose knot. Since this knot is not attached to the wood surrounding it, differential shrinkage in a piece of lumber often results in the branch stub falling from the surrounding wood, producing a knothole. Early death of branches results in small tight knots and rapid pruning of dead branches reduces the extent of particularly undesirable loose knots (Brown *et al.*, 1949).

Another commercially important defect in wood related to natural pruning is discoloration. Light colored wood of some hardwoods, chiefly *Fagus, Betula,* and *Acer,* is preferred for many products. A central column of discolored wood, which is distinct from heartwood, often occurs in these species and originates at dead branches and pruning wounds. Shigo (1965) found that mature trees that had grown approximately their first 25 ft with small side branches that naturally pruned early in the life of the trees had very small central columns of discoloration. However, large, persistent, dead branches usually resulted in extensive discoloration.

References

Addicott, F. T., Carns, H. R., Lyon, J. L., Smith, J. L., Smith, O. E., and McMeans, J. L. (1964). On the physiology of abscisins. *In* "Régulateurs naturels de la croissance végétale" (J. P. Nitsch, ed.), pp. 687–704. CNRS, Paris.

Åkebrand, V. (1957). Om sambandet mellan tallens kvistrensnig og vissa standorts forhallanden inom Sallskapets for praktisk skogsforadling inventeringsomrade 1952–1954. *Norrlands Skogsvardsforb. Tidskr.* **1957**(1), 95–109; from *Forest. Abstr.* **18,** 3807 (1957).

Alexander, R. R. (1958a). Silvical characteristics of engelmann spruce. *U. S., Forest Serv., Rocky Mt. Forest Range Exp. Sta., Sta. Pap.* **31.**

Alexander, R. R. (1958b). Silvical characteristics of subalpine fir. *U. S., Forest Serv., Rocky Mountain Forest Range Exp. Sta., Sta. Pap.* **32.**

Allen, R. M. (1955). Growth of planted longleaf pine on cutover land and old fields. *J. Forest.* **53,** 587.

Anderson, D. A., and Smith, W. A. (1970). "Forest and Forestry." Interstate Printers & Publ., Inc., Danville, Illinois.

Anonymous. (1930). Self-pruning in plants. *Amer. Bot.* **36,** 146–149.

Arnold, C. A. (1947). "An Introduction to Paleobotany." McGraw-Hill, New York.

Ashby, W. C. (1962). Bud break and growth of basswood as influenced by daylength, chilling, and gibberellic acid. *Bot. Gaz.* (*Chicago*) **123,** 162–170.

Bailey, L. H. (1933). "The Cultivated Conifers in North America." Macmillan, New York.

Baker, F. S. (1950). "Principles of Silviculture." McGraw-Hill, New York.

Barker, W. G., and Collins, W. B. (1963). Growth and development of the lowbush blueberry: Apical abortion. *Can. J. Bot.* **41**, 1319–1324.

Barnard, C. (1926). Preliminary note on branch fall in the Coniferales. *Proc. Linn. Soc. N. S. W.* **2**, 114–128.

Becker, D. A. (1968). Stem abscission in the tumbleweed, Psoralea. *Amer. J. Bot.* **55**, 753–756.

Bennett, F. A. (1961). Silvical characteristics of southern magnolia. *U. S., Forest Serv., Southeast. Forest Exp. Sta., Sta. Pap.* **139**.

Berntsen, C. M. (1958). Silvical characteristics of western hemlock. *U. S., Forest Serv., Pac. Northwest Forest Range Exp. Sta., Silvical Ser.* No. 3.

Bessey, C. E. (1900). Botanical notes—the annual shedding of cottonwood twigs. *Science* **12**, 650.

Bieniek, M. E., and Millington, W. F. (1967). Differentiation of lateral shoots as thorns in *Ulex europaeus*. *Amer. J. Bot.* **54**, 61–70.

Blackman, M. W. (1922). Mississippi bark beetles. *Miss., Agr. Exp. Sta., Tech. Bull.* **11**.

Boe, K. N. (1958). Silvics of western larch. *U. S., Forest Serv., Intermt. Forest Range Exp. Sta., Misc. Publ.* **16**.

Bold, H. C. (1957). "Morphology of Plants." Harper, New York.

Brinkman, K. A. (1957a). Silvical characteristics of black oak. *U. S., Forest Serv., Cent. States Forest Exp. Sta., Misc. Release* **19**.

Brinkman, K. A. (1957b). Silvical characteristics of black walnut. *U. S., Forest Serv., Cent. States Forest Exp. Sta., Misc. Release* **22**.

Broadhurst, J. (1917). Self-pruning in the American elm. *Torreya* **17**, 21–24.

Brown, H. P., Panshin, A. J., and Forsaith, C. C. (1949). "Textbook of Wood Technology." McGraw-Hill, New York.

Büsgen, M., and Münch, E. (1931). "The Structure and Life of Forest Trees" (T. Thomson, English Translation). Wiley, New York.

Campbell, R. A. (1957). Silvical characteristics of scarlet oak. *U. S., Forest Serv., Southeast. Forest Exp. Sta., Sta. Pap.* **86**.

Carmean, W. H. (1958). Silvical characteristics of yellow buckeye. *U. S., Forest Serv., Cent. States Forest Exp. Sta., Misc. Release* **29**.

Chamberlain, C. J. (1934). "Gymnosperms Structure and Evolution." Univ. of Chicago Press, Chicago, Illinois.

Chaney, W. R., and Leopold, A. C. (1972). Enhancement of twig abscission in white oak by ethephon. *Can. J. Forest. Res.* **2**, 492–495.

Chapman, H. H. (1953). Effect of thinning on yields of forest-grown longleaf and loblolly pines at Urania, La. *J. Forest.* **51**, 16–26.

Clark, F. B. (1958). Silvical characteristics of swamp white oak. *U. S., Forest Serv., Cent. States Forest Exp. Sta., Misc. Release* **25**.

Codaccioni, M. (1962). Recherches morphologiques et ontogenetiques sur quelques Cupuliferes. *Rev. Cytol. Biol. Veg.* **25**, 114–116.

Codaccioni, M. (1963). Le maintien en survie in vitro du meristeme terminal chez le Castanea sativa. *C. R. Acad. Sci.* **257**, 2319–2321.

Cook, O. F. (1903). The culture of the Central American rubber tree. *U. S., Dep. Agr. Bur. Plant Ind.* (now *Plant Sci. Res. Div. ARS*) *Bull.* No. 49.

Cooper, R. W. (1957). Silvical characteristics of slash pine. *U. S., Forest Serv., Southeast. Forest Exp. Sta., Sta. Pap.* **81**.

Cooperrider, C. K. (1938). Recovery processes of ponderosa pine reproduction following injury to young annual growth. *Plant Physiol.* **13**, 5–27.

Corner, E. J. H. (1951). "Wayside Trees of Malaya." Government Printing Office, Singapore.

Cornforth, J. W., Milborrow, B. V., Ryback, G., and Wareing, P. F. (1965). Chemistry and physiology of Dormins in Sycamore. *Nature* (*London*) **205,** 1269.

Curtis, J. D., and Lynch, D. W. (1957). Silvics of ponderosa pine. *U. S., Forest Serv., Intermt. Forest Range Exp. Sta., Misc. Publ.* **12.**

Dahms, W. G. (1958). Silvical characteristics of mountain hemlock. *U. S., Forest Serv., Pac. Northwest Forest Range Exp. Sta., Silvical Ser.* No. 11.

Dallimore, W., and Jackson, A. B. (1967). "A Handbook of Coniferae and Ginkgoaceae" (revised by S. G. Harrison). St. Martin's Press, New York.

Eames, A. J., and MacDaniels, L. H. (1947). "An Introduction to Plant Anatomy," 2nd ed. McGraw-Hill, New York.

Ecklund, P. R., and Moore, T. C. (1968). Quantitative changes in gibberellin and RNA correlated with senescence of the shoot apex in the Alaska pea. *Amer. J. Bot.* **55,** 494–503.

Ecklund, P. R., and Moore, T. C. (1969). RNA and protein metabolism in senescent shoot apices of 'Alaska' peas. *Amer. J. Bot.* **56,** 327–334.

Esau, K. (1960). "Anatomy of Seed Plants." Wiley, New York.

Everett, T. H. (1968). "Living Trees of the World." Doubleday, New York.

Fenton, R. H., and Bond, A. R. (1964). *U. S., Forest Serv., Northeast. Forest Exp. Sta., Res. Pap.* **NE-27.**

Foster, A. S., and Gifford, E. M., Jr. (1959). "Comparative Morphology of Vascular Plants." Freeman, San Francisco, California.

Fowells, H. A. (1965). Silvics of forest trees of the United States. *U. S. Dep. Agr., Agr. Handb.* **271.**

Fowells, H. A., and Schubert, G. H. (1956). Silvical characteristics of sugar pine. *U. S., Forest Serv., Calif. Forest Range Exp. Sta., Tech. Pap.* **14.**

Funk, D. T. (1957). Silvical characteristics of honeylocust. *U. S., Forest Serv., Cent. States Forest Exp. Sta., Misc. Release* **23.**

Garrison, R., and Wetmore, R. (1961). Studies in shoot-tip abortion: *Syringa vulgaris. Amer. J. Bot.* **48,** 789–795.

Geiger, R. (1965). "The Climate Near the Ground" (Scripta Technica., Inc., English Translation). Harvard Univ. Press, Cambridge, Massachusetts.

Gilbert, A. M. (1960). Silvical characteristics of yellow birch. *U. S., Forest Serv., Northeast. Forest Exp. Sta., Pap.* **134.**

Goebel, K. (1905). "Organography of Plants." Oxford Univ. Press (Clarendon), London and New York.

Guilkey, P. C. (1957). Silvical characteristics of American elm. *U. S., Forest Serv., Lake States Forest Exp. Sta., Pap.* **54.**

Harlow, W. M. (1946). "Fruit Key and Twig Key to Trees and Shrubs." Dover, New York.

Harlow, W. M., and Harrar, E. S. (1968). "Textbook of Dendrology." McGraw-Hill, New York.

Harris, R. W. (1972). High-temperature limb breakage. *Proc. 48th Int. Shade Tree Conf.* **48,** 133–134.

Hart, A. C. (1959). Silvical characteristics of red spruce. *U. S., Forest Serv., Northeast. Forest Exp. Sta., Pap.* **124.**

Heath, O. V. S. (1969). "The Physiological Aspects of Photosynthesis." Stanford Univ. Press, Stanford, California.

Heinselman, M. L. (1957). Silvical characteristics of black spruce. *U. S., Forest Serv., Lake States Forest Exp. Sta., Pap.* **45**.

Herman, F. R. (1958). Silvical characteristics of rocky mountain juniper. *U. S. Forest Serv., Rocky Mt. Forest Range Exp. Sta., Sta. Pap.* **29**.

Hickman, M. (1903). Notes on the cleavage-plane in stems and falling leaves. *Proc. Indiana Acad. Sci.* pp. 93–95.

Holsoe, T. (1950). Profitable tree forms of yellow poplar. *W. Va., Agr. Exp. Sta., Bull.* **341**.

Hooker, J. D. (1877). Report of the scientific committee of the Royal Horticulture Society. *Gard. Chron.* **7**, 506.

Hough, A. F. (1960). Silvical characteristics of black cherry. *U. S., Forest Serv., Northeast. Forest Exp. Sta., Pap.* **139**.

Isaac, L. A., and Dimock, E. J. (1958). Silvical characteristics of Douglas-fir var. menziesii. *U. S., Forest Serv., Pac. Northwest Forest Range Exp. Sta., Silvical Ser.* No. 9.

Kidston, R. (1885). On the relationship of Ulodendron, Lindley and Hutton, to Lepidodendron, Sternberg; Bothrodendron, Lindley and Hutton; Sigillaria, Brongniart; Rhytidodendron, Boulay. *Ann. Mag. Natur. Hist.* **16**, 123–139.

Kittredge, J. (1948). "Forest Influences." McGraw-Hill, New York.

Klebs, G. (1917). Uber das Verhaltnis von Wachstum und Ruhe bei den pflanzen. *Biol. Zentralbl.* **37**, 373–415.

Köstler, J. (1956). "Silviculture" (M. L. Anderson, English Translation). Oliver & Boyd, Edinburgh.

Kotok, E. S. (1951). Shall we prune to provide peeler logs for the future. *Timberman* **52**, 104, 106, and 108–109.

Kozlowski, T. T. (1964). Shoot growth in woody plants. *Bot. Rev.* **30**, 335–392.

Kozlowski, T. T. (1971a). "Growth and Development of Trees," Vol. 1. Academic Press, New York.

Kozlowski, T. T. (1971b). "Growth and Development of Trees," Vol. 2. Academic Press, New York.

Kozlowski, T. T., and Clausen, J. J. (1966). Shoot growth characteristics of heterophyllous woody plants. *Can. J. Bot.* **44**, 827–843.

Kramer, P. J. (1958). Photosynthesis of trees as affected by their environment. *In* "The Physiology of Forest Trees" (K. V. Thimann, ed.), pp. 157–186. Ronald Press, New York.

Labyak, L. F., and Schumacher, F. X. (1954). The contribution of its branches to the main stem growth of loblolly pine. *J. Forest.* **52**, 333–337.

Langdon, O. G. (1958). Silvical characteristics of baldcypress. *U. S., Forest Serv., Southeast Forest Exp. Sta., Sta. Pap.* **94**.

Leak, W. B. (1958). Silvical characteristics of sweet birch. *U. S., Forest Serv., Northeast. Forest Exp. Sta., Pap.* **113**.

Licitis-Lindbergs, R. (1956). Branch abscission and disintegration of the female cones of *Agathis australis* Salisb. *Phytomorphology* **6**, 151–167.

Liese, W. (1957). Orientierende Untersuchungen zur chemischen Astung von Pappeln. *Forstwirt.-Holzwirt.* **12**, 297–298; from *Forest Abstr.* **19**, 432 (1958).

Lindsey, M. (1915). The branching and branch shedding of *Bothrodendron*. *Ann. Bot.* (*London*) **29**, 223–230.

Little, S. (1959a). Silvical characteristics of Atlantic white-cedar. *U. S., Forest Serv., Northeast. Forest Exp. Sta., Pap.* **118**.

Little, S. (1959b). Silvical characteristics of pitch pine (*Pinus rigida*). U. S., *Forest Serv., Northeast. Forest Exp. Sta., Pap.* 119.

Lockhart, J. A., and Gottschall, V. (1961). Fruit-induced and apical senescence in *Pisum sativum. Plant Physiol.* 36, 389–398.

Long, W. H. (1924). The self-pruning of western yellow pine. *Phytopathology* 14, 336–337.

Lotti, T. (1957). Silvical characteristics of cherrybark oak. U. S., *Forest Serv., Southeast. Forest Exp. Sta., Sta. Pap.* 88.

Lyashenko, N. I. (1967). Differentiation of the apex of sympodially branching shoots of certain woody plants. *Dokl. Akad. Nauk SSSR* 177, 714–716.

Lynch, R. I. (1878). On the disarticulation of branches. *J. Linn. Soc. London, Bot.* 16, 180–183.

MacConnell, W. P., and Bond, R. S. (1961). Application of herbicides with mist blowers. *J. Forest.* 59, 427–432.

MacConnell, W. P., and Bond, R. S. (1962). Herbicide tests with shoulder mounted mist blowers in Massachusetts and New Hampshire. *Northeast. Loggers' Handb.* 11, 18–19, 30–31, and 44.

MacConnell, W. P., and Kenerson, L. (1964). Chemi-pruning northern hardwoods. *J. Forest.* 62, 463–466.

Martindale, D. L. (1958). Silvical characteristics of sweetgum. U. S., *Forest Serv., Southeast. Forest Exp. Sta., Sta. Pap.* 90.

Mattoon, W. R. (1915). The southern cypress. U. S., *Dep. Agr., Bull.* 272.

Maul, D. C. (1958). Silvical characteristic of white fir. U. S., *Forest Serv., Calif. Forest Range Exp. Sta., Tech. Pap.* 25.

Merz, R. W. (1957a). Silvical characteristics of Ohio buckeye. U. S., *Forest Serv., Cent. States Forest Exp. Sta., Misc. Release* 16.

Merz, R. W. (1957b). Silvical characteristics of shellbark hickory. U. S., *Forest Serv., Cent. States Forest Exp. Sta., Misc. Release* 18.

Merz, R. W. (1958). Silvical characteristics of American sycamore. U. S., *Forest Serv., Cent. States Forest Exp. Sta., Misc. Release* 26.

Millington, W. F. (1963). Shoot tip abortion in Ulmus americana. *Amer. J. Bot.* 50, 371–378.

Millington, W. F. (1966). The tendril of *Parthenocissus inserta*: Determination and development. *Amer. J. Bot.* 53, 74–81.

Minckler, L. S. (1957a). Silvical characteristics of pin oak. U. S., *Forest Serv., Cent. States Forest Exp. Sta., Misc. Release* 20.

Minckler, L. S. (1957b). Silvical characteristics of white oak. U. S., *Forest Serv., Cent. States Forest Exp. Sta., Misc. Release* 21.

Mirov, N. T. (1967). "The Genus *Pinus*." Ronald Press, New York.

Mollenhauer, W. (1938). Tools and methods in an experimental pruning of white pine. *J. Forest.* 36, 588–599.

Muhldorf, A. (1925). Uber den Ablosungsmodus der Gallen nebst einer kritischen Uebersicht uber die Trennungserscheinungen in Pflanzenreiche. *Beih. Bot. Zentralbl., Abt. 1* 42, 1–110.

Neely, D. (1970). Healing of wounds on trees. *J. Amer. Soc. Hort. Sci.* 95, 536–540.

Nelson, T. C. (1965). Silvical characteristics of the commercial hickories. U. S., *Forest Serv., Southeast. Forest Exp. Sta., Hickory Task Force Rep.* No. 10.

Neville, P. (1961). Etude expérimentale de l'influence des ébauches foliares sur la morphogénèse végétative de *Gleditsia. C. R. Acad. Sci.* 253, 1121–1123.

Neville, P. (1968). Morphogenese chez *Gleditsia triacanthos* L. I. Mise en évidence

expérimentale de correlations jouant un role dans la morphogénèse et la croissance des bourgeons et des tiges. *Ann. Sci. Natur.: Bot. Biol. Veg.* [12] **9**, 433–
510.

Neville, P. (1969). III. Etude histologique et expérimentale de la senescence des
bourgeons. *Ann. Sci. Natur.: Bot. Biol. Veg.* [12] **10**, 301–324.

Neville, P. (1970). Immunité des bourgeons de rameaux courts vis-à-vis de la senescence chez le *Maclura pomifera*. *Bull. Soc. Bot. Fr.* **117**, 49–54.

Nitsch, J. P. (1957). Photoperiodism in woody plants. *Proc. Amer. Soc. Hort. Sci.*
70, 526–544.

Olson, D. F. (1969). Silvical characteristics of yellow-poplar. *U. S., Forest Serv.,
Southeast. Forest Exp. Sta., Res. Pap.* **SE-48**.

Ovington, J. D., and Madgwick, H. A. I. (1955). A comparison of light in different
woodlands. *Forestry* **28**, 141–146.

Paille, G. (1966). Elagage naturel du sapin. *Forest Conserv.* **37**, 28–29.

Paul, B. H. (1933). Pruning forest trees. *J. Forest.* **31**, 563–566.

Paul, B. H. (1938). Knots in second-growth pine and the desirability of pruning.
U. S., Dep. Agr., Misc. Publ. **307**.

Paul, B. H. (1947). Lumber grades vs. site quality of second-growth Douglas Fir.
U. S., Forest Prod. Lab., Rep. **R1688**.

Paul, B. H., and Smith, D. M. (1950). Summary on growth in relation to quality of
southern yellow pine. *U. S., Forest Serv., Forest Prod. Lab., Rep.* **R1751**.

Petschke, K. (1962). (Studies in green pruning beech.) *Arch. Forstw.* **11**, 839–849;
from *Forest. Abstr.* **24**, 3672 (1963).

Phillips, I. D. J. (1971). "Introduction to the Biochemistry and Physiology of Plant
Growth Hormones." McGraw-Hill, New York.

Priestley, J. H., and Scott, L. I. (1932). Branch scars on trees. Their recording and
interpretation. *Naturalist (London)* pp. 275–278.

Putnam, J. A., and Bull, H. (1932). The trees of the bottom lands of the Mississippi
River delta regions. *U. S., Forest Serv., S. Forest Exp. Sta., Occas. Pap.* **27**.

Putnam, J. A., Furnival, G. M., and McKnight, J. S. (1960). Management and in-
ventory of southern hardwoods. *U. S., Dep. Agr., Agr. Handb.* **181**.

Quisenberry, J. E., and Kohel, R. J. (1971). Abortive terminal, a cytoplasmically
inherited character in cotton, *Gossypium hirsutum*. *Crop. Sci.* **11**, 128–129.

Rapraeger, E. F. (1939). Development of branches and knots in western white pine.
J. Forest. **37**, 239–245.

Read, R. A. (1958). Silvical characteristics of plains cottonwood. *U. S., Forest Serv.,
Rocky Mt. Forest Range Exp. Sta., Sta. Pap.* **33**.

Reifsnyder, W. E., and Lull, H. W. (1965). Radiant energy in relation to forests.
U. S. Dep. Agr., Tech. Bull. **1344**.

Renshaw, J. F. (1961). Silvical characteristics of white basswood. *U. S. Forest
Serv., Southeast. Forest Exp. Sta., Sta. Pap.* **136**.

Reynolds, R. R. (1967). Natural pruning of loblolly pine. *Forest Farmer* **26**, 8–9
and 18.

Roach, B. A. (1958). Silvical characteristics of black locust. *U. S., Forest Serv.,
Cent. States Forest Exp. Sta., Misc. Release* **30**.

Roe, E. I. (1957). Silvical characteristics of tamarack. *U. S., Forest Serv., Lake
States Forest Exp. Sta., Pap.* **52**.

Roe, E. I. (1958a). Silvical characteristics of balsam poplar. *U. S., Forest Serv.,
Lake States Forest Exp. Sta., Pap.* **65**.

Roe, A. L. (1958b). Silvical characteristics of black cottonwood. U. S. Forest Serv., Intermt. Forest Range Exp. Sta., Misc. Publ. 17.

Romberger, J. A. (1963). Meristems, growth, and development in woody plants. U. S., Dep. Agr., Tech. Bull. 1293.

Roth, E. R. (1948). Healing and defects following oak pruning. J. Forest. 46, 500–504.

Roy, D. F. (1957). Silvical characteristics of tanoak. U. S., Forest Serv., Calif. Forest Range Exp. Sta., Tech. Pap. 22.

Rudolf, P. O. (1957). Silvical characteristics of red pine. U. S., Forest Serv., Lake States Forest Exp. Sta., Pap. 44.

Rudolf, P. O. (1958). Silvical characteristics of jack pine. U. S., Forest Serv., Lake States Forest Exp. Sta., Pap. 61.

Rushmore, F. M. (1961). Silvical characteristics of beech. U. S., Forest Serv., Northeast. Forest Exp. Sta., Pap. 161.

Ruth, R. H. (1958). Silvical characteristics of sitka spruce. U. S., Forest Serv., Pac. Northwest Forest Range Exp. Sta., Silvical Ser. No. 8.

Ruth, R. H., and Muerle, G. F. (1958). Silvical characteristics of big leaf maple. U. S., Forest Serv., Pac. Northwest Forest Range Exp. Sta., Silvical Ser. No. 13.

Sacher, J. A. (1955). Dwarf shoot ontogeny in Pinus lambertiana. Amer. J. Bot. 42, 784–792.

Sander, I. L. (1957). Silvical characteristics of northern red oak. U. S., Forest Serv., Cent. States Forest Exp. Sta., Misc. Release 17.

Schaffner, J. H. (1902a). The self-pruning of woody plants. Ohio Natur. 2, 171–174.

Schaffner, J. H. (1902b). Observations on self-pruning and the formation of cleavage planes. Ohio Natur. 3, 327–330.

Schaffner, J. H. (1906). Additional observations on self-pruning. Ohio Natur. 6, 450–451.

Schaffner, J. H., and Tyler, F. J. (1901). Notes on the self-pruning of trees. Ohio Natur. 1, 29–32.

Scholz, H. F. (1957). Silvical characteristics of rock elm. U. S., Forest Serv., Lake States Forest Exp. Sta., Pap. 47.

Scholz, H. F. (1958). Silvical characteristics of American Basswood. U. S., Forest Serv., Lake States Forest Exp. Sta., Pap. 62.

Schubert, G. H. (1957a). Silvical characteristics of giant sequoia. U. S., Forest Serv., Calif. Forest Range Exp. Sta., Tech. Pap. 20.

Schubert, G. H. (1957b). Silvical characteristics of incense-cedar. U. S., Forest Serv., Calif. Forest Range Exp. Sta., Tech. Pap. 18.

Schulz, J. P., and Vink, A. T. (1966). Observations on the effect of early pruning on branch development of young soemaroeba (Simaruba amara Aubl.) Turrialba 16, 81–83.

Scott, D. H. (1962). "Studies in Fossil Botany." Hafner, New York (originally published by Black, London, 1900).

Seward, A. C. (1959). "Plant Life Through the Ages." Hafner, New York.

Seward, A. C. (1963). "Fossil Plants." Hafner, New York.

Shattock, S. G. (1888). On the scars occurring on the stem of Dammara robusta, C. Moore, J. Linn. Soc. London 24, 441–450.

Shigo, A. L. (1965). The pattern of decays and discolorations in Northern hardwoods. Phytopathology 55, 648–652.

Sitton, D., Itai, C., and Kende, H. (1967). Decreased cytokinin production in the roots as a factor in shoot senescence. Planta 73, 296–300.

Smith, C. (1963). Shoot apices in the family Moraceae with a seasonal study of *Maclura pomifera* (Raf.) Schneid. *Bull. Torrey Bot. Club* **90**, 237–258.

Smith, D. M. (1962). "The Practice of Silviculture," 7th ed. Wiley, New York.

Smith, J. H. G. (1956). Even 75-year-old fir can be pruned for profit. *Brit. Columbia Lumberman* **40**, 24 and 28.

Sowder, J. E., and Mowat, E. L. (1958). Silvical characteristics of western juniper. *U. S., Forest Serv., Pac. Northwest Forest Range Exp. Sta., Silvical Ser.* No. 12.

Spaulding, P., and Hansbrough, J. R. (1944). Decay of logging slash in the Northeast. *U. S., Dep. Agr., Tech. Bull.* **876.**

Spector, W. S., ed. (1956). "Handbook of Biological Data." Saunders, Philadelphia, Pennsylvania.

Splettstosser, A. (1957). Asten von Eichen mit Wuchsstoffen. *Forstwirtiz Holzw.* **12**, 127–130; from *Forest. Abstr.* **18**, 4075 (1957).

Staebler, G. R. (1958). Silvical characteristics of noble fir. *U. S., Forest Serv., Pac. Northwest Forest Range Exp. Sta., Silvical Ser.* No. 5.

Stark, J. (1876). On the shedding of branches and leaves in Coniferae. *Trans. Roy. Soc. Edinburgh* **27**, 651–660.

Stein, W. I. (1958). Silvical characteristics of California-laurel. *U. S., Forest Serv., Pac. Northwest Forest Range Exp. Sta., Silvical Ser.* No. 2.

Streets, R. J. (1962). "Exotic Forest Trees in the British Commonwealth" (H. G. Champion, ed.). Oxford Univ. Press (Clarendon), London and New York.

Strothmann, R. O., and Zasada, Z. A. (1957). Silvical characteristics of quaking aspen. *U. S., Forest Serv., Lake States Forest Exp. Sta., Pap.* **49.**

Swarbrick, T. (1928). The abscission of black currant shoots during the growth season. *Annu. Rep. Agr. Hort. Res. Sta. Long Ashton, Bristol* pp. 39–41.

Tackle, D. (1959). Silvical characteristics of lodgepole pine. *U. S., Forest Serv., Intermt. Forest Range Exp. Sta., Misc. Publ.* **19.**

Tarbox, E. E., and Reed, P. M. (1924). Quality and growth of white pine as influenced by density, site, and associated species. *Harvard Forest Bull.* **7.**

Thomas, M. (1933). The natural abscission of twigs in Canadian black poplar (*Polulus serotina* Hartig). *Naturalist (London)* pp. 79–86.

Titman, P. W., and Wetmore, R. H. (1955). The growth of long and short shoots in *Cercidiphyllum. Amer. J. Bot.* **42**, 364–372.

Toole, E. R. (1961). Rot entrance through dead branches of southern hardwoods *Forest Sci.* **7**, 218–226.

Toumey, J. W., and Korstian, C. F. (1947). "Foundations of Silviculture," 2nd ed. Wiley, New York.

Trelease, W. (1884). When the leaves appear and fall. *Univ. Wis., Annu. Rep. Agr. Exp. Sta.* p. 59.

van der Pijl, L. (1952). Absciss-joints in the stems and leaves of tropical plants. *Proc., Kon. Ned. Akad. Wetensch., Ser. C* **55**, 574–586.

van der Pijl, L. (1953). The shedding of leaves and branches of some tropical trees. *Madj. Ilmu Alam Untuk Indones.* **109**, 11–25.

Vezina, P. E., and Paille, G. (1967). Natural pruning of spruce and balsam fir in Quebec. *Commonw. Forest. Rev.* **46**, 125–132.

Vines, R. A. (1960). "Trees, Shrubs and Woody Vines of the Southwest." Univ. of Texas Press, Austin.

Wahlenberg, W. G. (1960). "Loblolly Pine." School of Forestry, Duke University, Durham, North Carolina.

Wareing, P. F. (1954). Growth studies in woody species. VI. The locus of photo-periodic perception in relation to dormancy. *Physiol. Plant.* **7**, 261–277.

Wareing, P. F., and Ryback, G. (1970). Abscisic acid: A newly discovered growth-regulating substance in plants. *Endeavour* **29**, 84–88.

Wareing, P. F., Eagles, C. F., and Robinson, P. M. (1964). Natural inhibitors as dormancy agents. *In* "Régulateurs naturels de la croissance végétale" (J. P. Nitsch, ed.), pp. 377–386. CNRS, Paris.

Watson, D. M. S. (1914). Structure and origin of the Ulodendroid scar. *Ann. Bot. (London)* **28**, 481–498.

Weidman, R. H. (1939). Evidence of racial influence in a 25-year test of ponderosa pine. *Agr. Res.* **59**, 855–887.

Weintraub, R., Brown, J., Nickerson, J., and Taylor, K. (1952). Studies on the relation between molecular structure & physiological activity of plant growth. regulators. I. Abscission inducing activity. *Bot. Gaz. (Chicago)* **113**, 348–362.

Wellner, C. A. (1962). Silvical characteristics of western white pine. *U. S., Forest Serv., Intermt. Forest Range Exp. Sta., Misc. Publ.* **26**.

Wenger, K. F. (1958). Silvical characteristics of loblolly pine. *U. S., Forest Serv., Southeast. Forest Exp. Sta., Sta. Pap.* **98**.

Whiting, A., and Murray, M. (1948). Abscission and other responses induced by 2,3,5-triiodobenzoic acid in bean plants. *Bot. Gaz. (Chicago)* **109**, 447–473.

Whitmore, T. C. (1966). "Guide to the Forests of the British Solomon Islands." Oxford Univ. Press, London and New York.

Wiesner, J. (1889). Der absteigende physiologische Bedeutung. *Bot. Ztg.* **47**, 1–9.

Williamson, M. J. (1957). Silvical characteristics of eastern red cedar. *U. S., Forest Serv., Cent. States Forest Exp. Sta., Misc. Release* **15**.

Wilson, R. W., Jr., and McQuilkin, W. E. (1963). Silvical characteristics of eastern white pine. *U. S., Forest Serv., Northeast. Forest Exp. Sta., Res. Pap.* **NE-13**.

Wright, J. W. (1959a). Silvical characteristics of green ash. *U. S., Forest Serv., Northeast. Forest Exp. Sta., Pap.* **126**.

Wright, J. W. (1959b). Silvical characteristics of white ash. *U. S., Forest Serv., Northeast. Forest Exp. Sta., Pap.* **123**.

Zimmermann, M. H., and Brown, C. L. (1971). "Trees: Structure and Function." Springer-Verlag, Berlin and New York.

. 6 .

Development
and Shedding of Bark

G. A. Borger

I. Introduction

As currently used, "bark" is a nontechnical term that describes all tissues exterior to the vascular cambium (Esau, 1965; Srivastava, 1964). Thus, the term bark applies to the aggregation of epidermis, cortex, and phloem in the primary plant body, as well as to the complex, secondarily thickened tissue commonly called bark (Esau, 1965; Kozlowski, 1971a). Earlier authors have used the term bark in a technical sense to describe all dead tissues exterior to a deep-seated periderm (DeBary, 1884; Büsgen and Münch, 1929). This aggregrate of dead tissues, consisting of alternating layers of periderms and tissues cut off by these periderms, is presently termed "rhytidome," often used interchangeably with the term "outer bark" (Eames and MacDaniels, 1947; Esau, 1965). The remaining, living tissues of the bark (phloem and the innermost phellogen and phelloderm) are collectively termed the "inner bark" (Eames and MacDaniels, 1947; Whitmore, 1963).

The terms rhytidome and inner bark cannot be used in the description of all barks. For example, in some trees the original periderm survives for the life of the tree, remaining in an external, superficial position. In other species, new periderms are formed at regular intervals, but arise in the inner layer (phelloderm) of the previously existing periderm. The bark of these trees technically cannot be divided into rhytidome and inner bark. Thus, the term outer bark (rather than rhytidome) should be applied to all dead tissues exterior to the innermost phellogen. The term outer bark would then be synonomous with rhytidome in species forming such a layer, but would still be applicable to species lacking a rhytidome.

Bark functions as a conducting tissue and a protective covering of the stem and roots (DeBary, 1884; Esau, 1965). Actively conducting sieve elements are juxtaposed to the vascular cambium, and only after their senescence do they become incorporated into the exterior protective layer. Depending on the structure of the mature bark, these nonfunctional sieve elements may be shed as part of the rhytidome or remain part of the inner bark.

Since bark contains cork cells (which are impervious to water), it prevents desiccation of underlying tissues (Esau, 1965). For example, interxylary bark develops in the stem, root, or rhizome of many species of shrubs and herbaceous plants and cuts off the current year's tissue from that formed in prior years (Diettert, 1938; Kingsley, 1911; Lemesle, 1927; Moss, 1934, 1936, 1940; Moss and Gorham, 1953). The interxylary bark mollifies the rate of desiccation of stem and root tissues and the plants are able to survive in arid environments that would otherwise be unfavorable to growth (Diettert, 1938; Moss, 1936).

Many pathogens are unable to penetrate the corky tissues of bark and become established only in wounds extending to living tissue (Boyce, 1961). In addition, outer bark often forms in response to pathogen invasion of the inner bark and walls off the pathogen (Bramble, 1936; Boyce, 1961; Kaufert, 1937). Kaufert (1937) reported that the furrowed bark of *Populus tremuloides* is atypical, forming in response to invasion of the cortex by *Macrophoma tumefaciens*.

The many air-filled cells of bark provide excellent thermal insulation and prevent rapid temperature changes at the cambial zone (Cooke, 1948; Hare, 1965; Martin, 1963; Stickel, 1941). DeZeeuw (1941) reported that deep-seated barks (those formed in the outer phloem) arose earlier in several species of trees when the stems were exposed to direct sunlight than when the stems were shaded by the forest canopy. Earlier bark formation deep in the stem may have been a response to decreased relative humidity or higher temperatures around the stem or to increased light intensity. Illumination of the hypocotyl of *Fraxinus pennsylvanica* early in ontogeny appears to enhance activity of the first phellogen (Borger and Kozlowski, 1972e).

Bark is formed beneath wounds by establishment of periderms that become confluent with the normally developed periderms. Bloch (1941, 1952) has written excellent reviews of wound repair in higher plants.

Cortical cells in the smooth bark of *P. tremuloides* are photosynthetically active (Pearson and Lawrence, 1958; Strain and Johnson, 1963). The green cortical cells of other smooth-barked species and young stems probably are photosynthetic also. Gregory (1889b) has suggested that the aliform periderm of *Euonymous alatus* is a contrivance for increased circumferential growth without appreciable loss of these assimilatory cortical cells.

Polyderm, a bark specialized for storage, is formed in the roots of species of Hyperaceae, Myrtaceae, Onagraceae, and Rosaceae (Esau, 1965) (see Section III,E).

This chapter will consider the development and shedding of barks that contain a periderm. The term outer bark will be used to describe all dead tissues exterior to the innermost phellogen. Tissues interior to the outer bark will be termed inner bark. The terms bark and rhytidome will be used as defined above.

II. Bark Tissues

A. Periderm

Periderm is a secondary tissue that replaces or reinforces the epidermis as a protective tissue (DeBary, 1884; Esau, 1965; Eames and MacDaniels, 1947). Roots, stems, and branches of gymnosperms, most dicotyledons, and a few monocotyledons develop periderm (DeBary, 1884; Esau, 1965). In monocotyledons, periderm replaces the epidermis of roots and stems of species of Pandanae, epiphytic Aroidae (*Philodendron, Monstera, Anthurium, Tornelia*), and roots and rhizomes of species of *Strelitzia, Dioscorea*, Dracaeneae, and Zingeberaceae (DeBary, 1884). Woody monocotyledons such as species of *Cocos* may develop a rhytidomelike layer resulting from repeated formation of new periderms beneath older ones (Eames and MacDaniels, 1947).

Herbaceous dicotyledons may form periderm, usually in the roots or oldest portions of the stem (Esau, 1965). For example, Moss (1936) reported extensive periderm development in the perenniating roots and pseudorhizome of *Epilobium angustifolium,* whereas the annual herbaceous stem lacked periderm.

Periderm normally develops beneath abscission scars of leaves and branches (Eames and MacDaniels, 1947; Esau, 1965) and beneath injured or diseased tissue in roots, stems, or fruits (DeBary, 1884; Esau, 1965).

Fruits such as the tropical sapodilla (*Achras zapota*) and sapote (*Calocarpum mammosum*) normally develop periderm; russet apples (*Malus* spp.) frequently develop one, also. Usually periderm forms in tubers such as those of potato (*Solanum tuberosum*) (Eames and Mac-Daniels, 1947).

Development of periderm in phyllomes is largely restricted to the scales of winter buds of several coniferous and dicotyledonous trees (DeBary, 1884; Esau, 1965), e.g., *Aesculus glabra* (Stover, 1951), *A. hippocastanum, Abies excelsa, Ulmus montana,* and species of *Populus, Carpinus,* and *Corylus* (DeBary, 1884). Rarely, as in petioles of several species of *Ficus,* periderm develops in leaves (Eames and MacDaniels, 1947).

In cryptogams, periderm has been found only on the surface of the rhizome of some species of Ophioglosseae (DeBary, 1884).

Periderm includes the phellogen (cork cambium), phelloderm, and phellem (DeBary, 1884; Esau, 1965). In transverse and radial sections, phellogen cells generally appear oblong. In tangential section these cells appear polygonal or irregularly shaped (Srivastava, 1964; Esau, 1965; Schneider, 1955). Phellogen cells are characteristically thin-walled, have protoplasts, are vacuolated to varying degrees, and may contain tannins, starch, and chloroplasts (Esau, 1965; Srivastava, 1964). The phellogen may consist of only 1 cell layer, as in *Fraxinus pennsylvanica,* or may consist of a zone of meristematic cells, as in *Ailanthus altissima* (Borger and Kozlowski, 1972a).

Phelloderm cells are living at maturity and resemble cortical parenchyma cells in shape and content. The walls may be thickened. Phelloderm cells, like those of the phellem, are radially arranged, hence easily distinguished from cortical cells. Intercellular spaces are usually present in the phelloderm (DeBary, 1884; Esau, 1965). Phelloderm cells of green-barked trembling aspen (*Populus tremuloides*) contain chlorophyll and are photosynthetically active (Pearson and Lawrence, 1958; Strain and Johnson, 1963).

Cells of the phellem may have suberized walls (cork cells) or lignified walls (phelloid cells) (Esau, 1965; DeBary, 1884; Gregory, 1888a). Cork cells usually are radially shortened with thick walls; phelloid cells generally are radially elongate with thin walls (Bowen, 1963; Gregory, 1889a; McNair, 1930). Cork and phelloid cells are dead at maturity and generally lack intercellular spaces (DeBary, 1884; Esau, 1965; Stover, 1951). The occurrence of "parenchymatous" phellem cells has been reported in tree species of the tribe Leptospermoidae (Myrtaceae). These cells are found in addition to phelloid and cork cells and are believed to provide abscission zones in the bark (Bamber, 1962). The arrangement of phellem cells in bark varies from species to species (Bamber, 1962; Esau, 1965).

B. CORTEX

Cortical tissues form an integral part of young bark. The first periderm in the stems of most species arises in the cortex (DeBary, 1884; Esau, 1965). In species such as *Citrus limon,* cortical tissues may persist for many years beneath the superficial periderm (Schneider, 1955). Usually, however, the cortex is shed as periderms arise deeper and deeper in the stem, until they are forming in the outer phloem (Büsgen and Münch, 1929; DeBary, 1884; Eames and MacDaniels, 1947; Esau, 1965). Since the first periderm in roots generally arises in the pericycle, the cortex of roots is important only in the bark of the primary plant body (Eames and MacDaniels, 1947; Esau, 1965).

C. PHLOEM

Phloem tissues are intimately involved in development of bark structure. The inner bark of smooth-barked species consists largely of phloem tissues; in addition, patterns of phloem element deposition in conjunction with particular patterns of periderm development are responsible for the structure of ring barks, scale barks, and furrowed barks (Kozlowski, 1971b). Deep-seated periderms arise in living cells of the outer phloem.

III. Bark Development

A. PERIDERM FORMATION

The first periderm arises in the vicinity of the root–shoot confluence (Borger and Kozlowski, 1972a). Smith (1958) reported that the first periderm in the hypocotyl of *Pseudotsuga menziesii* was a continuation of the periderm that originated earlier in the pericycle of the primary root. The exact locus of origin of the periderm on the root axis, relative to the root–shoot juncture, was not described. In a few species, periderm develops first as lenticels (DeBary, 1884) or hyperhydric tissue (Liphschitz and Waisel, 1970) and spreads over the stem.

The first periderm originates at radially dissimilar positions in the shoots of different species. In *Nerium oleander,* species of Pomaceae, many species of *Salix,* several species of *Viburnum* (DeBary, 1884), several species of *Pyrus* (Esau, 1965), and others, the first periderm originates in the epidermis. Epidermal and subepidermal cells give rise to the first periderm in the stems of *Citrus limon* (Schneider, 1955) and several species of *Pyrus* (Esau, 1965). Cortical cells immediately beneath the epidermis are the usual site of origin of the first periderm in the stem. *Pelargonium hortorum* (Lier, 1955), species of *Acer, Quercus, Tilia, Catalpa, Fraxinus, Syringa, Prunus, Populus,* and many others are examples of plants in which the first periderm is formed in the subepidermal position (DeBary, 1884; Solereder, 1908). In stems of *Robinia pseudoacacia, Gleditsia triacanthos,* some other legumes, and several species of *Aristolochia, Pinus,* and *Larix,* the first periderm usually arises in the second or third cortical layer (Esau, 1965). Phloem parenchyma or cortical cells near the stele give rise to the first periderm in species of Caryophyllaceae, Ericaceae, *Berberis, Camellia,* and others (Esau, 1965; DeBary, 1884). In the hypocotyl of *Pseudotsuga menziesii* (Smith, 1958) and *Pinus resinosa* (Borger and Kozlowski, 1972a), the outer layer of the pericycle gives rise to the first periderm.

The pericycle is the usual site of origin of the first periderm in roots (DeBary, 1884; Eames and MacDaniels, 1947; Esau, 1965; Stover, 1951); however, many exceptions occur. In *Citrus sinensis,* the first periderm in the root originates in subepidermal cells (Hayward and Long, 1942). In *Quercus agrifolia, Frankenia grandiflora,* and species of *Monstera* and *Salicornia,* the periderm in the root originates in cortical cells (Stover, 1951). In roots of *Delphinium scaposum* the first periderm replaces the endodermis (Kingsley, 1911).

The first periderm usually forms in the stem during the first year of growth after extension and primary differentiation have occurred. However, in some species (species of *Negundo, Ilex, Sophora*) the first periderm is not formed until the second or later years of growth. In *Euonymus alatus,* formation of the corky wings occurs during the first season; formation of a circumfluent periderm does not occur until the third or fourth season (Bowen, 1963).

Once initiated, the first periderm develops acropetally. During its development in many species, periderm appears to arise simultaneously over the entire stem circumference (Esau, 1965; Srivastava, 1964). However, close inspection reveals that periderm extends toward the apex further on some portions of the stem than on others (Borger and Kozlowski, 1972a).

B. LENTICEL FORMATION

The majority of species forming periderm also form lenticels (DeBary, 1884; Esau, 1965; Eames and MacDaniels, 1947). Lenticel formation slightly precedes or occurs concomitantly with periderm formation, and involves activity of the lenticel phellogen, which becomes continuous with phellogen of the periderm. Lenticels may form on areas of the stem or root lacking stomata (DeBary, 1884; Esau, 1965), but most often develop beneath stomata.

Activity of the lenticel phellogen produces a large mass of complementary cells centrifugally and usually an equal number of phelloderm-like cells centripetally. In face view these masses of cells appear lentil-shaped, hence the name lenticel. Lenticels of smooth-barked species such as those of *Prunus* and *Betula*, become horizontally elongate (often exceeding several centimeters) as the tree ages, whereas those of furrowed bark or scale bark species remain small, rarely exceeding several millimeters in horizontal or vertical dimensions (DeBary, 1884; Esau, 1965).

The arrangement of complementary cells in lenticels varies with species. In gymnosperms, the complementary cells are suberized and closely associated, but show intercellular spaces. In dicotyledons, 3 types of lenticels (based on arrangements of complementary cell types) are recognized. In the first type, the complementary cells are suberized, rather compact, and grade from thin-walled cells that are produced in the spring into thick-walled cells that are produced in the autumn (*Pyrus, Prunus,* etc.). Lenticels of species of *Quercus, Fraxinus,* and others, which consist of loose, nonsuberized complementary cells and annual bands of heavily suberized, compact "closing cells," form the second type. In lenticels of the third type (*Prunus, Betula,* etc.), several layers each of loosely arranged, nonsuberized complementary cells and compactly arranged closing cells are produced each season. Intercellular spaces separate the complementary cells of each of the lenticel types (Esau, 1965).

C. Effects of Environment on Early Periderm Development

Several authors have recognized that bark development is environmentally influenced. DeZeeuw (1941) reported that saplings of several species of trees that were exposed to direct solar radiation formed deepseated periderms sooner than saplings of the same species that were grown under a forest canopy. Guttenberg (1951) has reported a method of determining overall tree vigor from bark characteristics.

Beyond these observations, data are lacking on the complex physiological processes involved in development of bark types. However, the effects of several environmental factors on development of the first periderm have been investigated. It is probable that these factors also influence formation of subsequent periderms, and thus influence development of bark types.

1. Light Intensity

Although the time of formation of the first periderm and subsequent phellogen activity vary directly with light intensity, there appears to exist a minimum intensity requirement for this development. This requirement varies with species. *Fraxinus pennsylvanica* exhibits an extremely low light intensity requirement (around 1.6×10^3 ergs/cm^2/sec) for periderm initiation, whereas *Pinus resinosa* requires light energy at about 1.3×10^4 ergs/cm^2/sec before periderm is formed (Borger and Kozlowski, 1971a, 1972c).

2. Photoperiod

Day length also influences periderm development. Periderm is formed sooner and phellem cells produced at a greater rate as day length is increased up to continuous illumination. For example, whereas *Fraxinus pennsylvanica* seedlings grown on 8 hr of light and 16 hr of darkness at an intensity of 2.3×10^4 ergs/cm^2/sec for 30 days developed an average of 2.3 cell layers in the periderm, seedlings grown on continuous illumination at the same energy level developed an average of 3.9 cell layers (Borger and Kozlowski, 1971a, 1972f). Arzee *et al.* (1968) reported that phellogen initiation was retarded in seedlings of *Robinia pseudoacacia* grown on 17-hr days at 28°C and suggested that long days or high temperatures retarded phellogen formation. This retarding effect was probably a result of the temperature treatment and not the photoperiod treatment.

Although the foliage appears to be the receptor of the light stimulus, illumination of the stem apparently also influences periderm development (Borger and Kozlowski, 1972e).

3. Temperature

Phellogen initiation and activity increase with a rise in temperature until maximum activity is attained, then decrease with any further temperature increase. The temperature at which periderm formation is maximum varies with species. *Ailanthus altissima* and *Robinia pseudoacacia* exhibit maximum periderm development at 25°C, whereas *Fraxinus pennsylvanica* shows maximum periderm development at 30°C (Borger and Kozlowski, 1971c, 1972d).

4. Soil Moisture

Excessive soil moisture has been reported to prevent maturation of cork cells in tubers of *Solanum tuberosum* (Mylius, 1913). In fact, suberization may be entirely suppressed and callus tissue may replace cork (Kuster, 1925). These effects of excess soil water probably result from inadequate soil aeration which has been shown to induce lenticel hypertrophy in roots of *Gossypium* (Templeton, 1926) and *Nyssa sylvatica* var. *biflora* (Hook *et al.*, 1970). Internal water deficits induced by soil drying (Borger and Kozlowski, 1970) or hydroponically administered solutions of polyethylene glycol 4000 (PEG) have been shown to increase time to phellogen formation and decrease subsequent phellogen activity in seedlings of *Fraxinus pennsylvanica* when compared to controls. This decrease, however, was not as marked as the decrease in xylem production induced by the same water deficits. For example, seedlings of *F. pennsylvanica* grown for 35 days in a −2.2 bar solution of PEG formed an average of 6.6 cell layers in the periderm, whereas those grown in a −14.3 bar solution of PEG formed an average of 4.2 cell layers. Xylem increments were 46.5 and 20.2 cell layers, respectively (Borger and Kozlowski, 1971b, 1972b). Drying of tubers of *Solanum tuberosum* (i.e., establishing a severe internal water deficit) has been shown to prevent formation of wound periderm when the tubers were subsequently sliced (Shapovalov and Edson, 1919; Weimer and Harter, 1921).

5. Relative Humidity

Olufsen (1903) and Shapovalov and Edson (1919) reported that low relative humidity (RH) was best for wound periderm formation in tubers of *Solanum tuberosum*. However, later workers (Artschwager,

1927; Artschwager and Starrett, 1931, 1933; Weimer and Harter, 1921) showed that high RH was necessary for wound periderm formation, and very low RH prevented wound periderm formation (Weimer and Harter, 1921). The effects of RH on normal periderm development have not been investigated.

The influence of environmental factors on periderm formation appears to be closely associated with their effects on leaf development (Borger and Kozlowski, 1971a, 1972c,e). Although the apical meristem does not appear to affect periderm initiation directly, the ratio of phellem to phelloderm cells produced is apparently controlled by the apical meristem (Borger and Kozlowski, 1972c).

6. Exogenous Growth Regulators

Phellogen activity is strongly affected by exogenous application of ethylene gas. Stone (1913) reported abnormal proliferation of tissues of lenticels on roots of *Salix* grown in water charged with illuminating gas. Roots of *Catalpa* seedlings grown in soil aerated with illuminating gas (ethylene concentration of 1%) for 21 days and those of *Ailanthus* seedlings grown in soil aerated with ethylene gas at a concentration of 0.4% for 15 days showed rapid proliferation at the phellogen layer (Harvey and Rose, 1915). Doubt (1917) reported proliferation of cells of lenticels on roots of *Hibiscus* and *Sambucus,* tissues of leaf scars of *Lycopersicum,* and bark tissues of stems of several plants when they were treated with traces of illuminating gas or with ethylene. Woffenden and Priestly (1924) treated stems of *Sambucus nigra* with coal gas (about 2% ethylene) and found that proliferation of lenticel tissues occurred only during the period immediately succeeding phellogen formation. This abnormal activity resulted in excessive development of complementary cells.

Initiation of a peridermlike layer in roots of *Ophioglossum petiolatum* by 2 ppm benzyladenine (a cytokinin) has been reported (Peterson, 1971). This species of *Ophioglossum* normally does not develop periderm. Promotion of periderm development by formation of wound periderm has been reported in *Fraxinus pennsylvanica* (Borger and Kozlowski, 1972g,h) and other species (Haberlandt, 1928; Liphschitz and Waisel, 1970).

Isopropyl-N-(3-chlorophenyl)carbamate at concentrations as low as 25 ppm (Audia *et al.,* 1962) and commercial mixtures of naphthalene-acetic acid (methyl ester) and 2,3,5,6-tetrachloronitrobenzene (Cunningham, 1953) have been reported to prevent wound periderm formation in tubers of *Solanum tuberosum.* When 2,4-D, IAA, kinetin, or gibberellic acid was applied at 100 or 1000 ppm in lanolin paste to wounded hypocotyls of *Fraxinus pennsylvanica,* only 1000 ppm 2,4-D prevented wound periderm formation. However, all treatments retarded or prevented normal periderm development in hypocotyl tissues basipetal to the wounded area, indicating differences in physiological controls of wound and normal periderm development (Borger and Kozlowski, 1972h). Gibberellin and naphthaleneacetic acid have been reported to retard periderm development in *Robinia pseudoacacia* (Arzee *et al.,* 1968).

D. Development and Shedding of Characteristic Bark Types

Development and shedding characteristics of bark are related to the radial position of the initial periderm on the stem or root, patterns of formation of subsequent periderms, and arrangement of cells in the phloem. Smooth barks, furrowed barks, scale barks, ring barks, and winged barks (true winged barks: aliform; false winged barks: pseudo-aliform) have been recognized. The bark of *Quercus suber* is recognized as a distinct form of smooth bark (DeBary, 1884).

1. *Smooth Barks*

Smooth barks result when the periderm remains in a superficial position. This condition may persist for the life of the tree, as in *Populus tremuloides, Fagus sylvatica* (Fig. 1), and *Citrus limon*. In *Populus tremuloides* and *Fagus sylvatica* the original periderm persists for the life of the tree, increasing in girth through anticlinal divisions (Kaufert, 1937). In *Citrus limon,* periderms are formed subsequent to the original periderm. However, these arise in the phelloderm of the previous periderm, thus maintaining the periderm in a superficial position (Schneider, 1955). Several other species retain a smooth bark for many years, e.g., species of *Betula* and *Prunus* (DeBary, 1884; Eames and MacDaniels, 1947; Srivastava, 1964).

The outer bark of smooth barks consists of phellem cells in the case of *Populus tremuloides* and *Fagus sylvatica,* or of phellem and old phellogen cells in the case of *Citrus limon.* The number of layers of phellem cells usually increases for several years, then remains relatively constant. For example, Rees and Shiue (1957–1958) reported that the average number of phellem cells per radial row in the outer bark of *Populus tremuloides* increased from 5.1 in first year twigs to 10.0 in 5-year-old twigs, then remained nearly constant as twigs increased in age. The outer bark of species such as *Betula pendula* may show annual rings of cork due to the presence of thin-walled phellem cells produced early in the growing season that grade into thick-walled phellem cells produced at the close of the growing season (Scott, 1950).

The inner bark of smooth-barked species consists of cortical remnants

FIG. 1. Smooth bark of *Fagus grandifolia*. U. S. Forest Service Photo.

and phloem tissues. The cortex increases in girth through anticlinal divisions (Eames and MacDaniels, 1947) and tangential expansion of cortical cells (Rees and Shiue, 1957–1958). In *Populus tremuloides,* the cortex has been reported to increase in radial dimensions from an average of 29.6 μm for 2-year-old twigs to an average of 41.7 μm for 4-year-old twigs; the average number of cortical cells in the circumference increased from 25 for 1-year-old twigs to 452 in 4-year-old twigs (Rees and Shiue, 1957–1958).

Increase in girth of the phloem is accomplished by formation of dilation tissue, which may develop before the first phellogen has formed, e.g., *Citrus limon* (Schneider, 1955), or not until some time after initiation of the first phellogen, e.g., *Populus tremuloides* (Rhees and Shiue, 1957–1958). Dilation tissue results from anticlinal divisions and tangential enlargement of phloem ray cells. These divisions apparently occur in response to tangential stresses induced by activity of the vascular cambium. Since greater stresses are exerted near the exterior of the stem, more divisions of the phloem ray cells occur in this region, giving the dilation tissue a wedge shape in cross section. These wedges of dilation tissue eventually distort the arrangement of phloem cells, separating them into patches (Rees and Shiue, 1957–1958; Schneider, 1955).

Owing to the superficial position of the periderm, lenticels of smooth-barked trees such as *Prunus, Betula,* and *Fagus* may expand tangentially as the tree ages, eventually appearing as tangential bands. In other smooth-barked species the tangentially stressed lenticels may become branched as dense periderm forms within the lenticels (*Pyrus malus, Rhamnus frangula*) or may remain unchanged in tangential dimensions owing to formation of periderm at the edges of the lenticels (*Fraxinus excelsior, F. ornus*) (DeBary, 1884).

Attrition of tissue from smooth bark occurs slowly and is usually commensurate with the rate of formation of phellem cells. In *Populus tremuloides,* outer phellem cells are sloughed off individually or in small clusters, giving the bark a powdery appearance (Rees and Shiue, 1957–1958). In species such as *Betula pendula,* sheets of cork several cell layers thick are lost as tangential pressures induce tearing in the outer phellem. These sheets of cork usually separate along the thin-walled cork cells formed in the early part of the growing season (Scott, 1950).

2. *Bark of Quercus suber*

The first periderm in the stem of *Q. suber* arises in the epidermis (Eames and MacDaniels, 1947). The highly active phellogen produces a mass of cork cells but few phelloderm cells. Phellogen activity is seasonal:

FIG. 2. Cross section of the cork of *Quercus suber*. Note the annual growth rings (arrow) and the lenticels which run perpendicular to these rings. (×5.)

radially elongate, thin-walled cork cells produced in the spring and summer grade into radially shortened, thick-walled cells produced during the autumn. This seasonal activity results in annual bands of cork which are clearly discernible in the bark (Fig. 2) (DeBary, 1884). Tangential growth stresses soon tear the outer surface of the first-formed cork and the bark becomes furrowed (Eames and MacDaniels, 1947). As the tree ages, new periderms occasionally form internal to the first periderm, separating small patches of cortex between broad bands of cork (DeBary, 1884).

Commercial harvesting of cork alters development of the bark of *Q. suber*. When the tree has attained an age of 12–15 years, the first-formed periderm (male cork) is mechanically removed. This is accomplished by slitting the periderm vertically and separating it from the cortex along the cortex–periderm interface. The living cells on the exterior of the cortex die and become dry and brittle. A circumfluent wound periderm soon develops 1–2 mm beneath this cicatrice. The phellogen of the wound periderm is more active than the original phellogen and produces more cork than the original phellogen. Cork cells of the wound periderm (female cork) are of a finer grade than the male cork. Harvesting of female cork is performed every 9–10 years with the result that periderms eventually arise in the outer phloem (DeBary, 1884; Eames and Mac-Daniels, 1947).

The cork of *Betula alba* has been harvested in the U.S.S.R., using methods similar to those employed in harvesting cork of *Q. suber* (DeBary, 1884).

In *Q. suber*, the lenticels are found scattered over the surface of the stem. As phellem is produced, the outer complementary cells dry out, turning brown and forming a loose, powdery mass. In transverse section the lenticels appear as irregularly constricted, cylindrical columns running at right angles to the annual increments of phellem (Fig. 2) (DeBary, 1884).

3. Scale Barks

In species developing scale barks, the first periderm may remain in a superficial position for several years. However, as the stem ages, new periderms eventually arise beneath the original. These periderms are not formed as continuously circumfluent cylinders as is the original periderm, but are formed in small patches. In *Eucalyptus* spp. new periderms arise beneath cracks in the previously formed periderm, defining the crack in transverse and longitudinal dimensions (Chattaway, 1953). Seen in transverse or radiolongitudinal section, these patches appear as overlapping arcs or lunes; in face view they are irregular in shape. Once development of these irregular patches of periderm begins, it continues for the life of the tree. New periderms are formed annually or at irregular intervals beneath the previous periderms. The new periderms arise deeper and deeper in the inner bark until they eventually arise in cells of the outer phloem (DeBary, 1884; Eames and MacDaniels, 1947; Esau, 1965; Scott, 1950; Srivastava, 1964).

This pattern of periderm development produces a thick outer bark that consists of alternating layers of periderms and tissues cut off by these periderms. Since phellem cells of the periderms are dead at maturity, all cells exterior to the innermost phellogen eventually die. The outer bark of mature scale barks, therefore, is a true rhytidome; the inner bark consists of phloem.

Tissues exfoliating from scale barks take the form of flakes (*Pinus sylvestris, P. ponderosa*), sheets (*Acer pseudoplatanus*), or platelike strips (*Carya ovata*). In *Pinus sylvestris* the phloem consists of alternating bands of sieve elements and parenchyma cells; sclerenchyma cells are noticeably absent. The lunes of periderm are formed in parenchyma cells of the outer phloem and produce a phellem consisting of several layers of tough cork cells, a few layers of thin-walled and weak cork cells, and toward the end of the growing season, several more layers of tough cork cells. In the young stem, where even small, annual xylem increments produce large increases in stem circumference, the tangential stresses of growth soon tear the phellem tissue through the thin-walled cork cells and cracks develop in the bark. Further tangential stresses coupled with weathering, e.g., temperature changes (Chattaway, 1953) or moisture variations (Eames and MacDaniels, 1947), bring about exfoliation of the bark flakes. The tangential stresses induced by diameter growth decrease as the tree ages and the rate of loss of bark tissues decreases. The adherent flakes form the thick rhytidome of older stems. However, even in older bark, the flakes are easily separated and outer flakes are continuously being shed (Dittmann, 1931). Bark on older portions of the stem may also become furrowed (Fig. 3) (Chang, 1954; Scott, 1950).

The phloem of *Acer pseudoplatanus*, like that of *Pinus sylvestris*, shows a paucity of sclerenchyma (Esau, 1965). Periderms arise in the living cells of the outer phloem and produce several layers of phelloid cells followed by several layers of cork cells. However, unlike those of *Pinus sylvestris*, the lunes of periderm in *Acer pseudoplatanus* cover extensive regions of the stem. Tangential growth stresses tear the phellem through the phelloid cells and large sheets of bark are desquamated. The cork cells remain on the stem, forming a fairly smooth covering (DeBary, 1884).

FIG. 3. Scale bark of *Pinus ponderosa*. Note the shape of the bark flakes. The flakes have beveled edges. U. S. Forest Service Photo.

FIG. 4. Scale bark of *Carya ovata*. U. S. Forest Service Photo.

The phloem of *Carya ovata* contains large amounts of sclerenchyma, especially fibers, in addition to other phloem components. Because of the banded, interlocking arrangement of fibers in the phloem, periderms arise as long, narrow strips rather than well-defined lunes evident in the flakey bark of *Pinus sylvestris*.

Tangential growth stresses eventually induce splitting of the phellem tissues and vertical cracking of the rhytidome. However, because of the tenacity of fiber bands and the manner in which periderm is formed, exfoliation is incomplete and the rhytidome persists as long, narrow, vertical strips attached at their upper ends where tangential stresses are least (Fig. 4) (Eames and MacDaniels, 1947).

Lenticel development in scale barks reflects the continual shedding of large areas of rhytidome. In *Acer pseudoplatanus, Carya ovata,* and the young actively shedding stem of *Pinus sylvestris,* new lenticels appear on the surface exposed by the exfoliation of bark flakes. In the old furrowed bark of *Pinus sylvestris* lenticels occur in the bark fissures (DeBary, 1884; Esau, 1965).

4. Furrowed Barks

Furrowed barks of species such as *Sequoia sempervirens* are formed in much the same manner as the bark of *Pinus sylvestris*. However, the phloem of furrowed bark species contains large amounts of sclerenchyma, especially fibers, which interlock tissues of the rhytidome. In addition, cells of the periderm adhere tightly to other cells of the rhytidome. The result is a tough adherent bark (Chattaway, 1953, 1955a,b; Esau, 1965; Scott, 1950).

In *Sequoia sempervirens,* the phloem contains, in addition to sieve elements and parenchyma cells, bands of fibers forming a continuous network throughout the phloem. As the phloem is forced outward by activity of the vascular cambium, the sieve elements collapse and phloem parenchyma cells enlarge enormously to compensate for the increased circumference (Chang, 1954; Isenberg, 1943; Scott, 1950). Lunes of periderm arise in the enlarged parenchyma of the outer phloem, and cells exterior to these lunes die and collapse. The flakes of bark thus formed do not separate, however, owing to the interlocking system of fibers. The resultant rhytidome is a furrowed, loose, fibrous mass. Because of its nonflammable, decay-resistant nature, this fibrous rhytidome may accumulate to a depth of 2 ft on older trees (Scott, 1950). As in other furrowed bark species, lenticels are formed at the bottom of the bark fissures (Esau, 1965).

Several species of *Eucalyptus,* e.g., *E. baxteri* and *E. globoides,* are known as "stringy barks," because of their characteristic, loose, fibrous barks. Primary phloem of the stringy barks consists of a band of sieve elements and parenchyma next to the vascular cambium; then a layer of dead, crystalliferous parenchyma; and finally a band of fibers from 1 to 5 cell layers deep. This pattern of cell differentiation is repeated in the secondary phloem.

The first periderm in stringy barks arises in the outer cortex and produces a phellem consisting of narrow bands of cubical cork cells alternating with wide bands of tangentially flattened phelloid cells with heavily lignified walls. As the stem ages, attrition of the phelloid cells occurs by abrasive weathering, and the layers of cork cells split. Suberin is initially deposited around these cracks, sealing them temporarily, but further tangential stresses eventually produce cracks that penetrate the periderm. Once these cracks penetrate the periderm, lunes of periderm, confluent with the original periderm, are formed beneath the crack. As the stem ages and fissuring continues, the lunes of periderm form deeper and deeper in the stem, eventually arising in parenchyma of the outer phloem (Chattaway, 1953).

Cortical parenchyma, and later phloem parenchyma cut off by the lunes of periderm, undergo radial enlargement, often attaining a radial dimension 30 times their original size. The network of heavily lignified phelloid cells and phloem fibers binds the successive periderms together. The expanded parenchyma cells lend a spongy character to the cohesive, but loose, fibrous rhytidome (Chattaway, 1953, 1955a). The rhytidome accumulates to great thicknesses on older stems and is shed only gradually by abrasion (Chattaway, 1953).

Furrowed barks of *Salix, Quercus, Ulmus, Eucalyptus, Fraxinus, Juglans*, etc., develop in much the same manner as those of *Sequoia sempervirens* and the stringy barked species of *Eucalyptus*. However, large amounts of radially enlarged parenchyma are lacking and the rhytidome of these species owes its characteristics to the presence or absence of large amounts of cork cells or sclerified tissues. Soft barks such as those of *Celtis occidentalis* contain large amounts of cork cells and little sclerified tissue, whereas those of *Quercus* or *Fraxinus* (Fig. 5) contain few cork cells and a preponderance of sclerified cells.

In the group of eucalypts known as "ironbarks," e.g., *Eucalyptus crebra* and *E. jenseni*, periderms are formed in nearly every band of phloem parenchyma, forming an anastomosing network. These periderms usually remain active for a considerable period, producing large amounts of thin-walled cork cells. Tangential stresses induce lysigenous fissure formation in the rhytidome as the tree ages. However, the fissures are broad and widely separated since the rhytidome of the ironbarks con-

FIG. 5. Furrowed bark of *Fraxinus americana*. U. S. Forest Service Photo.

225

tains large amounts of hard, brittle kino which prevents rhytidome rupture (Chattaway, 1953, 1955b).

Bark of the "peppermints," e.g., *Eucalyptus andreusii*, is more finely furrowed than that of the ironbarks. The rhytidome is rather soft and contains kino deposits only along the margins of the schizogenously formed bark furrows (Chattaway, 1955b).

The pattern of fissuring of the rhytidome is to a great degree dependent on orientation of phloem fibers which, in turn, reflect the orientation of the xylem elements. A symmetrically fissured bark, such as that of species of *Juglans* or *Fraxinus* (Fig. 5) which shows vertically elongate, diamond-shaped meshes would suggest straight-grain timber. A marked spiral fissuring of the bark would indicate spiral-grained timber (Scott, 1950).

5. Ring Barks

The mode of periderm development in ring-barked species (Cupressaceae, *Vitis*, *Clematis*, and others) is markedly different from that of the other bark types discussed above. Rather than forming in a superficial position, the first periderm of ring barks forms as a circumfluent cylinder in the outer phloem. Subsequent periderms arise annually beneath the first one, usually concentric with the previously formed periderms. This pattern of periderm development results in exfoliation of hollow cylinders (rings) of rhytidome from the stem (DeBary, 1884; Esau, 1965; Eames and MacDaniels, 1947).

Separation of the cylinders of rhytidome occurs through thin-walled phellem elements. However, the exfoliation is incomplete and the rhytidome remains attached to the stem by numerous groups of phloem fibers that form an anastomosing network in the phloem. Thus, the rhytidome of ring barks assumes its characteristic, shaggy appearance (Fig. 6) (DeBary, 1884).

Since the formation of annual rings of periderm in ring barks brings about necrosis of the previous year's bark tissue, new lenticels are formed in each new periderm (Esau, 1965).

6. Wing Barks

Wing barks are formed by the activity of localized periderms (aliform bark) as in *Euonymous alatus* or by the longitudinal fissuring of circumfluent periderms (pseudoaliform bark) as in *Acer campestre*, *Ulmus × hollandica*, *Quercus macrocarpa*, and *Liquidambar styraciflua*.

The periderm of *E. alatus* is initiated in cortical cells underlying vertical grooves in the stem. The grooves extend from slightly above each lateral bud to slightly past the point of insertion of the decussate leaves at the immediate node above. The vertical grooves appear acropetally in the

FIG. 6. Ring bark of *Juniperus virginiana*. U. S. Forest Service Photo.

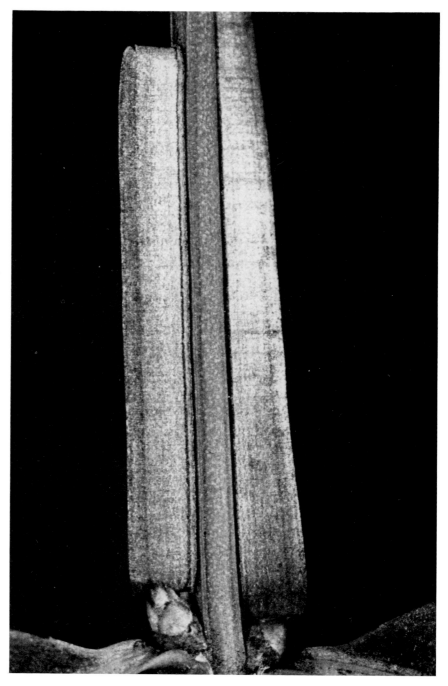

FIG. 7. Aliform bark of *Euonymous alatus*. Note the vertical band of closing cells in the wing on the left. Photo courtesy of William R. Bowen.

internodes 3–4 days after stem emergence in the spring and become wider as stem elongation continues. Periderm initiation begins in the lowermost internode and proceeds acropetally. Only 4–5 days are required for formation of periderm throughout the current year's growth. With formation of periderm, stem elongation ceases (Bowen, 1963).

During the growing season, the phellogen produces a vast number of radially elongate phellem cells which form the wing (Bowen, 1963). Circumferential growth of the stem is reflected by anticlinal divisions in the phellogen which broaden the base of the wing. At the end of the growing season, thick-walled, oblong closing cells are formed. These closing cells can be seen as annual bands in the aliform periderm (Fig. 7) (Gregory, 1888c, 1889a).

The position of formation of aliform bark in *Euonymus alatus* is related to the arrangement of primary vascular tissues. In any internode, the wings are aligned with the leaf gaps of the leaves inserted at the node below and are at right angles to the leaf gaps of the leaves inserted at the node above, reflecting the opposite, decussate phyllotaxy. Thus, the wings in each internode are opposite and decussate to the wings in the internodes above and below (Bowen, 1963).

Aliform periderm is the only periderm formed during the first 3 years of stem growth. In the fourth year, a circumfluent periderm arises in tissues between the wings, and the phellogen of the aliform periderm ceases to function. The wings are gradually obliterated by activity of the normal periderm and weathering (Bowen, 1963).

In the pentagonal stem of *Quercus macrocarpa*, the first periderm arises circumfluently from subepidermal, cortical cells. As activity of the phellogen and vascular cambium proceeds, tangential stresses occur and the epidermis and phellem rupture in vertical lines along the angles of the stem. Phellogen activity becomes pronounced beneath these fissures, producing large amounts of radially elongate phelloid cells. The edges of the sections of periderm on the 5 faces of the stem are pushed out by these abnormally active regions of the phellogen, forming pseudoaliform bark (Fig. 8). Increased activity of the phellogen at the angles of the stems occurs for several years, but eventually furrowed bark is formed (Gregory, 1888a).

In *Acer campestre*, longitudinal splits in the initial periderm occur at the angles of the hexagonal stem. Again as in *Quercus macrocarpa*, localized, increased phellogen activity occurs beneath the cracks. However, unlike that of *Quercus macrocarpa*, the phellogen on the 6 faces of the stem also becomes markedly active. Thus, the wings of the bark are formed on the faces of the stem, rather than on the angles (Gregory, 1888b; Smithson, 1952). The pseudoaliform bark of *Ulmus × hollandica*

is formed in the same manner as that of *Acer campestre* (Smithson, 1954).

In the pseudoaliform bark of *Liquidambar styraciflua,* the wings are formed only on the upper sides of the branches. The first circumfluent periderm arises in the second layer of cortical cells beneath the epidermis. Numerous lenticels are formed concurrently with the periderm. Lenticels on the upper sides of the branches are larger, better developed, and more numerous than those on the lower side of the branch. Some time after periderm initiation, increased activity of the phellogen is apparent around the lenticels on the upper side of the branch. After a short time, the interlenticular phellogen also becomes very active and the lenticels on the top of the branch are eventually connected by a cork ridge or wing. Later, additional wings may be formed on the top of a branch in a similar manner (Gregory, 1888b). The wings of bark thus formed consist of thin-walled, radially elongate cork cells formed during the spring and summer and several layers of thick-walled closing cells formed at the end of the growing season. Tangential walls of the thin-walled cells meet the radial walls of adjacent cells near the center, whereas the tangential walls of adjacent closing cells are confluent. This arrangement gives the wing a banded appearance, each band representing a year's growth (Gregory, 1889a).

E. Bark Development in Roots

In dicotyledons and gymnosperms, roots that undergo secondary thickening normally develop a bark. In roots of this type, the first periderm usually develops in the pericycle; the cortical tissues and epidermis are subsequently shed. In many species the first periderm persists indefinitely, forming a smooth, brown covering on the root. Well-developed lenticels are commonly present in the periderm. In other species, a rhytidome is formed by the progressive formation of new periderms deeper and deeper in the phloem. Little accumulation of rhytidome tissues occurs in roots, however, because of the rapid decay of the dead tissues in the humid soil atmosphere (Eames and MacDaniels, 1947).

Dicotyledon roots that show little or no secondary thickening may retain their cortex indefinitely. In this case, periderm may not be formed or a superficial periderm may form in the outer cortex (Esau, 1965). In roots of *Citrus sinensis* that show little secondary growth, a superficial periderm is formed in the outer cortex and later replaced by a deep-seated periderm originating in the pericycle (Hayward and Long, 1942).

A periderm specialized for storage, the polyderm, occurs in underground stems and roots of Hyperaceae, Myrtaceae, Onagraceae, and Rosaceae (Esau, 1965). The polyderm of adventitious roots of *Fragaria virginiana* × *chiloensis* contains normal phellogen and phelloderm cells, but the phellem is modified, consisting of alternating bands of living cork and phelloid cells with an outer layer of dead cork cells. The phelloid bands are typically 2- to 3-cell layers thick; the cork bands typically 1 cell layer thick. Living cells of the phellem may attain a thickness of 20 or more layers. In the fall and winter, the living phelloid cells store large quantities of starch. These reserves are mobilized during active vegetative and reproductive growth (Mann, 1930; Nelson and Wilhelm, 1957).

In roots of herbaceous, perennial dicotyledons such as *Epilobium angustifolium, E. latifolium, Gaura coccinea,* and *Artemisia dracunuloides,* periderms may form between the annual xylem increments. These periderms are in addition to the external periderm that arises in the pericycle. The formation of periderms external and internal to the annual xylem increments probably decreases desiccation and pathogen entry associated with the annual dying down of the shoot and facilitates survival of the plant (Moss, 1936).

F. BARK VOLUME

The volume of bark produced during the life of a plant is difficult to determine accurately since bark tissues, unlike xylem tissues, are shed. In addition, the shedding process is continuous and imperceptible. However, bark thickness and volume have been determined for both diameter and age classes of many commercially valuable timber species. These volumes are used as an aid in estimating wood volumes of standing timber.

Bark thickness generally increases with stem age and diameter. In *S. sempervirens* a straight-line relationship exists between bark thickness and stem diameter. This relationship probably results from the resistance of bark to weathering and decay and to the persistent nature of the rhytidome (Dittman, 1931). In most other species, however, the relationship between diameter and bark thickness is curvilinear, owing to shedding of bark tissues. Hale (1955) determined bark thickness by age and diameter classes for *Abies balsamea, Picea marinana, P. glauca, Pinus banksiana, Populus tremuloides,* and *Betula papyrifera.* For all species, a plot of bark thickness against stem diameter produced an upcurving line, regardless of stem age. When grouped by 20-year age classes, the same species showed an increase in bark thickness with age class, regardless of diameter class. For example, in 7-inch-diameter *Populus tremuloides* stems in age classes of 41–60 years, 81–100 years, 101–120 years, and 150 years and older, the bark thickness were 0.19, 0.27, 0.30, and 0.61 inches, respectively. In the 81–100-year age class, bark thicknesses were 0.24, 0.25, 0.27, 0.30, 0.33, and 0.36 inches for stem diameters of 5, 6, 7, 8, 9, 10, and 11 inches, respectively.

Bark volume as percent of total stem volume, however, decreased as stem diameter increased, but increased with stem age. For *Populus tremuloides* stems with an average diameter of 7 inches, the bark represented approximately 10, 16, 18, and 32% of stem volume for trees in age classes of 41–60 years, 81–100 years, 101–120 years, and 150 years and older, respectively. In the 81–100-year age class, bark composed approximately 18% of the volume of 5-inch stems; in 11-inch stems this value had fallen to 12%. However, as stem diameter increases, actual bark volume also increases. A 100-inch stem section of *Populus tremuloides* 5 inches in diameter would yield approximately 353 cubic inches of bark, whereas a 100-inch section of stem 11 inches in diameter would yield approximately 1140 cubic inches of bark.

In addition to varying with stem age and diameter, bark thickness varies with tree vigor. Hale (1955) reported that the average thickness of bark on vigorously growing stems of *Picea glauca* in the age class 41–60 years was greater than that of all age classes of less vigorous trees.

References

Artschwager, E. (1927). Wound periderm formation in the Irish potato as affected by temperature and humidity. *J. Agr. Res.* **35**, 995–1000.

Artschwager, E., and Starrett, R. C. (1931). Suberization and wound–periderm formation in sweet potato and gladiolus as affected by temperature and relative humidity. *J. Agr. Res.* **43**, 353–364.

Artschwager, E., and Starrett, R. C. (1933). Suberization and wound–cork formation in the sugar beet as affected by temperature and relative humidity. *J. Agr. Res.* **47**, 669–674.

Arzee, T., Liphschitz, N., and Waisel, Y. (1968). The origin and development of the phellogen in *Robinia pseudoacacia* L. *New Phytol.* **67**, 87–93.

Audia, W. V., Smith, W. L., Jr., and Craft, C. C. (1962). Effects of isopropyl N-(3-chlorophenyl) carbamate on suberin, periderm and decay development by Katadin potato slices. *Bot. Gaz. (Chicago)* **123**, 255–258.

Bamber, R. K. (1962). The anatomy of the barks of Leptospermoidae. *Aust. J. Bot.* **10**, 25–54.

Bloch, R. (1941). Wound healing in higher plants. *Bot. Rev.* **7**, 110–146.

Bloch, R. (1952). Wound healing in higher plants. II. *Bot. Gaz. (Chicago)* **18**, 655–679.

Borger, G. A., and Kozlowski, T. T. (1970). Effect of water deficits on periderm formation in *Fraxinus pennsylvanica* seedlings. *Plant Physiol.* **46**, Suppl., 18.

Borger, G. A., and Kozlowski, T. T. (1971a). Effects of light intensity and photoperiod on periderm development in tree seedlings. *Plant Physiol.* **47**, Suppl., 12.

Borger, G. A., and Kozlowski, T. T. (1971b). Effects of PEG-induced water deficits on periderm development in *Fraxinus pennsylvanica* seedlings. *Plant Physiol.* **47**, Suppl., 36.

Borger, G. A., and Kozlowski, T. T. (1971c). Temperature and periderm development in woody angiosperm seedlings. *Bull. Ecol. Soc. Amer.* **52**, 34–35.

Borger, G. A., and Kozlowski, T. T. (1972a). Early periderm ontogeny in *Fraxinus pennsylvanica, Ailanthus altissima, Robinia pseudoacacia,* and *Pinus resinosa* seedlings. *Can. J. Forest Res.* **2**, 135–143.

Borger, G. A., and Kozlowski, T. T. (1972b). Effects of water deficits on first periderm and xylem development in *Fraxinus pennsylvanica* seedlings. *Can. J. Forest Res.* **2**, 144–151.

Borger, G. A., and Kozlowski, T. T. (1972c). Effects of light intensity on early periderm and xylem development in *Pinus resinosa, Fraxinus pennsylvanica,* and *Robinia pseudoacacia. Can. J. Forest Res.* **2**, 190–197.

Borger, G. A., and Kozlowski, T. T. (1972d). Effects of temperature on first periderm and xylem development in *Fraxinus pennsylvanica, Robinia pseudoacacia,* and *Ailanthus altissima. Can. J. Forest Res.* **2**, 198–205.

Borger, G. A., and Kozlowski, T. T. (1972e). Effects of cotyledons, leaves, and stem apex on early periderm development in *Fraxinus pennsylvanica* seedlings. *New Phytol.* **71**, 691–702.

Borger, G. A., and Kozlowski, T. T. (1972f). Effects of photoperiod on early periderm and xylem development in *Fraxinus pennsylvanica, Robinia pseudoacacia,* and *Ailanthus altissima* seedlings. *New Phytol.* **71**, 703–708.

Borger, G. A., and Kozlowski, T. T. (1972g). Wound periderm ontogeny in *Fraxinus pennsylvanica* seedlings. *New Phytol.* **71**, 709–712.

Borger, G. A., and Kozlowski, T. T. (1972h). Effects of growth regulators and herbicides on normal and wound periderm ontogeny in *Fraxinus pennsylvanica* seedlings. *Weed Res.* **12**, 190–194.

Bowen, W. R. (1963). Origin and development of winged cork in *Euonymous alatus. Bot. Gaz.* (*Chicago*) **124**, 256–261.

Boyce, J. S. (1961). "Forest Pathology," 3rd ed. McGraw-Hill Book, New York.

Bramble, W. C. (1936). Reaction of chestnut bark to invasion by *Endothia parasitica. Amer. J. Bot.* **23**, 89–94.

Büsgen, M., and Münch, E. (1929). "The Structure and Life of Forest Trees." Wiley, New York.

Chang, Y. P. (1954). Bark structure of North American conifers. *U. S., Dep. Agr., Tech. Bull.* **1095**.

Chattaway, M. M. (1953). The anatomy of bark. I. The genus *Eucalyptus. Aust. J. Bot.* **1**, 402–433.

Chattaway, M. M. (1955a). The anatomy of bark. V. *Eucalyptus* species with stringy bark. *Aust. J. Bot.* **3**, 165–169.

Chattaway, M. M. (1955b). The anatomy of bark. VI. Peppermints, boxes, ironbarks, and other eucalypts with cracked and furrowed barks. *Aust. J. Bot.* **3**, 170–176.

Cooke, G. B. (1948). Cork and cork products. *Econ. Bot.* **2**, 393–402.

Cunningham, H. S. (1953). A histological study of the influence of sprout inhibitors on *Fusarium* infection of potato tubers. *Phytopathology* **43**, 95–98.

DeBary, A. (1884). "Comparative Anatomy of the Vegetative Organs of the Phanerogams and Ferns." Oxford Univ. Press (Clarendon), London and New York.

deZeeuw, C. (1941). Influence of exposure on the time to deep cork formation in three northeastern trees. *N. Y. State Coll. Forest., Syracuse Univ. Tech. Publ.* **56**.

Diettert, R. A. (1938). The morphology of *Artemisia tridentata* Nutt. *Lloydia* **1**, 3–74.

Dittmann, C. P. (1931). The relation of bark thickness to the diameter of western yellow pine. *Idaho Forester* **14**, 39 cont. to 44.

Doubt, S. L. (1917). Responses of plants to illuminating gas. *Bot. Gaz.* (*Chicago*) **63**, 209–224.

Eames, A. J., and MacDaniels, L. H. (1947). "Introduction to Plant Anatomy," 2nd ed. McGraw-Hill, New York.

Esau, K. (1965). "Plant Anatomy," 2nd ed. Wiley, New York.

Gregory, E. L. (1888a). Development of cork-wings on certain trees. I. *Bot. Gaz.* (*Chicago*) **13**, 249–258.

Gregory, E. L. (1888b). Development of cork-wings on certain trees. II. *Bot. Gaz.* (*Chicago*) **13**, 281–287.

Gregory, E. L. (1888c). Development of cork-wings on certain trees. III. *Bot. Gaz.* (*Chicago*) **13**, 312–316.

Gregory, E. L. (1889a). Development of cork-wings on certain trees. IV. *Bot. Gaz.* (*Chicago*) **14**, 5–10.

Gregory, E. L. (1889b). Development of cork-wings on certain trees. V. *Bot. Gaz.* (*Chicago*) **14**, 37–44.

Guttenberg, S. (1951). Listen to the bark. S. *Lumberman* **183**, 220–222.

Haberlandt, G. (1928). Zur Entwicklungsphysiologie des Periderms. *Sitzungsber. Preuss. Akad. Wiss., Phys.-Math. Kl.* pp. 317–318.

Hale, J. D. (1955). Thickness and density of bark; trends of variation for six pulpwood species. *Pulp Pap. Mag. Can.* **56**, 113–117.

Hare, R. C. (1965). Contribution of bark to fire resistance of some southern trees. *J. Forest.* **63**, 248–251.

Harvey, E. M., and Rose, C. R. (1915). The effects of illuminating gas on root systems. *Bot. Gaz.* (*Chicago*) **60**, 27–44.

Hayward, H. E., and Long, E. M. (1942). The anatomy of the seedling and roots of the Valencia orange. *U. S., Dep. Agr., Tech. Bull.* **786**.

Hook, D. D., Brown, C. L., and Kormanik, P. P. (1970). Lenticel and water root development of swamp tupelo under various flooding conditions. *Bot. Gaz.* (*Chicago*) **131**, 217–224.

Isenberg, I. H. (1943). The anatomy of redwood bark. *Madroño, San Francisco* **7**, 85–91.

Kaufert, F. (1937). Factors influencing the formation of periderm in aspen. *Amer. J. Bot.* **24**, 24–30.

Kingsley, M. A. (1911). On the anomalous splitting of the rhizome and root of *Delphinium scaposum. Bull. Torrey Bot. Club* **38**, 307–318.

Kozlowski, T. T. (1971a). "Growth and Development of Trees," Vol. 1. Academic Press, New York.

Kozlowski, T. T. (1971b). "Growth and Development of Trees," Vol. 2. Academic Press, New York.

Kuster, E. (1925). "Pathologische Pflanzenanatomie," 3rd ed. Fischer, Jena.

Lemesle, R. (1927). De l'existence d'un liege intraligneaux chez une Labiée (*Hymenocrater bituminous* Fisch. et Mey.) *Bull. Soc. Bot. Fr.* [5] **9–10**, 904.

Lier, F. G. (1955). The origin and development of cork cambium cells in the stem of *Pelargonium hortorum. Amer. J. Bot.* **42**, 929–936.

Liphschitz, N., and Waisel, Y. (1970). Phellogen initiation in the stems of *Eucalyptus camaldulensis* Dehnh. *Aust. J. Bot.* **18**, 185–189.

McNair, G. T. (1930). Comparative anatomy within the genus *Euonymous. Kans. Univ. Sci. Bull.* **19**, 221–273.

Mann, C. E. T. (1930). Studies in the root and shoot growth of the strawberry. V. The origin, development, and function of the roots of the strawberry (*Fragaria virginiana* × *chiloensis*). *Ann. Bot.* (*London*) **44**, 55–86.

Martin, R. E. (1963). Thermal and other properties of bark and their relation to fire injury of tree stems. *Diss. Abstr.* **24**, 1322–1323.

Moss, E. H. (1934). Rings of cork in the wood of herbaceous perennials. *Nature* (*London*) **133**, 689.

Moss, E. H. (1936). The ecology of *Epilobium angustifolium* with particular reference to rings of periderm in the wood. *Amer. J. Bot.* **23**, 114–120.

Moss, E. H. (1940). Interxylary cork in *Artemisia* with a reference to its taxonomic significance. *Amer. J. Bot.* **17**, 762–768.

Moss, E. H., and Gorham, A. L. (1953). Interxylary cork and fission of stems and roots. *Phytomorphology* **3**, 285–294.

Mylius, G. (1913). Das Polyderm. Eine vergleichende Untersuchung über die physiologischen Scheiden. Polyderm, Periderm, und Endodermis. *Bibl. Bot. Stuttgart* **18**(79), 1–119.

Nelson, P. E., and Wilhelm, S. (1957). Some anatomical aspects of the strawberry root. *Hilgardia* **26**, 631–642.

Olufsen, L. (1903). Untersuchungen über Wunderperiderm-bildung an Kartoffel-knollen. *Beih. Bot. Zentralbl.* **15**, 269–308.

Pearson, L. C., and Lawrence, D. B. (1958). Photosynthesis in aspen bark. *Amer. J. Bot.* **45**, 383–387.

Peterson, R. L. (1971). Induction of a 'periderm-like' tissue in excised roots of the fern *Ophioglossum petiolatum* Hook. *Ann. Bot.* (*London*) [NS] **35**, 165–167.

Rees, L. W., and Shiue, C. (1957–1958). The structure and development of the bark of quaking aspen. *Minn. Acad. Sci.* **25–26**, 113–125.

Schneider, H. (1955). Ontogeny of lemon wood bark. *Amer. J. Bot.* **42**, 893–905.

Scott, L. I. (1950). The changing surface of the tree. *The Naturalist* **832**, 1–10.

Shapovalov, M., and Edson, H. A. (1919). Wound-cork formation in the potato in relation to seed-piece decay. *Phytopathology* **9**, 483–496.

Smith, F. H. (1958). Anatomical development of the hypocotyl of Douglas fir. *Forest Sci.* **4**, 61–70.

Smithson, E. (1952). Development of winged cork in *Acer campestre* L. *Proc. Leeds Phil. Lit. Soc., Sci. Sect.* **6**, 97–103.

Smithson, E. (1954). Development of winged cork in *Ulmus* × *hollandica* Mill. *Proc. Leeds Phil. Lit. Soc., Sci. Sect.* **6**, 211–220.

Solereder, H. (1908). "Systematic Anatomy of the Dicotyledons," Vols. I and II. Oxford Univ. Press, (Clarendon), London and New York.

Srivastava, L. M. (1964). Anatomy, chemistry, and physiology of bark. *Int. Rev. Forest. Res.* **1**, 204–277.

Stickel, P. W. (1941). The relation between bark character and resistance to fire. *Pulp Pap. Mag. Can.* **42**, 420.

Stone, G. E. (1913). Effect of illuminating gas on Vegetation. *Rep. Bot., 25th Annu. Rep., Mass., Agr. Exp. Sta.* pp. 45–56.

Stover, E. L. (1951). "An Introduction to the Anatomy of Seed Plants." Heath, Boston, Massachusetts.

Strain, B. R., and Johnson, P. L. (1963). Corticular photosynthesis and growth in *Populus tremuloides*. *Ecology* **44**, 581–584.

Templeton, J. (1926). Hypertrophied lenticels on the roots of cotton plants. *Egypt, Min. Agr., Tech. Sci. Bull.* **59**.

Weimer, J. L., and Harter, L. L. (1921). Wound-cork formation in the sweetpotato. *J. Agr. Res.* **21**, 637–647.

Whitmore, T. C. (1963). Why do trees have different sorts of bark? *New Sci.* **16**, 330–331.

Woffenden, L. M., and Priestley, J. H. (1924). The toxic action of coal gas upon cork and lenticel formation. *Ann. Appl. Biol.* **11**, 42–53.

. 7 .

Shedding of Roots

G. C. Head

I. Introduction

Root shedding is a natural part of development of healthy perennial plants. While the roots of perennial herbaceous plants probably have a limited life span, some tree roots live for many years and may be as old as the tree itself. However, only a few tree roots survive for many years;

FIG. 1. Half the root system of a 16-year-old Cox's Orange Pippin apple tree on M.1 rootstock exposed by excavation. (Photo courtesy of East Malling Research Station.)

although seedlings or rooted cuttings of woody species are usually well supplied with roots at the time of planting, mature trees often have relatively few main roots spreading from the base of the trunk (Fig. 1).

Detailed studies of growth of tree roots have shown that while some long roots continue growing in length for several years and eventually commence secondary thickening, other roots, apparently similar in external appearance, do not thicken at all and soon die. By far the greatest natural loss of roots from the tree root systems occurs among the short rootlets that grow to only a few millimeters in length. The development of these is favored by soils rich in humus, and the rootlets are often infected with mycorrhizae. As a result of frequent branching, the short rootlets form dense networks and these are constantly replaced as new areas of soil are explored.

In addition to natural death of roots, all tree roots shed tissue and sometimes excrete water and complex materials in the course of their development, and a considerable amount of organic matter is deposited in this way in the soil each year. Together with litter from the aerial parts of plants, the decaying root material forms the basis for the complex biological cycles in the soil that include bacteria, fungi, and soil animals. Accurate measurement of the total amount of dry matter deposited in the soil each year is virtually impossible, but some investigators have attempted to estimate annual root production, which, for instance, in mature forests may be equated with annual loss of roots. Bray (1963) reviewed some of the estimates of annual root production and these ranged from 1.2 tons/ha in *Pinus sylvestris* to 2.2 tons/ha in *Betula verrucosa;* root production in tropical rain forests was estimated as 2.6–2.8 tons/ha/year. These figures are of the same order of magnitude as those given by Orlov (1955) for *Picea excelsa* forests of different ages. Orlov estimated that between 0.3 and 2.0 tons/ha of roots less than 3 mm in diameter died off each year.

Apart from natural losses of root material from healthy trees, severe losses sometimes occur as a result of unfavorable soil conditions, root disease, or activities of soil animals feeding directly on roots.

II. Life History of Roots

A. Morphology and Structure

Root systems vary widely in their morphology and show adaptations to markedly varied life cycles. Most monocotyledonous plants have fibrous root systems. The roots that develop from the embryo in grasses, known as seminal roots, are later supplemented by adventitious roots that arise from the base of the stem. Those species of monocotyledons that have bulbs, corms, or rhizomes also have mainly adventitious roots.

The root systems of dicotyledonous plants, on the other hand, are usually based on a taproot and its branches. The taproot often does not persist, however, and then the framework of the root system consists of several of the lateral branches.

In longitudinal section it can be seen that the meristematic region of a root tip gives rise to a root cap that is often considered to protect the meristematic region. The cells of the root cap are living parenchymatous cells and many of them, particularly those on the outside of the root cap, have mucilaginous cell walls. In the aerial roots of many tropical plants the root cap may become dry and crusty.

The cortex of primary roots of dicotyledonous plants and gymnosperms that develop secondary growth is generally simple in structure and consists of relatively homogeneous cells. In the roots of monocotyledons, which have a persistent cortex, a more complex structure is found.

In the primary root the cortex is surrounded by an epidermis which typically bears unicellular root hairs. In some plants all epidermal cells are capable of initiating root hairs, in others only certain of the epidermal cells have this capacity.

The innermost layer of the cortex in primary roots differentiates into an endodermis, the cells of which have Casparian strips on their anti-

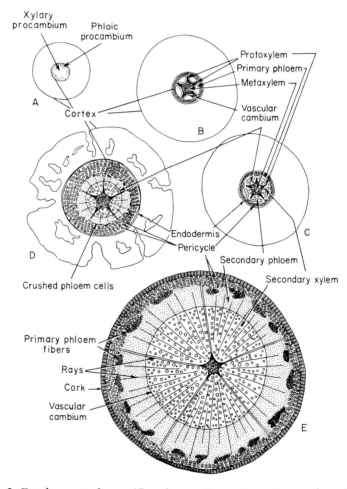

FIG. 2. Development of pear (*Pyrus*) root. Transections: A, vascular cylinder in procambial state; B, primary growth completed; C, strips of vascular cambium between phloem and xylem have produced some secondary vascular tissues; D, vascular cambium, now a cylinder, has produced additional secondary tissues; pericycle has undergone periclinal divisions; endodermis partly crushed; cortex breaking down; and E, secondary growth has progressed further; periderm has appeared; cortex has been shed. Cambium opposite protoxylem poles has formed wide rays (D, E). (All, ×29.) [From Esau (1965).]

clinal walls. The origin and function of Casparian strips are controversial, but it is generally held that they form a barrier to free diffusion of water through cell walls of the endodermis, both from the outside to the inside of the root and vice versa.

The endodermis encloses the vascular cylinder of the root but a full description of the development of the vascular elements will not be given here. The layer of cells immediately inside the endodermis forms the pericycle which, in flowering plants and gymnosperms, has a meristematic function. Lateral roots arise in the pericycle and, in roots that show secondary growth after the eventual shedding of the cortex and the endodermis, the pericycle gives rise to part of the vascular cambium and also to the phellogen that forms cork tissue on the outside of the root (Fig. 2).

B. Root Growth

The growth of individual roots is very difficult to study under entirely natural conditions, and several investigators have resorted to the use of root observation boxes (Rogers, 1939b; Wilcox, 1962) or root observation trenches (Rogers, 1939a; Richardson, 1953a; Rogers and Head, 1963; Lyr and Hoffman, 1967; Hilton *et al.*, 1969) in which roots growing in soil can be viewed through glass windows. These methods are valuable for studies of root life history and of seasonal trends in root growth on small numbers of trees, but as an unknown proportion of the roots of the tree is visible, observation methods are not suitable for some quantitative studies.

Most observations of root life history have been made on fruit trees or bushes or on forest trees. In the two underground root observation laboratories at East Malling Research Station in England the roots of apple (*Pyrus malus*), pear (*Pyrus cydonia*), plum (*Prunus cerasifera*), cherry (*Prunus avium*), black currant (*Ribes nigrum*), raspberry (*Rubus idaeus*), strawberry (*Fragaria* sp.), and hops (*Humulus lupulus*) have been studied. In some cases observations have been made on roots from the same trees for more than 12 years.

The root systems of trees extend into new areas of soil, or reoccupy areas that were explored earlier, by means of fairly thick fast-growing roots. On apple trees these extension roots, or long roots, are 1–2 mm thick and at East Malling they grew 4–6 cm per week. The extension roots of cherry trees are thicker and grow more rapidly (up to 7–8 cm per week) and Lyr and Hoffman (1967) reported that roots of *Robinia pseudoacacia* grow up to 5–6 cm per day.

The growth of the extension roots is very variable, however. For instance, Wilcox (1962) found that the roots of incense cedar seedlings (*Libocedrus decurrens*) were not all growing at any one time. In this species the individual roots had growth cycles that appeared to be internally regulated. The roots that can be seen through glass observation panels do not always grow against the glass for long periods, but Head (1966b) followed the growth of a single apple root over an uninterrupted period of 24 weeks. During this period many other roots apparently similar in size and external appearance ceased growth after only a few days and subsequently remained dormant.

Growth of the extension roots is usually more or less straight, and Wilson (1967) found that roots of red maple (*Acer rubrum*) tended to continue growing in the original direction after negotiating artificial barriers. Head (1966b), however, reported sudden unexplained deviations in the direction of growth of apple roots.

C. Root Hair Development

The density and length of root hairs vary considerably between species and as a result of environmental influences (Kramer and Kozlowski, 1960). Some trees such as avocado (*Persea*) and pecan (*Carya*) have been reported not to have root hairs at times (Woodroof and Woodroof, 1934; Smith and Wallace, 1954). Dittmer (1949) reported considerable differences in root hair length between species, the lengths ranging from 0.17 mm in *Celtis occidentalis* to 0.37 mm in *Fraxinus lanceolata*. Through the glass panels in the East Malling root observation laboratory the root hairs of apple were measured and found to be about 0.1 mm long (Head, 1964) and those of cherry about 0.3 mm long (Head, 1966b). The root hairs of black currant were much longer, measuring about 1.0 mm (Head, 1966b) (Fig. 3). In plum the distribution of root hairs over the root surface was often uneven and appeared to be influenced by the proximity of soil particles (Head, 1968c). In rapidly growing roots the root hairs are first seen several millimeters behind the growing root tip but when growth slows down or ceases root hair development occurs within a millimeter or so of the tip.

D. Root Branching

Root branching varies with variety and species as well as with soil conditions, but is always much more extensive than the branching of the

FIG. 3. Root of black currant showing typical long thin root hairs. (By courtesy of East Malling Research Station.)

aerial parts of a tree. Figure 4 shows an extension root of apple growing behind a glass observation panel (etched with ½-inch squares) with lateral roots emerging 4–10 cm behind the growing tip. The root had been growing at a rate of 4–6 cm per week. Lateral roots may emerge much nearer the tip in slower growing or dormant roots.

The lateral roots themselves may branch again giving rise to laterals

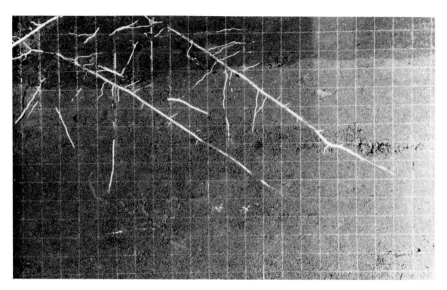

FIG. 4. Roots of James Grieve/M.VII, about 9–14 inches below the soil surface in June. The lines engraved on the glass are approximately ½ inch apart. (By courtesy of East Malling Research Station.)

of higher orders. Kolesnikov (1966) found that apple seedlings produced up to 5–7 orders of lateral roots in one growing season with a predominance of third and fourth order roots. Kolesnikov found that in the conditions of the Moscow region a 1-year-old Chinese apple (*Malus*

TABLE I

ORDERS, NUMBERS, AND LENGTHS OF ROOTS OF ONE 6-MONTH-OLD *Cornus florida* AND ONE *Pinus taeda* SEEDLING GROWN IN THE GREENHOUSE IN THE ABSENCE OF COMPETITION[a]

Order	*Cornus florida*		*Pinus taeda*	
	No. of roots	Lengths (cm)	No. of roots	Lengths (cm)
First	1	44.1	1	32.2
Second	93	859.7	71	187.5
Third	1035	2714.4	496	146.1
Fourth	1336	1357.0	199	21.2
Fifth	191	168.1	0	0.0
Sixth	1	0.6	0	0.0
	2657	5143.9	767	387.0
Length (ft)		168.76		12.70

[a] From Kozlowski and Scholtes (1948).

prunifolia) seedling grew up to 40,000 roots in 1 season with a total length of 230 m, and he estimated that a mature apple tree had several million roots with a total length of several kilometers.

Similar counts and measurements were carried out by Kozlowski and Scholtes (1948) and showed the variability that exists between species. Four-month-old seedlings of *Pinus taeda* and *Robinia pseudoacacia* had 419 and over 7000 roots, respectively, with total root lengths of 161 cm and 32,500 cm. They also compared 6-month-old seedlings of *Pinus taeda* and *Cornus florida* grown in the absence of competition and found them to have 767 and 2657 roots, respectively, with total lengths of 387 and 5144 cm, respectively. The breakdown of numbers and lengths of roots by order of branching is shown in Table I. Because of root competition forest grown seedlings had many fewer roots.

The growth potential of roots of higher order is generally less than that of the extension roots described above and on apple trees few of them grow more than a few centimeters in one growth cycle. Repeated branching occurs in apple tree roots, particularly in soils rich in organic matter, and this leads to formation of dense networks of absorbing rootlets (Fig. 5). Repeated branching in some species is associated with mycorrhizal infection.

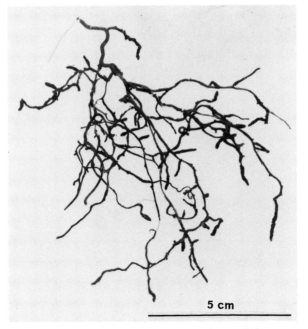

FIG. 5. Finely branched rootlets of M.VII rootstock sampled in winter from a tree growing in sandy soil. (By courtesy of East Malling Research Station.)

E. MATURATION AND SECONDARY THICKENING

The extension roots are white or yellowish white when young but after
a short time the outer tissues, the epidermis, and cortex turn brown as a
result of internal changes and eventually rot away (see Section IV,C).
Some of the extension roots subsequently commence secondary thickening and eventually become the thick structural roots of the tree. Secondary thickening starts after the cortex has turned brown and may start
during the year of primary growth or in a subsequent year. The proportion of roots that thicken varies among species and cultivars of the
same species and is one of the factors that gives a root system its particular character. Apple trees grown on certain varieties of rootstock may
have only a few main roots, whereas the same variety on a different
rootstock may have many roots.

FIG. 6. The roots that grew across one of the glass panels in 1 year in the underground root observation laboratory at East Malling Research Station are shown in
thin lines. The few thicker lines show which of the original roots had become
secondarily thickened after 3 additional years. (By courtesy of East Malling Research Station.)

Root thickening could be observed and measured through the glass panels of the underground root observation laboratories at East Malling. The small porportion of the roots that grew across one of the glass panels that showed secondary thickening during the following 3 years is illustrated in Fig. 6.

Relatively little is known about the timing and distribution of secondary thickening on tree roots. Cockerham (1930) with *Acer pseudoplatanus* and Wight (1933) with *Pinus sylvestris* found that there was considerable root thickening close to the base of the trunk but that the roots tapered sharply beyond 50–60 cm from the trunk. Further away from the trunk the roots tapered only very gradually and often appeared to be more or less uniform in thickness for considerable distances. In the thickened basal parts of *Pinus sylvestris* roots Wight (1933) reported that cambial activity could be regarded almost as an extension of that in the base of the trunk. In the more distal parts of the roots, secondary thickening started more or less uniformly about 6 weeks later than cambial activity in the trunk. In apple and cherry roots Head (1968b) found that thickening was irregular along the length of the root. Sometimes thickening started first in more distal parts of the root and, in some years, roots did not thicken at all. Thus, in these species the number of rings of secondary wood cannot be used to estimate the age of a root.

III. Control of Root Growth

A. Internal Factors

Growth of the aerial and subterranean parts of plants is very closely interrelated and some of the principles of this relationship are illustrated by a series of experiments conducted in Holland by Richardson on seedlings of *Acer saccharinum* L. in their first year of growth. The root growth of seedlings grown in gravel culture under controlled conditions of illumination and of root and shoot temperature was very sensitive to changes that affected the rate of photosynthesis of the foliage (Richardson, 1953a,b). Reductions in light intensity or in leaf temperature affected the rate of root elongation within a few hours. However, after plants had been exposed to a low level of illumination or to low leaf temperature for several days, the rate of root growth increased again to between 50 and 60% of its original level. Defoliation at this stage, however, stopped root growth completely. Richardson interpreted this response as indicating that root growth continued at the expense of carbohydrate reserves as long as contact with the leaves was retained. Further experiments involving disbudding and defoliation showed that during the growing season root elongation was controlled by a stimulus supplied by the leaves and that root initiation was determined by the presence of an active terminal meristem (Richardson, 1957). The influence of the terminal meristem could, after excision, be replaced by indoleacetic acid (IAA) (Richardson, 1958) but the role of the leaves in controlling root elongation could not be replaced by IAA.

Thus it is clear that carbohydrates and hormonal materials supplied by the aerial parts have a profound and, at least in tree seedlings, a fairly direct effect on root activity. In larger trees with considerable carbohydrate reserves, the relationship with current photosynthesis is more tenuous, but may be important at times of stress.

Experimentation with excised roots in artificial media has confirmed that roots need a supply of a soluble sugar and materials such as thiamin, niacin, and pyridoxin. This type of experiment has also shown that internal growth regulation occurs within the roots themselves (Street, 1969). In a suitable culture medium, with a sucrose concentration that caused some restriction of root growth rate, the same root meristem would continue growing for 40–50 weeks if transferred periodically into fresh culture medium. On the other hand, a glucose concentration that led to maximum root growth rate brought about earlier aging of the meristem and eventual cessation of growth. Street interpreted these results as indicating the accumulation at the meristem of an auxinlike growth factor that was essential to meristem function but which was inhibitory to growth at higher concentrations.

B. Seasonal Factors

The times of year at which new root extension and the emergence of new lateral roots take place most rapidly are controlled by external factors, such as temperature and soil moisture and also by internal physiological factors and it must be remembered that, at least as far as the internal regulating factors are concerned, the 2 types of new root growth may respond differently.

Most studies of root growth periodicity have been carried out by observations of growing roots through glass observation panels. Lyr and Hoffmann (1967) showed that considerable variation occurred in the season of maximum root growth among 7 tree species at the Eberswalde root laboratory (Fig. 7). In general, maximum root growth, both as regards the number of growing roots and increment in length, occurred in most species in June and July (early summer) and growth decreased

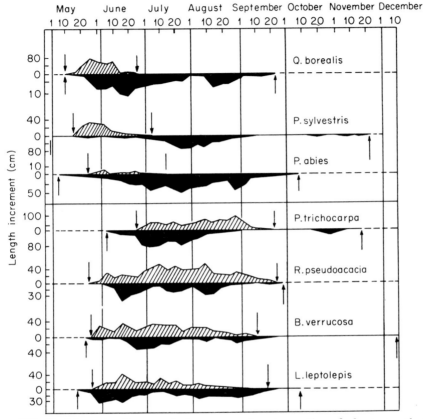

FIG. 7. Schematic diagram of the course of root growth and shoot growth of 7 tree species in the Eberswalde root laboratory. Shading indicates shoot growth; solid black represents root growth; start and termination are indicated by arrows. [From Lyr and Hoffmann (1967).]

in August and September. In *Pinus sylvestris* little root growth occurred during the period when new shoots and needles were being produced but root growth increased after the needles had expanded.

The first root growth measurements carried out in the root observation laboratory at East Malling were made on trees that were irrigated regularly and deblossomed. The experiments were designed to investigate the ways in which the seasonal cycle of root growth could be modified by cultural treatments. The measurements were of the total length of white root present at weekly intervals and as the rate at which the white root turned brown remained fairly constant during most of the growing season, the figures were a good measure of new growth (Head, 1966a).

In experiments concerned with the effects of pruning of branches on

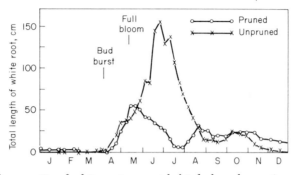

FIG. 8. The quantity of white root present behind glass observation panels through the months (Jan., Feb., etc.) of 1 season from pruned and unpruned trees of Cox's Orange Pippin on MM.111 (pruning March 11). [From Rogers and Head (1969).]

root growth, large differences in timing of root growth occurred. Figure 8 shows a comparison of the pattern of root growth on one unpruned apple tree and one pruned tree. Other replicated experiments confirmed that pruning had a profound effect on root growth and Head (1967, 1968a) interpreted this as being related to differences in the intensity of shoot growth (rate of shoot growth and number of growing shoots). Further experiments showed that the amount of new root growth in the late summer and autumn was much less if the trees were carrying a crop of fruit or if the trees were artificially defoliated (Head, 1969a). The amount of new root growth in the spring was reduced in trees growing in a grass sward when some degree of nitrogen deficiency was present (Head, 1969b), and this reduction was corrected by fertilizer application.

Root growth in the spring was also less on trees that had shown weak shoot growth in the previous season (Head, 1966b) or that had been defoliated in the previous autumn (Head, 1969a).

Many other workers studying different tree species, usually without artificial irrigation, have reported 2 or more periods of root growth each year separated by periods of lesser activity. Some have interpreted this pattern as being a result of soil moisture conditions (Rogers, 1939a; Eremeev, 1960; Orlov, 1957; Kolesnikov, 1959), and others considered that the periods of inactivity were due to competition between root and shoot development (Waynick and Walker, 1930; Reed and MacDougal, 1937; Monselise, 1947; Marloth, 1950; Sakanishi and Hayashi, 1959; Hiroyasu, 1961).

The experiments at East Malling reported above indicate quite clearly that the seasonal distribution of root growth on trees of the same variety may show large variations in the same year if they are pruned differently. Thus it is clear that internal regulating factors play a strong part in controlling the seasonal activity of the root system, although in unirrigated orchards or woodlands the level of soil moisture may have an overriding influence.

C. Soil Factors

As with growth of other parts of plants, growth of roots is profoundly affected by the medium in which they grow. Growth is the result of metabolic processes and is thus regulated by soil temperature. Ladefoged (1939) made short-term studies of root growth of *Fagus sylvatica*, *Fraxinus excelsior*, *Picea abies*, and *Abies alba* and found the maximum growth rate for these species to be at temperatures between 24° and 32°C. Under natural conditions, however, the optimum temperature is likely to vary with such factors as carbohydrate balance and mineral nutrition.

Minimum temperatures for root growth vary considerably with species from the values of 2°–4°C quoted by Ladefoged (1939) for *Picea abies* and *Abies alba* to values as high as 11°C for citrus (Muromcev, 1962). The range of temperatures suitable for root growth is also an important specific characteristic (Table II). Muromcev (1962) found that citrus roots would grow over a range of only 7°C whereas those of strawberry grew over a range of 16°C.

Soil temperature also affects root morphology and the pattern of root branching. Excised roots of *Robinia* showed inhibition of lateral root formation below 19°C and inhibition of growth in length of the main root at 33°C, favoring the greater development of laterals. Lindemann (1956) also found that high temperatures of the rooting medium promoted formation of lateral roots on cuttings of a wide range of shrubs, whereas lower temperatures gave rise to longer unbranched roots.

Nightingale (1935) grew young Elberta peach trees in containers at controlled temperatures and found marked differences in morphology and vigor of the new roots produced. After 45 days at 45°F no new roots had been produced, at 50°F there were a few roots but no branching, whereas at 55° and 65°F new root growth was more extensive with many primary branches. At temperatures above 65°F the roots were thinner and fewer but very much branched. No new roots were produced at 95°F. However, experiments with redwood (*Sequoia sempervirens*) seedlings did not give quite the same results (Hellmers, 1963). Roots of seedlings grown in soil at 18°C appeared to be the healthiest but they were not always the longest. Roots of plants grown at a soil temperature of 28°C were thinner and darker and had fewer and shorter white root tips. At a soil temperature of 8°C, the roots were short and

TABLE II

CARDINAL VALUES OF TEMPERATURE FOR ROOT GROWTH[a]

Tree species	Temperature (°C)			Authority
	Minimum	Optimum	Maximum	
Pinus strobus L.	5–6	—	—	Engler (1903)
Pinus cembra L.	5–6	—	—	Engler (1903)
Pinus mugo Turra.	5–6	—	—	Engler (1903)
Pinus taeda L.	5	25	35	Barney (1951)
Pinus sylvestris L. (in culture)	—	19	25	Slankis (1949)
Picea abies (L.) Karst.	2–4[b]	26[b]	—	Ladefoged (1939)
Picea abies (L.) Karst.	3–4	>10	—	Busarova (1961)
Abies alba Mill.	2–4[b]	32[b]	—	Ladefoged (1939)
Larix decidua L.	5.7	—	—	Ladefoged (1939)
Fagus sylvatica L.	>0[b]	24[b]	—	Ladefoged (1939)
Fagus sylvatica L.	2–3	—	—	Engler (1903)
Acer pseudoplatanus L.	2–3	—	—	Engler (1903)
Fraxinus excelsior L.	4–6[b]	29[b]	—	Ladefoged (1939)
Betula sp.	5	—	—	Vorobieva (1961)
Malus sp.	7.2[b]	18.3[b]	29.4[b]	Nightingale (1935)
Malus sp.	4–5	—	—	Bodo (1926)
Prunus domestica L.	2–4	—	—	Bodo (1926)
Prunus persica (L.) Batsch.	7.2[b]	18.3[b]	29.4[b]	Nightingale (1935)
Prunus persica (L.) Batsch.	—	24[b]	35[b]	Proebsting (1943)
Citrus sp.	—	24–27	—	Girton (1927)
Citrus sp.	11	—	—	Muromcev (1962)
Carya pecan (Marsh.) Engl. & Graebn.	1–3[b]	30[b]	36[b]	Woodroof and Woodroof (1934)
Robinia pseudoacacia L. (in culture)	—	22.5[b]	—	Seeliger (1959)

[a] From Lyr and Hoffmann (1967).
[b] Experimentally determined values.

thick except for those on older seedlings grown at warmer air temperatures.

The water content of the soil affects root growth both when the soil is too wet and when it is too dry. In dry soils many workers have reported that root growth is restricted although it is obviously difficult to conduct controlled experiments under natural conditions while ensuring that all other factors remain constant. Ladefoged (1939) found that root growth of most species stopped when the soil moisture content was reduced to 12–14% on an oven-dry basis, and that increase of soil moisture above 40% gave no further improvement in root growth. Decreases in rates of root growth with decreasing soil moisture levels were reported for *Pinus banksiana* by Kaufman (1945).

The effects of excess soil moisture are largely due to changes in the gaseous environment in the soil. This is demonstrated by the fact that tree roots will grow well when immersed in water if the culture solution is well aerated. In wet soils the level of available oxygen decreases and there is also a buildup of carbon dioxide levels; opinions differ as to which factor is more important in reducing root growth. Appreciable differences probably exist between tree species. Boynton (1940) found that production of new roots on apple trees was reduced in the presence of 5% of carbon dioxide. However, it has been reported that concentrations of carbon dioxide as high as 45% can be tolerated by *Salix* roots (Cannon and Free, 1925) in the presence of an adequate oxygen supply. For *Pinus* and *Picea*, the same authors found that root growth was noticeably inhibited at oxygen levels around 10%. Leyton and Rousseau (1958) presented data on the rate of root growth of several tree species at different oxygen concentrations (Fig. 9). Some species such as *Salix*, *Alnus*, and *Betula* are thought to have internal systems of intercellular spaces to conduct oxygen to their roots to enable them to grow in poorly aerated soils.

Soil fertility affects the spread of tree root systems and also the root : shoot ratio. Apple trees excavated from a good loam soil at East Malling showed a root weight : shoot weight ratio of about 2:1, whereas in trees excavated from a poor sandy soil the root : shoot ratio was about 1:1. The spread of the root system was about 3 times that of the branches on the sandy soil compared with 1.5 times on the fertile loam (Rogers and

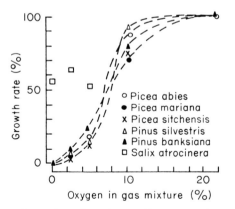

FIG. 9. Influence of aeration of rooting medium on root growth. Root growth expressed as percentage of that with normal aeration. [From Leyton and Rousseau (1958).]

Vyvyan, 1934). Apple trees grown in pots with water and nutrients liberally supplied have been found to have root:shoot ratios of 7:1 or more (Rogers and Head, 1969).

Although it is clear that mineral nutrition influences both root growth and root morphology, the response is complex and it is not always clear whether it is a direct response to a particular mineral, a result of other, perhaps biological, changes in the growth medium, or whether it is associated with changes within the plants resulting from a general improvement in growth.

In soil layers rich in nutrients, growth of the main root is usually reduced in favor of development of more lateral roots (Lyr and Hoffmann, 1967). Otto (1964) found that the rooting quotient (total root length/number of root tips) was very high on trees in sandy soil but was lower after fertilization.

The type of nitrogen fertilizer has also been reported to affect root morphology. Both Leyton (1952) and Smith (1957) reported that nitrate fertilizer produced a stronger root system than did ammonia nitrogen. The root system of *Alnus glutinosa* was dense and more branched when nitrate fertilizer was used, whereas ammonia fertilizer tended to produce longer but less branched roots.

IV. Loss of Tissue from Healthy Roots

A. Root Cap

The root cap is usually thought of as protecting the meristem as the root pushes its way through the soil and Haberlandt (1914) suggested that the mucilaginous walls that occur in the outer layers of the root cap might help ease the passage of the root tip through the soil. Certainly the outermost cells of the root cap are constantly being sloughed off as new cells are produced from within.

Jenny and Grossenbacher (1962) showed that root caps and sometimes the terminal few millimeters of roots of many species are covered by a film of mucilage that may be quite thick, and others have shown that the mucilage layer on the aerial roots of some tropical plants may be several millimeters thick and that it sometimes dries into a hard crust. Mucilage sheaths were found to be a particularly striking feature of roots of species of Ericaceae (Leiser, 1968). Sheaths were present on all species examined and on healthy roots the sheath was often large enough to be seen with the naked eye, especially *Arctostaphylos* and *Pernettya*. On inactive, diseased, or poorly growing roots the sheath was absent. *Azalea*

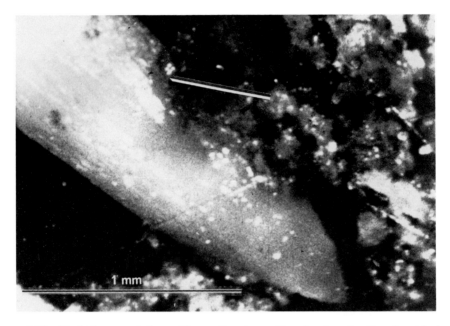

FIG. 10. Enlargement from 16-mm time-lapse film showing the tip of a black currant (*Ribes nigrum*) root. A layer of mucilage (arrowed) is being scraped off the upper side of the root as it pushes past the soil particles. (By courtesy of East Malling Research Station.)

roots had a transparent gelatinous or mucilaginous sheath external to the root cap, the meristematic region, and the zone of elongation that varied with the microenvironment. The sheath could be drawn out if touched, thus indicating that it had a very sticky or viscous consistency and contained sloughed root cap cells and oil droplets. Leiser concluded that the sheath was not derived from breakdown products of a senescent root cap.

Studies of root growth of fruit plants by time-lapse cinematography at East Malling Research Station (Head, 1968c) showed clearly that mucilage was scraped off the tip of a black currant root as it grew past some soil particles (Fig. 10). No attempts have been made to estimate the quantity of dry matter deposited in the soil in this way from the roots of trees but Samtsevich (1965) described the large mucilage caps that may form on the tips of roots of some agricultural crops and attempted to estimate the amount of dry matter lost from the roots in this way. The tips of wheat (*Triticum* sp.) roots have particularly prominent mucilage caps that can be seen easily with the naked eye when they are grown in humid chambers (Head, 1970). Figure 11 shows that these caps are full of cells that have become detached from the root cap. Samtsevich meas-

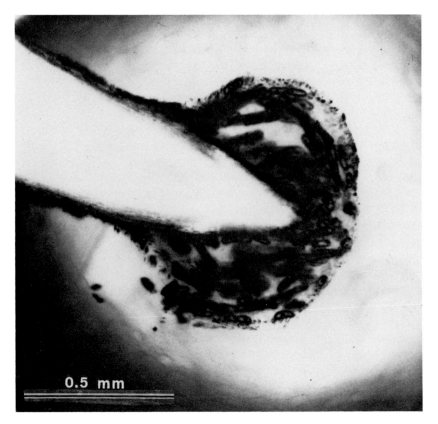

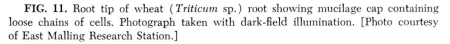

FIG. 11. Root tip of wheat (*Triticum* sp.) root showing mucilage cap containing loose chains of cells. Photograph taken with dark-field illumination. [Photo courtesy of East Malling Research Station.]

ured the diameters of many root tips and of their mucilage caps and thus calculated the volume of mucilage on each root tip. Using the product of this calculated figure and an estimate of the number of root tips per hectare and taking into account the dry matter content of the mucilage, he estimated that the amount of dry matter deposited in the soil in this way during the development of a crop of wheat was about equal to the dry weight of the grain removed from the same area.

B. Root Hairs and Secretion

The young roots of most species of dicotyledons that show secondary thickening of roots are covered with root hairs less than 1 mm long (see Section II,C), and these are shed when the root cortex sloughs off.

The density of root hairs over the root surface varies with species and with environmental conditions. Kozlowski and Scholtes (1948) estimated the number of root hairs on 7-year-old tree seedlings grown in the greenhouse and found that on average seedlings of *Robinia pseudoacacia* had 11,000 root hairs while those of *Pinus taeda* had less than 600 root hairs (Table III).

According to Dittmer (1949), approximately 75–80% of the surface area of the entire root system of dicotyledons exhibiting secondary thickening is covered with root hairs, but this proportion would obviously vary with root growth activity of the tree, this being affected greatly by the time of year and the soil conditions.

Although it is clear that all the root hairs are shed when the cortex separates from the central part of the root when secondary thickening begins, it appears that some root hairs may die earlier. Kozlowski (1971) reported that the majority of root hairs live for only a few hours or weeks, and certainly time-lapse cinematography has shown that although, in general, the root hairs of black currant are quite long-lived some of them shrivel after only a few days (Head, 1968c). Dittmer (1949) found root hairs of some species to be quite persistent structures, however. On one species, *Gleditsia triacanthos*, Dittmer observed the epidermis separating from the root surface in strips with the root hairs still undamaged. McDougall (1921) also found that the root hairs of *Gleditsia* and related species may become very thick-walled and persist for many months.

Organic materials are lost from the roots into the soil by excretion.

TABLE III

NUMBERS, LENGTHS, AND SURFACE AREA OF ROOTS, AND SURFACE AREA OF ROOT HAIRS OF 7-WEEK-OLD *Robinia pseudoacacia* AND *Pinus taeda* SEEDLINGS[a,b]

Order	Root length (cm)	Root surface area (cm²)	No. of root hairs	Root hair surface area (cm²)
Robinia pseudoacacia				
Primary	16.20	3.4466	1,166	3.6346
Secondary	115.62	15.7167	8,321	25.2172
Tertiary	30.60	3.1151	2,081	5.1759
	62.42	22.2784	11,568	34.0277
Pinus taeda				
Primary	6.45	1.7341	215	0.7973
Secondary	5.93	0.9683	371	2.0770
	12.38	2.7024	586	2.8743

[a] Data represent averages for 12 seedlings of each species.
[b] From Kozlowski and Scholtes (1948).

The phenomenon of excretion is very well known and forms the basis of the rhizosphere effect, but it is imperfectly understood as it is difficult to study under natural conditions. Ivarson and Sowden (1969) sampled rhizosphere soil and nonrhizosphere soil near roots of different species of plants and found that the rhizosphere soil contained larger quantities of amino acids than nonrhizosphere soil, the most prominent amino acids being aspartic acid, threonine, serine, glutamic acid, glycine, and alanine; asparagine, glutamine, and citrulline were also present. These amino acids were also present in the roots but their proportions in the soil were different from those in roots and the authors concluded that the presence of amino acids in the soil was not a consequence of root breakdown. The composition of the mixture of amino acids in the soil varied with plant species.

The exudation of liquid from apple tree roots is an unusual phenomenon that has been studied in the East Malling root observation laboratories. Root hairs form on apple tree roots in the region a few millimeters behind the root tip soon after cell elongation has ceased. This process was studied by time-lapse cinematography in a place where an apple root crossed one of the many air spaces in the soil. On the film small liquid droplets appeared on the tips of all root hairs just when

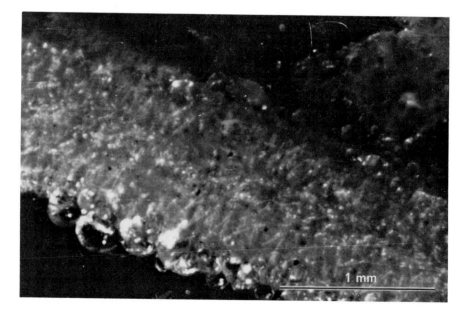

FIG. 12. Droplets of exudation on the surface of 3½-day-old apple root. Some of the individual droplets have coalesced to form large globules. (By courtesy of East Malling Research Station.)

they had grown to their full length. As the quantity of liquid increased, the individual droplets coalesced to form larger globules (Fig. 12) that were easily visible with the naked eye (Head, 1964). Some of the larger globules were then seen to shrink when the root was 4½ days old and it was suggested that some of the liquid might be reabsorbed by the root. After disappearance of the liquid the root hairs appeared to have lost their turgidity.

Exudation of liquid, as described above, occurs regularly on all apple tree roots of all the varieties so far observed at East Malling as they pass through the particular stage of development, and the droplets may be seen on white roots dug from the orchard (Fig. 13) or on roots of potted trees. In the root observation laboratory, the presence of exudation was used as one of the distinguishing features of apple tree roots as nothing similar was observed on any of the other fruit species studied. Richardson (1953a) observed similar droplets on the roots of *Acer pseudoplatanus*. This type of exudation can be observed only where a root crosses an air space in the soil and it is not known whether liquid is released by root hairs that are embedded in the soil.

C. Cortex and Bark

The shedding of the cortex of roots of those species that show secondary thickening of roots seems to be a natural process that occurs about the time of the formation of the endodermis. The outer tissues of apple tree roots, initially yellowish white, begin to turn brown under natural soil conditions after 1–3 weeks in the summer in England (Head, 1966a); in the cooler weather of spring, autumn, and winter, cortical browning is slower to commence.

Information is not available on the fate of the cortex of roots grown in sterile culture, but in natural soil the initial browning is patchy, indicating that soil microorganisms may play a part in the first changes. Patchy browning was also noted by Richardson (1953a) for roots of *Acer pseudoplatanus* and by Bhar *et al.* (1970) for plum (*Prunus salicina*) roots. On the other hand, Mason *et al.* (1970) found an orderly progression of cortical browning along the roots of gymnosperm species.

In the studies at East Malling, it was noticed that soon after the initial brown patches had joined up and the root had become uniformly brown, small patches of white woolly fungus were visible on the root surface. There was, however, no apparent relation between the location of the first brown patches and subsequent development of the fungus (Head, 1966b).

Initiation of cortical browning in some cases appeared to be asso-

FIG. 13. An apple root carefully removed from orchard soil, showing large globules of exudation. (By courtesy of East Malling Research Station.)

ciated with contact with the soil, as those regions of the root that crossed air spaces in the soil were often last to change color.

The process of cortical browning and decay has been studied in detail on apple roots by time-lapse cinematography at East Malling, the changes occurring on the surface of one root over a period of several weeks was filmed in color (Head, 1968c). On the second day of filming, a group of nematodes was seen to feed close to the root tip and, when the region of the root being filmed was 6 days old, a marked increase in activity of the soil microfauna was evident. At first several mites appeared, then various types of nematodelike worms, and on day 13 collembola appeared for the first time. After 14 days, the region of the root on which the eelworms had fed on day 2 turned a rich brown as did a similar small area at the other side of the field of view. No further browning occurred, however, until the root was about 30 days old when the first brown patches started extending over the root surface across the air space in the soil.

The brown areas were sharply delimited and the boundary extended in irregular jumps proceeding along the root about 2–3 mm within the hour between successive exposures, the new brown area gradually darkening during the following 3–4 hr (Head, 1968c). One or more days later, the brown patch extended rapidly again to include another adjacent area of cortical tissue.

Although the root was brown at either edge of the field of view, where it was in close contact with the soil, at the age of 2–3 weeks, it was more than 5 weeks before the tissue bridging the air space in the soil became completely brown.

Several different species of soil fauna were active close to the root surface throughout the period of filming, but activity increased greatly once the cortical tissue had turned completely brown. Nematodes and enchytraeid worms fed on the degenerating tissue, opening channels which allowed collembola and mites of various species to help the process of breakdown and dissemination of the cortical remains. As a result of continued feeding over a period of months, the rotted cortex was removed completely, leaving the healthy central vascular cylinder exposed.

The rate of cortical browning, i.e., the first stage of cortical degeneration, in apple roots studied through glass observation panels, remained fairly constant throughout the period from May to September in England (Head, 1966a) but the process was considerably slower in autumn, winter, and spring. Wilcox (1968b) found that the cortex of some of the dormant roots of *Pinus resinosa* degenerated in the winter from December to March, but that in many long roots degeneration was delayed by the presence of mycorrhizal fungi in the cortex.

Rogers (1968) attempted to estimate the quantity of cortical material deposited in the soil per unit length of apple root by separating the cortex from the central vascular cylinder and weighing the parts. The dry weight of the cortex that would eventually have been sloughed naturally was approximately equal to the weight of the central part that remained.

Leiser (1968) found that genera of Ericaceae having fine roots (*Calluna* and *Rhododendron*) had a root cortex that was only 2 cells thick, consisting of an outer parenchyma layer and an endodermis. Genera of *Arctostaphylos*, however, had thicker roots. Leiser also found that roots of plants with nutritional problems showed deviations from the structure of healthy roots. Among the signs of root problems were premature sloughing of large strips of epidermis and development of an epidermis with oval shaped loose-fitting cells.

V. Natural Death of Roots

A. Trees and Bushes

The death of many roots may be regarded as natural, although as the soil is an environment shared by so many other living organisms it is impossible to be certain that the shedding of roots, as distinct from death due to disease or insect attack, is not affected by the fungi, bacteria, or animals that live on and around root surfaces.

The root systems of all forest trees contain a fair proportion of dead roots. Copeland (1952) examined apparently healthy trees of loblolly pine (*Pinus taeda*) and shortleaf pine (*Pinus echinata*) of different ages and found that older trees had a larger proportion of dead roots. Under 15-year-old trees of loblolly and shortleaf pine, 3.7 and 2.9%, respectively, of the roots were dead; with older trees (35 years old) the proportions of dead roots were 6.3 and 18%, respectively. Copeland classified the roots into 4 diameter classes—<¼ inch, ¼–½ inch, ½–1 inch, and >1 inch —and found that in the younger trees most of the dead roots were in the smallest diameter class. On the older trees some of the thicker roots were dead, including some in the largest diameter class on shortleaf pine trees on poorly drained soils.

Although roots do not have a predetermined length of life as leaves do, Kolesnikov (1966), summarizing much of the work done on fruit tree root systems in the U.S.S.R. since the 1920's, has likened the regular shedding of roots to the annual leaf drop that occurs in deciduous trees. He made the point, often neglected by ecologists, that roots are the principal source of organic matter in the deeper layers of the soil and that the activity of microorganisms and soil animals is to a large extent regulated by root activity. Kolesnikov (1966) stated that the tip of the main root of most seedlings (fruit tree species) dies by the time the seedling is 2 months old. Subsequently the tips of roots of higher orders continue to die throughout the growing season. Apple, pear, and sour cherry (*Cerasus vulgaris*) seedlings up to 3 years old had 2.0–4.8% of their root tips dead; even seedlings in their first year might loose thousands of rootlets in the course of one growing season. Kolesnikov also reported that root shedding increased in older trees.

Referring to other work in U.S.S.R., Kolesnikov (1968) showed that the intensity of root shedding may vary in response to external conditions. Z. N. Voronchkhina carried out studies on the roots of gooseberry (*Grossularia reclinata*), which do not decay as quickly as those of tree fruit species, and demonstrated considerable variation in root shedding from year to year, from 25 to 72% of the roots being shed in different years. In a very dry year, irrigation reduced root shedding to some extent: irrigated bushes had 25% living and 75% dead roots 10 days after irrigation, while unirrigated bushes had 15% living and 85% dead roots.

Orlov (1955) investigated the quantity of dead roots in *Picea excelsa* forests by examining minutely the whole of large soil blocks with a pair of forceps, a needle, and a lens. In 25-year-old and 50-year-old forests, respectively, he found 4.0 and 1.2 tons/ha (dry weight) of living rootlets less than 3 mm in diameter, and he estimated that 50 and 20%, respectively, of the mass of small rootlets died off each year.

Remezov (1959) used similar methods in a 50-year-old oak forest and found a decrease in dry weight of rootlets from 0.8 tons/ha in July to 0.5 tons/ha in August. He considered that this confirmed Orlov's conclusion that up to 50% of the small rootlets may die each year.

Although roots do not have a fixed length of life, an average life span can be established after very painstaking and detailed studies. For fir trees and pine trees Orlov (1966) reported the average length of life of the roots to be 3.5 years and 4.0 years, respectively, and he used these figures to calculate the mass of organic matter entering the soil each year. He assumed that the average annual increase in the mass of the feeding roots would be equal to the quantity of feeding roots that died off. His calculations showed that 75–300 kg of roots died off each year per hectare in piceeta of the forest steppes and 350 kg/ha in pineta of the taiga zone. In dense young piceeta, the amount of root dying off each year reached 1100 kg/ha. In the taiga forests with well-drained soils the annual quantity of dead roots represented $\frac{1}{20}$ and $\frac{1}{10}$ of the total amount of litter from the above-ground parts of piceeta and pineta, respectively.

In all studies of root shedding, the criteria of root death are, of course, all important, and it may be easier to determine with certainty which roots are really dead if the roots are separated from the soil and can be handled than if they are viewed through glass observation panels. All roots undergo a color change as they mature, and after the cortex has separated many roots remain apparently unchanged for many months.

In apple the smallest rootlets lose their cortex much more slowly than the fatter extension roots and it is very difficult through glass to be sure when they are dead. Figure 14 illustrates this point with photographs of the same group of apple rootlets taken through the observation panels at East Malling over a period of almost 2 years.

In observation box studies, however, Childers and White (1942) recorded the fact that apple roots 1–1.5 mm in diameter died after 1–2 weeks even though conditions seemed favorable for growth. Death of the roots they studied was confirmed when the roots disintegrated a short time later.

There are undoubtedly considerable differences between species in longevity of roots. Kolesnikov (1968) reports that the roots of gooseberry (*Grossularia reclinata*) do not decay as quickly as those of tree fruits and Head (1966b) found that black currant (*Ribes nigrum*) roots remained white and apparently healthy for more than 12 months. Bosse (1960) found that the small absorbing roots of apple began to decay after 80–100 days, whereas many of the absorbing roots of *Picea excelsa* lived for 3–4 years.

Much of the annual loss of small absorbing rootlets appears to take place in the winter months. Bode (1959) found that 90% of the absorbing roots of *Juglans regia* were lost during the winter. In fact, the production of new leaves in the spring was delayed until the soil had warmed up sufficiently for new root growth to begin. In the tea plant (*Camellia thea*) Voronkov (1956) also found a considerable loss of active roots during the winter. The proportions of dead roots in the soil layer 30–50 cm deep were 4.3% in August, 4.2% in October, 5.3% in December, and 9.3% in February.

The patterns of growth of roots and the dying off of roots of different orders were studied by Wilcox (1968a) in *Pinus resinosa*. Earlier studies (Wilcox, 1962) had shown that not all the roots of any one tree were growing at the same time and that growth of individual roots tended to be cyclic. In *Pinus resinosa* the primary roots grew more rapidly and for longer periods than first-order branch roots, and these more rapidly than second-order branches. About two-thirds of the second-order branches

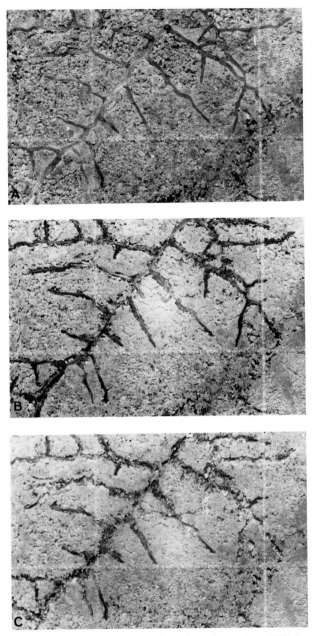

FIG. 14. (A) Much-branched system of small apple rootlets. (B) Same rootlets 12 months later. The cortex of all rootlets has turned dark brown but in most places has not rotted away. (C) Same area photographed 23 months after (A). The central root and many lateral rootlets have rotted away completely. (Engraved lines are approximately ½ inch apart.) [From Head (1970).]

became dormant after their first cycle of growth and did not resume growth. The failure of many of them to renew their growth activity was associated with a characteristically high degree of natural pruning. Many of the successful second-order laterals were those that had been initiated early in the season during the period of growth of the parent root, or alternatively those initiated near the end of the growing season. The late-formed laterals often became dormant shortly after emerging from the parent root but resumed their growth in the following spring. The laterals that were most likely to abort after one growth cycle were those initiated in the middle of the growing season. Wilcox (1968a) thought that the cause of the abortion of these second-order laterals might be related to an internal hormonal correlative mechanism.

Further information concerning the identification of those roots that survive and become part of the permanent woody root system was provided by Horsley and Wilson (1971). By examining the primary xylem diameter (PXD) of the initial seedling roots of *Betula papyrifera* and that of the subsequent branches and also the PXD's of roots 6–12 m long on forest trees, they established that to have a chance of becoming a permanent root the PXD of the new root tip must be at least 25% of the parent root PXD. All roots with PXD's smaller than this were ephemeral.

B. Grasses

Although this chapter concerns mainly the roots of trees and bushes, any discussion of root longevity should include some mention of the work of Weaver and his associates on the roots of prairie plants in Nebraska (Weaver, 1968).

Grasses possess 2 distinct root systems. The primary or seminal root system starts to develop with germination of the seed and consists of one or several main roots and their branches: the number varies with species and is often only 1 to 3. Weaver remarked that earlier workers had often mistakenly regarded the seminal roots as temporary roots.

In root box experiments Weaver found that the seminal roots were all alive and functioning at a time when the abundant tillers were one-half to two-thirds as tall as the parent culm. The seminal roots were usually very fine, being 0.3 mm or less in diameter and little or no decortication had occurred on most species after 21 days. Microscopical examination after 90–123 days showed that about half of the seminal roots appeared normal and Weaver considered that in natural field conditions they would probably remain alive and active for at least the first season of growth. Adventitious roots grow later from each new tiller.

To study the annual increases in roots and tops, Weaver grew plants of three grass species in drums 34 inches high and 22.5 inches in diameter. Very similar amounts of underground materials were produced in 3 consecutive years and the amounts were very similar to those found in typical mature stands of mature prairie. The data for annual production of tops and roots in the 3 years were as shown in the following tabulation.

	Dry weight (gm) of tops			Dry weight (gm) of roots		
	1943	1944	1945	1943	1944	1945
Andropogon gerardi	306	518	804	152	261	321
Andropogon scoparius	388	790	906	89	166	160
Bouteloua gracilis	283	394	441	76	118	92

Weaver calculated that the roots alone of the 3 species mentioned above yielded approximately 5.5, 2.7, and 1.6 tons/acre, respectively, after 3 years' growth.

To study longevity of the adventitious roots Weaver grew seedlings in special drums with removable sections at the top. These sections were removed and the soil was washed away from around the upper portions of the roots when it was desired to examine them. During the examination the roots were sprayed with water and eventually dry earth was replaced around them. Before the earth was replaced, however, small aluminum tags were placed around certain roots so that they could be identified at a later date.

Survival of the banded roots of *Bromus inermis* at 3 consecutive annual examinations was 92, 84, and 36%; for *Panicum virgatum* and *Agropyron smithii* the survival after 3 year's growth was 100 and 42%, respectively. On the plants of *Agropyron cristatum* 75% of the tagged roots survived the first year, and on *Andropogon gerardi* 81% survived for 3 years. On the less stable *Elymus canadensis*, however, none survived for 3 years.

Grass species were broadly divided by Stuckey (1941) into those that replaced their roots annually, such as perennial ryegrass (*Lolium perenne*), and those that possessed mainly perennial roots, such as cocksfoot (*Dactylis glomerata*). Williams (1969) considered that the variation among grass species in this respect was related to the ecological habitat to which they were adapted.

According to Jacques and Schwass (1956), 60% of the roots of perennial ryegrass and tall fescue (*Festuca arundinacea*) were replaced annually. Garwood (1967) found that longevity of perennial ryegrass roots was

TABLE IV

AVERAGE LENGTH OF ACTIVE LIFE OF ADVENTITIOUS ROOTS
OF PERENNIAL RYEGRASS, S.23[a]

Month of origin	Longevity (days)	Month of origin	Longevity (days)
January	111	July	68
February	98	August	70
March	95	September	168
April	64	October	188
May	61	November	154
June	79		

[a] From Garwood (1967).

shorter than others had found and that it varied with the time of year at which the roots were laid down. Roots initiated in September–November lived longest with an average length of life of 5–6 months. The maximum recorded for an individual root was 8 months. Adventitious roots initiated in April and May survived for the shortest period, many of them not living for more than a week or so. Garwood's results, shown in Table IV, indicate that all roots were replaced in less than 12 months.

C. MEASUREMENT OF ROOT LOSSES

Although the studies of Orlov (1966) and Remezov (1959) enable a rough estimate to be made of the quantity of root shed every year, more precise measurements must await development of new techniques. Preliminary investigations using radioactive tracers have been made on a laboratory scale by Racz et al. (1964). They developed a method of labeling completely with ^{32}P the root systems of cereal plants growing in containers. The amount of radioactivity remaining in the growth medium after the removal of all the living roots of the experimental plant indicates the amount of root material that had become separated as a result of natural death, shedding of root cortex, feeding of predators, exudation, etc. Radioactive phosphorus can only be used for short-term experiments, however, and a tracer with a longer half-life would be preferable. The main problem with this type of technique, however, is the difficulty of ensuring a uniform distribution of the tracer.

Another approach to the problem of estimating root losses has been by the use of $^{14}CO_2$ feeding (Bartholomew and McDonald, 1966; Shamoot et al., 1969). In these experiments plants of 11 annual and perennial herbaceous species were enclosed in gas-tight growth chambers and supplied with measured quantities of $^{14}CO_2$ in the atmosphere. At the end of these experiments, some of which continued for 212 days, the calculated quantities of organic material left in the soil after all the living roots had been removed, ranged from 20.2 to 49.4 gm organic debris per 100 gm of roots removed. The authors stated that the quantities of organic debris measured at the time of harvest underestimated the total because much of the root debris would have been extensively decomposed by that time and the ^{14}C would have been released as $^{14}CO_2$ into the atmosphere. Further refinement of this method would permit measurement of the CO_2 released by respiration in the soil but this would include the products of both root respiration and microbial respiration and it would then be desirable to determine these 2 components separately.

VI. Death of Roots Associated with the Nature of the Soil and Fertilization

The fact that the nature of the soil and type of fertilizer applied affect root growth and morphology was mentioned earlier (Section III,C) and it is clear that unfavorable conditions may lead to root death. In several cases various authors have referred to a reduction of root growth in unfavorable conditions, but for the purposes of this account it has been assumed that this is equivalent to root death. The absence of root growth on otherwise healthy trees at suitable temperatures must indicate that the root tips are being killed as fast as they are produced.

Lack of oxygen or accumulation of carbon dioxide in the soil is a frequent cause of root death and this state of affairs may be brought about either by the nature of the soil or by flooding. Childers and White (1942) experimented with roots of young apple trees grown in root observation boxes, submerging the roots in water in the spring. Submersion inhibited formation of new roots but did not cause an immediate reduction in the linear growth of the roots already present. After 18 days of submersion all the roots visible against glass panels were dead, but when the free water was drained away it became clear that the older woody roots had not been killed as new roots appeared at the glass close to the soil surface within 8 days. Such results are in agreement with those of Boynton and Compton (1943) who concluded that a higher level of oxygen may be necessary for production of new rootlets than for survival of existing roots.

The critical level of oxygen supply will certainly vary with species. With 3 varieties of avocado (*Persea gratissima*—varieties Scott, Duke, and Topa Topa) Valoras *et al.* (1964) found that root growth did not take place when the oxygen diffusion rate was less than 0.20 gm/cm²/min. This value was dependent on temperature.

Heinicke (1933) had previously submerged the roots of 1-year-old McIntosh apple trees grown in 3-gal galvanized buckets at different times of the year. He showed that roots could be submerged from late autumn, before the ground was frozen, to late spring without causing noticeable injury. However, if any leaves were present on the trees while the roots were submerged, the leaves and the flowers showed severe damage, especially on hot sunny days. After submersion, the roots were black and dying back from the extremities.

There are many references in the literature to death of roots as a result of fluctuations in the water table in orchards, particularly in the autumn and winter. Harris (1926), for instance, observed that a rise of the water table in the autumn was associated with cessation of apple root growth and eventual death of newly formed roots. Rogers and Vyvyan (1934) also ascribed the sharp cutoff in the vertical sinker roots of an apple tree they had excavated to a fluctuating water table.

Copeland (1952), investigating the root systems of shortleaf (*Pinus echinata*) and loblolly (*Pinus taeda*) pines in areas of South Carolina subject to littleleaf disease, found that on the healthy trees on 5 soil series 2.9–3.7% of the roots were dead. The highest percentages were found with both species on the poorly drained Cataula and Mecklenburg soil series. On these soils more roots of a thicker category were affected than on well-drained soils.

Variable top growth of Washington Navel orange and Balady mandarin trees in surface-irrigated groves in the United Arab Republic was related to different water table levels (Minessy *et al.*, 1971), and soil sampling showed that the lateral and vertical spread of the roots was greatly reduced on the trees in the areas with a high water table.

Roots may also be damaged if soils are too dry, particularly in the surface layers (Buchholz and Neumann, 1964) and Kramer (1950) found that suberization, or browning of the cortex, occurred sooner in dry soils than in wet ones. In parts of Quebec where white birch (*Betula papyrifera*) trees suffer from a form of dieback that has so far not been associated with any particular pathogen, Pomerleau and Lortie (1962) found that the most seriously affected trees were always on shallow soils. Both the quantity of healthy roots and the proportion of dead roots were related to soil depth (Table V).

Redmond (1955), investigating dieback of yellow birch (*Betula alleghaniensis*), found that rootlet mortality in a normal 55-year-old stand was 6% and that this increased to 19 and 60% if the soil temperature was raised by 1° and 2°C, respectively, in one summer.

TABLE V

ROOT DEPTH AND ROOTLET MORTALITY IN *Betula papyrifera* TREES
DIFFERING IN PERCENTAGE OF LIVING FOLIAGE[a]

Tree no.	Age (stump)	D.b.h. (inches)	% Living foliage	Mean depth of rooting (inches)	Average % of rootlet mortality
1	83	8	98 (1B)[b]	20	14
2	85	8	80 (2B)	13	18
3	86	8	15 (4A)	9	69
4	85	7.5	2 (4B)	7	97

[a] From Pomerleau and Lortie (1962).
[b] Numbers and letters within parentheses refer to crown injury class, after Hawboldt and Skolko (1948).

A high rate of fertilization with ammonium nitrate was found to reduce the concentration of feeder roots in orange groves in Florida (Ford *et al.*, 1957). Rates of application had been gradually increased over a period of 6 years before root sampling was undertaken, up to a maximum rate of 3.5 lb N per tree. Trees that had received the highest level of N had a weight of roots (<2 mm diameter) 37% less than the trees on the low N plots. The reduction in root concentration extended down to 5 ft (Table VI). There was no visible evidence that the feeder roots from any of the plots were stubby, misshapen, or off-color. Similar reductions in the amount of root were found on plots where the same total annual quantity of nitrogenous fertilizer had been applied in 3 split doses. Thus it appears that the effect on the roots was not a direct effect of high salt concentration.

VII. Death of Roots as a Result of Factors Affecting the Aerial Parts of the Plant

As it has been established that there are internal correlative factors controlling root growth (see Section III,A) in terms of both hormonal materials and available carbohydrates, it is clear that disturbance of the natural growth pattern of the aerial parts will have a strong influence on root growth. Apart from inhibition of root growth in early summer as a result of the stimulation of vigorous shoot growth by branch pruning (Head, 1967), the studies in the root observation laboratory at East Malling on apple trees also showed that fruiting resulted in a considerable reduction in root development in late summer and autumn, and that

TABLE VI
ROOT DISTRIBUTION OF VALENCIA ORANGE TREES[a,b,c]

| | Total feeder roots in 5-foot depth zone[d] | | |
	1953	1954	1955[e]
Nitrogen rate			
1.8 lb N/tree	15.5	19.9	23.3
3.6 lb N/tree	15.3	14.2	19.4
Significance	N.S.	f	g
L.S.D. at 0.05	—	3.3	3.5
Timing of N application			
3 equal applications	15.3	17.6	20.1
All N in fall	16.5	19.3	18.6
All N in spring	14.4	14.9	25.3
Significance	N.S.	f	g
L.S.D. at 0.05	—	4.1	4.3
Ratios (N:K:Mg)			
4:3:6	15.7	16.7	21.2
8:3:6	15.0	17.4	21.4
Significance	N.S.	N.S.	N.S.
L.S.D. at 0.05	—	—	—

[a] Data from a 3-year study at Montverde, Florida.
[b] From Ford et al. (1957).
[c] N.S., not significant; L.S.D., least significant difference.
[d] Expressed as grams dry weight in a column one foot square and 5 feet deep. Timing is expressed as the mean of 24 trees. Rate and ratio are expressed as the mean of 36 trees.
[e] Roots were sampled with a 2-inch diameter tube in contrast to an 8-inch diameter auger used in 1954. Direct weight comparisons between the 2 years are not entirely valid.
[f] Statistical significance at 1% level.
[g] Statistical significance at 5% level.

defoliation 6 weeks before the date of natural leaf fall rapidly reduced the amount of white root present in the autumn (Head, 1969a).

Removal of several large branches from citrus trees, or skeletonization, in March resulted 5 months later in a considerable decrease in the quantity of feeder roots found at all soil levels down to 5 ft in comparison with untreated trees (Biely et al., 1958). Heavy cropping caused severe damage to the root system of coffee (Coffea) trees (Nutman, 1933), and Chandler (1923) found a 50% decrease in the root system of Prunus after a heavy crop.

The effects of defoliation by the spruce budworm (Choristoneura fumiferana Clem) on the root system of balsam fir trees (Abies balsamea L. Mill) were studied by Redmond (1959). From a preliminary examination of about 60,000 rootlets from trees that had not been defoliated

TABLE VII

MORTALITY OF ROOTLETS OF *Abies balsamea* AS RELATED TO AMOUNTS OF DEFOLIATION BY THE SPRUCE BUDWORM[a]

Plot number	History of defoliation (% current year's foliage removed)[b]							Size of trees (inches d.b.h.)	Mortality of rootlets[c] Normal Medium Heavy Com-plete (No. of trees)			
	1950	1951	1952	1953	1954	1955	1956		Normal	Medium	Heavy	Complete
1	40	25	15	15	30	15	20	7 to 12	9	1	—	—
2 (June/56)	5	5	5	5	30	65	90	3 to 6	4	4	2	—
2 (Sept./56)	5	5	5	5	30	65	90	3 to 6	—	—	10	—
3	10	60	100	100	95	10	0	4 to 6	—	3	6	—
4	0	5	5	5	50	100	80	3 to 8	—	—	10	—
5	5	5	5	50	70	100	80	4 to 6	—	1	18	—
6 (June/56)	20	45	90	98	98	35	90	4 to 6	—	—	9	—
6 (Aug./56)	20	45	90	98	98	35	90	4 to 6	—	9	—	—
7	65	100	100	100	100	80	40	8 to 12	—	—	3	10
8	10	80	100	95[d]	35	75	80[d]	3 to 4	—	19	—	—
9	30	80	100	95[d]	30	75	80[d]	6 to 12	—	1	17	—

[a] From Redmond (1959).

[b] Information on defoliation was supplied by D. G. Mott, Forest Biology Laboratory, Fredericton, New Brunswick, Canada.

[c] Normal, 0 to 15%; medium, 31 to 50%; heavy, 76 to 99%; and complete, 100%.

[d] Aerial spraying with DDT was carried out during the spring or early summer.

it was established that not more than 15% of the roots were dead. On the affected trees the degree of defoliation on different trees was classified as: light, <20%; medium, 30–60%; and heavy, >70%. Rootlet mortality was classified as: normal, <15%; medium, 31–50%; and heavy, >75%. The data showed that trees with only light defoliation had <15% root mortality, even if the defoliation had been repeated annually for several years. On trees with defoliation approaching 100%, root mortality in June varied between trees, but by late September all trees in this category had suffered heavy rootlet mortality (Table VII).

When defoliation ceased, root initiation commenced on young trees but these trees succumbed to a second defoliation. Mature trees seemed incapable of producing new rootlets following total defoliation of new shoots over 4 or 5 successive years. Coyne (1968) simulated attacks by the pine tip moth (*Rhyaciona frustrana*) and showed reduced root growth after 2 years. No rootlet mortality was observed, but it was thought that severe attacks would almost certainly have this effect.

Visser (1969) tested different pruning techniques to bring tea (*Camellia thea*) plants into bearing and found that the bushes always tended to retain the same root:shoot ratio. If the shoots were hard pruned some roots died, presumably because of a reduction in carbohydrate reserves. Death of roots was less severe in plantations at higher altitudes, possibly because carbohydrate reserves are known to be higher in tea bushes grown in the cooler temperatures found at higher altitudes (Tubbs, 1935, 1937). Parker and Houston (1971) analyzed the roots of sugar maple (*Acer saccharum*) trees that had been defoliated in June or July and found the level of "root extractives"[1] (as a percentage of the dry weight) in roots harvested 2–6 weeks later to be lower than in roots of nondefoliated trees. Although after defoliation in July for 3 years in succession, the level of extractives in the roots was not significantly lower than in roots of nondefoliated trees, a definite downward trend was associated with the number of defoliations.

VIII. Damage to Roots Associated with Soil Animals

Damage to roots resulting from feeding of soil animals is not so well documented as that caused by fungi. De Fluiter (1941) studied the feeding habits of species of white grubs and decided that they could be divided into 3 categories: (1) grubs that fed only on dead organic matter, (2) grubs that normally fed on dead organic matter but which would feed on living roots in its absence (Rutelineae and Dynastinae), and

[1] Roots were extracted in 80% ethanol by the method of Priestley (1965, 1969).

(3) grubs feeding by preference on living roots of crop plants (Melolonthinae). If this type of classification were applied to soil animals in general, it would be found that few species live for preference on living roots.

In Java, De Fluiter (1941) found that larvae of *Lepidiota, Holotrichia,* and *Psilopholis* fed on roots of rubber (*Hevea* sp.) trees, and Nirula *et al.* (1952) reported that *Leucopholis* larvae injured the roots of coconut palm (*Cocos nucifera*) in India. In Japan and Israel, larvae of *Anomala albopisola* and *Pentodon* damaged roots of pine windbreaks and apple trees, respectively (Nitto and Tachibana, 1955; Plant, 1953). Ritcher (1958) reviewed the biology of the Scarabaeidae and mentioned that in some parts of the world numerous members of the Rutelinae, such as the Japanese beetle, *Popillia japonica* Newman, and the Chinese rose beetle, *Adoretus sinicus* Burmeister, feed on living roots by preference.

In Europe, Ritcher (1958) recorded that larvae of Melolonthinae could cause extensive damage to the roots of small fruit plants, shrubs, and trees.

Periodical cicadas have also been observed to cause considerable damage to apple trees in the United States (Banta, 1960; Hamilton, 1961; Hamilton and Cleveland, 1964), and it has been reported that nymphs of this species suck the juices from the roots of forest trees and may be responsible for considerable loss of primary production (Lloyd and Dybas, 1966), particularly in the years immediately preceding emergence (Edwards *et al.*, 1970). Dybas and Davis (1962) estimated the live weights of mature periodical cicada nymphs in Illinois at between 230 and 428 kg/ha in an upland forest stand and between 1913 and 3655 kg/ha in a floodplain area.

Root-boring insects have also been found to cause considerable damage to tree roots. Warren (1956) found that the spruce root borer (*Hypomolyx piceus*) caused considerable damage to the main roots or root collars of white spruce (*Picea glauca*) and noted that damage had also been reported on pines and tamarack (*Larix laricina*). Adults of this species feed on the bark of small roots and twigs but the most serious damage is done by the larvae which bore into the bark and along the cambium of roots. They may be found in roots more than 1 inch in diameter or in root collars, but they seem to prefer root crotches. Small trees may be completely girdled or one or more of the main roots may be girdled on larger trees. Damage was worst on trees in damp situations, either on wet soils or on dry soils where fallen leaves covered living tree roots. Whitney (1961) also described and illustrated damage caused by *Hylobius* sp. (*Hypomolyx* sp.). Figure 15A shows tunnels made by larvae to the depth of the cambium or slightly into the wood. Figure 15B–D shows that roots smaller than 1 inch in diameter are commonly girdled

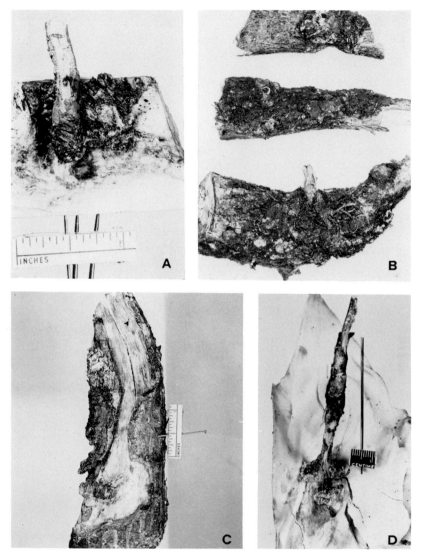

FIG. 15. (A) One-quarter inch root girdled at the base by *Hylobius* several years earlier and now infected. (B) *Hylobius* damage on roots 3–4 inches in diameter; the rough appearance is due to an accumulation of soil and debris soaked with resin and larval tunnels. (C) *Hylobius* wound in spruce root, from which the resin-soaked debris has been removed. (D) Root girdled by *Hylobius* several years earlier, and now infected by *Armillaria mellea* which has spread into the main root from which the bark has been removed. [From Whitney (1961).]

or killed. Whitney reported that roots less than ⅛ inch in diameter sometimes were chewed off completely.

Farrar and Kerr (1968) found that the broad-necked root borer
(*Prionus laticollis,* Drury) was injurious to the roots of many trees including apple in Rhode Island. He quoted others who had found that this
species damaged apple, grape, linden, oak, pine, poplar, aspen, black-
berry, cherry, chestnut, dogwood, maple, and rhododendron.

The part played by soil animals in the breakdown and dissemination
of the root cortex was described earlier (Section IV,C). In the root
observation laboratory at East Malling, heavy infestations of root aphids
were found on black currant roots; no estimate of the damage to the
bush was possible, but Dixon (1970) discussed the energy intake and
honeydew or wax production of root aphids.

Considerable damage is done to roots by both ecto- and endoparasitic
nematodes. The ectoparasitic forms feed on the root hairs and cortex
generally near the root tip. Species such as *Paratylenchus* and *Tylen-
chorhynchus* penetrate the epidermal cells with their stylet and leave
little evidence of their feeding. However, heavy infestations of *Paratylenchus hamatus* in figs in California caused the infested roots to become enlarged and spongy and suppressed further growth (Thorne and
Allen, 1950).

Migratory ectoparasitic nematodes of the species *Trichodorus* also
appear to leave little evidence of their feeding but cause a marked change
in tissue differentiation that leads to loss of meristematic activity. The
feeding of *Trichodorus christei* on cranberry (*Vaccinium* sp.) roots was
described by Zuckerman (1961) as resulting in aborted lateral roots and,
frequently, swollen tips. Pitcher (1967) studied the massing of *Trichodorus viruliferus* on the roots of Worcester Pearmain apple (Fig. 16)
by using time-lapse cinematography and described typical scarring of
the root surface and stunting of root growth.

Another group of migratory ectoparasites causes death of epidermal
and cortical cells. This group includes the genera *Xiphinema* and *Cri-
conemoides.* The damage is probably not due to mechanical injury but
to certain enzymes secreted into the host. Galling of the root tips of *Rosa*
spp. was found to be associated with *Xiphinema diversicaudatum* infestation (Schindler, 1957; Davis and Jenkins, 1960).

Surface lesions are also caused by other nematodes such as the migratory endoparasites *Pratylenchus* and *Radopholus.* They penetrate the
host by puncturing and entering the epidermal cells and eventually they
colonize the cortex. Once again the damage is thought to be largely due
to secretion of enzymes by the nematodes. *Pratylenchus penetrans* is one
of the most pathogenic species causing serious damage to roots of many

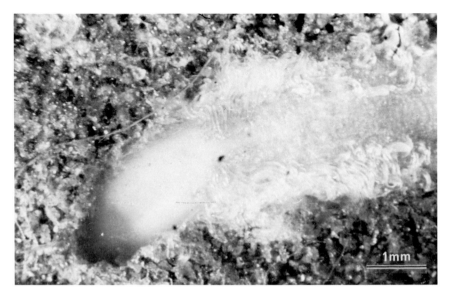

FIG. 16. *Trichodorus viruliferus* massed on an extending apple root. [From Pitcher (1967).]

host species. Mountain and Patrick (1959) showed that this species, free of bacteria and fungi, produced extensive lesions on peach roots and sometimes killed them. The damaged area was ahead of the part colonized by the nematode thus indicating that toxic products were involved.

Radopholus spp. are endoparasites that cause extensive damage to the root cortex. In banana (*Musa* sp.), *Radopholus similis* caused extensive necrosis of both roots and rhizomes and gave the entire root system a darkened and unhealthy appearance (Blake, 1966). The same species causes one of the most important and dramatic plant diseases, spreading decline of citrus (Suit and DuCharme, 1953). In decline-affected orchards, feeder roots did not exist below 30 inches, but large secondary roots were present. Small roots were produced during periods of active growth, but they were quickly destroyed and affected trees had only one-half as many roots as healthy trees.

IX. Death of Roots Caused by Soil Fungi

Severe deterioration of roots has been consistently found in trees of shortleaf pine, *Pinus echinata* Mill., and loblolly pine, *Pinus taeda* L., affected by littleleaf disease. Siggers and Doak (1940) were the first to note that the roots were defective, the finer ones often being dead. Similar findings were reported by Jackson (1945) who gave more detail of the root lesions. He found that there was excessive exfoliation of bark and dieback of fine roots; extensive brown patches on the root surface were a feature of the roots of diseased trees. Large pitchy cankerlike lesions were also seen.

The brown patches were distributed on roots of all sizes on diseased trees, whereas on healthy trees they were prominent only on the larger roots; roots less than 1 inch in diameter usually had only small widely scattered patches. Roots of diseased shortleaf pines had nearly 6 times more brown patch than the healthy shortleaf trees within the littleleaf disease area and 32 times more than shortleaf pines growing outside the

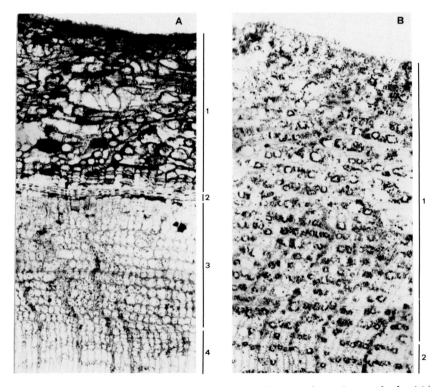

FIG. 17. Photomicrographs of shortleaf pine (*Pinus echinata*) root bark. (A) Transverse section showing (1) brown patch, (2) periderm layer, (3) secondary phloem, and (4) current secondary phloem layer. (B) Transverse section of healthy root bark showing accummulation of starch grains in (1) old secondary phloem and in (2) current secondary phloem layer. Starch grains stained black with iodine. [From Jackson (1945).]

littleleaf area. Dead or dying primary or lateral roots were 4 times more frequent on diseased than on healthy trees.

Histological examination showed that the brown patches consisted of older layers of secondary phloem (Fig. 17) that had abscised because of the development of periderm layers. The size and arrangement of the elements of the brown patches did not differ from that of those of the underlying live phloem, indicating that no pathological changes had occurred during their formation. Jackson found that dark hyphae closely resembling those of *Torula* spp. were abundant throughout the brown patches, and that inoculations of this fungus, or of *Poria cocos*, would induce formation of brown patches within 6–8 months.

Campbell (1951) made a study of the distribution of *Phytophthora cinnamoni* in different soil layers in areas with littleleaf diseased trees, and subsequently Copeland (1952) found close agreement between his data for root mortality and those of Campbell for distribution of *Phytophthora* (Fig. 18).

Copeland studied root mortality on shortleaf and loblolly pines on 5 different soil series. On the older shortleaf trees (35 years old) suffering

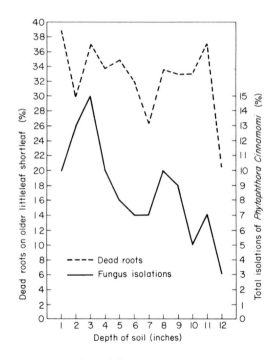

FIG. 18. Comparison of *Phytophthora cinnamomi* isolations with the percentage of dead roots on older littleleaf shortleaf trees. [From Copeland (1952).]

from littleleaf he found an average of 34.4% dead roots in comparison with 18% on healthy trees of the same age, and he considered that the critical proportion of dead roots lay somewhere between 18 and 34%. When a tree's root system deteriorated beyond this stage, he considered that nutrient absorption would be decreased to such an extent that the littleleaf symptoms would appear. Copeland (1952) found an increase in the proportion of dead roots on healthy shortleaf trees from 3.0% at the age of 15 years to 18% at 35 years and thought that this might explain why littleleaf disease usually occurred in trees over 20 years of age.

Copeland classified by diameter the dead roots found on healthy and littleleaf trees (Table VIII). Most of the dead roots that occurred on the healthy young and old loblolly pine trees were of the smallest size class, less than ¼ inch in diameter. On the young healthy shortleaf pine trees mortality was also highest in the smallest class of roots and in older healthy shortleaf pine trees the highest mortality was found in roots less than ½ inch in diameter. On the older littleleaf shortleaf pine trees mortality occurred to a great extent in roots of all size classes.

Another major root disease for which root mortality figures are available is the root rotting disease of white spruce (*Picea glauca*), sometimes called stand-opening disease, that is largely caused by *Polyporus tomentosus*. Whitney (1961) showed that 53.5% of the 1170 diseased roots examined had, in fact, been previously damaged. The types of damage that led to rotting were complete or partial girdling wounds caused by the weevil larvae, in particular by *Hylobius* sp. (*Hypomolyx* sp.), damage caused by excess moisture, compression wounds, damage caused by animal trampling, root cankers, and other unknown causes. The relation between the severity of *Hylobius* wounding and root rot in 630 living and 26 dead trees is shown in Table IX. Whitney (1962) also analyzed the advanced root decay that he found associated with the stand-opening disease (Fig. 19). Trees in early stages of the disease had lateral roots ¼–¾ inch in diameter infected with *Polyporus tomentosus* and 1–10% of them were dead. No discoloration was found in roots less than ⅛ inch in diameter. In the early middle stage of the disease up to 40% of the root wood was decayed but at this stage there were no above-ground symptoms. In late stages of the disease, less than 20% of the roots were living, and trees with this amount of root damage had many dead branches and reduced height increment. Whitney reported that *P. tomentosus* infected all regions of the root, the heartwood being most severely attacked. In small roots the fungus appears to invade wood and bark at equal rates. Although dead rootlets 1–3 mm in diameter were found, the primary invasion seemed to be in roots ¼–½ inch in diameter. No fungus could be cultured from roots smaller than 2.8 mm in diameter.

TABLE VIII

PERCENTAGE MORTALITY OF ROOTS OF DIFFERENT DIAMETERS ON TREES OF *Pinus echinata* AND *P. taeda*[a]

	Size classes							
	Less than $\frac{1}{4}$-inch diameter		$\frac{1}{4}$ to $\frac{1}{2}$-inch diameter		$\frac{1}{2}$ to 1-inch diameter		1 inch and larger in diameter	
Soils	Total no.	Dead (%)	Total no.	Dead (%)	Total no.	Dead (%)	Total no.	Dead (%)
Young healthy *Pinus taeda*								
Cecil	199	3	17	0	6	0	3	0
Davidson	151	5	14	0	2	0	0	0
Cataula	183	3	10	0	6	0	1	0
Vance	130	2	10	0	4	0	0	0
Mecklenburg	217	7	15	0	11	0	1	0
Young healthy *Pinus echinata*								
Cecil	184	3	16	0	4	0	0	0
Davidson	176	1	13	0	1	0	0	0
Cataula	177	5	11	0	4	0	0	0
Vance	151	2	11	9	5	0	0	0
Mecklenburg	164	5	7	0	1	0	1	0

Older healthy *Pinus taeda*							
Cecil	282	9	10	0	14	0	6	0
Davidson	214	8	16	0	13	0	2	0
Cataula	210	4	18	11	7	0	4	0
Vance	280	8	24	0	7	0	6	0
Mecklenburg	288	6	25	0	15	0	5	0
Older healthy *Pinus echinata*							
Cecil	489	17	23	4	6	0	7	0
Davidson	374	14	26	12	6	0	4	0
Cataula	265	24	18	11	8	13	4	0
Vance	392	19	25	20	16	19	5	20
Mecklenburg	563	21	21	5	8	0	5	20

[a] From Copeland (1952).

TABLE IX

NUMBER OF TREES CLASSIFIED[a] BY SEVERITY OF *Hylobius* WOUNDING
AND SEVERITY OF ROOT ROT[b]

Severity of *Hylobius* wounding[c]	Number of trees						
	Severity of root rot[d]						
	None	Trace	Light	Moderate	Heavy	Severe	Total
None	102	80	44	17	6	1	250
Trace	20	31	8	3	2	2	66
Light	11	41	31	13	8	9	113
Moderate	2	7	36	27	16	13	101
Heavy	0	3	10	17	28	25	83
Severe	0	0	1	8	11	23	43
	135	162	130	85	71	73	656

[a] Number includes 630 living trees and 26 dead trees.

[b] From Whitney (1961).

[c] The severity of *Hylobius* wounding is assessed on the basis of the number of roots girdled or damaged, and the proportion of the root collar damaged, according to the technique of Warren (1956).

[d] The severity of root rot was assessed by visually estimating the percentage of root and root crown wood decayed in the dissected roots and butts. The ratings were as follows: none, 0% decayed; trace, 1–20% decayed; light, 21–40% decayed; moderate, 41–60% decayed; heavy, 61–80% decayed; severe, 81–100% decayed.

Whitney found that dead rootlets for which the cause of death could not be determined occurred commonly on both healthy and diseased trees. The frequency of these was determined on the living parts of the root systems of 6 diseased trees and on 3 adjacent healthy trees (Table X). The average proportion of dead rootlets was higher on the diseased trees, and Whitney thought that some weakly pathogenic fungi might have been responsible for death of the rootlets.

An interesting interaction between rootlet mortality as a result of fungus attack and fertilization was found by Rowan (1971). Seedlings of *Pinus elliottii* grown at 35°C in soil heavily infested with *Macrophomina phaseolina* and *Fusarium oxysporum* showed more severe root rot when fertilized with ammonium nitrate at the rate of 112.1 kg/ha.

Root growth is severely hindered on some kinds of trees planted on sites where the same species has been grown previously. This effect, of which the cause is not known, has been extensively reviewed by Savory (1966).

Apple, cherry, and peach are particularly affected by specific replant diseases, as the troubles are called, and it has been observed in the East

FIG. 19. (A) Rotted roots of an 80-year-old diseased white spruce (*Picea glauca*) in a stand opening. (B) White pockets, or advanced stage of decay caused by *P. tomentosus* in root wood. (C) Early stage of stand-opening disease showing the outer portions of roots (stained reddish brown) infected with *P. tomentosus*. (D) Root and butt of a diseased tree (early middle stage) with bark removed. The entire center main root and part of the root collar (dark area) were infected with *P. tomentosus*. (E) Tree in the late middle stage of disease. Decay caused by *P. tomentosus* extended from the infected roots to the butt. (F) Decayed roots of trees in the late stage of disease. Only the root on the extreme right was alive at the time of excavation; all roots were infected with *P. tomentosus*. [From Whitney (1962).]

TABLE X

PERCENTAGES OF ROOTLETS DEAD ON HEALTHY *Picea glauca* TREES AND
ON LIVING, DISEASED ROOT BRANCHES FROM DISEASED TREES[a]

	Tree no.	No. of rootlets examined	Dead (%)
Healthy trees	1	2325	32.8
	2	1229	25.2
	173	130	26.2
Average			30.1
Diseased trees	43	227	45.4
	44	123	80.4
	62	58	38.9
	67	65	31.8
	93	74	42.9
	171	259	39.6
Average			47.8

[a] From Whitney (1962).

Malling root observation laboratories that extension growth and branching of apple and cherry roots are greatly reduced in soils in which apple and cherry, respectively, have been grown previously. The root cortex was found to turn brown more quickly in replant soil.

So far, these problems, the basis of which undoubtedly lies in the soil, have been recognized only where grubbing and replanting have occurred but it is possible that the same effect occurs within the root system of an established tree. In the East Malling root observation laboratories, the root growth of apple trees in the first years after the trees were planted in fresh soil was vigorous and branching was extensive. In subsequent years, however, not only was the quantity of new growth less, but the new roots looked less healthy. It is clear that the roots under established trees grow in a somewhat hostile biological environment and at East Malling, once the glass panels had become fully exploited by roots, few of the new roots seemed to develop their full growth potential.

References

Banta, E. S. (1960). Apple orchard decline. *Proc. Ohio State Hort. Soc., 13th Annu. Meet.* pp. 80–90.

Barney, C. W. (1951). Effects of soil temperature and light intensity on root growth of loblolly pine seedlings. *Plant Physiol.* **26,** 146–163.

Bartholomew, W. V., and McDonald, I. (1966). Measurement of the organic material deposited in soil during the growth of some crop plants. *In* "The Use of Isotopes in Soil Organic Matter Studies," FAO/IAEA Tech. Meet., Brunswick-Völkenrode, 1963, pp. 235–242. Pergamon, Oxford.

Bhar, D. S., Mason, G. F., and Hilton, R. J. (1970). *In situ* observations on plum root growth. *J. Amer. Soc. Hort. Sci.* **95**, 237–239.

Biely, M. I., Wallace, A., and Kimball, M. H. (1958). Some factors influencing feeder-root weights and distribution in citrus. *Univ. Calif., Los Angeles, Spec. Rep.* No. 1, pp. 16–29.

Blake, C. D. (1966). The histological changes in banana roots caused by *Radopholus similis* and *Helicotylenchus multicinctus. Nematologica* **12**, 129–137.

Bode, H. R. (1959). The relationship between leaf development and the formation of new absorbing roots in Juglans. *Ber. Deut. Bot. Ges.* **72**, 93–98.

Bodo, F. (1926). Untersuchungen auf dem Gebiet des Wurzelwachstums des Apfels und der Zwetschge. *Fortschr. Landw.* **24**, 768–773.

Bosse, G. (1960). The root development of apple clones and apple seedlings during the first three years. *Erwerobstbau.* **2**, 26–30.

Boynton, D. (1940). Soil atmosphere and the production of new rootlets by apple tree root systems. *Proc. Amer. Soc. Hort. Sci.* **37**, 19–26.

Boynton, D., and Compton, O. C. (1943). Effect of oxygen pressure in aerated nutrient solution on production of new roots and on growth of roots and tops by fruit trees. *Proc. Amer. Soc. Hort. Sci.* **42**, 53–58.

Bray, J. R. (1963). Root production and the estimation of net productivity. *Can. J. Bot.* **41**, 65–72.

Buchholz, F., and Neumann, E. (1964). The influence of cultivation and fertilization on the soil moisture of, and root penetration into, sandy woodland soils in Brandenburg. *Albrecht-Thaer-Arch.* **8**, 525–536.

Busarova, E. I. (1961). Vliyanie nekotorykh faktorov gredy na rost eli i sosny v zabolochennykh tipakh lesa [Effect of certain environmental factors on the growth of spruce and pine in swampy forests] *Les. Khoz.* **5**, 11-14. *Ref. Zh. Biol.,* 1961, No. 17633.

Campbell, W. A. (1951). Relative abundance of *Phytophthora cinnamoni* in the root zones of healthy and littleleaf-diseased shortleaf pine. *Phytopathology* **39**, 752–753.

Cannon, W. A., and Free, E. E. (1925). Physiological features of roots, with special reference to the relationship of roots to aeration of the soil. *Carnegie Inst. Wash. Publ.* **368**, 1–168.

Chandler, W. H. (1923). Results of some experiments in pruning fruit trees. *N. Y., Agri. Exp. Sta., Ithaca, Bull.* **415**, 1–75.

Childers, N. F., and White, D. G. (1942). Influence of submersion of the roots on transpiration, apparent photosynthesis, and respiration of young apple trees. *Plant Physiol.* **17**, 603–618.

Cockerham, G. (1930). Some observations on cambial activity and seasonal starch content in sycamore (*Acer pseudoplatanus*). *Proc. Leeds Phil. Lit. Soc., Sci. Sect.* **2**, Part 2, 64–80.

Copeland, O. L., Jr. (1952). Root mortality of shortleaf and loblolly pine in relation to soils and littleleaf disease. *J. Forest.* **50**, 21–25.

Coyne, J. F. (1968). Simulated Nantucket pine tip moth attacks reduce root development of shortleaf pine seedlings. *J. Econ. Entomol.* **61**, 319–320.

Davis, R. A., and Jenkins, W. R. (1960). Nematodes associated with roses and the root injury caused by *Meloidogyne hapla* Chitwood, 1949, *Xiphinema diversi-*

caudatum (Micoletzky, 1927) Thorne, 1939, and *Helicotylenchus nannus* Steiner, 1945. *Md. Agri. Exp. Sta., Bull.* A-106, 1–16.

De Fluiter, H. J. (1941). Mededeeling van het Besoekisch Proefstation: Waarneminger Omtrent Engerlingen (Oerets) en hun Bestrijding in Hevea-Aanplantingen. *Arch. Rubbercult. Ned.-Indie* 25, 167–227.

Dittmer, H. J. (1949). Root hair variations in plant species. *Amer. J. Bot.* 36, 152–155.

Dixon, A. F. G. (1970). Aphids as root-fluid feeders. *In* "Methods of Study in Soil Ecology," pp. 243–247. UNESCO, Paris.

Dybas, H. S., and Davis, D. D. (1962). A population census of seventeen-year periodical cicadas (Homoptera:Cicadidae:Magicicada). *Ecology* 43, 432–443.

Edwards, C. A., Reichle, D. E., and Crossley, D. A., Jr. (1970). The role of soil invertebrates in turnover of organic matter and nutrients. *In* "Ecological Studies. Analysis and Synthesis" (D. E. Reichle, ed.), Vol. 1, pp. 147–172. Springer-Verlag, Berlin and New York.

Engler, A. (1903). Untersuchungen über das Wurzelwachstum der Holzarten. Mitteil. d. schweizer. *Zentralanstalt forstliche Versuchswesen* 7, 247–317.

Eremeev, G. N. (1960). The growth of the absorbing roots of fruit trees in relation to soil conditions. *Dokl. Akad. Nauk SSSR* 130, 678–681.

Esau, K. (1965). "Plant Anatomy." Wiley, New York.

Farrar, R. J., and Kerr, T. W. (1968). A preliminary study of the life history of the broad-necked root borer on Rhode Island. *J. Econ. Entomol.* 61, 563–564.

Ford, H. W., Reuther, W., and Smith, P. F. (1957). Effect of nitrogen on root development of Valencia orange trees. *Proc. Amer. Soc. Hort. Sci.* 70, 237–244.

Garwood, E. A. (1967). Seasonal variation in appearance and growth of grass roots. *J. Brit. Grassland Soc.* 22, 121.

Girton, R. E. (1927). The growth of citrus seedlings as influenced by environmental factors. *Univ. Calif. Publ. Agric. Sci.* 5, 83–117.

Haberlandt, G. (1914). "Physiological Plant Anatomy" (translated from 4th German ed. by M. Drummond). Macmillan, New York.

Hamilton, D. W. (1961). Periodical cicadas, *Magicicada* spp., as pests in apple orchards. *Proc. Indiana Acad. Sci.* 71, 116–121.

Hamilton, D. W., and Cleveland, M. L. (1964). Periodical cicadas in 1963, brood 23. *Proc. Indiana Acad. Sci.* 72, 167–170.

Harris, G. H. (1926). The activity of apple and filbert roots especially during the winter months. *Proc. Amer. Soc. Hort. Sci.* 23, 414–422.

Hawboldt, L. S., and Skolko, A. J. (1948). Investigation of yellow birch dieback in Nova Scotia in 1947. *J. Forest* 46, 659–671.

Head, G. C. (1964). A study of "exudation" from the root hairs of apple roots by time-lapse cine-photomicrography. *Ann. Bot. (London)* [N.S.] 28, 495–498.

Head, G. C. (1966a). Estimating seasonal changes in the quantity of white unsuberized root on fruit trees. *J. Hort. Sci.* 41, 197–206.

Head, G. C. (1966b). Studies on growth and development of roots of fruit plants in relation to environment, season and growth of the aerial parts. Ph. D. Thesis, University of London.

Head, G. C. (1967). Effects of seasonal changes in shoot growth on the amount of unsuberized root on apple and plum trees. *J. Hort. Sci.* 42, 169–180.

Head, G. C. (1968a). Seasonal changes in the amounts of white unsuberized root on pear trees on quince rootstock. *J. Hort. Sci.* 43, 49–58.

Head, G. C. (1968b). Seasonal changes in the diameter of secondarily thickened roots of fruit trees in relation to growth of other parts of the tree. *J. Hort. Sci.* **43**, 275–282.

Head, G. C. (1968c). Studies of growing roots by time-lapse cinematography. *Trans. Int. Congr. Soil Sci., 9th, 1968* pp. 751–758.

Head, G. C. (1969a). The effects of fruiting and defoliation on seasonal trends in new root production on apple trees. *J. Hort. Sci.* **44**, 175–181.

Head, G. C. (1969b). The effects of mineral fertilizer on seasonal changes in the amount of white root on apple trees in grass. *J. Hort. Sci.* **44**, 183–187.

Head, G. C. (1970). Methods for the study of production in root systems. *In* "Methods of Study in Soil Ecology," pp. 151–157. UNESCO, Paris.

Heinicke, A. J. (1933). The effects of submerging the roots of apple trees at different seasons of the year. *Proc. Amer. Soc. Hort. Sci.* **29**, 205–207.

Hellmers, H. (1963). Effects of soil and air temperatures on growth of redwood seedlings. *Bot. Gaz. (Chicago)* **124**, 172–177.

Hilton, R. J., Bhar, D. S., and Mason, G. F. (1969). A rhizotron for *in situ* root growth studies. *Can. J. Plant Sci.* **49**, 101–104.

Hiroyasu, T. (1961). Nutritional and physiological studies on the grapevine. I. Growth studies of the underground and aboveground parts of the vines. *J. Jap. Soc. Hort. Sci.* **30**, 77–81 (in Japanese).

Horsley, S. B., and Wilson, B. F. (1971). Development of the woody portion of the root system of *Betula papyrifera*. *Amer. J. Bot.* **58**, 141–147.

Ivarson, K. C., and Sowden, F. J. (1969). Free amino acid composition of the plant root environment under field conditions. *Can. J. Soil Sci.* **49**, 121–127.

Jackson, L. W. R. (1945). Root defects and fungi associated with littleleaf disease in southern pines. *Phytopathology* **35**, 91–105.

Jacques, W. A., and Schwass, R. H. (1956). Root development of some common New Zealand pasture plants. 7. Seasonal root replacement in perennial ryegrass, Italian ryegrass, and tall fescue. *N. Z. J. Sci. Technol., Sect. A* **37**, 569.

Jenny, H., and Grossenbacher, K. (1962). Root-soil boundary zones as seen by the electron microscope. *Calif. Agr.* **16**, 7.

Kaufman, C. M. (1945). Root growth of jack pine on several sites in the Cloquet Forest, Minnesota. *Ecology* **26**, 10–23.

Kolesnikov, V. A. (1959). The growth of axial and absorbing roots of top and small fruit plants throughout their life cycle. *Izv. Timiryazev. Sel'skokhoz. Akad.* No. 1, pp. 127–148.

Kolesnikov, V. A. (1966). "Fruit Biology." Mir Publishers, Moscow.

Kolesnikov, V. A. (1968). Dynamics of the growth of root systems in fruit plants. *Indian J. Hort.* **25**, 37–40.

Kozlowski, T. T. (1971). "Growth and Development of Trees," Vol. 2. Academic Press, New York.

Kozlowski, T. T., and Scholtes, W. H. (1948). Growth of roots and root hairs of pine and hardwood seedlings in the Piedmont. *J. Forest.* **46**, 750–754.

Kramer, P. J. (1950). Effects of wilting on the subsequent intake of water by plants. *Amer. J. Bot.* **37**, 280–284.

Kramer, P. J., and Kozlowski, T. T. (1960). "Physiology of Trees." McGraw-Hill, New York.

Ladefoged, K. (1939). Investigations on the periodicity of branching and growth in length of the roots of some of our most common woodland trees. *Forstl. Forsogsv. Danm.* **16**, 1–256.

Leiser, A. T. (1968). A mucilaginous root sheath in Ericaceae. *Amer. J. Bot.* **55,** 391–398.

Leyton, L. (1952). The effect of pH and form of nitrogen on the growth of Sitka spruce seedlings. *Forestry* **25,** 32–40.

Leyton, L., and Rousseau, Z. (1958). Root growth of tree seedlings in relation to aeration. *In* "The Physiology of Forest Trees" (K. V. Thimann, ed.), pp. 467–475. Ronald Press, New York.

Lindemann, A. (1956). Rooting factors in cuttings. *Deut. Baumsch.* **8,** 12–16.

Lloyd, M., and Dybas, H. S. (1966). The periodical cicada problem. I. Population ecology. *Evolution* **20,** 133–149.

Lyr, H., and Hoffmann, G. (1967). Growth rates and growth periodicity of tree roots. *Int. Rev. Forest. Res.* **2,** 181–236.

McDougall, W. B. (1921). Thick-walled root hairs of *Gleditsia* and related genera. *Amer. J. Bot.* **8,** 171–176.

Marloth, R. H. (1950). Citrus growth studies. I. Periodicity of root growth and top growth in nursery seedlings and budlings. *J. Hort. Sci.* **25,** 50–59.

Mason, G. F., Bhar, D. S., and Hilton, R. J. (1970). Root growth studies on Mugho pine. *Can. J. Bot.* **48,** 43–47.

Minessy, F. A., Barakat, M. A., and El-Azab, E. M. (1971). Effect of some soil properties on root and top growth and mineral content of Washington Navel orange and Balady mandarin. *Plant Soil* **34,** 1–15.

Monselise, S. P. (1947). The growth of citrus roots and shoots under different cultural conditions. *Palestine J. Bot., Rehovot. Ser.* **6,** 43–54.

Mountain, W. B., and Patrick, Z. A. (1959). The peach replant problem in Ontario. VII. The pathogenicity of *Pratylenchus penetrans* (Cobb, 1917) Filip. & Stek. 1941. *Can. J. Bot.* **37,** 459–470.

Muromcev, I. A. (1962). Soil temperature and growth of roots. *Fiziol. Rast.* **9,** 419–424.

Nightingale, G. I. (1935). Effects of temperature on root growth, anatomy and metabolism of apple and peach roots. *Bot. Gaz.* (*Chicago*) **96,** 581–639.

Nirula, K. K., Anthony, J., and Menon, K. P. V. (1952). Investigations on the pests of the coconut palm—the rhinoceros beetle (*Oryctes rhinocerus* L.); life history and habits. *Indian Coconut J.* **5,** 57–70.

Nitto, M., and Tachibana, K. (1955). Decision on the number of the instar and the generation on the basis of measurements on the larval head of the May beetle. *J. Jap. Forest. Soc.* **37,** 326–333 (in Japanese).

Nutman, F. J. (1933). The root system of *Coffea arabica*. II. The effect of some soil conditions in modifying the "normal" root system. *Emp. J. Exp. Agr.* **1,** 285–296.

Orlov, A. J. (1955). The role of feeding roots of forest vegetation in enriching soils with organic matter. *Pochvovodenie* No. 6, pp. 14–20.

Orlov, A. Ja. (1957). Observations on the absorbing roots of *P. abies* in natural conditions. *Bot. Zh.* (*Leningrad*) **42,** 1172–1181.

Orlov, A. Ja. (1966). Significance of dying feeding tree roots on the organic cycle of the forest. *Zh. Obshch. Biol.* **27,** 40–46.

Otto, G. (1964). A coefficient for determining the branching density of the fibrous roots of woody plants in relation to problems of the rhizosphere. *Tagunsber., Deut. Akad. Landwirtschafts-wiss. Berlin* **65,** 259–264.

Parker, J., and Houston, D. R. (1971). Effects of repeated defoliation on root and root collar extractives of sugar maple trees. *Forest Sci.* **17,** 91–95.

Pitcher, R. S. (1967). The host–parasite relations and ecology of *Trichodorus viru-*

liferus on apple roots, as observed from an underground laboratory. *Nematologica* 13, 547–557.

Plant, N. (1953). A new white grub on apple in Israel. *FAO Plant Prot. Bull.* 1, 171.

Pomerleau, R., and Lortie, M. (1962). Relationships of dieback to the rooting depth of white birch. *Forest Sci.* 8, 219–224.

Priestley, C. A. (1965). A new method for the estimation of the resources of apple trees. *J. Sci. Food Agr.* 16, 717–721.

Priestley, C. A. (1969). Some aspects of the physiology of apple rootstock varieties under reduced illumination. *Ann. Bot. (London)* [N.S.] 33, 293–300.

Proebsting, E. (1943). Root distribution of some deciduous trees in a California orchard. *Proc. Amer. Soc. hort. Sci.* 43, 1–4.

Racz, G. J., Rennie, D. A., and Hutcheon, W. L. (1964). The P32 injection method for studying the root system of wheat. *Can. J. Soil Sci.* 44, 100–108.

Redmond, D. R. (1955). Rootlets, mycorrhiza, and soil temperatures in relation to birch dieback. *Can. J. Bot.* 33, 595–627.

Redmond, D. R. (1959). Mortality of rootlets in balsam fir defoliated by the spruce budworm. *Forest Sci.* 5, 64–69.

Reed, H. S., and MacDougal, D. T. (1937). Periodicity in the growth of the orange tree. *Growth* 1, 371–373.

Remezov, N. P. (1959). Methods of study of the biological turnover of elements in the forest. *Pochvovodenie* No. 1, pp. 71–79.

Richardson, S. D. (1953a). Root growth of *Acer pseudoplatanus* L. in relation to grass cover and nitrogen deficiency. *Meded. Landbouwhogesch., Wageningen* 53, 75–97.

Richardson, S. D. (1953b). Studies on root growth in *Acer saccharinum* L. I. The relation between root growth and photosynthesis. *Proc., Kon. Ned. Akad. Wetensch., Ser. C* 56, 185–193.

Richardson, S. D. (1953c). Studies on root growth in *Acer saccharinum* L. II. Factors affecting root growth when photosynthesis is curtailed. *Proc., Kon. Ned. Akad. Wetensch., Ser. C* 56, 346–353.

Richardson, S. D. (1957). Studies on root growth in *Acer saccharinum* L. VI. Further effects of the shoot system on root growth. *Proc., Kon. Ned. Akad. Wetensch., Ser. C* 60, 624–629.

Richardson, S. D. (1958). The effect of IAA on root development of *Acer saccharinum* L. *Physiol. Plant.* 11, 698–709.

Ritcher, P. O. (1958). Biology of Scarabaeidae. *Annu. Rev. Entomol.* 3, 311–334.

Rogers, W. S. (1939a). Root studies. VIII. Apple root growth in relation to rootstock, soil, seasonal, and climatic factors. *J. Pomol. Hort. Sci.* 17, 99–130.

Rogers, W. S. (1939b). Root studies. IX. The effect of light on growing apple roots: A trial with root observation boxes. *J. Pomol. Hort. Sci.* 17, 131–140.

Rogers, W. S. (1968). Amount of cortical and epidermal tissue shed from roots of apple. *J. Hort. Sci.* 43, 527–528.

Rogers, W. S., and Head, G. C. (1963). A new underground root-observation laboratory. *Annu. Rep., East Malling Res. Sta., Kent for 1962*, 55–57.

Rogers, W. S., and Head, G. C. (1969). Factors affecting the distribution and growth of roots of perennial woody species. *Proc. Easter Sch. Agr. Sci., Univ. Nottingham* 15, 280–295.

Rogers, W. S., and Vyvyan, M. C. (1934). Root studies. V. Rootstock and soil effect on apple root systems. *J. Pomol.* 12, 110.

Rowan, S. J. (1971). Soil fertilization, fumigation, and temperature affect severity of black root rot of slash pine. *Phytopathology* 61, 184–187.

Sakanishi, Y., and Hayashi, K. (1959). Studies on the year's growth of tops and roots of roses. *Bull. Univ. Osaka Prefect., Ser. B* 9, 49–59 (in Japanese).

Samtsevich, S. A. (1965). Active excretions of plant roots and their significance. *Fiziol. Rast.* 12, 837–846.

Savory, B. M. (1966). Specific replant diseases. *Commonw. Bur. Hort. Plant. Crops (Gt. Brit.), Res. Rev.* 1.

Schindler, A. F. (1957). Parasitism and pathogenicity of *Xiphinema diversicaudatum*, an ectoparasitic nematode. *Nematologica* 2, 25–31.

Seeliger, I. (1959). Über die Bildung wurzelbürtiger Sprosse und das Wachstum isolierter Wurzeln der Robinie (*Robinia pseudoacacia* L.). *Flora, Jena,* 148, 218–254.

Shamoot, S., McDonald, I., and Bartholomew, W. V. (1969). Rhizo-deposition of organic debris in soil. *Soil Sci. Soc. Amer., Proc.* 32, 817–820.

Siggers, P. V., and Doak, K. D. (1940). The littleleaf disease of shortleaf pine. *S. Forest Exp. Sta., Occas. Pap.* 95, 1–5.

Slankis, V. (1949). Einfluss der Temperatur auf das Wachstum der isolierten Wurzeln von *Pinus silvestris*. *Physiol. Plant., Copenhagen* 2, 131–137.

Smith, P. F. (1957). Studies on the growth of citrus seedlings with different forms of nitrogen in solution cultures. *Plant Physiol.* 32, 11–15.

Smith, R. L., and Wallace, A. (1954). Preliminary studies on some physiological root characteristics in citrus and avocado. *Proc. Amer. Soc. Hort. Sci.* 63, 143–145.

Street, H. E. (1969). Factors influencing the initiation and activity of meristems of roots. *Proc. Easter Sch. Agr. Sci., Univ. Nottingham* 15, 20–41.

Stuckey, I. H. (1941). The seasonal growth of grass roots. *Amer. J. Bot.* 28, 481.

Suit, R. F., and DuCharme, E. P. (1953). The burrowing nematode and other parasitic nematodes in relation to spreading decline of citrus. *Plant Dis. Rep.* 37, 379–383.

Thorne, G., and Allen, M. W. (1950). *Paratylenchus hamatus* n.sp. and *Xiphinema index* n.sp., two nematodes associated with fig roots, with a note on *Paratylenchus anceps* Cobb. *Proc. Helminthol. Soc. Wash.* 17, 27–35.

Tubbs, F. R. (1935). Report of the Plant Physiologist for 1934. *Bull. Tea Res. Inst., Ceylon* 12, 32–39.

Tubbs, F. R. (1937). Investigations on the planting, pruning and plucking of the tea bush. *Bull. Tea Res. Inst. Ceylon* 15, 1–59.

Valoras, N., Letey, J., Stolzy, L. H., and Frolich, E. F. (1964). The oxygen requirements for root growth of three avocado varieties. *Proc. Amer. Soc. Hort. Sci.* 85, 172–178.

Visser, T. (1969). The effect of root and shoot damage on the growth of tea plants. *Neth. J. Agr. Sci.* 17, 234–240.

Vorobieva, T. G. (1961). The dynamics of growth of birch roots under various ecological conditions. *Obsc. Ispyt. Prir., Bjll. Otd. Biol., Moscow* 66, 89–96. (Russian.)

Voronkov, V. V. (1956). The dying-off of the feeder root system in the tea plant. *Dokl. Vses. Akad. Sel'skokhoz. Nauk* 21, 22–24.

Warren, G. L. (1956). The effect of some site factors on the abundance of *Hypomolyx piceus* (Coleoptera:Curculionidae). *Ecology* 37, 132–139.

Waynick, D. D., and Walker, S. J. (1930). Rooting habits of citrus trees. *Calif. Citrogr.* 15, 201 and 238–239.

Weaver, J. E. (1968). "Prairie Plants and Their Environment." Univ. of Nebraska Press, Lincoln.

Whitney, R. D. (1961). Root wounds and associated root rots of white spruce. *Forest. Chron.* **37**, 401–411.

Whitney, R. D. (1962). Studies in forest pathology. XXIV. *Polyporus tomentosus* Fr. as a major factor in stand-opening disease of white spruce. *Can. J. Bot.* **40**, 1631–1658.

Wight, W. (1933). Radial growth of the xylem and the starch reserves of *Pinus sylvestris:* a preliminary survey. *New Phytol.* **22**, 77–95.

Wilcox, H. (1962). Growth studies of the root of incense cedar, *Libocedrus decurrens.* II. Morphological features of the root system and growth behavior. *Amer. J. Bot.* **44**, 237–245.

Wilcox, H. (1968a). Morphological studies of the root of red pine, *Pinus resinosa.* I. Growth characteristics and patterns of branching. *Amer. J. Bot.* **55**, 247–254.

Wilcox, H. (1968b). Morphological studies of the roots of red pine, *Pinus resinosa.* II. Fungal colonization of roots and the development of mycorrhizae. *Amer. J. Bot.* **55**, 686–700.

Williams, T. E. (1969). Root activity of perennial grass swards. *Proc. Easter Sch. Sci., Univ. Nottingham* **15**, 270–279.

Wilson, B. F. (1967). Root growth around barriers. *Bot. Gaz. (Chicago)* **128**, 79–82.

Woodroof, J. G., and Woodroof, N. C. (1934). Pecan root growth and development. *J. Agr. Res.* **49**, 511–530.

Zuckerman, B. M. (1961). Parasitism and pathogenesis of the cultivated cranberry by some nematodes. *Nematologica* **6**, 135–143.

. 8 .

Shedding of Pollen and Seeds

Robert G. Stanley and E. G. Kirby

I. Introduction

Patterns of pollen and seed shedding are among the principal factors influencing the distribution and survival of plants. Variations in the gene pool facilitating selective evolution are maintained in part by the diversity of dispersal patterns of pollen and seeds. Insects, birds, and mammals are often directly involved in the distribution of pollen and seeds of many species. Such animal vectors may dislodge and carry pollen or seeds to receptive tissues or growing sites as by-products of their normal feeding patterns. In other cases, pollen and seeds have evolved truly beautiful and intricate adaptations for wind dispersal.

This chapter focuses primarily on pollen shedding mechanisms and distribution patterns. Factors relating the shedding of pollen to seed development are also discussed. Processes and patterns of seed shedding are reviewed with due consideration to the extensive publications on this subject by Fahn and Werker (1972) and van der Pijl (1969).

II. Pollen

A. GENERAL DEVELOPMENT

1. Different Patterns

Gymnosperm pollen is formed on microsporophylls, which are spirally joined together forming the strobilus, a portion of which is shown in Fig. 1A. Following meiosis of the microspore mother cells, young pollen grains arise as spore tetrads. During development pollen cells are nourished by substrates derived from the tapetal layer. The resulting pollen grains are enclosed in a two-layered wall, the outer wall often being expanded to form two air sacs which influence the rate of fall of these wind-dispersed pollens. Mature gymnosperm pollen contains the tube or vegetative cell, a generative cell, and vestiges of two nonfunctional prothallial cells (Fig. 1B).

While the general patterns of differentiation leading to mature pollen are somewhat similar in gymnosperms and angiosperms, the pollen-bearing structures differ.

Angiosperm microspores develop in anthers, usually a pair of double-lobed pollen sacs, termed microsporangia. The anther lobes are joined together by vegetative connective tissue containing a vascular bundle, and the anthers are supported by a filament which attaches them to the flower receptacle tissue. The anther and filament combined comprise the stamen (Fig. 2).

Three different patterns of angiosperm pollen development are commonly recognized based on whether or not microspore enlargement precedes nuclear division (Maheshwari, 1950). Microspores, with their haploid chromosomes, go through one to three mitotic divisions. Pollen, with its contained vegetative and generative cells, then rests until it is shed.

Mature pollens are classified by the exine patterns of their exterior walls. These are genetically controlled and often highly symmetrical. The study and classification of pollen, called palynology (Wodehouse, 1935; Erdtman, 1969), requires an understanding of factors influencing pollen development and dispersal.

2. Influencing Factors

External environmental conditions and the internal nutritional status of the plant may influence the genetic control of pollen development.

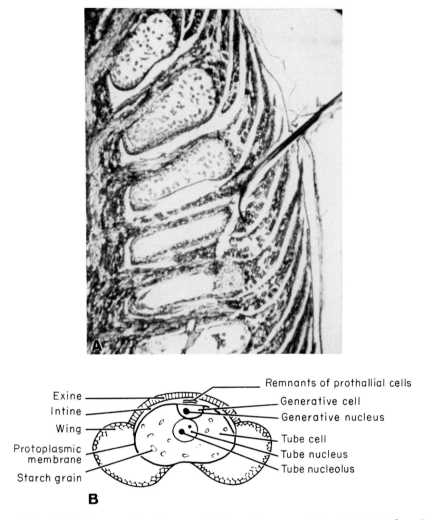

Exine
Intine
Wing
Protoplasmic membrane
Starch grain

Remnants of prothallial cells
Generative cell
Generative nucleus
Tube cell
Tube nucleus
Tube nucleolus

B

FIG. 1. (A) Portion of male cone of *Pinus*. (Courtesy of Dr. H. F. Linskens.) (B) *Pinus* pollen grain.

Such factors determine the viability of mature pollen. If the pollen aborts, the male structures do not mature but abscise without discharging their pollen.

Temperature and photoperiod are probably the most critical environmental factors influencing pollen development. Goss (1971) has shown that pollen production decreases when leaves of *Ornithogalum caudatum* are shaded. He has suggested that pollen may be induced to differentiate

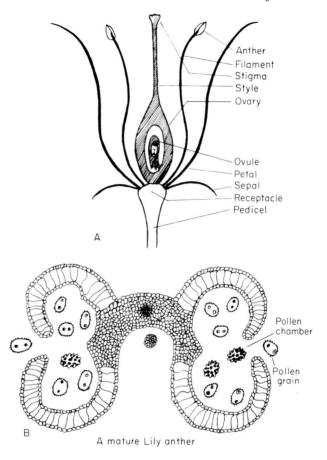

A

B

A mature Lily anther

FIG. 2. (A) Angiosperm flower. (Redrawn from Porter and Livingston, 1956.) (B) *Lilium* anther sac, transverse section. (Redrawn from Russell, 1958.)

in response to a chemical produced in the leaves, rather than in the flowers. However, shading may cause reduced or modified levels of available carbohydrates which could result in a decrease in the amount of pollen produced.

Low light intensities or short day lengths during microspore development may result in sugar pools insufficient to support new cell synthesis. In *Beta vulgaris* the disruption of sugar-flavonoid metabolism results in the production of anthers containing an abundance of red pigments. Such atypical pigmentation may be indicative of a metabolic condition resulting in the production of sterile pollen.

Either low or high temperatures at meiosis and during wall formation can block further pollen development. Eriksson (1968) has attributed chromosome aberrations produced during microsporogenesis to temperature disturbances. Partial or total sterility will occur if low temperatures prevail during the critical stages of reduction division and wall differentiation. Temperature ranges can be established for most species outside of which pollen development is blocked. Poggendorff (1932) determined that 21°–39°C is the critical temperature range in rice. However, in some rice strains night temperatures above 15°C and day temperatures above 19°C may decrease production of sterile pollen. Kiyosawa (1962) related these observations to starch and sugar metabolism in anthers.

In breeding programs it is often desirable to hasten pollen development. Factors which decrease the time required for flowering, hasten maturity, and increase the abundance of flowers include practices such as grafting, fertilizing, ringing, stem constricting, and root pruning (Stanley, 1958). Such practices can also affect the quantity of pollen produced. Once a plant is producing pollen, simple methods can often be used to accelerate maturation and anthesis.

Hastening pollen development during a growing season is commonly accomplished by removing branches bearing microsporangia from the plant and placing the cut ends in water in a warm room. Normally developed microspores from angiosperms, such as *Pyrus* and *Alnus*, can be obtained as much as 1 month prior to pollen shed in nature (Barner and Christensen, 1958). Table I lists examples of early maturation and pollen dehiscence produced under greenhouse conditions with controlled light and temperature. The time at which the flowering branches are excised is not as critical in angiosperms as in gymnosperms. Development of microspores in gymnosperms such as pine is usually inhibited if branches are removed from the tree before reduction division has begun (Chira, 1967).

TABLE I

ACCELERATED POLLEN DEHISCENCE[a,b]

Genus	Weeks detached before natural dehiscence
Alnus	10
Fraxinus	3
Abies	3
Cedrus	4
Chamacyparus	3
Picea	8
Pinus	4

[a] 600 W daily for 20–22 hr at 27°C.
[b] From Worsley (1959).

Some attempts have been made to facilitate maturation by adding nutrients including purines, pyrimidines, nucleic acids, auxins, and growth factors to the water–buffer solution into which the branches bearing developing microspores are placed (Vasil, 1959). The demonstration of positive effects of nutrients and growth factors in enhancing pollen development *in vitro* have been primarily limited to angiosperms. Taylor (1950) correlated requirements for specific substrates in excised *Tradescantia* flowers with the stage of nuclear division of the microsporocytes at the time the flowers are removed from the plant. He thus provided a method for studying nutrient requirements of developing pollen. In gymnosperms, Konar (1963) found that occasionally viable pollen will develop *in vitro*, even though strobili do not reach normal size.

Conversely, the same technique as described above with excised anthers can be used to determine the inhibiting action of external factors on pollen development. In this way, the influences of salts, radiation, and purine analogs (e.g., 5-fluorodeoxyuridine) have been assessed (Rushton, 1969). Experiments aimed at chemical emasculations have shown that of many chemicals tested, sodium-2,2-dichloropropionate applied at 250 ppm most effectively blocked morphological changes leading to anther dehiscence (Hirose, 1969).

B. SHEDDING

1. *Dehiscence Mechanisms*

The Greek word dehiscence means to open wide, as a mouth yawns; it describes the opening of the cavity in which microspores mature.

In gymnosperms, at pollen maturity, microsporophylls retract at their distal ends and separate. The opening of the microsporangial chamber walls is induced by dehydration of the bract scales. Pollen grains are shed mechanically by the shaking of strobili in the wind.

Among angiosperms, dehiscence occurs by one of several mechanisms involving disruption of the wall of the anther sac. In the mature anther the locule wall generally consists of two layers of cells, the epidermis and an underlying endothecial layer. The endothecium, also called the fibrous layer, usually develops specific bandlike patterns of lignified thickenings on their inner walls as the anthers mature. These genetically controlled wall thickenings have been used for taxonomic classification in the Compositae (Dormer, 1962). The orientation of these bands is related to the function of the fibrous layer. The cells shrink hygroscopically, splitting the cell walls; if drying continues, the edges may retract, opening the anther locule. The edges may curve back so far as to turn the pollen sacs inside out. Under conditions of alternating high and low humidity, the anther sacs may open and close. In anthers which do not dehisce, the endothecial cell walls are not modified; the thickening bands are absent.

In some species part or all of the epidermis of the anther sac may disintegrate leaving the fibrous layer exposed. Shedding of pollen from dehiscent anthers may occur through one of several different types of openings resulting from rupture of the fibrous layer. Longitudinal and poricidal dehiscence are the two most common forms, with the poricidal form being less frequently reported. Drawings of Kerner (1904) clearly illustrate different opening mechanisms. A few are reproduced to clarify the common anther dehiscence patterns (Fig. 3). Some anthers are curved in a spiral (e.g., *Erythraea*) or U shape (e.g., *Bixa*). Venkatesh (1956) has suggested that the U curvature with a slit opening in the middle, represents a dehiscent form intermediate between the pore and longitudinal slit. Anthers in the same flower may dehisce by different mechanisms depending on their location on the flower receptacle. In *Cassia*, the outer whorl of anthers dehisces by incomplete slits, whereas those of the inner whorl dehisce by pores. This suggests an evolutionary transition between these two anther opening mechanisms.

Longitudinal dehiscence varies with the arrangement of sporangia. In some stamens with sunken sporangia, each locule has its own endothecial cap. When anther sacs are joined laterally in pairs, the opening slit may develop between the two sporangia permitting pollen grains from both sacs to mix on dehiscence. When anthers are four-cornered, dehiscence occurs separately in each microsporangium. All pollen may be released at once or the anther sacs may only partially empty depending on moisture and wind conditions at dehiscence.

Richter (1929) compared dehiscence in 38 species of angiosperms and concluded that anthers of entomophilous flowers generally dehisce by a longitudinal contraction exposing pollen to visiting insects. Pollen in

FIG. 3. Stamen forms with dehiscence mechanism indicated (from Kerner, 1904). (A) *Calandrinia compressa* (longitudinal slit). (B) *Calla palustris* (distal slit), early stage. (C) *Calla palustris* (distal slit), late stage. (D) *Garcina* sp. (circumscissile).

such anthers is often associated with adhesive mucilage or viscin, thread-like residues of the tapetum. The transition to anemophily has been cor-related with a reduction in longitudinal dihescence patterns and domi-nance of poricidal openings. Anthers opening by pores undergo little or no change in shape at dehiscence and the pollen surface is usually dry. Richter suggested that enzymes may function to break the endothecial cell walls in *Cryptocoryne griffithii*, in which there is no well-defined fibrous cell layer. Becquerel (1932) concluded that anther dehiscence in many species, including *Lilium*, resulted from enzymatic breakdown of connecting wall of the anther sacs. In many cases, dehiscence probably results from a combination of physical forces preceded by enzyme-induced changes.

Some anthers which do not dehisce or release pollen in response to air currents have evolved symbiotic relationships with animals. Insects usu-ally transmit pollen from anthers which have already dehisced. As they strive to reach the nectar, which may be located in perianth modifications or on the receptacle, insects brush against dehisced anthers whose surfaces are covered with pollen. Beetles, wasps and moths, or birds and bats may be involved in transfer of pollen. Some angiosperm flowers have truly nondehiscent anthers and depend on biological pollinating vectors to transfer pollen, or in some cases, the entire anther or its modification to the female tissue (Faegri and van der Pijl, 1966). Anthesis does not occur in all angiosperm flowers; cleistogamous flowers remain closed to most external pollinating agents.

With the exception of the cycads, all gymnosperms are wind pollinated.

Patterns of pollination and behavior of pollinating vectors have been described as "pollination ecology" by Faegri and van der Pijl (1966). The volume of Meeuse (1961) colorfully describes many of the pollen conveying mechanisms. Anther pores of some tropical species are actually squeezed open by insects. In *Centaurea* the insect irritates the filament causing a contraction and anther opening, resulting in pollen exposure. The female yucca moth, *Pronuba yuccasella*, pollinates the yucca plant by collecting a ball of pollen in the maxillary tentacles and depositing it on the stigma after an act of oviposition. While this is hardly an active form of pollen shedding, it is an interesting example of symbiotic evolution. Another rather interesting adaptive pattern involving wind currents occurs in water plants such as *Vallisneria* and *Zostera*. Male flowers, when released, float to the surface and move by currents on the water surface until striking a female flower at which time pollen is catapulted onto the stigma. In other water plants, e.g., *Ceratophyllum*, pollen discharge and pollination take place under water via a balloonlike attachment which brings the released pollen into contact with the stigma.

Hagerup (1950) has observed a rain-triggered pollen flotation mechanism in *Ranunculus* and *Narthecium* flowers on the Faroe Islands. As rain accumulates in these cuplike flowers, pollen floats on the surface of the water until it reaches the level of the stigma, at which time pollination takes place.

The ability of many flowers to close in overcast or inclement weather, or in darkness, conveniently restricts pollen exposure to periods when pollination has the greatest probability of success. The number of cleistogamous flowers in wheat florets generally increases with low temperatures, rain, dry heat, or drought. Such conditions decrease pollen dehiscence and seed yields. Highly bred varieties of Durum wheat, however, express their flower opening pattern as a stable characteristic, independent of prevailing meteorological factors (De Vries, 1971). In wheat varieties, the percentage of florets open during anthesis varies from a low of 20% to a high of 100% (male sterile types).

2. Sampling Methods

Determining quantity, shedding time, and dispersal patterns of pollen is important for ecologists, plant breeders, fruit orchard growers, allergists, palynologists, and farmers. Ecologists are interested in the range of viable pollen flight as a contributing factor to interspecific variation; plant breeders are concerned with the distance necessary to isolate seed production plots from contaminating pollens. Orchard managers must consider factors such as probable pollen mix and the number of male trees or pollinator varieties required to pollinate dioecious female trees of compatible varieties. Allergists evaluate air-borne pollens capable of inducing allergies in humans; palynologists analyze the distributions and selective survival patterns of pollens in ground and water sources.

Each particular interest in pollen dispersal patterns requires a specific sampling method. While standard assay methods have not been established, certain techniques which afford reproducible data have been perfected and are often recommended for workers in specific areas. Often commercial survey instruments are available; however, sampling techniques generally suffer inherent potential errors. For a plant breeder, the number of viable pollen grains arriving at a specific site in a certain period of time is a meaningful measurement. From an allergist's viewpoint, the density of allergenic pollen in a given air volume may be used as an index of the potential influence of such pollens on sensitive human subjects. Therefore, two main types of assays are used. The deposition method, the most common type, involves counting the number of pollen grains deposited per unit area. The other, the volumetric method, assesses the density of pollen grains per unit of air volume (Koski, 1967). Both methods involve counting pollen microscopically and extrapolation from a small sample to a larger total environment. Data from either method can be interconverted.

Methods of varying degrees of complexity are available utilizing both of these concepts. The simplest deposition assay exposes a glass slide coated with Vaseline, silicone grease, or another sticky substance to the air. Rain or heavy dew may wash pollen off the sticky collecting surfaces, but can be avoided by using a rather cumbersome, but workable, water trap pollen sampling device utilizing an electric pump (Kenady, 1968). To maintain a slide surface directed into the wind, mountings similar to weather vanes are used. Some techniques expose slides for fixed periods of time by mounting them on rotating drums which sequentially expose one slide for a given time interval (Voisey and Bassett, 1961). The single slide, or mounted slides, can be used at ground level or specific elevations corresponding to the level of the receptive flowers. Within a stand of trees, air movements result in different pollen distribution patterns at different heights. Slides have been exposed from airplanes to detect pollen levels in the upper air masses (Polunin, 1951).

The efficiency of glass slides, or any sampling device in catching airborne particles, is measured as the "efficiency of impact," or E (Gregory, 1960). E values vary from 0.01 to 60%. Moving a slide from a horizontal to a vertical position increases the E value; wind speeds above 0.5 m/sec increase E. Pohl (1937) found that by exposing glass rods with sticky surfaces he could obtain higher E values than could be obtained with flat slides. This circular deposition method has since been further developed.

A relatively crude but reasonably effective assay method consisting of a modified circular trapping device has been developed by Grano (1958) utilizing polyester stick tape applied to a 4-sided slide mount. Sarvas (1962) developed a sophisticated and efficient globe sampler (Fig. 4) which avoids the problem of rain washing pollen grains from coated surfaces. The globe, generally exposed at the level of the tree crowns, retracts into a small shelter when rain threatens. Exposed globes are removed and transferred to a special mount for viewing under the microscope. Premarked sample areas of the globe are counted, rather than the entire surface.

An automatic volumetric spore trap designed by Hirst (1952) is widely used. Air is pumped through a narrow slit over a glass slide; the slide moves at a rate of 2 mm/hr and yields data covering a 24-hr period. Sarvas (1955, 1962) also has devised a volumetric pollen collection device that operates on a block mechanism with vane adapters to maintain the

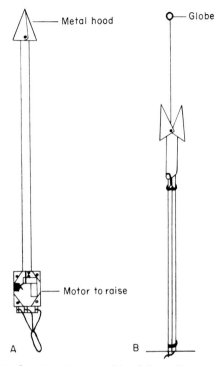

Metal hood

Globe

Motor to raise

A

B

FIG. 4. Schematic drawing of retractable globe pollen sampler (Sarvas, redrawn from Koski, 1967). (A) Globe in shelter. (B) In working position.

exposure slit to the wind. Such applications, while affording sequential metering of air volumes, still yield relatively low E values.

To circumvent all instrumentation, McDonald (1962) collected and counted pollen washed from the air in raindrops. By computer analysis, he established the size range of pollen grains that are carried by specific sizes of raindrops. McDonald also has suggested that the scavenging action of raindrops in removing pollen from the air may be a greater cause of reduced pollination than the mere reduction of anther dehiscence by rain. His suggestions overlook the complex and often detrimental physiological effects of rainwater in reducing pollen viability in some species, even when they are not washed from the surface of the stigma. McDonald's method is a novel concept. although it currently has little application.

The use of radioisotopes to determine pollen distribution patterns is feasible, although of limited application. Initial studies by Colwell (1951) using ^{32}P as a marker and subsequent reports utilizing ^{32}P and [131]I (Turpin and Schlising, 1971) have shown that extracted pollen may be labeled with the isotope and then redispersed as if by normal anther dehiscence. Supplying the isotopes to developing pollen results in inefficient use of radioactive labels because of high dilution and absorption of the isotopes by vegetative tissues. Isotopes are useful in determining the range and pattern of pollen distribution from a point source. However, cost of the isotope and the associated detection equipment and essential safety precautions limit the technique's applicability.

Plants with dominant mutations will produce pollen carrying such mutations. Thus, genetic mutants can serve as natural pollen markers, particularly when associated with leaf pigments. A yellow needle variant found in *Picea abies* (Langner, 1952) can provide an index of pollen flow, as can some virescent chlorophyll gene markers in *Pinus elliottii* (A. E. Squillace, personal communication). Visible leaf pigments are easier to assess than pollen markers as are the "sugary" and "starchy" endosperm in *Zea mays* (Mangelsdorf, 1932). While affording delayed results, pollen markers have many advantages over pollen isotope labeling for determining pollen flow.

Collecting total pollen samples for quantitative determinations and breeding applications may be done by enclosing or "bagging" developing flowers while on the plant, or by placing cut branches with maturing flowers or strobili in enclosed areas or containers prior to dehiscence. Maturing strobili or male flowers can be isolated by covering them with sealed bags and total pollen released collected by shaking the flowers or cones into the bag or container after dehiscence. If other plant parts are collected along with the pollen, then pure pollen may be obtained by the use of sieve screens.

Nienstaedt and Kriebel (1955) tested 9 different types of collecting bags on *Tsuga canadensis* trees in the eastern United States. White cloth bags proved the most adequate for collecting pollen, whereas black and transparent bags were least effective. Plastic polyethylene bags generally accumulated excess moisture and were therefore detrimental to collection of dry pollen.

In vitro drying techniques for flowers and strobili can also be critical in affecting total pollen yield. If anemophilous anthers are dried too rapidly they will only partially open, if at all. Unopened strobili are also a common phenomenon in material which is improperly dried or cut too early. Entomophilous pollen can conveniently be collected from individual dehisced anthers by a vacuum tube or medicine dropperlike extension operating off a vacuum pump. The vacuum line is separated from the vacuum pump by a removable frittered screen (Barrett, 1969) or nylon cloth disk which catches the pollen. Occasionally, other flower parts are sucked into the pollen mass. If a quantitative collection of pure pollen is desired, it must be removed with sieves or by hand selection. This method is fairly rapid and useful with anthers that dehisce, but whose pollen is not wind dispersed.

3. *Quantity Shed*

The amount of pollen shed varies with species, age of plant, and environmental conditions during development. For most purposes the quantity of pollen produced is measured in cubic centimeters (cc). Examples of pollen yields have been compiled by Snyder and Clausen (1973; Table II).

Obviously grain size, presence or absence of wings, and characteristics of the pollen exine influence packing volume. The quantities of pollen developed in pine strobili (Table II) depend indirectly on nutrition and directly on exposure, i.e., side of the tree, position on the tree, and age of the tree on which the strobili form. Maximum quantities of pollen occur in strobili located at the top part of the crown in *Pinus elliottii* (Chira, 1966). Strobili growing near the top and outside of the tree are generally larger and heavier than those growing in lower or inside areas. In plants flowering over a large part of the growing season, the number of anthers dehiscing often decreases toward the end of the growing season. This may reflect the decreasing availability of nutrients.

Merely comparing volume yields of pollen per strobilus or per flower distorts the assay of pollen productivity by removing it from the functional context of plant reproduction (Kugler, 1970). From the ecological viewpoint, a reasonable way of comparing pollen yields is to measure numbers of grains produced per viable ovule. Pohl (1937) found that anemophilous species almost always out produce entomophilous species.

TABLE II

POLLEN YIELDS[a]

Genus	Source	Quantity (cc)
Alnus	100 Catkins	4
Betula	100 Catkins	5–20
Fagus	100 Flowers	1–1.6
Larix	100 Strobili	1.0–1.4
Pinus	100 Strobili	7–33
Pinus	1 Liter of strobili	100–200
Populus	100 Catkins	50–100
Pseudotsuga	100 Strobili	0.7–3.7
Ulmus	100 Flowers	0.1–0.4

[a] After Snyder and Clausen, 1973.

However, when considered on the basis of pollen grains per ovule the ratios in both groups are quite similar. The amount of pollen produced on an individual plant varies in successive years. Thus, an assay in any one year may not be a valid assay of the pollen production potential of a given plant.

The capacity to produce pollen is primarily under genetic and physiological control. Among apple varieties, counts of pollen grains per anther show Winesap consistently low, about 400 grains per anther, while Delicious usually produces about 7000 grains per anther. Oberle and Goertzen (1952) found that in anthers of Delicious apple 85% of the microspores matured, while 15% aborted before being shed; the statistics prevailed in Winesap. While the potential to produce higher quantities of pollen probably initially exists in Winesap and other low yielding varieties, a genetic control mechanism or metabolic block apparently inhibits production of more viable pollen.

Many meteorological factors also influence quantities of pollen produced. While it is possible to select parent plants that are consistently high pollen producers, it is preferable to make final selections after examining several years' data. Hyde (1951) analyzed pollen yields over 7- and 10-year intervals at different locations in Great Britain. Some trees, such as *Fraxinus* and *Ulmus*, follow 3-year cycles, with a year of high pollen production every third year; *Fagus sylvatica* shows alternate year highs. Changing patterns of quantities in 7 genera, assayed by impact slide counts (Table III), reveal dissimilar yields in any two successive years.

Patterns of relative productivity, noted in the preceding discussions (Section II,B,2), depend partly on the size of the area sampled and

TABLE III

YEARLY VARIATION IN POLLEN PRODUCTIVITY[a]

Genus	Average of catch 1943–1948 (= 100%)	Catch by year % of 1943–1948 averages					
		1943	1944	1945	1946	1947	1948
Alnus	385	45	226	120	55	34	118
Fraxinus	675	330	25	24	159	54	8
Fagus	273	23	140	11	143	8	275
Betula	620	57	160	42	106	22	214
Ulmus	4579	84	87	180	50	48	150
Quercus	2776	58	90	42	280	10	117
Pinus	386	92	121	72	127	40	146

[a] Hyde, 1951.

method of pollen collection. Percent distribution of pollen from forest species, based on patterns observed at the forest floor or in deposits in lakes within the forest, can be very misleading. A stand comprised of 82% *Fagus* and 4% *Picea* yielded pollen counts of 49 and 10%, respectively, at an opening in the stand. Lake mud samples from a similarly composed stand yielded pollen counts of 8 and 67% for the same two species (Anderson, 1970). Obviously, pollen counts can be totally misleading when the distribution of trees or total pollen yields from different genera in the flora are extrapolated. This finding also indicates that paleobotanists may make errors in describing past forest compositions from residual pollen found in particular geologic formations.

Pollen yields of a particular plant may vary with specific morphological modifications. Ornduff (1970) studied pollen production by direct count in the two different heterostylic forms of *Jepsonia parryi*. Each plant population had flowers of two types, one with short stamens and long style (pin type) and the other with long stamens and short styles (thrum type). Pin produced four times the number of pollen grains per flower as the thrum type. In addition, pin had a mean pollen size about 33% smaller than thrum. However, both flowers produce an equal number of seeds, primarily due to effective pollinating insects.

4. Dispersal Patterns

Considerable contributions to the literature concerning time of pollen shedding have come from plant breeders and allergists. Plots of pollen concentrations paralleled with climatic variables, such as temperature and rainfall reveal the influence of such factors on pollen shedding. Romashov (1957) analyzed shedding of oak pollen and associated climatic factors in the Khar'kov region of the U.S.S.R. Plots of dissemination of oak pollen with wind velocities greater than 1 m/sec demonstrated that shedding increased with rising temperatures and decreased with falling temperatures. Rain, usually accompanied by decreased air temperatures, delayed the time of pollen shedding until late May and shortened the period of dehiscence. Matthews (1963), therefore, concluded that pollen dispersal in anemophilous species was not completely dependent upon weather during flowering; pollen shedding occurred quickly in brief periods when conditions were optimal. However, pollen shedding might be delayed beyond the time of receptivity of female flowers.

In the American tropics, many plants flower and develop fruit in the wet season, but the peak of flowering in many entomophilous species and development of flower and fruit by species requiring wind dispersal of pollen occur in the dry season (Croat, 1969; Janzen, 1967). In the temperate zone, the dominant pollen in the spring air is from trees, while summer air frequently contains high percentages of grass pollen. Pollens from herbaceous plants, particularly from members of the Compositae, dominate the autumn air. Table IV lists a seasonal distribution pattern for dominant pollens. A more complete list of pollens shed in various months for the United States is given by Wodehouse (1945), for southern Europe by Cefulu and Smiraglia (1964), and for central Europe by Stix and Grosse-Brauckmann (1970).

Periods of pollen shedding in pines depend largely on species, location, and climate. *Pinus radiata* growing in a mild climate in Australia sheds pollen over a 6-week period, with the peak release time ranging from the second to fourth week (Fielding, 1957). Maximum pollen shed occurs in the early afternoon; strobili release no pollen at night. Pines along a mountain elevational transect shed their pollen earlier at lower elevations. Duffield (1953) compiled data on the influences of elevation, latitude, and longitude on shedding of pollen. Mean daily temperature was determined to be the most significant factor influencing shedding by different species along mountain transects. Species obtained from different elevations and planted in an arboretum show considerably less variance in dates of pollen shedding although genetically established differences and temperature requirements may persist.

Peaks of pollen production by *Pinus palustris* in southwest Alabama varied from as early as February 23 to as late as April 3 over a 10-year period. Related field experiments by Boyer (1970) showed that temperature was the most critical factor controlling the time of pollen shedding.

TABLE IV

MONTHLY PATTERNS OF POLLEN SHEDDING IN TEMPERATE CLIMATES

Species	Period of dehiscence
Corylus avellana	February–March
Salix repens	March–April
Salix caprea	March–May
Taraxacum officinale	March–September
Pyrus malus	April–May
Brassica oleracea	May
Pinus sylvestris	May
Acer pseudoplatanus	May–June
Robinia pseudoacacia	May–July
Fagopyrum esculentum	June–July
Ambrosia artemisiifolia	June–September
Solidago virgaurea	July–October
Aster spp.	August–November

Boyer also found that temperature during developmental phases in November influenced the time of strobili maturation the following spring. Air temperature in pine stands may control the duration of pollen shedding as well. Cool weather prolongs the period of pollen shedding.

Pine pollen, as previously mentioned, is shed in a diurnal pattern. Certain peak times for shedding are characteristic for each species. Recognizing diurnal patterns of pollen availability is important in understanding the pollination activity of animals. Table V lists approximate maximum and minimum times of pollen dehiscence in 5 plant species. Some plants shed their pollen in the evening or at night, to the benefit of nocturnal pollinators such as moths and bats. However, pollen of most species is shed in a 3- to 4-hour period each day.

The volume, density, and shape of wind-dispersed pollen grains determine the distances to which they are disseminated. Palynologists generally describe pollen by size, recognizing normal statistical variations. About 90% of all pollen observed ranges between 20 and 60 μm in diameter, but within individual species size can vary as much as 100%. Anemophilous pollens are usually small, 20–30 μm in diameter, or occasionally measure up to 100 μm; entomophilous pollens range in diameter up to 300 μm (Faegri and van der Pijl, 1966). Some anemophilous pollens are quite light (e.g., *Betula* and *Ulmus* weigh approximately 1×10^{-9} gm per grain). Conifer pollens are large, 50–150 μm in diameter, and weigh between 50 and 270×10^{-9} gm.

Buoyancy is the most critical factor influencing dissemination of anemophilous pollens. Because of air moisture, weight alone is a poor index of buoyancy potential. In the large conifer pollens, buoyancy is achieved by the presence of air sacs. Dispersal of anemophilous pollens is also facilitated by smooth and dry surfaces; thus, anemophilous pollen grains rarely adhere to each other. Koski (1967) maintained that pollens of all anemophilous species are "lifted up and transferred with equal

TABLE V

DIURNAL CHANGES IN POLLEN AVAILABILITY[a]

Species	Time	
	Maximum	Minimum
Papaver somniferum	0600–0930	1100
Verbascum phlomoides	0600–0930	1400
Rosa multiflora	0630–1030	1330
Verbena officinalis	0700–1130	1400
Convolvulus tricolor	0800–1400	1600

[a] Kleber, 1935.

efficiency in spite of differences in their sizes." This generalization is based largely on the rate at which air movements carry pollen. As discussed below, free fall velocities of pollen grains strongly influence the distances that pollen will travel.

Individual species of plants may produce pollen which varies greatly in size. Pollen grain size is determined by many factors including inherited characteristics and environmental conditions during pollen development. Size variations associated with ploidy levels are well recognized in mature pollen. The average diameters of pollen from diploid, tetraploid, and hexaploid varieties of wheat, *Triticum vulgare*, have been calculated to be 45.83, 50.44, and 58.87 μm, respectively (De Vries, 1971). Pollen grains of tetraploid rye, *Secale cereale*, are significantly larger than those of diploid varieties. However, the quantity of pollen produced by the diploids has been shown to be 2.5 times greater than that produced by the tetraploids. In addition, there are fewer anthers produced on tetraploid rye plants. Giant tetraploid pollen grains have been artificially produced by colchicine treatment in plum and cherry trees (Olden, 1954). Thus, there appears to be a positive correlation between ploidy level of a pollen grain and its size.

In nature, size variations within species are often related to ecotypes. Pollens of *Quercus macedonica* from 11 locations in Yugoslavia show a size range of 25–45 μm in diameter with the variation at each location being only 3–4 μm (Jovančević, 1962). Other oak species from different locations in Yugoslavia varied considerably less than *Q. macedonica*. A geographic cline was found in the pollen grain size of *Pinus echinata*, ranging from 42–70 μm in diameter, with the larger mean values found in pollen from trees in the northern part of the range.

Influences of such factors as age of the tree and position of the male catkins on the tree may indeed be slight since greater pollen size variations are seen from year to year. Clausen (1960) established that *Betula* pollen does not statistically vary in size in different parts of the catkin or from catkins located at different areas of the tree. Szwabowicz (1971) studied pollen production of *B. oycoviensis* over 3 years. Both young and old trees produced pollen of approximately the same size. There was no difference in amount of pollen produced by trees growing in the wild or in cultivation. However, variation in pollen size from year to year was about 10%. Sources of such variation may lie in differences in growing conditions during successive years.

Growing conditions can influence pollen size. For example, low night or high day temperatures result in small pollen in tomatoes (Kurtz and Liverman, 1958). Schoch-Bodmer (1940) has shown that the amount of water available during development had a profound influence on pollen size in the purple loosestrife, *Lythrum salicaria*. The effect of water in solubilizing soil nutrients may relate Schoch-Bodmer's work to Bell's detailed growth studies. Bell (1959) has shown that tomato yields significantly different sized pollen (20% variation) when experiencing nutrient deficiencies. Bell's studies on three additional genera yielded data consistent with those for tomato pollen.

Pollen grains produced on the same plant, or on several plants of the same species grown under identical conditions, can vary greatly in size with no change in ploidy. A classic example of this phenomenon is observed in the heteromorphic pollen of corn (*Zea mays*). The large sugar-containing pollen grains of corn are so different in size from the smaller starch-containing pollen grains that the two types may be mechanically separated (Mangelsdorf, 1932). The early studies of Breslavets (1935) correlating sexual expression in hemp with different pollen sizes have been further verified by Herich (1961). Herich has shown that nuclear size and volume are reduced when pollen size is reduced. The significance of sexual dimorphism in hemp is still not understood.

Probably the most common source of size differences in anemophilous pollens is the tendency of certain pollens to clump together. The number of grains per clump can vary from 2 to 10. Anderson (1970) reported that up to 85% of wind-dispersed pollens of some species have their sedimentation velocities and distribution patterns affected by clumping (Table VI). This indicates that the size of a single pollen grain alone is not a valid assay when considering dissemination patterns. The percentage of pollen adhering together at the time of shedding must also be considered.

The free fall rate, or sedimentation velocity, of pollen is usually measured by determining the average time required for pollen to fall a given

TABLE VI
POLLEN SHED IN CLUMPS[a]

Species	% of all pollen clumped	Average number grains/clump	Terminal velocity (cm/sec)
Alnus glutinosa	34	3.2	1.7
Betula verrucosa	34	2.9	2.4
Carpinus betulus	—	—	6.8
Corylus avellana	37	3.9	2.5
Fagus sylvatica	24	2.4	5.5
Quercus spp.	50	2.7	2.9
Tilia cordata	85	9.0	3.2
Ulmus montana	58	3.1	3.2

[a] Anderson, 1970.

TABLE VII

POLLEN SEDIMENTATION VELOCITIES[a]

	Rate in cm/sec		
Species	Dyakowska, 1937	Knoll, 1932	Eisenhut, 1961
Larix decidua	12.3	9.9	12.6
Picea abies	6.8	8.7	5.6
Pinus sylvestris	3.7	2.5	3.7
Betula verrucosa	2.9	2.4	2.6
Carpinus betulus	6.8	4.5	4.2
Quercus robur	4.0	2.9	3.5

[a] Eisenhut, 1961.

distance. Eisenhut (1961) compared values reported in the literature for sedimentation in calm air. A few values are listed in Table VII.

Differences encountered for the same species may be due to experimental technique or different pollen moisture contents. Since the sedimentation rate is a function of volume and weight, water content strongly affects the density of pollen and sedimentation velocity. The falling rate is also influenced by pollen form and shape. However, it is incorrect to conclude that gymnosperm pollens with air sacs have a slower sedimentation rate than angiosperm pollens which lack air sacs. The approximate velocities of pollen sedimentation have been calculated according to Stokes' law, assuming sedimentation rates in pollen are comparable to rates for equal sized water droplets (Koski, 1967). In Stokes' law both drop size and density affect sedimentation rate. Conifer pollen density,

TABLE VIII

POLLEN DISPERSION AT A WIND SPEED OF 0 AND 10 m/sec[a]

Species	Average weight of pollen (10^{-9} gm)	Average of maximum diameter (μm)	Sedimentation velocity at 0 wind (cm/sec)	Distance transported at wind 10 m/sec (km)
Picea excelsa	93.20	162.0	6.8	22.2
Pinus sylvestris	30.08	59.0	3.7	267.8
Alnus spp.	9.37	24.6	2.8	546.7
Corylus avellana	9.45	24.2	2.9	267.8
Dactylis glomerata	21.85	33.3	3.1	174.0
Quercus robus	18.10	24.8	4.0	199.0

[a] Knoll, 1932; Dyakowska and Zurzycki, 1959.

which is about 50% that of water, suggests that its sedimentation velocity will be one-half that of a water droplet of equal size. However, distinct external morphology of gymnosperm pollen introduces deviations from the computed values of Stokes' law. Slight differences in wing size of conifer pollens can markedly affect buoyancy. A variety of wing sizes exists within a given species. In *Pinus resinosa* 4% of the pollen will sediment within 10 min, while 9% requires 120 hr prior to sedimentation (Hopkins, 1950). Damaged wings probably contribute to the initial rapidly settling 4%. Sedimentation rates are important in understanding and predicting the distribution of anemophilous pollens. Once pollen is shed, the important factor is to be able to predict dispersal patterns in different natural environments.

Comparing pollen dispersal distances at a constant wind speed, Table VIII, demonstrates that weight alone is not the only factor that determines the distance pollen is transported. Size and density, correlated with sedimentation rates at zero air speed are better indices of the distance pollen will move in an air stream.

Differences in dispersal distances of *Alnus* and *Corylus* pollens are, in part, related to their surface properties which may result in increased grain clumping in *Corylus* (Table VI). Pollens of *Abies alba* and *Zea mays* with weights of about 330 and 480 ng per grain, respectively, have even higher sedimentation velocities and in general travel even shorter distances than pollen of *Picea excelsa* (Table VIII). However, most anemophilous pollens are light weight and readily transported for long distances. In addition to pollen volume and density, rate of fall, and presence of rain or moisture, the following factors also markedly influence pollen dispersal distances: wind velocity and direction; land topography (i.e., dispersal from a pollen source is greater in the downhill direction than in the uphill direction); air turbulence; and height of the pollen source (Wright, 1952).

Analyses of dispersal at different distances from the pollen source (Fig. 5) show that dispersal and settling often follow a discontinuous pattern. Low pollen levels may result near the tree, with increasing amounts further away from the source (Fig. 5A), reaching a maximum deposition at a considerable distance from the source. Morphologically similar pollens from closely related species shed at identical locations may not follow similar flight patterns. Anemophilous gymnosperm pollens usually travel shorter distances than lighter anemophilous angiosperm pollens. Data reported for *Pinus edulis* pollen can help detect variations caused by prevailing wind patterns, but, as seen in Fig. 5B, wind patterns are not always significant causes of differences in the pattern of pollen shedding.

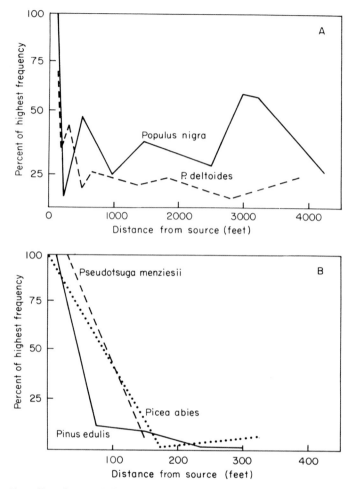

FIG. 5. Pollen dispersal distances from source trees (from Wright, 1952). (A) Poplar. (B) Conifers (source heights indicated).

Most pollens settle from the air by rain or water droplets. Pollen settled in rainy weather is dispersed for shorter distances and exhibits decreased viability. In rainy weather, pine pollen is dispersed only about 60 km, as compared to 1700 km in dry weather (Busse, 1926). Busse postulated that low levels of pollen produced by forest trees do not drastically limit pollination potential since the mix and dispersal of pollen in forest stands are still proportionately high per unit of female receptive tissue. This concept has not been adequately explored.

Evidence suggests that the critical factor is not simply the quantity of pollen shed, but also potential contaminating pollen sources. Determining distances that must be maintained free of related species in an isolation zone is essential in establishing forest seed orchards or hybrid "certified" seed production areas. Volume of pollen in the air differs seasonally and with location. Chances of pollen contamination in Douglas fir (*Pseudotsuga menziesii*) seed orchards are greater in valley locations. Silen and Copes (1972) reported seed orchards in level flat areas several miles from contaminating trees had counts of 3000 contaminating pollen grains per inch^2; however, counts reached 7500 grains per inch^2 in narrow valley orchards.

In determining the essential distances required to isolate seed orchards, studies of the impact of contaminating pollens are often preceded by quantitative assessment of pollen present based on air volume counts (see Section II,B,2) or by theoretical calculations of the amount of pollen shed certain distances from the orchards. Analysis of sedimentation by Stokes' law is limited to predicting velocity of descent of a pollen grain. Numerous factors including clumping, alternating wind patterns, and interfering objects all modify the validity of any predictive value. Koski (1967) tested a mathematical model for estimating the dispersal range of tree pollens based on sedimentation velocity in air. Recognizing that different sedimentation rates occur in calm air versus air moving at different velocities permits the use of a mean sedimentation velocity based on measured wind speeds.

$$\text{Probable range of pollen flight} = X = 0.91 \left(\frac{AV}{\delta c^2} \right) \tag{1}$$

where A = coefficient of exchange or shedding rate of pollen in gm/cm/sec; V = wind velocity (cm/sec); δ = air density (cm^{-3}); c = sedimentation velocity (cm/sec). Koski calculated probable ranges of flight, X values, for pollen of *Betula* and *Pinus* species. From crude measurements, he determined A, the rate of pollen shedding, to be approximately 100 gm/cm/sec. Koski's calculated values for X are as follows:

Betula verrucosa $X = 43.8$ km

Pinus sylvestris $X = 21.6$ km

Strong upward air currents are particularly helpful in distributing pollen over great distances. Formula 2 derived in relation to Formula 1 may be used to determine the altitude of a pollen cloud, Z_{max},

$$Z_{max} = 0.227 \left(\frac{A}{\delta c} \right) \tag{2}$$

The probable flight duration, Y, the period of time required for 50% of the pollen shed in a given time interval to return to the altitude from which they were shed, is calculated as follows:

$$Y = 0.91 \left(\frac{A}{\delta c^2} \right) \tag{3}$$

Koski indicated that Y is a difficult term to calculate because of great variation in meteorological conditions. A term which accounts for changes in wind turbulence must generally be included in such calculations to yield values which approximate empirical observations.

The amount of pollen shed, or quantity of male flowers, is often correlated with seed set. A good seed year is generally considered to follow heavy pollen dispersion, however, measurements show that this is not always the case. Receptivity of the female tissue as well as pollen dispersal pattern influence the level of successful pollinations and seed set.

III. Pollination

In gymnosperms, pollination takes place by deposition of the air-borne pollen on a droplet of liquid secreted by the micropylar tissues and located at the tip of the ovule. As the pollination drop dries, pollen deposited on it is drawn into the pollen chamber formed by the breakdown of the apical region of the nucellus. In angiosperms, pollen begins its functional role on receptive stigmata (Fig. 2). To facilitate the adherence of deposited pollen, the surfaces of some angiosperm stigmata are sticky or covered with fatty substances. Growth of pollen on receptive tissues and release of growth substances generally delay shedding of the pistil and ovary (angiosperms) or the megasporangiate strobilus (gymnosperms).

The time of pollen dehiscence does not always correspond to the time at which female tissues are receptive. In fact, settling of "clouds" of wind-borne pollen may take place long after shedding, depending, of course, on atmospheric conditions. The only pollen of reproductive significance is that which lands on receptive tissues and is able to grow and release functional male cells destined for fertilization.

Size of exposed receptive tissue has been correlated with pollen grain size (Koski, 1967). Small angiosperm pollens have, as landing targets, stigmata which are considerably smaller than the landing surfaces available to larger conifer pollens. In pine, stacks of receptive openings located on large female strobili are able to catch large pollen grains with greater efficiency than small pollen grains. However, closeness of the pollen source to the female receptive tissue is probably the most critical factor in pollination, regardless of the mechanism of pollen deposition. In the catkin inflorescences of trees or spadix inflorescences, pollen-bearing structures are often located below receptive female tissues. Upward air currents, or flying insects, may concentrate pollen in the zones where stigmatic surfaces await pollination.

The amount of pollen settling on a given unit of stigmatic surface may vary from year to year. However, in trees, flowers from different locations in the crown usually receive, on a yearly basis, consistent percentages of the total amount of pollen available. Romashov (1957) compared pollen counts on oak stigmata, in 3 successive years (Table IX). Maximum deposition occurred in the upper parts of the crown. Concentrations of pollen deposited on stigmatic surfaces and, consequently, the number of seeds produced were greater in flowers from trees on the margin of the stand, where downward drafts deposited greater amounts of pollen as compared to the interior of the stand, where upward drafts were mainly responsible for pollen circulation. It has often been reported that trees along the edge of a forest stand receive more sunlight, grow more vigorously, and, therefore, produce more seed. This explanation may be an oversimplification which does not consider the aerodynamics of pollen flight.

The percentage of ovules fertilized is determined, in part, by the quantity and physiological condition of pollen reaching receptive female tissues. In some flowers dead pollen can stimulate ovary development. This phenomenon, parthenocarpic fruit development, yields seedless fruit, without viable embryos. In fact, dead tomato pollen has been shown to set as many seeds as fresh pollen (Visser, 1955). In nature, parthenocarpic development may function to increase fruit yields in

TABLE IX

POLLEN PER MM^2 OF STIGMA OF *Quercus* AFTER 4 DAYS RECEPTIVE CONDITION[a]

Height (m)	Year		
	1951	1952	1953
22–23	5.7	15.4	6.0
18–19	2.3	20.5	8.7
17	1.7	9.9	2.7
14–15	1.0	6.9	2.4
11	0.9	4.3	1.2
Mean ($\bar{x}$)	2.3	11.4	4.2

[a] Romashov, 1957.

circumstances where pollen is continually present, even though it may not be sexually compatible or viable. McQuade (1952) applied different noncompatible pollens to *Petunia* stigmata and obtained parthenocarpic fruit from 2 of 5 Solanaceae pollens. Obviously, some pollens may interact with nonrelated species inducing parthenocarpy.

Foreign pollen may release hormones or stimulate production of hormones in receptive female tissues. Rapidly diffusing chemicals are known to be released from pollens as soon as they are placed in a solution or on moist surfaces (Linskens and Schrauwen, 1969; Stanley and Search, 1971; Knox *et al.,* 1972). Dead pollen can, in many cases, also supply a chemical factor which initiates auxin activity, resulting in development of mature fruits and seeds as Visser (1955) has shown.

A. Time of Receptivity

In most trees, the time at which flowers are receptive depends on the time at which growth is resumed in the spring and on the rate of shoot and flower development. Factors which impede development, in particular low temperature, also hinder pistil maturation. However, such factors may not affect stigma receptivity. Once the stigma becomes receptive, it usually remains so for several days, depending on species. Anther dehiscence is generally more responsive to weather conditions which control the operation of anatomical dehiscence mechanisms (see Section II,B,1).

The standard procedure for determining the time of stigma receptivity is to apply freshly extracted pollen to stigmata by hand at given time intervals. Generally, maximum receptivity occurs at the time of anthesis or shortly thereafter. Rice stigmata tested for receptivity from 1 to 7 days after emasculation have shown a decreasing capacity to set seed the third day after opening with stigmata being essentially nonresponsive after 5 days.

Pollen dispersion is generally more prolonged than stigma receptivity. In *Pinus pinaster,* pollen shedding occurs over a 3-week period, but the female cones are receptive for only 8 days of that period (Illy and Sopena, 1963). In *Tilia,* the stigmata of flowers located in the center of the crown are receptive several days before the stigmata of laterally positioned flowers, although pollen is dispersed freely throughout the entire period of receptivity (Wettstein and Onno, 1949).

An effective mechanism ensuring cross-pollination has evolved in flowers which shed their pollen either prior to or after the period of receptivity of the stigmata, such species usually being self-sterile. Avocado (*Persea*) flowers normally open twice (Peterson, 1955). During the first opening the female tissues are receptive. The flowers then close and reopen to shed pollen. Individual avocado flowers do not set fruit after self-pollination. If flowers such as *Hibiscus*, which displays a diurnal pattern of opening and closing, are not pollinated during their normal receptive period, they may remain open and receptive for several additional days, thus ensuring pollination. However, postpollination conditions may cause failure of seed development and result in abscission of the ovary and developing fruit. It is frequently observed that good flowering years and abundant pollen shedding are more frequent than years of good seed yield. This suggests that seed development is also influenced by factors subsequent to the period of receptivity; stimulatory or inhibitory factors may influence either pollination or fertilization and decreased seed production.

B. PHYSIOLOGICAL AND ENVIRONMENTAL FACTORS

Factors reducing seed yields include the failure of pollination, incompatible reactions between pollen and stigmatic tissues, and attacks by insects and fungi. Clouds or rain, which usually delay pollen shedding, generally extend stigma receptivity. In millet Aziz *et al.* (1959) have found that receptivity may be extended from 2 to 6 days during inclement weather. When clouds restrict bee activity, pear pollination is drastically reduced. However, Stephen (1958) has reported that although pear flowers remain receptive for 19 days, the pollinating activities of bees on only one day of good weather are sufficient to ensure a good pear crop. When inclement weather prevails throughout the entire 19-day period of stigma receptivity, less than 8% of the flowers develop fruit.

Artificial methods of pollination may be utilized to insure seed set. Often, supplemental pollinations are performed with mixtures of several pollens, the advantage being that various chemicals released by the different pollens may interact with maternal tissue to insure fertilization. Under these conditions the pistil tissue may continue to function and formation of an abscission layer at the base of the style may be delayed. However, pear and apple pollen interact to inhibit both pollen growth and seed set (Saitan, 1952). Levels of artificially applied pollen may, in some cases, reduce the percentage of seed development. Excess supplemental pollen, when applied to *Juglans* species, results in decreased seed set (Kavetskaya and Tokar, 1963). These studies have also shown that 10–18 pollen grains per stigma produce maximum fruit yields. Higher densities of pollen on the sigma may result in fruit abscission.

Metabolic activities, as evidenced by respiratory rates, are at peak levels in stigmata at the point of optimum receptivity. These high metabolic rates are reflected in changes in water potential of the stigma as time of maximum receptivity approaches. As suggested by Stanley (1964), the increased potential for pollen to affect pollination when placed on corn stigmata in the morning, as compared to the afternoon, may be related to diurnal variations in water potential of the stylar tissues. A large number of pollen grains germinating on a single stigma may function to withdraw water from the female tissues and upset the delicate metabolic balance of the stigma and style. In addition, many pollen grains germinating on a single stigma may afford a potential for excess auxin or chemical secretion. Such chemicals at abnormally high levels may result in premature shedding of flowers prior to seed development. Metabolic processes leading to normal seed maturation and seed or fruit shedding are extremely vulnerable to environmental conditions as well as physiological factors initiated by pollination.

IV. Seeds

The topic of seed dispersal is one that has received considerable attention in the past, partly because of its importance in agriculture and plant biogeography. In fact, van der Pijl (1969) stated ". . . next to pollen transport, seed dispersal is the most important single factor promoting the gene flow in populations." Excellent reviews of seed dispersal may be found in Guppy (1912), Ridley (1930), van der Pijl (1969), and Fahn and Werker (1972). The latter review provides considerable detail on anatomical mechanisms of seed dispersal.

A. SHEDDING PATTERNS

Fruits may be classified by gross morphology as dry or fleshy fruits. Dry fruits may be further divided into two groups. In dry indehiscent fruits, the pericarp decays after the fruit reaches the ground exposing usually a single seed; examples are nuts and achenes. Dry dehiscent fruits remain attached to the parent plant and split or dehisce, usually releasing many seeds. Such fruits include legumes, in which a single carpel splits along both margins, and capsules, which contain several carpels and may dehisce by means of fracture lines or pores.

Fleshy fruits, on the other hand, all are indehiscent, the main types being the berry, pome, and drupe. Fleshy fruits may involve the elaboration of tissues in addition to the ovary and thus may be termed accessory fruits, e.g., strawberry.

B. Physiological Basis for Shedding of Fruits and Seeds

The shedding of seeds and fruits requires formation of a special abscission zone where cell walls dissolve and separate or are torn apart at maturity. Most research has been on fleshy fruits, such as apples and citrus, which are of economic importance. The abscission of fruits and leaves has been well reviewed (Carns, 1966; Leopold, 1971). Chapter 12 of this volume and Fahn and Werker's (1972) treatment of anatomical mechanisms of abscission provide access to the current literature on this subject.

Auxin has been shown to effectively retard the process of abscission by preventing structural weakening of the abscission zone; however, once weakening of this specialized layer of cells has begun, auxin enhances the abscission process (Rubinstein and Leopold, 1963). By preventing the beginning of the weakening process, auxin can effectively block the action of ethylene.

Ethylene has been shown to act on weakened cells in the abscission zone by stimulating secretion of hydrolyzing enzymes, such as cellulase and pectinase, thus causing breakdown of cell walls in the abscission zone (Abeles, 1969; Morré, 1968). In fact, the accumulation of calcium oxalate crystals in the abscission zone may be indicative of the breakdown of pectic materials in the middle lamella and the simultaneous release of associated calcium ions (Scott *et al.*, 1948).

In general, cytokinins inhibit abscission, whereas gibberellins promote abscission. Abscisic acid has been found to promote the abscission process (Liu and Carns, 1961). However, at the present time, auxin and ethylene appear to be the important chemicals controlling both leaf and fruit abscission (Leopold, 1971).

C. Dispersal

The evolution of specific seed shedding and dispersal mechanisms appears concomitant with the presence of effective dispersal agents (Ridley, 1930). In the development of lower vascular plants, ferns, and mosses, wind has by far played the greatest role in dispersal. Evolution of terrestrial herbivores such as early reptiles and mammals and birds was accompanied by a preponderance of plants with fleshy and edible fruits. As a result of competition among species for dispersal agents,

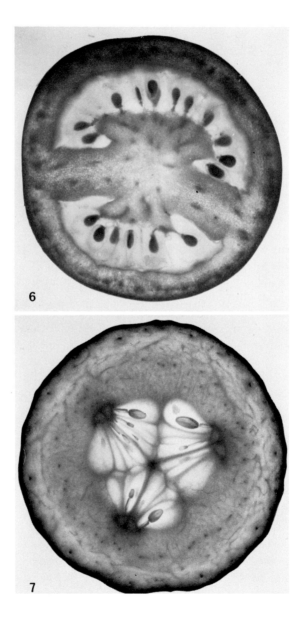

FIG. 6. Fruit of *Lycopersicon esculentum*. Photograph courtesy of Dr. C. Stumm.
FIG. 7. Fruit of *Cucurbita pepo*. Photograph courtesy of Dr. C. Stumm.

attractive and colorful fruits were evolved. Water and rainfall have also proven to be effective dispersal agents. However, plants have also evolved autochory, or dispersal of seeds without the help of specific dispersal agents.

The presence of animals in the biosphere has offered plants many possible methods for seed dissemination. Seed dispersal by animals is termed zoochory. There are several ways which animals function in seed dispersal. Animals may ingest seeds and excrete them, unharmed, some distance from the parent plant. In fact, digestive enzymes and acids can function to hasten seed germination once seeds are excreted (Krefting and Roe, 1949). As previously mentioned, some fruits have colorful fleshy elaborations of the female tissues making them attractive to hungry dispersal agents. However, inconspicuous fruits may be ingested by herbivores while eating foliage. Native species of tomato (*Lycopersicon* sp.) have attractive berry fruits specifically adapted for dispersal by birds and other herbivores (Fig. 6). Cucumbers (*Cucurbita pepo*) also are dispersed by birds (Fig. 7). Currently, man is playing a very significant role in the distribution of these plants. The fruit of *Cyclamen neapolitanum* (Primulaceae) has a fleshy testa containing edible material in a diffuse form (Fig. 8). Ants are attracted to these fruits and by carrying the fruit to their colony the insects not only disperse the seeds, but also plant them (Ridley, 1930).

The relative abundance of food in a given environment may determine whether an animal is a seed predator or a dispersal agent (Janzen, 1971). If few seeds are available, the majority are eaten by predator-dispersal agents. However, if the food supply is plentiful, many seeds may be removed from the vicinity of the parent plant and not eaten; these seeds are effectively dispersed. Little does the squirrel realize how many oak trees he plants in years of good acorn yields.

Another method of zoochory is adhesion of fruits and seeds to the feathers of birds and fur of mammals. Species of the genus *Caucalis* (Umbelliferae) have fruits armed with processes and hooks by which they can be attached to fur (Fig. 9). The fruit of the beggar's tick (*Bidens* sp.) has a hooked pappus that very effectively becomes attached to the fur of animals and the clothing of man (Fig. 10), thus transporting the seed for considerable distances.

Fruits and seeds may also become attached to the feet of animals that have been trampling in the mud. The pond plant *Trapa natans* is dispersed by seeds adhering to water birds and other animals in the water (Fig. 11).

It appears that almost all groups of animals from earthworms to man are involved in seed dispersal (van der Pijl, 1969). However, it is widely

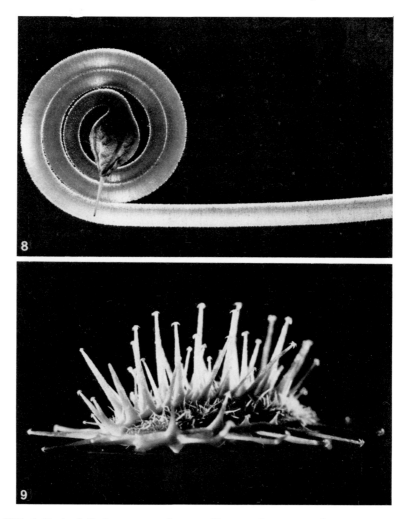

FIG. 8. Fruit of *Cyclamen neopolitanum.* Photograph courtesy of Dr. C. Stumm.
FIG. 9. Fruit of *Cauculis lappula.* Photograph courtesy of Dr. C. Stumm.

held that birds, more than any other group, are responsible for dissemination of higher plants throughout the world, as a result of the very long distances they can fly. Well documented discussions of seed dispersal and predation by animals may be found in the works of Ridley (1930), van der Pijl (1969), and Janzen (1971).

Action of water in plant distribution (hydrochory) is well described in the plant kingdom. Moreover, water provides perhaps the most primitive means of dispersal for higher plants, since the earliest plants

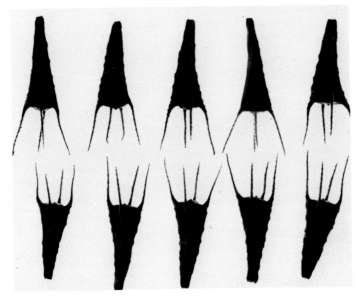

FIG. 10. Fruit of *Bidens cernua*. Photograph courtesy of Dr. C. Stumm.

FIG. 11. Fruit of *Trapa natans*. Photograph courtesy of Dr. C. Stumm.

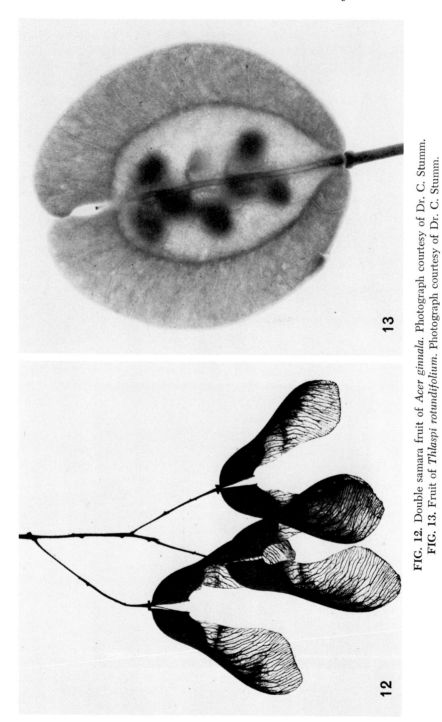

FIG. 12. Double samara fruit of *Acer ginnala*. Photograph courtesy of Dr. C. Stumm.
FIG. 13. Fruit of *Thlaspi rotundifolium*. Photograph courtesy of Dr. C. Stumm.

FIG. 14. (A) *Taraxacum officinale* receptacle with seeds removed. (B) *Taraxacum officinale* single achene. Photograph courtesy of Dr. C. Stumm.

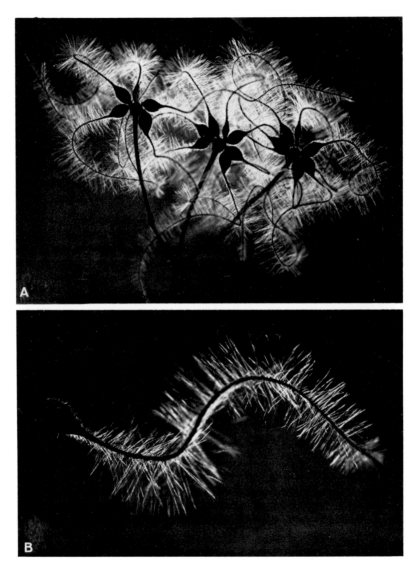

FIG. 15. (A) *Clematis vitalba* seeds removed exposing receptacle. (B) *Clematis vitalba* single achene. Photograph courtesy of Dr. C. Stumm.

were undoubtedly aquatic. Dispersal by water may be by rain wash, ice, streams, rivers, floods, and seas. Further discussions of water dispersal mechanisms in higher plants may be found in Ridley (1930) and van der Pijl (1969).

TABLE X
SEED WEIGHTS[a]

Species	Weight (gm $\times$ 10^{-4})
Nepenthes phyllamphora	3.50
Rhododendron verticillatum	0.28
Aeschynanthus sp.	0.20
Dendrobium attenuatum	0.06

[a] After Ridley, 1930.

Wind dispersal, or anemochory, is also a relatively primitive mechanism for seed distribution. This mode of dispersal seems more significant in colonizing plants than in plants of mature communities, which are characterized by a high degree of symbiosis. Seeds and fruits may be dispersed by wind in several ways. Some seeds are extremely light in weight

FIG. 16. Fruit of *Eriophorum angustifolium*. Photograph courtesy of Dr. C. Stumm.

and are easily carried by the wind for considerable distances without requiring structural modifications. The weights of some such seeds are given in Table X. Ridley (1930) has termed these very light weight seeds "dust seeds."

In order to accomplish effective wind dispersal of the heavier fruits and seeds, many structural modifications have been evolved. Winged samaras such as in the Aceraceae (Fig. 12) are indehiscent fruits, the wings being dry, elongate extensions of the ovary. An unusual fruit of this type is seen in the genus *Thlaspi* (Cruciferae) (Fig. 13) in which the carpels dilate when ripe forming wide, thin wings, resulting in a

FIG. 17. Fruit of *Vicia sativa*. Photograph courtesy of Dr. C. Stumm.
FIG. 18. Fruit of *Robinia pseudoacacia*. Photograph courtesy of Dr. C. Stumm.

nearly circular papery indehiscent fruit containing one to several seeds.

Many seeds have developed silky or woolly appendages to insure staying aloft long enough to be effectively transported by wind. In *Taraxacum officinale* (Compositae) the sepals are modified, forming a pappus which functions to keep the tiny achenes bouyant in the air (Fig. 14a and b). Seeds of *Clematis* (Ranunculaceae) have silky plumules derived from the styles which function in wind dispersal (Fig. 15). The seeds of genus *Eriophorum* (Cyperaceae) are characterized by hypogenous bristles which make the seeds woolly and plumed and thereby aid in wind dispersal (Fig. 16). Thus, in such diverse groups of angiosperms (Ranunculaceae, Compositae, and Cyperaceae) functionally similar adaptations for wind dispersal have arisen.

FIG. 19. Fruit of *Geranium* sp. Photograph courtesy of Dr. C. Stumm.

Autochory, or dispersal mechanisms exerted by the plant itself without the help of a dispersal agent, is not a recent innovation in the plant kingdom. Many fungi, bryophytes, vascular cryptograms, and ferns possess autochorous, or "active" mechanisms, for spore discharge. Among the higher plants, van der Pijl (1969) has distinguished several types of autochory. Plants that disperse their seeds by tension changes in dead hygroscopic tissues, balists, include many members of the Leguminosae. In *Vicia sativa* the carpels of the dry fruit split open and release seeds (Fig. 17). In turn, these seeds may be eaten by birds or cattle. In *Robinia pseudoacacia* (Fig. 18), pods dehisce after being shed. The pods are very light in weight and may be blown along the ground for considerable distances, aided by thin valves formed by dehiscence. External agents for seed dispersal may accompany autochory, as seen in the examples of *Vicia sativa* and *Robinia pseudoacacia*. This condition in which more than one process is involved in dispersal has been termed polychory (van der Pijl, 1969).

Van der Pijl's second group of autochorous dehiscence mechanisms involves active balists by tension in living tissues. Seed dispersal of the explosive fruits of *Geranium* (Fig. 19) involves the violent upward coiling of the beaklike, elongated mericarps, thus ejecting the seeds from the ovary (Fahn and Werker, 1972).

FIG. 20. Fruit of *Avena sterilis* with hygroscopic appendages. Photograph courtesy of Dr. C. Stumm.

TABLE XI
VARIATION IN METHODS OF DISPERSAL IN THE LEGUMINOSAE[a]

Wind-dispersed fruits and seeds
    Dalbergia
    Albizzia
    Trifolium
Water-dispersed fruits and seeds
    Colutea
    Pongamia
    Neptunia
    Lathyrus maritimus
Animal-dispersed fruits and seeds
    Acacia spp.
    Tamarindus
    Cassia goratensis
    Adenanthera
    Medicago
    Desmodium
    Vicia

[a] After Ridley, 1930.

Some seeds and fruits possess hygroscopic bristles or hairs that are capable of movements under alternating dry and humid conditions. Seeds of *Avena sterilis* (Fig. 20) possess awns which are capable of such movements and function in penetration of the seeds into the soil.

D. CONSEQUENCES OF SEED DISPERSAL

The underlying concepts involved are competition and reestablishment. Effective dispersal is achieved when seeds of a plant are transported away from the parent plant, thus avoiding competition for resources with the parent plant and other germinating seeds. Older diverse families of vascular plants may display high diversity of dispersal mechanisms. Such is the case in the Leguminosae (Table XI).

V. Conclusions

The shedding of pollen and seeds displays the intricate interrelationships between plant species and the biotic and abiotic components of their immediate environments. Moreover, dispersal and pollination function to bring about gene exchange between populations and are therefore significant from the standpoint of population genetics and speciation. To ensure maximum productivity of his agricultural lands, man has endeavored to understand and influence pollen and seed shedding patterns. Increased knowledge of the environmental factors and physiological mechanisms controlling pollen and seed shedding can further the efforts to develop plants with selected genetic characteristics.

References

Abeles, F. B. (1969). Abscission: Role of cellulase. *Plant Physiol.* **44**, 447–452.

Anderson, S. T. (1970). The relative pollen productivity and pollen representation of north Engvar trees and correction factors for tree pollen spectra. *Dan. Geol. Unders.* [*Afh.*], *Raekke 2* **96**, 1–99.

Aziz, M. A., Bajwa, M. A., and Shah, S. (1959). Pollen viability and stigma receptivity in pearl millet (*Pennisetum typhoideum* Rich.). *Agriculture* (*London*) **10**, 457–462.

Barner, H., and Christiansen, H. (1958). On the extraction of forest tree pollen from inflorescences forced in a specially designed house. *Silvae Genet.* **7**, 19–24.

Barrett, H. C. (1969). A new pollen collection technique for plant breeders. *Hort. Res.* **9**, 153–155.

Becquerel, P. (1932). La déhiscence de l'anthère du lis blanc. *C. R. Acad. Sci.* **195**, 165–167.

Bell, C. R. (1959). Mineral nutrition and flower to flower pollen size variation. *Amer. J. Bot.* **46**, 621–624.

Boyer, W. D. (1970). Influence of temperature on data and duration of pollen shed by longleaf pine. Ph.D. Thesis, Duke University, Durham, North Carolina.

Breslavets, L. P. (1935). Differential fertilization of the hemp plant. *Dokl. Akad. Nauk SSSR* **2**, 297–302.

Busse, J. (1926). Kiefernpollen und forstliche Saatgutanerkennung. *Tharandter. Forstl. Jahrb.* **77**, 225–231.

Carns, H. R. (1966). Abscission and its control. *Annu. Rev. Plant Physiol.* **17**, 295–314.

Cefulu, M., and Smiraglia, C. B. (1964). "I Pollini Aerodiffusi e le Pollinosi." G. Denaro, Rome.

Chira, E. (1966). The effect of age and position in the crown on the pollen quality of *Pinus sylvestris*. *Sb. Vys. Sk. Zemed. Brne* **35**, 339–343.

Chira, E. (1967). Acceleration of the pollen development of *Picea excelsa* under laboratory conditions. *Biologia* (*Bratislava*) **22**, 260–265.

Clausen, K. E. (1960). A survey of variation in pollen size within individual plants and catkins of three taxa of Betula. *Pollen Spores* **2**, 299–364.

Colwell, R. N. (1951). The use of radioactive isotopes in determining spore distribution patterns. *Amer. J. Bot.* **38**, 511–523.

Croat, T. B. (1969). Seasonal flowering behavior in central Panama. *Ann. Mo. Bot. Gard.* **56**, 295–307.

De Vries, A. P. L. (1971). Flowering biology of wheat, particularly in view of hybrid seed production—a review. *Euphytica* **20**, 152–170.

Dormer, K. J. (1962). The fibrous layer in the anthers of Compositae. *New Phytol.* **61**, 150–153.

Duffield, J. W. (1953). Pine pollen collection dates—Annual geographic variation. *Calif. Forest Range Exp. Sta., Forest Res. Notes* **85**.

Dyakowska, J. (1937). Researches on the rapidity of the falling down of pollen of some trees. *Bull. Acad. Pol. Sci., Ser. B* **8/10**, 155–168. 155–168.

Dyakowska, J., and Zurzycki, J. (1959). Gravimetric studies on pollen. *Bull. Acad. Pol. Sci.* **7**, 11–16.

Eisenhut, G. (1961). The morphology and ecology of the pollen of indigenous and exotic forest trees. *Forstwiss. Forsch.* **15**, 1–68.

Erdtman, G. (1969). "Handbook of Palynology." Hafner, New York.

Eriksson, G. (1968). Temperature response of pollen mother cells in *Larix* and its importance for pollen formation. *Stud. Forest. Suec.* **63**, 5–131.

Faegri, K., and van der Pijl, L. (1966). "The Principles of Pollination Ecology." Pergamon, Oxford.

Fahn, A., and Werker, E. (1972). Anatomical mechanisms of seed dispersal. *In* "Seed Biology" (T. T. Kozlowski, ed.), Vol. 1, pp. 151–221. Aacdemic Press, New York.

Fielding, J. M. (1957). Notes on the dispersal of pollen by Monterey pine. *Aust. Forest.* **21**, 17–22.

Goss, J. A. (1971). Effect of light intensity on pollen production in *Ornithogalum caudatum. Amer. J. Bot.* **58**, 476.

Grano, C. X. (1958). A time saving slide for trapping atmospheric pollen. *Forest Sci.* **4**, 94–95.

Gregory, P. H. (1960). "Microbiology of the Atmosphere." Wiley (Interscience), New York.

Guppy, H. P. (1912). "Studies in Seeds and Fruits." Williams & Norgate, London.

Hagerup, O. (1950). Rain pollination. *Kgl. Vidensk. Selsk. Biol. Medd.* **18**(5), 1–19.

Herich, R. (1961). Gerschlechtsausprägung und Grösse der Zellkerne sowie der Pollenkörner bei monözischen Hanf verschiedener Formen. *Zuechter* **31**, 48–51.

Hirose, T. (1969). Studies on chemical emasculation in pepper. I. Suppression of anther dehiscence and other morphological changes caused by Na 2,2-dichloropropionate. *J. Jap. Soc. Hort. Sci.* **38**, 29–35.

Hirst, J. M. (1952). An automatic spore trap. *Ann. Appl. Biol.* **39**, 257–265.

Hopkins, J. S. (1950). Differential flotation and deposition of coniferous and deciduous tree pollen. *Ecology* **31**, 633–641.

Hyde, H. A. (1951). Pollen output and seed production in forest trees. *Quart. J. Forest.* **45**, 172–175.

Illy, G., and Sopena, J. (1963). The dispersal of *Pinus pinaster* pollen. *Rev. Forest. Fr.* **15**, 7–18.

Janzen, D. H. (1967). Synchronization of sexual reproduction of trees within the dry season in Central America. *Evolution* **21**, 620–637.

Janzen, D. H. (1971). Seed predation by animals. *Annu. Rev. Ecol. Syst.* **2**, 465–492.

Jovančević, M. (1962). Determining the germinative potential of forest tree pollen by the size, form and color of pollen grains. *Nar. Sumar.* **16**, 493–502.

Kavetskaya, A. A., and Tokar, L. O. (1963). The vegetative action of a large quantity of pollen during pollination of *Juglans regia. Bot. Zh.* (*Leningrad*) **48**, 580–585.

Kenady, R. M. (1968). A water-trap pollen sampler. *Forest Sci.* **14**, 105–106.

Kerner, A. J. (1904). "The Natural History of Plants." Gresham Publ. Co., London.

Kiyosawa, S. (1962). Influence of some environmental factors on the development of pollen in rice plant. *Proc. Crop Sci. Soc. Jap.* **31**, 37–40.

Kleber, E. (1935). Hat das Zeitgedächthis der Bienen biologische Bedeutung? *Z. Vergl. Physiol.* **22**, 221–262.

Knoll, F. (1932). Über die Fernverbreitung des Blütenstaubes durch den Wind. *Forsch. Fortschr.* **8**, 301–302.

Knox, R. B., Willing, R. R., and Ashford, A. E. (1972). Role of pollen wall proteins as recognition substances in interspecific incompatibility in poplars. *Nature (London)* **237**, 381–383.

Konar, R. N. (1963). In vitro studies of *Pinus roxburghii* Sar. *Plant Tissue Organ Cult., Symp., 1961* pp. 224–229.

Koski, V. (1967). Pollen dispersal and its significance in genetics and silviculture. *Forest Res. Inst., Final Rep.* **P.L.-480-No. E8-FS50**, 1–117.

Krefting, L. W., and Roe, E. I. (1949). The role of some birds and mammals in seed germination. *Ecol. Monogr.* **19**, 269–286.

Kugler, H. (1970). "Blütenökologie," 2nd ed. Fischer, Stuttgart.

Kurtz, E. B., and Liverman, J. L. (1958). Some effects of temperature on pollen characters. *Bull. Torrey Bot. Club* **85**, 136–138.

Langner, W. (1952). Eine Mendelspaltung bei Aura-Formen von *Picea abies* (L.) Karst als Mittel zur Klärung der Befruchtungsverhältnisse im Walde. *Z. Forstgenet.* **2**, 49–51.

Leopold, A. C. (1971). Physiological processes involved in abscission. *HortScience* **6**, 376–378.

Linskens, H. F., and Schrauwen, J. (1969). The release of free amino acids from germinating pollen. *Acta Bot. Neer.* **18**, 605–614.

Liu, W. C., and Carns, H. R. (1961). Isolation of abscisin, an abscission accelerating substance. *Science* **134**, 384–385.

McDonald, J. E. (1962). Collection and washout of air-borne pollen and spores by raindrops. *Science* **135**, 435–436.

McQuade, H. A. (1952). The induction of parthenocarpy in Petunia. *Ann. Mo. Bot. Gard.* **39**, 97–112.

Maheshwari, P. (1950). "An Introduction to the Embryology of Angiosperms." McGraw-Hill, New York.

Mangelsdorf, P. C. (1932). Mechanical separation of gametes in maize. *J. Hered.* **28**, 288–295.

Matthews, J. D. (1963). Factors affecting the production of seed by forest trees. *Forest. Abstr.* **24**, i–xiii.

Meeuse, B. J. D. (1961). "The Story of Pollination." Ronald Press, New York.

Morré, D. J. (1968). Cell wall dissolution and enzyme secretion during leaf abscission. *Plant Physiol.* **43**, 1545–1559.

Nienstaedt, H., and Kriebel, H. B. (1955). Controlled pollination of eastern hemlock. *Forest Sci.* **1**, 115–120.

Oberle, G. D., and Goertzen, K. L. (1952). A method for evaluating pollen production of fruit varieties. *Proc. Amer. Soc. Hort. Sci.* **59**, 263–265.

Olden, E. J. (1954). Giant pollen grains in fruit trees from colchicine treatment in vacuum. *Hereditas* **40**, 526–529.

Ornduff, R. (1970). Incompatibility and the pollen economy of *Jepsonia parryi*. *Amer. J. Bot.* **57**, 1036–1041.

Peterson, P. A. (1955). The avocado flower cycle as related to pollination and fruit set. *Citrus Leaves* **35**, 8–9.

Poggendorff, W. (1932). Flowering, pollination, and natural crossing in rice. *Agr. Gaz. N. S. W.* **43**, 898–904.

Pohl, F. (1937). Die Pollenkorngewichte einiger windblütiger Pflanzen und ihre ökologische Bedeutung. *Bot. Zentralbl.* **57**, 112–172.

Polunin, N. (1951). Arctic aerobiology: Pollen grains and other spores observed on sticky slides exposed in 1947. *Nature (London)* **168**, 718–721.

Porter, R. H., and Livingston, C. H. (1956). "Laboratory Manual for Botany and Seed Plants." Burgess, Minneapolis, Minnesota.

Richter, S. (1929). Über den Öffnungsmechanismus der Antheren bei einigen Vertretern der Angiospermen. *Planta* **8**, 154–184.

Ridley, H. N. (1930). "The Dispersal of Plants Throughout the World." Reeve, Kent, England.

Romashov, N. V. (1957). Laws governing fruiting in oak. *Bot. Zh. (Leningrad)* **42**, 41–56.

Rubinstein, B., and Leopold, A. C. (1963). Analysis of the auxin control of bean leaf abscission. *Plant Physiol.* **38**, 262–267.

Rushton, P. S. (1969). Effects of 5-fluorodeoxyuridine on radiation-induced chromatid aberrations in *Tradescantia* microspores. *Radiat. Res.* **38**, 404–413.

Russell, N. H. (1958). "An Introduction to the Plant Kingdom." Mosby, St. Louis, Missouri.

Saitan, I. M. (1952). Mutual interaction between the individual components of pollen mixtures. *Agrobiologiya* **5**, 108–111.

Sarvas, T. (1955). Investigations into the flowering and seed quality of forest trees. *Commun. Inst. Forest. Fenn.* **45**, 7.

Sarvas, T. (1962). Investigations in the flowering and seed crop of *Pinus sylvestris*. *Commun. Inst. Forest. Fenn.* **53**, 4.

Schoch-Bodmer, H. (1940). The influence of nutrition upon pollen grain size in *Lythrum salicaria*. *J. Genet.* **40**, 393–401.

Scott, F. M., Schroeder, M. R., and Turrell, F. M. (1948). Development, cell shape, ruberigation of internal surface, and abscission in the leaf of the Valencia orange, *Citrus sinensis. Bot. Gaz. (Chicago)* **109**, 381–411.

Silen, R. R., and Copes, D. L. (1972). Douglas-fir seed orchard problems—a progress report. *J. Forest.* **70**, 145–147.

Snyder, E. B., and Clausen, K. E. (1973). Woody-plant seed manual. *U. S., Dep. Agr., Misc. Publ.* **654** (in press).

Stanley, R. G. (1958). Methods and concepts applied to a study of flowering in pine. *In* "The Physiology of Forest Trees" (K. V. Thimann, ed.), pp. 583–599. Ronald Press, New York.

Stanley, R. G. (1964). Physiology of pollen and pistil. *Sci. Progr. (London)* **52**, 122–132.

Stanley, R. G., and Search, R. W. (1971). Pollen protein diffusates. *In* "Pollen Development and Physiology" (J. Heslop-Harrison, ed.), pp. 174–176. Butterworth, London.

Stephen, W. P. (1958). Pear pollination studies in Oregon. *Oreg., Agr. Exp. Sta., Tech. Bull.* **43**.

Stix, E., and Grosse-Brauckmann, G. (1970). Der Pollen- und Sporengehalt der Luft und seine tages- und jahreszeitlichen Schwankungen unter mitteleuropäischen Verhältnissen. *Flora (Jena), Abt. B* **159**, 1–37.

Szwabowicz, A. (1971). Badamia ned pytkiem polomstova krzozy ojcowskiej *Betula oycovensis. Acta Soc. Bot. Pol.* **40**, 91–121.

Taylor, J. H. (1950). The duration of differentiation in excised anthers. *Amer. J. Bot.* **37**, 137–143.

Turpin, R. A., and Schlising, R. A. (1971). A new method for studying pollen dispersal using iodine-131. *Radiat. Bot.* **11**, 75–78.

van der Pijl, L. (1969). "Principles of Dispersal in Higher Plants." Springer-Verlag, Berlin and New York.

Vasil, I. K. (1959). Cultivation of excised anthers in vitro—effect of nucleic acids. *J. Exp. Bot.* **10**, 399–408.

Venkatesh, C. S. (1956). The form, structure, and special modes of dehiscence in anthers of Cassia. I. Subgenus Fistula. *Phytomorphology* **6**, 168–176.

Visser, T. (1955). Problemen bij stuifmeel van fruit gewassen. *Meded. Dir. Tuinbouw* (*Neth.*) **18**, 856–865.

Voisey, P. W., and Bassett, I. J. (1961). A new continuous pollen sampler. *Can. J. Plant Sci.* **41**, 849–853.

Wettstein, W., and Onno, M. (1949). Blütenbiologische Untersuchungen an Koniferen und bei Tilia. *Oesterr. Bot. Z.* **95**, 475–478.

Wodehouse, R. P. (1935). "Pollen Grains." Hafner, New York.

Wodehouse, R. P. (1945). "Hayfever Plants." Chronica Botanica, Waltham, Massachusetts.

Worsley, R. G. F. (1959). The processing of pollen. *Silvae Genet.* **8**, 143–148.

Wright, J. (1952). Pollen dispersion of some forest trees. *Northeast. Forest Exp. Sta., Pap.* **46**.

. 9 .

Shedding of Reproductive Structures in Forest Trees

G. B. Sweet

I. Introduction

Until recently, relatively few botanists with the freedom to choose the plant material on which to carry out research chose forest trees. Thus a great deal of the general forest tree literature originated from researchers without that freedom, viz., those employed by forestry interests. This fact has influenced the type of research material published on forest trees: first it is somewhat problem oriented, and second it tends to be rather species limited. To date forest management (and consequently forestry research) has been mainly concentrated in the temperate and sub-temperate zones of the world where the genera of commercial importance are largely coniferous, and of the conifers probably the most important genus in terms of area of forest under management is *Pinus*. Thus a great deal of the material reviewed in this chapter relates to problems associated with seed production in conifers, and particularly in *Pinus*.

How important to forestry is the premature death or abscission of

341

reproductive structures? Clearly it is less important than in horticulture where the end product of growth is frequently the fruit. In forestry the production of seed is essentially a means to the end of establishing a new crop for the production of wood, and thus one person to whom it is important is the forest manager. To him the premature abscission of a flower crop may cause a seed shortage which has a major effect on forest establishment, not only for the year in which it occurs but possibly also for subsequent years, as many forest tree species are periodic flowerers. To the tree breeder a high level of flower abscission may mean a costly delay in completing progeny tests or in bringing genetically improved material into production; or it may mean the physical inability to produce seed of a potentially important hybrid. Certainly it will result in an increase in the cost of seed produced in seed orchards. To the geneticist the abortion of ovules after self-pollination may be important in its effect on the constitution of both present and future populations. Researchers in all these fields have contributed to the literature on abscission in reproductive structures in forest trees, as have plant physiologists, pathologists, ecologists, and economists. As stated previously, much of this research is problem oriented but, because it has been concentrated on a limited number of species some of which have botanically interesting characteristics, the literature does in addition make a useful contribution to botanical knowlege per se.

It is necessary at this stage to define some of the terminology used in the chapter. The word flower is classically defined as "a determinate sporogenous shoot, bearing carpels" (Hillman, 1963). Such a definition excludes the strobili of gymnosperms, a fact which is inconvenient when discussing gymnosperms and angiosperms jointly. To overcome this a flower will therefore be redefined for the purpose of this chapter simply as "a determinate sporogenous shoot" (Jackson and Sweet, 1972). Thus, as used here the terms flower and strobilus are synonymous when applied to the gymnosperms. In *Pinus* a female strobilus between the time of anthesis and the time of fertilization is frequently referred to as a conelet and subsequent to fertilization as a cone. These terms also will be used in the chapter.

Before discussing the incidence of premature abscission of reproductive structures, it is useful to have some data on the *total* amount of reproductive material shed in a forest stand and its relation to the total dry matter production of the stand. Bray and Gorham (1964) cite references listing from 1 to 17% of the total dry weight of litter on the forest floor as composed of reproductive parts: the genera examined included *Pinus, Picea, Betula, Quercus,* and *Eucalyptus.* Rodin and Bazilevich (1967) similarly list values between 1 and 6% for *Pinus sylvestris* and from 2 to

30% for *Picea abies.* Some of the variability in these data may reflect differences in the relative rates of breakdown of reproductive and other material on different sites.

Fielding (1960) estimated the total dry matter expended per annum on sexual reproduction in a plantation of *Pinus radiata* in Australia to be approximately 16% of the total mean annual increment of the stand. In actual terms, Sarvas (1962) estimated the seed production of *Pinus sylvestris* in Finland to be approximately half a million seeds per hectare per annum; this compares with nearly three million seeds for *Picea abies,* of which almost one million were filled (Sarvas, 1968). These numbers, however, are relatively low compared with the peak shedding of one-third of a million viable seeds per hectare per week recorded for *Eucalyptus pilularis* (Florence, 1964) and the ten million viable seeds per hectare per annum reported for *Eucalyptus delegatensis* by Grose (1960).

Sarvas (1962, 1968) reported an average production of some 50 kg dry weight ha/annum of pollen and pollen strobili for *Pinus sylvestris* and *Picea abies* in Finland, while Fielding (1960) estimated some 900 kg/ha/annum for *Pinus radiata* in Australia. Some of the differences between these values for *Pinus* relate to variations in strobilus weight; if weights of pollen only are compared, Fielding reported some ten times the annual production shown by Sarvas. The *number* of pollen grains produced was estimated by Sarvas to average $120/mm^2/annum$ ($= 12 \times 10^{11}/ha/annum$). An estimate based on Fielding's data suggests some four times as many grains as that.

The premature abscission of flowers is as common in forest tree species as in other parts of the plant kingdom. With *Eucalyptus pilularis* in Australia, for example, Florence (1964) recorded a loss of a potential 20–25 million viable seeds per hectare through premature abscission of flower buds and immature capsules. In *Eucalyptus delegatensis,* Grose (1960) recorded a loss in one year of 40 million potentially viable seeds per hectare through premature dropping of primordia, flower buds, and immature capsules; the actual seed crop in that year was six million viable seeds per hectare. Florence reported that although shedding occurred throughout the period of reproductive development, there were peaks which correlated with stages of major increases in the growth of the reproductive structures.

Williamson (1966) recorded that more than 90% of flowers initiated on trees of *Quercus alba* in Kentucky abscised prematurely in the 4 months subsequent to anthesis; some 65% of this abscission occurred between anthesis and fertilization, with the remainder occurring during fertilization, embryo development, and acorn maturation (see Fig. 1). Comparable abscission occurs in the European species of *Quercus* and has also

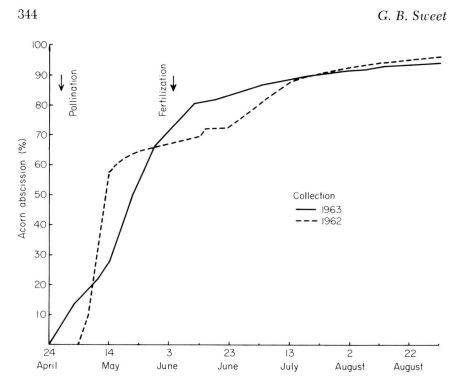

FIG. 1. Percentage abscission of flowers and fruit of *Quercus alba* in Kentucky, U. S. A. during the 4-month period from anthesis to acorn maturity (from Williamson, 1966).

been described for teak (*Tectona grandis*) (Bryndum and Hedegart, 1969). Neither is premature abscission restricted to female or bisexual flowers, it also occurs in male strobili (Silen, 1967).

In *Pinus* there are a large number of references to premature abscission presenting very similar information. Snyder and Squillace (1966) found that the successful development of cones from controlled pollinations of several *Pinus* species averaged over a 10-year period was only about 40%. Forbes (1971) found that cones developed from only 40% of strobili control pollinated in *P. echinata*, and Bingham and Rehfeldt (1970) obtained some 60% survival of control-pollinated strobili of *P. monticola*. Similarly, Wang (1970) obtained 52% survival with *P. ponderosa*; Brown (1970b) reported survivals ranging from 30 to 80% in *P. sylvestris* and Katsuta and Satoo (1965) with *P. thunbergii* obtained strobilus survivals around 50%. Ebel (1964) described conelet drop in wind-pollinated strobili of *P. palustris*, and Sweet and Thulin (1969) and Forbes (1971) showed that survival under these conditions is no greater than in control-pollinated strobili, with only some half of all strobili initiated in *P. radiata* and *P. echinata* surviving past the first year. S. L. Krugman (unpublished data) has shown this to be typical of a number of other *Pinus* species also. Sarvas (1962) related premature conelet drop to site quality and observed an incidence of drop in *P. sylvestris* plantations of between 20 and 100%; Bramlett (1972) with *P. echinata* observed mortalities ranging from 35 to 97%.

A number of cases of premature abscission have been shown to result from damage; frost has been implicated by many authors (e.g., Hard, 1963; Hutchinson and Bramlett, 1964; Krugman, 1966) as well as drought, high temperatures, and attacks by insects, fungi, birds, and mammals. The incidence of such damage in all stages of development is very great, as evidenced by the fact that *Forestry Abstracts* has cited some 235 papers published between 1951 and 1971 that are concerned primarily with injury to flowers, cones, fruits, and seeds of forest trees. While this subject is clearly an important one, it tends to be very much a local one in terms of the pathogenic and other agencies responsible; for this reason it will not be dealt with further in this chapter.

The majority of the references cited so far relate to the death or abscission of reproductive structures that have no visible signs of injury or in which any occurrence of injury is secondary. Such drop has been called physiological drop (Bramlett, 1972) to distinguish it from that caused by physical injury. It is with the physiological abscission, death, or abortion of reproductive structures that this chapter will be essentially concerned.

II. Abscission or Death prior to Anthesis

Death or abscission during this period has not been extensively recorded in forest trees. In *Pinus radiata* (G. B. Sweet, unpublished data) there is an estimated 3–5% of female strobili initiated which abort during the 3 months prior to anthesis. Such strobili are not abscised but remain attached to the stem, and it appears that the meristem remains alive in at least some of them. D. L. Bramlett (unpublished) has also observed this phenomenon in some 5% of strobili in *Pinus echinata*, and Hashizume (1963) has recorded cessation of development prior to anthesis in strobili of *Chamaecyparis obtusa*. In *Eucalyptus pilularis*, preanthesis abortion was accompanied by abscission (Florence, 1964), and this was the time of the major loss of reproductive material.

Silen (1967) has recorded extensive cessation of growth of male strobili on Douglas fir (*Pseudotsuga menziesii*) in Oregon some 3½ months after

FIG. 2. A twig of *Pseudotsuga menziesii* in autumn showing male reproductive buds. The large buds have developed normally, the smaller ones have ceased development precociously (after Silen, 1967).

initiation. The buds did not abscise, they merely ceased growth (see Fig. 2) and the number of buds which ceased growth was positively correlated with the altitude at which the trees were growing. A meristem remained alive in the buds which could subsequently be forced into growth; such growth however was vegetative, not reproductive, although there was little doubt that the buds initially had male reproductive potential.

As far as is known there has been little discussion as to the cause of such early death, stagnation, or abscission of reproductive structures of forest trees. Florence has correlated heavy flower bud fall in *Eucalyptus* during this period with a major increase in rate of bud growth, implying perhaps that abscission resulted from competition for metabolites. In

Pinus lack of development frequently occurs to only one or two strobili in a cluster. The impression is that these strobili may initially have been smaller than the others, and that as the size gap widened the small ones simply ceased growth. This could be interpreted in terms of the inability of a small organ to attract metabolites successfully in competition with the larger ones.

Overall, the physiology of reproductive structures at an early development stage probably does not differ markedly from that at a later stage, except for the fact that early in reproductive development the status may still be reversible (see, e.g., Owens, 1969). Thus the more detailed physiological discussion presented in some of the later sections of this chapter may also be relevant to considerations of abscission or death prior to anthesis.

III. Abscission between Anthesis and Fertilization

A. INTRODUCTION AND DESCRIPTION

Between anthesis and fertilization is the period during which the major incidence of reproductive drop occurs in most species of forest trees which have been intensively studied. In *Quercus,* for example, more than 65% of the total abscission occurred during this period (Williamson, 1966), and for *Pinus radiata* the figure is more than 90% (G. B. Sweet, unpublished). This latter value is probably fairly typical for many species of *Pinus.* Ebel (1964) has given the following description of conelets destined to drop in *Pinus palustris* during this period: ". . . many (1st year) conelets, including their stalks, were still succulent. They were of comparable size and appearance to conelets still firmly attached to the shoots. But in most of the readily detached conelets internal brown discoloration was present in the stalks at the point of contact with the supporting shoots. When picked, these conelets exuded little or no resin from the stalks, in contrast to firmly attached conelets which produced a very noticeable resin flow." This description has been since repeated with only minor variations by a number of workers. In *Pinus radiata* Sweet and Thulin (1969) reported that conelets ready to drop were smaller in size than the remaining ones (see Fig. 3), and they frequently noted the presence of resin droplets on the scales, often associated with some necrosis; but the essential description by Ebel held true.

Investigators reported strong clonal differences in both the amount and the timing of drop, with possibly also a clonal effect of the pollen

FIG. 3. Female strobili of *Pinus radiata* 33 days after anthesis. The strobilus on the right will develop normally; the other two strobili which are smaller will abscise within a few days (from Sweet and Thulin, 1969).

parent (Sweet and Thulin, 1969; Forbes, 1971); also geographical differences exist in the incidence of drop (Sarvas, 1962) and a site–genotype interaction may occur (G. B. Sweet, unpublished). The overall incidence of drop varies strongly from year to year; Bramlett (1972), for example, quoted a 3% survival one year and a 65% survival the following year (although these data did include some losses due to insect and frost damage). In *P. radiata* the incidence of drop is similar after both wind and controlled pollinations (Sweet and Thulin, 1969), but Brown (1970a) with *P. sylvestris* has shown clonal variability in this respect. Within a tree, Sweet and Thulin showed some indication of a higher percentage drop in the lower than in the upper part of the crown of *Pinus radiata*, and Brown (1970b) reported a higher incidence of drop on the colder aspects of the tree. Although only limited work has been done on the anatomy of precocious shedding, it appears that in many cases an abscission layer is formed at the base of the pedicel, probably similar to that in an unpollinated strobilus (see later). Abscission layers have been reported by Bramlett (1972) and Brown (1970a) and have been observed by me (G. B. Sweet, unpublished). They are not always present, however, and in some cases flowers simply die and shrivel up without being shed.

B. THE ABORTED OVULE

1. *Abortion through Lack of Pollination*

Agamospermy (the development of viable seeds in the absence of pollen) can occur in some species of forest trees, e.g., *Acer saccharum* (Gabriel, 1967) and possibly *Pseudotsuga* (Allen, 1942; Orr-Ewing, 1957a). It is not common, however, and as a regular phenomenon it probably does not occur in gymnosperms (Dogra, 1966), in which normally the unpollinated ovule aborts. In genera such as *Pinus* in which ovule growth and development are slow, the seed coats of aborted ovules almost universally fail to develop. In genera such as *Pseudotsuga* and *Picea*, however, in which seed development is considerably more rapid, the nonpollinated ovule, while it aborts, still leads to the development of an empty seed (Sarvas, 1962).

a. *Pinus*. Sarvas (1962) in a detailed study of *Pinus sylvestris* showed that development of the unpollinated ovule was normal until after the meiotic division of the megaspore mother cell, some three weeks after anthesis. (Note that in *Pinus* as distinct from many other species, female meiosis occurs *after* anthesis.) From that time on the nonpollinated ovule gradually begins to degenerate. The process starts in the region of the megaspore: the cells of the spongy tissue enlarge, and first their contents and then their cell walls disappear, creating a gradually enlarging cavity around the megaspore. Within a month the spongy tissue changes entirely into a formless mass and most of the adjacent nucellus tissue decomposes. Finally the center of the ovule collapses entirely. B. S. Henderson (unpublished) has confirmed the essential pattern of this sequence in *Pinus radiata*, but found differences in timing and in some areas of detail. She also has found evidence to suggest that some of the unpollinated ovules in the fertile area of the strobilus of *P. radiata* may be unpollinated because of abnormalities present prior to pollination. Sarvas points out that although the megasporangium initially develops independently of pollination, the female gametophyte does not; and it is the collapse of the gametophyte which ultimately halts ovule development.

Physiologically it is of interest to determine just how pollination in *Pinus* prevents ovule abortion. In many angiosperms pollen germination and fertilization occur at a close time interval, and it is accepted that fertilization provides the key to ovule development. In gymnosperms, however, the time interval is frequently longer. In *Pinus*, fertilization does not occur for some 13 months after pollination (Ferguson, 1904), and thus fertilization cannot be necessary for early ovule development. Rather the trigger to ovule development must lie in the process of pollination alone. It therefore seems reasonable to postulate that the germinating pollen grains provide some substance(s) that assists normal development of the ovule. That there is some specificity to the compound is implied by work with interspecific crossing. McWilliam (1959), Krugman (1970), and Katsuta (1971), for example, describe patterns of ovule collapse following interspecific pollinations which are similar to those described for unpollinated ovules. This topic will be discussed in more depth later in the chapter.

b. GENERA WITH MORE RAPID REPRODUCTIVE DEVELOPMENT. The abortion of unpollinated ovules in *Picea* has been described by Sarvas (1968) and Mikkola (1969). In *Picea* meiosis occurs prior to anthesis, and while the degeneration of unpollinated gametophytes begins soon after anthesis, integument development continues relatively normally. However the development threshold of the gametophyte appears to be considerably lower in *Picea* than in *Pinus*, and gametophytes have been observed to develop in a small number of unpollinated ovules. Further, pollination with dead pollen, pollen of other species of *Picea*, and even with pollen of *Pinus* has been able to promote normal ovule development. All ovules, pollinated or not, which pass the threshold associated with pollination produce archegonia, but the unpollinated ones finally abort some weeks after the time fertilization should have occurred (Mikkola, 1969). In *Pseudotsuga* unpollinated ovules also abort, but not until some 80 days after anthesis (Orr-Ewing, 1957b). This is some weeks after the date on which fertilization would have occurred had the ovules been pollinated. The seed coat by that time is full sized. In *Alnus* and *Betula* also (Hagman, 1970), unpollinated ovules abort leaving empty seeds which are full sized. It is important to note that in the species discussed in this paragraph, in which unpollinated ovules form full sized (but empty) seed, ovule abortion does not normally lead to abscission of the developing conelet or fruit.

2. *Abortion through Other Causes*

In forest trees the abortion of ovules prior to fertilization can occur for a variety of reasons. In some cases it represents normal development, e.g., where there is only room for one of several ovules in an ovary to develop. In other cases ovules abort because morphologically they are not sufficiently developed to continue growth. For example, in *Populus trichocarpa* the pistils contain 50–70 ovules of which only 25–40 are potentially functional (Stettler and Bawa, 1971). Such incomplete development has been particularly well documented in conifers. In *Pinus sylvestris,* for example, Sarvas (1962) demonstrated that 78% of the scales in each cone were deficient or sterile, of these 92% were at the base of the strobilus and 8% were at the apex. In *Pinus radiata* in New Zealand some 60% of the scales cannot produce seed. The degree of deficiency increases in scales progressively further away from the functional zone represented by the distal part of the conelet. Those scales bordering the region of fully fertile scales may have ovules with a normal megaspore mother cell and spongy tissue, which may be pollinated in the normal way. Moving further away, however, ovules are present in which the micropylar arms become smaller, the nucellus rudimentary, and megaspore development does not occur (B. S. Henderson, unpublished). Beyond this point, the structures present are too rudimentary to be regarded as ovules and the scales may be classified as sterile. In *Pinus radiata* in New Zealand the ratio of deficient to fully sterile scales is approximately 1:5 (B. S. Henderson, unpublished data).

While slightly deficient ovules may be pollinated in the normal way, deficient ovules usually begin to deteriorate soon after anthesis and there is no further development of ovule or seed coat. In a study of *Pseudotsuga menziesii* in New Zealand the percentage of deficient ovules was shown to range from 13 to 52%, being influenced in part by geography and/or climate (Sweet and Bollmann, 1972). In *Picea abies* in Finland the figure is about 7% (Sarvas, 1968). In all these species, while the deficient ovules abort, their seed wings develop relatively normally with the amount of development being dependent largely on the rate of growth (Sarvas, 1968).

The question now arises as to whether all ovules which are functionally capable of development following pollination do in fact develop. There are frequently practical problems in determining directly whether or not ovules which abort have been pollinated, but Sarvas has shown by indirect means that in forest stands of *P. sylvestris* aborted ovules (which were not initially deficient) result almost exclusively from lack of pollination. While these data are convincing it is by no means certain that this situation is universally true. For example, B. S. Henderson (unpublished) has observed in *P. radiata* a number of well-pollinated ovules in the fertile portion of the strobilus in the process of abortion (see Fig. 4), and Burdon and Low (1973) believe the incidence of this type of abortion to be reasonably high. They have suggested (on the basis of limited

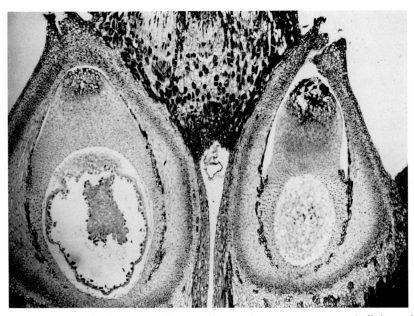

FIG. 4. Longitudinal section of two ovules on the same megasporophyll from the fertile portion of a strobilus of *Pinus radiata*. Well-developed pollen tubes can be seen in both ovules, but these are particularly clear in the right-hand one which is aborting, as evidenced by the shrunken nature of its tissues. The gap in the spongy tissue of the left-hand (healthy) ovule is due to collapse on sectioning. (Photograph by B. S. Henderson, Canterbury University, New Zealand.)

evidence) that competition for nutrients may be a factor leading to abortion of pollinated ovules, and Lyons (1956) has made a similar suggestion. Bramlett (1972) reported an average abortion of 69% of the fertile ovules of *Pinus echinata* and that two out of every three of these ovules aborted at a later stage than would normally be expected from unpollinated ovules. While it is possible that insects were implicated in some of this abortion, it seems likely that at least some of it represented physiological abortion of pollinated ovules. This is also evidenced indirectly by individual clones in well-pollinated seed orchards which show a high percentage of undeveloped ovules.

Krugman (1970) has reported frequent ovule abortion following retarded penetration of the nucellus by pollen tubes in interspecific crossings within the *Diploxylon* subgenus of *Pinus*. In the cross between *P. coulteri* and *P. ponderosa*, for example, pollen tube growth ceased 4 months after pollination, the developing megaspore collapsed, the surrounding spongy tissues enlarged and then collapsed, and the nucellus shrank. In *Picea* also, many interspecific crosses result in varying degrees of restricted pollen tube growth, and in angiosperms a number of genera in the Fagales show retarded pollen tube growth and ovule abortion following interspecific pollination (Hagman, 1970). In *Pinus* (Hagman and Mikkola, 1963) and *Pseudotsuga* (Orr-Ewing, 1957b) there has been no clear demonstration of selection on the nucellus against the germination of selfed pollen. An incompatibility mechanism does operate, however, in the styles of flowers of a number of genera such as *Alnus, Betula, Quercus,* and *Corylus* (Hagman, 1970) inhibiting the growth of tubes from selfed pollen and frequently leading to ovule abortion.

C. The Effect of Aborted Ovules on Flower Development

1. *The Unpollinated Flower*

The development of fruits or seed-bearing structures in the absence of pollination is referred to as agamocarpy. The ability of flowers to develop without pollination into fruit is present in widely different measure in different species of the plant kingdom, and variation in this factor is equally present in forest trees. *Fagus sylvatica,* for example, does show agamocarpy (Nielsen and Schaffalitzky de Muckadell, 1954) as does *Acer saccharum* (Gabriel, 1967) and *Alnus* and *Betula* (Hagman, 1970). *Castanea* species on the other hand do not (Nienstaedt, 1956). In gymnosperms, as probably in angiosperms, differences occur basically at the family level. In the *Abietaceae* (e.g., *Abies, Pseudotsuga, Picea*) unpollinated conelets develop essentially in the same way as do normal pollinated ones (Dogra, 1966; Orr-Ewing, 1957a), whereas in *Pinus* this happens most infrequently (Sarvas, 1962). On those occasions when it does happen there is a marked size difference in the cones at maturity (Katsuta, 1971). *Juniperus* (Sarvas, 1962) and perhaps some other members of the Cupressaceae behave like *Pinus* in this respect, and Sarvas (1962) has suggested that whether or not agamocarpy occurs in gymnosperms may relate to the rate of development of the strobilus. *Pseudotsuga,* for example, with a time of only 5 months between anthesis and seed maturation, contrasts strongly with *Pinus* (a minimum of 14 months) in this respect.

The abscission of unpollinated strobili has probably been studied more in *Pinus* than in any other forest tree genus, and in *Pinus* we now have a clear-cut picture of the anatomy and timing of drop of unpollinated strobili, even though our understanding of the physiology is limited. The generalized situation (M. Hagman, unpublished) is one in which development of the unpollinated strobilus is perfectly normal for some weeks following anthesis. Then frequently during a period of only 2 weeks or so 80 to 90% strobilus abscission occurs. At this time ovule degeneration is well advanced and is visible without sectioning. Thus the time sequence would suggest that the abscission of unpollinated strobili in *Pinus sylvestris* may be triggered by the failure of their ovules to develop. Katsuta (1971) has indicated a similar time sequence in *Pinus thunbergii*. Figure 5 shows a longitudinal section of an abscission layer in the pedicel of an unpollinated strobilus of *P. cembra*. The abscission appears typical of that described in much botanical literature (see, e.g., Addicott, 1965). It occurs at the proximal end of the pedicel in a zone of thin-walled cells lacking intercellular spaces. A considerable buildup of suberin exists above the zone, and abscission takes place by dissolution of the membranes of cells in a separation layer. The separation layer does not pass through the vascular tissue which is presumably blocked by tyloses and which ultimately fractures under wind or other pressure; it does, however, pass through the pith. External appearance suggests the formation of a periderm proximal to the separation zone, but this has not been formally described.

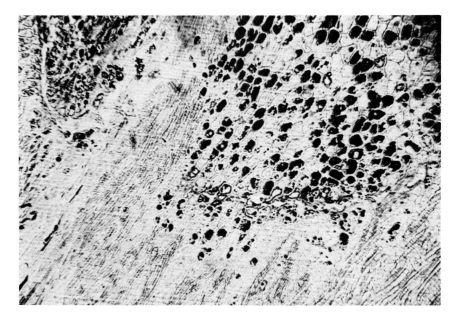

FIG. 5. Formation of the abscission layer in the pedicel of an unpollinated strobilus of *Pinus cembra*. The photograph, taken 30 days after anthesis, shows the abscission layer running across the pith and the tissues on one side of the vascular cylinder. (The photograph is from an unpublished study by M. Hagman, The Finnish Forest Research Institute, Helsinki.)

The length of time between anthesis and the start of abscission varies from species to species, being longer (in Finland) in *P. banksiana* and *P. contorta* than in *P. sylvestris,* which in turn takes longer than *P. cembra* and *P. peuce* (M. Hagman, unpublished). In Japan there is a marked difference between *Pinus thunbergii* and *Pinus densiflora* in this respect (Katsuta, 1971). There is some indication that the length of time varies from year to year, and that within any one species it also varies between clones (Sweet and Bollmann, 1970; Katsuta and Satoo, 1965; Katsuta, 1971). Almost all workers in this field have recorded some unpollinated strobili which, under some conditions, have been able to survive for longer than 1 year (Katsuta and Satoo, 1965; M. Hagman, unpublished; G. B. Sweet and M. P. Bollmann, unpublished) and in some cases to maturity (Sarvas, 1962; Brown, 1970b; Katsuta, 1971).

2. *The Pollinated Flower*

The probability that in *Pinus* conelet abscission in an unpollinated strobilus is a direct result of ovule abortion raises the question as to the

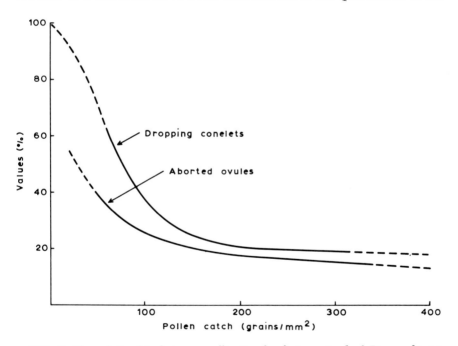

FIG. 6. The relationship between pollination levels in a stand of *Pinus sylvestris* in Finland and the percentage of aborted ovules in the female strobili (lower curve) and the percentage of strobili which abscise (upper curve). The figure combines data from separate figures in Sarvas, 1962.

influence that the proportion of aborting ovules may have on the development of a pollinated strobilus in that genus. Sarvas (1962), with *Pinus
sylvestris*, examined the relationship between the abortion of ovules in
strobili and the incidence of conelet drop. His data indicated a fairly
strong relationship between the two parameters (see Fig. 6) and he
regarded the relationship as causal. Sarvas also showed (Fig. 7) that
(a) the *percentage* of aborted ovules associated with dropping conelets
differed between species and (b) within the one species, it was lower in
larger strobili than in smaller ones. In fact Sarvas found very little variation between different sized strobili in the number of aborted ovules
associated with strobilus drop; thus he suggested tentatively that aborting
ovules might form substances which accumulate quantitatively to prevent cone development. If so, there must be quite a marked genotypic
variation in this character in that, under some conditions, some unpollinated strobili can develop to maturity. The clonal differences affecting
the ease with which unpollinated strobili abscise suggest also that there
may well be clonal differences in the number of aborted ovules leading
to conelet drop in pollinated strobili.

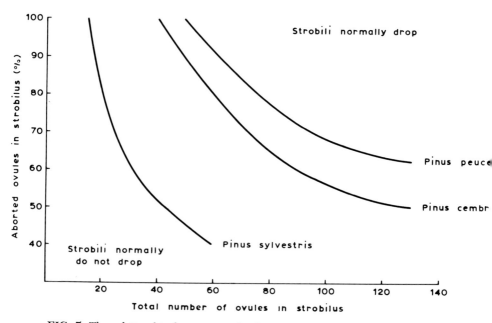

FIG. 7. The relationship between ovule abortion and conelet drop in three species
of *Pinus* in Finland. A strobilus with parameters above the curve for its species would
normally abscise, while one falling below the curve usually would not (after Sarvas,
1962).

While in *Pinus* most of the abscission of strobili with aborted, un-pollinated ovules occurs soon after meiosis, there are a number of refer-ences (e.g., McWilliam, 1959; Krugman, 1970) to the abscission of strobili with aborted ovules in the spring following pollination, a period of rapid increase in conelet size (Sweet and Bollmann, 1971b). Further, Brown (1970b) has reported premature abscission of mature *P. sylvestris* cones just prior to seed shed which appears to relate back to a high incidence of early ovule abortion. In coniferous genera such as *Pseu-dotsuga* and *Picea*, the requirements for strobilus development of a measure of ovule development are apparently not present.

D. The Major Natural Causes of Strobilus Abscission in *Pinus*

The first worker to provide data on the causes of naturally occurring conelet drop of *Pinus* in the forest was Sarvas (1962). Essentially on the basis of the data shown in Fig. 6, he postulated that some 80% of naturally occurring conelet drop in *P. sylvestris* was caused by inadequate pol-lination, and that a low level of pollen caused a high level of ovule abortion which in turn led to conelet drop. He regarded the remaining 20% of drop as probably resulting from damage of one type or another.

M. Hagman (unpublished) has aided our interpretation of the cause of conelet drop by his comparisons of conelet drop between pollinated and unpollinated strobili in a number of pine species (see Fig. 8). Using controlled pollinations of conelets which received adequate supplies of outcrossed pollen, he still found abscission levels of between 26 and 36% for *Pinus peuce*, *P. cembra*, and *P. sylvestris*. The pattern of this abscis-sion is shown in Fig. 8 to differ from that of the unpollinated strobili, where the drop began sharply between 36 and 48 days after anthesis according to the species. Why in these experiments the drop in pollinated strobili occurred prior to that of unpollinated ones is not known, but the data support Sarvas' (1962) contention that in *P. sylvestris* there is a certain basic level of drop, and above this, further drop is due to inadequate pollination. There can be no real doubt that in *P. sylvestris* in Finland low levels of pollen cause much of the naturally occurring conelet drop.

Figure 8 shows that the abscission of unpollinated strobili in *P. sylvestris* occurred some 3 weeks after meiosis, which in its turn was some 3 weeks after anthesis. In *P. radiata* meiosis does not occur for 6 to 7 weeks after anthesis (B. S. Henderson, unpublished) and the major abscission of *unpollinated* strobili does not occur until 10 or 12 weeks after that time (Sweet and Bollmann, 1970). The natural process of conelet drop how-

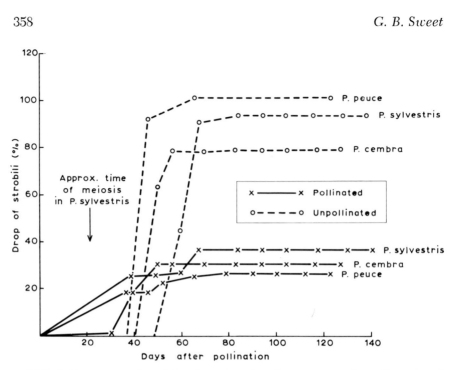

FIG. 8. Percentage of strobilus abscission plotted against time for pollinated and unpollinated strobili of *Pinus peuce, P. sylvestris,* and *P. cembra.* (From unpublished data by M. Hagman, Helsinki.)

ever, which occurs in some 50% of all strobili initiated in seed orchards, reaches a peak some 4 to 6 weeks after anthesis and then decreases in intensity. Conelet drop occurring after the first 10–12 weeks is usually less than 10% of the total (Sweet and Thulin, 1969). The temporal relationship between meiosis, ovule abortion, and conelet drop is thoroughly documented for *P. sylvestris,* and the drop in *P. radiata* does not fit that pattern. It thus seems highly improbable that most abscission of conelets in *P. radiata* seed orchards can be regarded as being caused by inadequate pollination. This is supported by the fact that there is virtually no evidence that pollen levels in *P. radiata* are limiting. As stated in the introduction to this chapter, pollen production in *P. radiata* in Australia is apparently some ten times as high as it is in *P. sylvestris* on a good site in Finland.

Sweet and Bollmann (1970) examined other possible causes of conelet drop in *P. radiata.* In New Zealand *P. radiata* grows vigorously in comparison with *P. sylvestris* in Finland. In the areas where the studies were made, for example, the predominant height of *P. radiata* was 30 m at age 20 years, compared with dominant heights ranging from 17–28 m at age 100 years for the *P. sylvestris* studied in Finland. In the seasonal development of *P. radiata* there is little terminal vegetative growth on a shoot prior to the strobili becoming receptive. Concurrently with, or immediately following anthesis, however, the shoot apex and lateral vegetative buds associated with the strobili increase in length very

FIG. 9. The leading shoot of a graft of *Pinus radiata* growing in New Zealand. Three clusters of strobili have been initiated in the current season. At the time of photography the apical cluster was receptive to pollen, while the basal cluster had been receptive some 5 weeks previously.

rapidly and on the leading shoot of the tree there may be three clusters of strobili produced in one season, each with a cluster of branches distal to it (see Fig. 9). On first-order branches there would be normally only one cluster of strobili. Figure 10 shows the change with time in the relative growth rate in length (RLGR) of the distal portion of strobilus-bearing shoots of *Pinus radiata* on two sites. On site 1, where the incidence of conelet drop is about 50%, RLGR was high after anthesis, but not at its seasonal maximum. On site 2 where conelet drop normally approaches 100%, shoot RLGR reached a peak soon after anthesis, which was much higher than at any other time examined. There is ample evidence in the literature to show that in woody plants, rapid vegetative growth imposes a high demand both for carbohydrates and mineral nutrients, and consideration was thus given to whether conelet drop in *P. radiata* might result from the inability of strobili to compete successfully with vegetative tissues, and with each other, for carbohydrate or other nutrients.

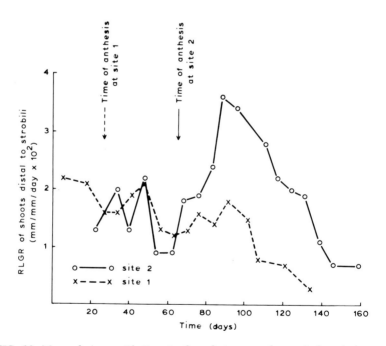

FIG. 10. Mean changes with time in the relative growth rate in length (measured from the pedicels of the strobili to the apex of the shoot) of strobilus-bearing shoots of *Pinus radiata*, on two different sites. On each site measurements were made on 27 branches, comprising 3 clones × 3 ramets per clone × 3 branches per ramet. (From G. B. Sweet, unpublished data.)

In a first investigation of this possibility Sweet and Bollmann (1970) showed by radioactive labeling that

(1) the demand for carbohydrates by strobili reached a peak at about the time of anthesis
(2) in the apical cluster of strobili, the removal of the apical vegetative bud above the strobili and the secondary branch buds in the same cluster as the strobili increased the amount of radioactive photosynthate reaching the strobili
(3) the above treatment significantly reduced the incidence of conelet drop during the first 10 weeks after receptivity and significantly increased the size of the strobili (see Table I)
(4) pollination did not affect the ability of the strobili to mobilize carbohydrate in the first few weeks after receptivity, nor did it affect the incidence of conelet drop at that time.

Having demonstrated that the pedicels of strobili have the ability to photosynthesize, Sweet and Bollmann in a second study (1970) examined the effect on conelet drop of covering pedicels with Vaseline and aluminum foil. The incidence of conelet drop was twice as high in the treated strobili as in uncovered controls. Thus there is strong presumptive evidence to implicate competition for carbohydrates in the process of drop in *P. radiata,* and Abbott (1960) has produced evidence to suggest a similar occurrence in apple. With the high requirement for photosynthates in developing strobili it is probably relevant that in *Pseudotsuga,* in which the strobilus reaches full size in 12 weeks from anthesis,

TABLE I

EFFECT OF TREATMENT[a] ON THE GROWTH AND SHEDDING OF FEMALE STROBILI OF *Pinus radiata*[b]

	Treatment number				Level of significance (%)	Least significant difference (1% level)
	1	2	3	4		
Conelet drop (%)	13.9	19.9	8.3	40.7	1.0	20.1
Strobilus mean diameter (cm)	1.82	1.72	1.82	1.57	0.1	0.13
Strobilus mean length (cm)	2.72	2.56	2.72	2.29	0.1	0.15

[a] Treatment number 1, removal of the apical vegetative bud distal to the strobili and the secondary branch buds in the same "cluster" as the strobili; 2, removal of the secondary lateral branches in the cluster immediately proximal to the strobili; 3, a combination of treatments 1 and 2; 4, control (no removal).

[b] Data assessed 10 weeks after anthesis.

the conelet bracts have a high ability to photosynthesize (Rook and Sweet, 1971).

Sweet and Bollmann (1970) further produced evidence to suggest that competition for mineral (as well as carbohydrate) nutrients might be important in affecting strobilus drop in *Pinus radiata*. In subsequent (unpublished) work by G. B. Sweet and M. P. Bollmann, individual applications of N, P, and K have all been effective on occasion in reducing drop, but on other occasions either singly or in combination they have been ineffective. Katsuta and Satoo (1964) have shown that N, expressed as a percentage of the dry weight of strobili in *P. thunbergii*, is higher at anthesis than at any other time in development, suggesting a high demand and probable competition for it at this stage. Dickmann and Kozlowski (1969b) with *P. resinosa* have similarly shown peak levels of P, K, and Ca at anthesis. Burdon and Low (1973) presented data which suggest that ovule abortion may be considerably higher on a phosphate-deficient than a nondeficient site. They do not, however, have corresponding data for conelet drop. It is also easy to imagine water stress being a factor which could affect competition within the plant and thus conelet drop. Data from Rehfeldt *et al.* (1971) suggest that this may in fact happen in *Pinus monticola*.

The proposal that carbohydrate and mineral nutrient levels affect conelet drop aids understanding of the year-to-year differences in the incidence of drop discussed earlier. At the time of year when the main level of drop occurs, seasonal levels of photosynthesis are still low and there must be a major drain on photosynthetic reserves. The level of these reserves (and thus probably the degree of drop) will be very much dependent on the weather of the previous summer. Thus a warm, moist summer could well mean high reserves and a low level of drop the following spring.

A summary of information on the major natural causes of strobilus abscission in *Pinus* follows:

(1) Lack of pollination causes ovule abortion, and when the incidence of ovule abortion in each strobilus reaches a certain level it is followed by conelet abscission

(2) In all species there is a level of conelet abscission not caused by lack of pollination. It occurs to a markedly greater extent in some species than others (this probably relating to differences in growth rate), and there is evidence to implicate competition for carbohydrate as a major cause of the factor. Competition for mineral nutrients may also be contributory, as also may physical damage

There may well be a physiological difference between these two types of abscission. In the case of the unpollinated strobilus, abscission follows ovule degeneration, but whether this happens with strobili aborting because of competition is not known. Certainly such strobili do have degenerating ovules at the time of shed, but tissues of the scales and axis are also abnormal and it has not proved possible to determine whether ovule degeneration precedes general degeneration or not.

Let us deal first with the unpollinated ovule in which the gametophyte is unable to develop and consider what the germinating pollen grain may supply which enables gametophytic development to proceed. A comparison of the pattern of abortion of unpollinated ovules with that following some interspecific pollinations is interesting. In the latter, although pollen germination usually occurs normally, there are many cases in which pollen tube growth is slow and ovule abortion is not prevented. Abortion may be delayed somewhat, but when it does occur, it does so in a manner very similar to that of an unpollinated ovule [compare, e.g., Sarvas (1962) with Krugman (1970)]. This suggests the possibility that ovule development in *Pinus* may be triggered by a recognition response resulting first from the presence of pollen, requiring not only that the pollen be present, but that it be an appropriate pollen. If this is so, then the recognition substance is fairly specific. M. Hagman (unpublished) has shown in a series of experiments that the subgenera *Haploxylon* and *Diploxylon* of *Pinus* (Mirov, 1967) are quite distinct in their ability to act successfully as pollen donors to one another. Within each of the subgenera the pollen of any one species appears relatively well able to limit the abscission of the strobili of another species, irrespective of taxonomic closeness; but between groups this is not so; strobili pollinated with pollen of the alternate subgenus generally behave as though they have not been pollinated. This is illustrated in Fig. 11 with the *Haploxylon* species, *P. cembra*, as the female parent. Similarly, *Picea* pollen is ineffective in preventing abscission in *Pinus* (Hagman and Mikkola, 1963), as is pollen of *Chamaecyparis* and *Cryptomeria* (Katsuta, 1971). Irradiated pollen has the same effect as normal pollen in preventing flower abscission in *Populus trichocarpa* (Stettler and Bawa, 1971), but in *Pinus* dead pollen merely delayed but did not prevent abscission (D. L. Bramlett, unpublished). Katsuta (1971) was not able to prevent abscission in *Pinus* by "pollinating" with fractions of an 80% methanol–pollen extract, but the residual pollen left after extraction was highly effective in preventing abscission, and heat-killed pollen also had some effect. The specificity implicit in many of these results suggests that it is unlikely that the recognition substance is one of the standard plant hormones.

In many plant and animal recognition systems the compounds involved are either proteinaceous in nature or have a protein component (as, e.g.,

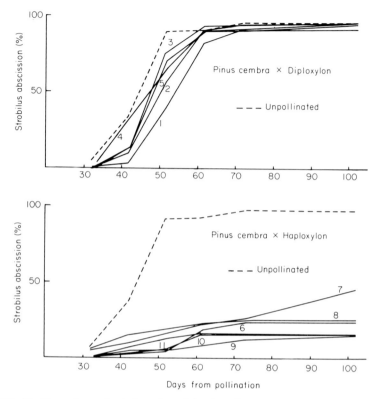

FIG. 11. Percentage abscission of strobili plotted against time. The female parent in all cases is *Pinus cembra*, a member of the subgenus *Haploxylon*. The pollen parents belong to the *Diploxylon* (top graph) and *Haploxylon* (bottom graph). The graphs are from unpublished data of M. Hagman. In his experiment one *Diploxylon* species, *P. ponderosa*, behaved similar to a *Haploxylon*. It has not been included in these diagrams.

glycoproteins); and Knox and Heslop-Harrison (1970) have shown the presence of proteins attached to the intine of the wall of pollen grains in a large number of species, including *Pinus banksiana*. They have shown (1971) that in one species of grass these compounds represent the main source of antigens released by the pollen grain and presumed (though not shown) to be concerned with the incompatibility process. Thus it is reasonable to query whether a protein or protein-containing substance in the wall of the pollen grain may constitute a recognition substance in *Pinus* concerned (a) with recognition by an ovule that it has been pollinated and (b) that it has been pollinated by pollen of an appropriate species.

First, although there is ample evidence of protein substances diffusing out of conifer pollens before and after germination (Stanley, 1967), it is perhaps unlikely that the substance concerned with recognition by the ovule that it has been pollinated is one of the pollen wall proteins reported by Knox and Heslop-Harrison. While pollen wall proteins diffuse very readily into aqueous solutions, it has not proved possible to prevent abortion of unpollinated ovules or strobili in *Pinus* by the application of diffusates from pollen grains (M. Hagman, unpublished). On the contrary, Katsuta's work in which the pollen residue left after a 70% methanol extraction prevented ovule abortion suggests that the recognition substance is very difficult to remove from the pollen grain; it also suggests that the substance is not solely a product of the germinated pollen.

There is no reason, however, to doubt an essential protein component to the substance. Hagman (1967) used disc electrophoresis to demonstrate protein differences between the pollens of pine species from the two subgenera and showed immunologically that some of the differences were antigenic. He has subsequently demonstrated (unpublished data) that at least some of the serologically active substances are present in the pollen tube. One serologically distinct substance detected in one subgenus, but not the other, had some of the attributes of a glycoprotein. Thus, on the basis of compounds present in *Pinus* pollen and the literature of other species, it can be expected with some degree of confidence that the substance triggering ovule development will, when isolated, prove to be a protein or to have a protein component.

Next it is necessary to examine how the pollinated, and thus developing, ovule serves to prevent abscission of the strobilus. As in almost all cases of abscission, so in *Pinus* it can be shown that auxin has a role to play. I (G. B. Sweet, unpublished) have applied to *P. radiata* the classic technique of removing the strobilus and applying lanolin with and without an auxin (in this case 2,4,5-T at 1000 ppm) to the distal end of the pedicel. All minus auxin pedicels abscised, while a high percentage of plus auxin pedicels remained firmly attached with no indication of development of an abscission layer.

In terms of the work of Luckwill (1959, 1970) it would be tempting to speculate that the gametophyte developing after pollination is a rich source of auxin and/or other plant hormones, and that if *sufficient* gametophytes develop, hormone levels are high enough in the strobilus as a whole to prevent abscission. Let us consider the evidence for such a suggestion. It is known from the work of Kopcewicz (1969), Ivonis (1969), and Bochurova (1970) that strobili of *Pinus* contain auxins, gibberellins, and inhibitors, and that the levels of these change (a) with time after anthesis and (b) with pollination. Some 20 days after pollination

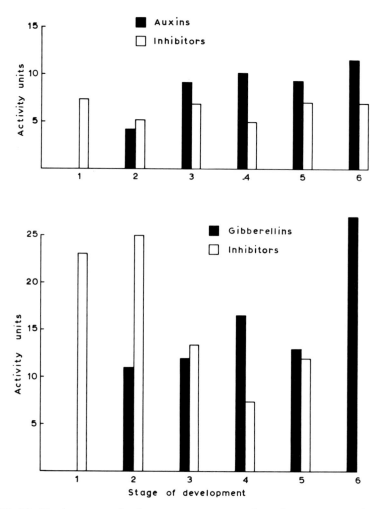

FIG. 12. Total activity of substances promotive and inhibitory in auxin bioassays
and substances promotive and inhibitory in gibberellin bioassays in extracts from
female strobili of *Pinus sylvestris*. Extraction was made at the following times:
(1) Eight days prior to pollination, (2) at the time of pollination, (3) 9 days after
pollination, (4) 19 days after pollination, (5) 4 months after pollination, (6) at the
time of fertilization (from Kopcewicz, 1969).

of *Pinus sylvestris*, gibberellin and auxin levels are high whereas levels
of inhibitor are low (Kopcewicz, 1969) (see Fig. 12). However, at this
stage neither ovule abortion nor conelet abscission has occurred. The
crucial time in terms of drop of unpollinated strobili in *P. sylvestris* is
some 6 to 7 weeks after anthesis. So far as is known the only work which

has examined auxin levels in pollinated and unpollinated strobili at that time is by Hagman. He has one study (unpublished) which suggests that just prior to the time of abscission of unpollinated strobili there was an increase in the level of at least one major indole compound in pollinated strobili. This did not occur in unpollinated strobili (Fig. 13), and thus constitutes some supporting evidence for the suggestion. However, our understanding of the timing of the processes is not sufficiently precise to be certain that the absence of the indole in the unpollinated strobili does not represent effect rather than cause.

If the role of pollen at pollination is to allow gametophytic development to occur, and if the developing gametophyte then prevents conelet drop by virtue of its high level of endogenous hormones (see Luckwill, 1970), then it should be possible to substitute for the gametophyte in conelet development by the application of plant growth substances. If this were possible, agamocarpy (but not agamospermy) would occur. Spraying of unpollinated strobili has been carried out by several workers with variable results. The only published work is by Hagman and Mikkola (1963) who sprayed strobili of *P. cembra* and *P. peuce* with solutions of indoleacetic acid (IAA) (100 ppm) and gibberellic acid (GA₃) (10 ppm). The only positive result was a delay in abscission caused by GA_3; as would be expected, however, there was no change in ovule abortion. I. R. Brown (unpublished data) has tested the effect of a number of plant growth substances (applied in a talc dust) to unpollinated strobili of *P. sylvestris* at anthesis. With 8–10% indolebutyric acid (IBA), despite the degeneration of ovule tissue, there was a reduction in conelet drop during the first season after anthesis; the effect did not last into the second season, however. 6-Benzyladenine (2–5%) on the other hand, while again not preventing the degeneration of ovule tissue, did reduce conelet drop down almost to the level of the pollinated control treatment. The cones developed to maturity, but of course contained no seed. S. L. Krugman (unpublished data) has also delayed the drop of unpollinated strobili with applications of IBA and GA_3.

These results, plus the fact that natural agamocarpy has been reported in *Pinus* and occurs regularly in other conifers, support the suggestion that the role of gametophyte development in terms of conelet development may be expressed at a relatively simple hormonal level. There is no firm evidence as to whether the control is exerted positively by developing ovules or, as suggested by Sarvas, negatively by aborting ovules.

It has been shown that the type of early conelet abscission predominating in *Pinus radiata* probably results from an inability to compete successfully for metabolites (as it does in apples; Abbott, 1960), and in contrast to abscission resulting from low levels of pollination the process

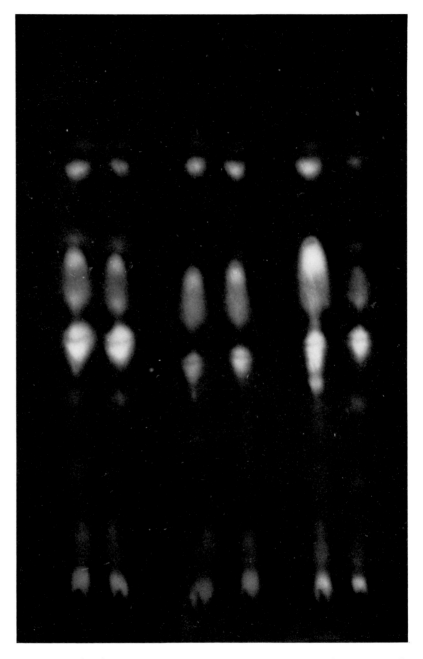

FIG. 13. A thin layer chromatogram of 6 ether extracts from first-year conelets of
Pinus cembra. The chromatogram was developed in chloroform:acetic acid (95:5),

may not begin with ovule breakdown. However, regardless of whether the process commences in the ovules or elsewhere in the strobilus, the conelets which survive must have a high ability to obtain metabolites. Luckwill (1970) has postulated that the phenomenon of hormone-directed transport plays a major role in the competition between shoots and fruits in apple trees, and there is every reason to expect that the same is true for *Pinus*. Auxins and cytokinins appear to be the principal hormones involved in hormone-directed transport, but gibberellins also have been implicated through their ability to increase both the production of auxin and its effects (see Luckwill, 1970 for references). Pine pollen is known to contain a large number of growth substances (Michalski, 1967; Sweet and Lewis, 1971) and when initially it was supposed that pollination did reduce the incidence of early conelet abscission in *Pinus radiata,* it was postulated that an auxin diffusing from germinating pollen might stimulate increased auxin production by the ovule. This diffusion, by attracting increased metabolites, would reduce ovule abortion and thus the level of strobilus drop (Sweet and Lewis, 1969). It was subsequently shown, however, (Sweet and Bollmann, 1970) that pollination did not cause an increased movement of ^{14}C to the ovule and neither did it reduce this early drop, and the growth substance when applied exogenously has not been shown to have any clear-cut effect on the incidence of drop in *P. radiata* (G. B. Sweet, unpublished data).

If hormone-directed transport does have a role in controlling the early drop of well-pollinated *Pinus* strobili, then exogenous hormone application could possibly be beneficial when the incidence of such drop is high. At least three and possibly more investigators have applied plant growth substances to normally pollinated *Pinus* strobili in the field. D. L. Bramlett (unpublished) tested naphthaleneacetic acid (NAA) (500 ppm); Sweet and Bollmann (1971a, and unpublished) used 2,4,5-T, NAA, GA_3, kinetin, and combinations of these compounds at concentrations of between 10 and 100 ppm; and S. L. Krugman (unpublished) has used IBA and GA_3 in a dust at 5 to 10% (w/w) concentration. The most clearly visible response was to auxin sprays (Sweet and Bollmann, 1971a; D. L. Bramlett, unpublished). These caused marked responses in growth rate of the strobili and also interfered with their normal geotropic

---

stained with Prochazkas reagent and fluoresced under UV light. From left to right the chromatogram shows extracts made on July 1, wind-pollinated and unpollinated; July 10, wind-pollinated and unpollinated; July 19, wind-pollinated and unpollinated. The first differences between the pollinated and unpollinated extracts appeared on July 19; all unpollinated strobili abscised soon after that date. (From unpublished data by M. Hagman, Helsinki.)

bending. The results generally, however, have been difficult both to interpret and repeat and are certainly less clear than when growth substances were applied to unpollinated strobili.

Overall, the morphological pattern of development from anthesis to fertilization has been well described for *Pinus* and for most conifers. There is a need, however, for parallel studies to examine quantitative and qualitative changes in the hormonal and nutritional status of strobili over the same period. Such data, of the type obtained by Luckwill (1959) for apple, could be expected to clarify considerably our understanding of conelet drop in well-pollinated strobili.

IV. Abscission or Death between Fertilization and Seed Maturity

A. INTRODUCTION

In a number of fruit trees, such as apple, fertilization occurs early in the reproductive cycle and provides a stimulus which is necessary both to the development of the seed and the fruit (Luckwill, 1959). In contrast, conelets and ovules of many conifers are well developed in size prior to fertilization. In *Pseudotsuga menziesii* for example, fertilization occurs some 8 weeks after anthesis, by which time the strobilus has made considerable growth and is within 4 weeks of being full sized (Owens and Smith, 1965). In *Pinus*, where fertilization occurs some 13 months after pollination, there is some variation between species in the size the cone has reached. In *P. radiata* the strobilus at fertilization is one-quarter of its final dry weight and more than 80% of its final width and length (Sweet and Bollmann, 1971b); in *P. sylvestris* (Brown, 1970b) and *P. thunbergii*

TABLE II

RELATIVE GROWTH RATES OF FEMALE STROBILI OF *Pinus radiata* DURING THE SECOND YEAR AFTER ANTHESIS IN ROTORUA, NEW ZEALAND[a]

Period	RGR (mg/gm/day)
July 19–Aug 1	3.4
Aug 1–Sept 3	16.2
Sept 3–Oct 2	24.9
Oct 2–Nov 6	20.1
Nov 6–Jan 8	19.2[b]
Jan 8–Nov 4	12.5
Mar 4–May 2	2.6
May 2–June 30	2.0

[a] After Sweet and Bollmann, 1971b.
[b] Fertilization occurred during this period.

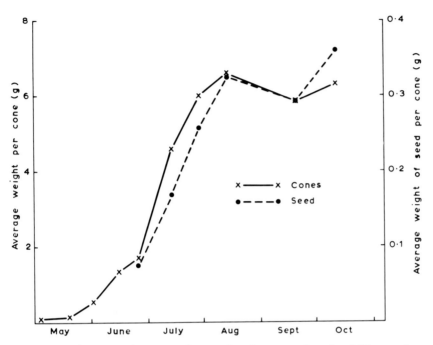

FIG. 14. Changes with time in dry weight of cones and seeds of *Pinus resinosa*, in Wisconsin, U.S.A., during the second growing season after anthesis (from Dickmann and Kozlowski, 1969b). Fertilization occurred in June (Dickmann and Kozlowski, 1969a).

(Katsuta, 1970) it is even closer to full size, and in *P. ponderosa* it is claimed to be full sized at fertilization (Wang, 1970). Table II suggests that, even in *P. radiata* where there is a considerable increase in cone dry weight following fertilization, syngamy does not lead to accelerated growth. Rather, the second-year patterns of cone growth in *Pinus* appear to follow seasonal patterns of vegetative growth (Katsuta and Satoo, 1964; Sweet and Bollmann, 1970). The parallel between cone growth and seed growth during the period subsequent to fertilization is very close (see Fig. 14).

B. Abortion in the Ovule and Its Effects on the Cone

Developmentally the first ovules to abort postfertilization time are those which do not become fertilized. Lack of fertilization has been reported after pollination with both self- and out-crossed pollen in *Pinus peuce* (Hagman and Mikkola, 1963), and Sarvas (1962) cites comparable examples. Fertilization in *Pinus* requires pollen tubes which have been dormant over winter to recommence growth, and in view of the fact that this does not happen early in the growing season when the peak of vegetative and strobilus growth is occurring (see Table II), it seems reasonable to suggest that this renewal of growth may be initiated hormonally, rather than by seasonal weather factors. Kopcewicz (1969) has shown that both gibberellin and auxin levels are high at the time of fertilization and inhibitor levels are low (see Fig. 12). The upsurge in gibberellin levels is particularly marked, though whether this precedes or succeeds syngamy is not known. Since Sweet and Lewis (1971) demonstrated a substance with some gibberellinlike properties which can stimulate pollen tube growth in *Pinus radiata,* it is tempting to speculate that gibberellins may be involved in the recommencement of pollen tube growth. However, it is also pertinent to consider that kinetin has a marked effect on pollen tube growth in *Pinus* (Konar, 1958) and that cytokinin levels have not been examined in strobili at this time.

Lack of fertilization has also been reported in interspecific crosses in the subgenus *Diploxyon* of *Pinus* (McWilliam, 1959; Krugman, 1970). It is probably not correct, however, to consider these as examples of a fertilization barrier to interspecific crossing. As indicated in a previous section, the abortion of ovules following interspecific pollination is progressive; the better the quality of pollen germination and pollen tube growth, the longer the ovule can remain without abortion. Probably ovules which abort (unfertilized) after the time at which fertilization would have occurred represent the ultimate example of this progression; Mikkola's (1969) studies with *Picea* would support such an interpretation.

Many conifer genera are polyzygotic, that is, each ovule is pollinated by a number of pollen grains which fertilize a number of archegonia. In *Pinus sylvestris,* for example, the number of pollen grains in each ovule ranges from 0 to 6 (mean 2.0) and the number of archegonia from 1 to 4 (mean 2.1) (see Sarvas, 1962). In *Picea abies* the figures are, respectively, 0 to 6 (mean 3.3) and 2 to 7 (mean 3.1) (Sarvas, 1968). The correlation between the number of pollen grains germinating in an ovule and the number of archegonia formed is very low in *Picea abies* ($r = 0.073$: Sarvas, 1968), and thus there must on many occasions be archegonia which develop but are not fertilized. Correspondingly there must be cases of surplus pollen. Dogra (1966), in a study of embryogeny of 24 conifer species, found no evidence for the continued development of unfertilized archegonia. With possibly an occasional exception (Orr-Ewing, 1957a), unfertilized ovules in conifers abort yielding empty seeds. Under optimal conditions of pollination all archegonia may be fertilized, but it is uncommon for more than one embryo to develop to maturity in each ovule. Thus, after successful fertilization, there is generally a considerable amount of proembryo or embryo abortion. Abortion following syngamy is a frequent cause of ovule breakdown following interspecific pollination in the subgenus *Haploxylon* of *Pinus* (Kriebel, 1970); it can, however, also occur in the *Diploxylon* (Krugman, 1970). It is believed that the critical point at which growth ceases occurs when the zygote nucleus begins to differentiate cell walls. Hagman and Mikkola (1963) speculate that this breakdown occurs as the result of a nucleus–cytoplasm incompatibility, perhaps as a result of the cytoplasmic organelles being unable to carry out the instructions of the hybrid nucleus. Krugman (1970) believes that this does not represent an incompatibility process in the normally accepted terminology and that it should properly be termed "embryonic inviability."

Sarvas (1962) regards polyzygotic polyembryony in conifers possibly as an opportunity for genetic selection to take place and certainly as an opportunity for the elimination of embryos which are homozygous for lethal, sublethal, or defective genes. He points out (1968) that the significance of polyembryony is that the abortion of embryos incapable of survival can largely take place without reducing the number of sound seeds produced. A similar type of occurrence exists in some angiosperms. In *Acer saccharum*, for example, four ovules are present in each fused pair of carpels but during the postfertilization period three or sometimes all four of these abort (Gabriel, 1967). A number of studies have been made of self-pollination in conifers and it seems clear that subsequently there is generally a very high incidence of abortion of developing embryos, with this occurring either at a late stage in proembryo or an early stage in embryo development (Orr-Ewing, 1957b; Sarvas, 1962, 1968; Hagman and Mikkola, 1963; King *et al.*, 1970). Postfertilization failures of this type are considered to provide the major mechanism to reduce homozygosity in a number of forest tree genera (Sarvas, 1962, 1968).

It has been suggested (Sarvas, 1968) that while much embryo abortion is absolute, resulting from homozygosity or lethal or defective genes, other abortions result from competitive effects. It seems reasonable to suggest, in terms of the fact that self-pollinated seedlings are generally less vigorous than out-crossed ones, that self-pollinated embryos containing some sublethal genes may also be less vigorous and may compete unsuccessfully in the embryo cavity with out-crossed ones. This idea was first advanced by Buchholz in the 1920's, and subsequently supported first by Stockwell and then by Sarvas (see Mikkola, 1969, for references). It has been criticized, however, as a concept lacking in evidence, and in a recent paper Mikkola (1969) was unable to show any evidence that such competition occurs in *Picea abies*. Mikkola did, however, report very large differences in the *timing* of fertilization of different archegonia, and he observed that these differences persisted into embryo development. The earliest fertilized embryo normally assumes the apical position in the embryo cavity, and Mikkola has observed that almost invariably the embryo in this position is the one that develops, while the remaining embryos abort. Dogra (1966) made a similar observation in *Pinus*. If indeed Mikkola's observations are correct, this still does not rule out a measure of genetic selection. Barnes *et al.* (1962) have suggested for *Pinus monticola* a correlation between vigor of the pollen tube and genetic vigor of the parent tree, and Mulcahy (1971) in work with maize has reported male gametophytic competition affecting sporophytic characteristics, with heavier seeds resulting from fertilization by faster-growing pollen grains.

It frequently happens, of course, that none of the proembryos in an ovule are fit to survive, and in that case complete embryo abortion occurs, resulting in an empty seed. This has been particularly well documented following interspecific crossing in *Pinus* (e.g., Kriebel, 1970) and *Picea* (Mikkola, 1969). A normal consequence of this in *Pinus* is the rapid degeneration of the developing endosperm and the transfer of any accumulated nutrients elsewhere, although in some interspecific crosses where embryo abortion occurs, endosperms do develop (Hagman and Mikkola, 1963).

Embryo abortion does not of necessity occur early in development. Sarvas (1962) lists three stages in which it is particularly severe in *Pinus*: (1) the rosette stage, (2) the stage at which the tip cell of the embryo is replaced by a number of embryonic initial cells, and (3) the stage of seed germination.

There is an extensive literature on factors affecting seed germination which has been well reviewed by Andersson (1965). It will not be considered in this chapter other than to indicate that embryo abortion at the time of seed germination in *Pinus* can result both from interspecific pollination (see Krugman, 1970) and from self-pollination (Wang, 1970). It can also, however, represent simply a delayed polyzygotic competition, as demonstrated by Ching and Simak (1971) with *Pinus* and *Picea* grown in subarctic climates. As at most stages of development, factors which are other than physiological and genetic can affect embryo development in the postfertilization period. Sarvas (1962), for one, cites examples of damage caused by weather and insects, and the literature contains a very large number of other comparable references.

The question now arises as to what extent embryo abortion leads to cone abscission. In contrast to the situation in many fruit trees, there is very little postfertilization drop of strobili in conifers. Some examples do occur in the literature (e.g., Brown, 1970b), but they are not common (Bramlett, 1972). Almost certainly this difference between fruit trees and conifers relates to the fact (discussed in the introduction to this section) that, in conifers, cones and seeds have approached full size by the time of fertilization, which in consequence does not provide a major stimulus to development. The drop of cones which does occur during this period is probably restricted to species which shed their cones naturally following seed-shed and which have therefore maintained a potential abscission zone. Rehfeldt *et al.* (1971) have described such abscission in *Pinus monticola* and have shown an influence of early summer water deficit on conelet drop at this stage. Drop of this type has not been observed in species such as *Pinus radiata* where cones normally remain closed and are not shed at maturity. However, cone death (probably due to drought) has been reported in *P. radiata* during this period (Pawsey, 1960). Cones have also been collected which contain empty seeds only, resulting from postfertilization abortion. They were considerably lighter, but little smaller in size than normal cones.

V. The Practical Implications of Abscission of Reproductive Structures in Forest Trees

With the advent of large-scale breeding projects in forestry, seed collection is frequently concentrated in special orchards grown specifically for seed production purposes. Typically these orchard areas may contain grafts or cuttings of selected and progeny-tested clones, with a number of ramets of each clone set out in a specific design aimed at minimizing self-pollination. The cost of such seed is high; the orchards need to be reasonably isolated from outside pollen of the species, yet in areas which are optimal for growth and thus flowering of the species. They need to be flat and easily worked by machinery and close to sources of labor for ease of establishment and protection purposes, as well as for cone collection. They also need to be close to seed extraction plants to minimize cone transport costs. The cost of seed is high both on account of these factors and also because it carries the compounded costs of an expensive breeding program. Examination of the costs of orchard seed in *Pinus taeda* in the United States has been made by Davis (1967) and Bergman (1968). Although their criteria for cost analysis vary to some degree, their final values are similar. Bergman in a 7-year-old orchard showed an average seed production of 38 kg/ha at a cost of some $31/kg. The curve relating cost per kilogram to production per hectare was hyperbolic; reducing seed yield per hectare to one-tenth increased the cost per kilogram tenfold.

How large in fact are seed losses in actual and percentage terms? Data from Sarvas (1962) can be used to examine the extent of potential seed loss in *P. sylvestris* on a relatively fertile site in Finland and to see where the major parts of this loss occur. The average strobilus of *P. sylvestris* has 57 scales of which 14 are fertile, bearing (potentially) two ovules each. One hundred strobili when initiated have therefore potentially $100 \times 14 \times 2 = 2800$ seeds (100%). Thirty % death or drop of strobili leaves potentially 1960 seeds (70%). Twenty-five % of the ovules will abort leaving potentially 1470 seeds (52.5%). Fourteen % of the developing embryos will abort leaving potentially 1264 seeds (45.1%). The exercise underlines the fact that losses of a given percentage of potential seed occurring early in the development cycle are economically much more important than losses of a similar magnitude occurring later. For this reason conelet drop is particularly important economically.

One other area of potential seed loss early in the cycle lies in the deficient ovules. B. S. Henderson (unpublished data) has classified approximately 20% of the sterile ovules of *Pinus radiata* as deficient; that is, while not in fact functional, they do not differ greatly in their morphology from functional ovules. If the calculation above is remade to initially include 20% of the sterile ovules (i.e., 10% of the sterile scales) as potentially fertile, one obtains the following calculations. One hundred strobili have potentially $100 \times (14 + 4.3) \times 2 = 3660$ seeds (100%). Thirty percent death or drop leaves potentially 2562 seeds (70%). "Loss" due to deficient ovules reduces the number of 1960 seeds (53.6%). Twenty-five percent of the fertile ovules will abort leaving potentially 1470 seeds (40.2%). Fourteen percent of the developing embryos will abort leaving potentially 1264 seeds (34.5%).

A comparison of the two calculations suggests that an increase in the percentage of fertile scales of the above order would boost the total seed production per strobilus by some 10%.

The possibility of such a prospect has received almost no attention but warrants examination. There are clearly big differences between species in the percentage of fertile ovules; in an unpublished study of *Pinus radiata* in New Zealand, for example, the mean percentage of fertile ovules was 36, compared with Sarvas' value of 25% for *Pinus sylvestris* in Finland. Lyons (1956) presents an even higher value (45%) for *Pinus resinosa* and of the sterile ovules he classified 87% as rudimentary in varying degree (as distinct from totally sterile). Sarvas (1962) showed year-to-year and tree-to-tree variation in the percentage of fertile ovules as well as an effect of geographic variation. The latter has also been shown in large degree for *Pseudotsuga menziesii* by Sweet and Bollmann (1972). Sarvas' data suggest a probable genetic component of quite large size, opening up the possibility of incorporating this character into selection programs for breeding purposes. There must however also exist prospects for chemical manipulation. Duff and Nolan (1958) have pointed out that in the species of *Pinus* growing in colder climates the female strobilus is literally the only portion of the tree actively growing throughout the entire winter. They point out that hormone levels must be extremely low during the time when much of this ovule development is occurring, and they postulate that hormone deficiency may control the level of fertile ovules in a strobilus. This suggestion warrants experimental examination. In terms of the cost of orchard seed, the exogenous application of a hormone that could increase the percentage of fertile ovules by even 10% would probably be highly economic.

The question of conelet drop, on the contrary, has received a great deal of examination. To the extent that it reflects inadequate pollination, it is not perhaps a major problem in seed orchards. Everything else being equal, seed orchards are established on fertile sites, and on these sites pollen production in *Pinus* is maximal (Sarvas, 1962). For genetic reasons also it is desirable to have high pollen levels in seed orchards (Sarvas, 1962) and techniques exist to increase the incidence of initiation of male strobili (e.g., see Jackson and Sweet, 1972). If despite these precautions pollen levels are still low, the use of "mentor" pollen offers the possibility of further restricting conelet drop (Stanley, 1967).

The proportion of the drop which is caused by factors other than lack of pollination is more difficult to control. Careful selection of seed orchard sites in climatic areas where the incidence of drop is low, and application of mineral fertilizers where nutrients are deficient, will be helpful. But the problem of increasing the ability of strobili to compete with the vegetative parts of the plant for carbohydrates has not been solved. In controlled pollinations made in New Zealand for the production of *Pinus radiata* seed for progeny testing, vegetative shoots which compete with strobili for nutrients are broken out as a routine measure; but this is not practicable in seed orchards in which optimal vegetative growth (for optimal flower production) is required. Many workers have examined alternative possibilities, e.g., that of spraying trees with growth substances to boost the growth of strobili (and their ability to attract metabolites) or alternatively to restrict the growth of competing vegetative shoots during the period when competition is critical. So far these techniques have not proved effective, one problem being the apparent difficulty in getting growth substances into strobili of conifers and another the wide diversity in strobilus development stage that exists in many species of conifer at any one time of spraying.

The other two factors contributing to loss of potential seed in orchards, viz., ovule abortion and embryo abortion, generally cause less than 10% of the total loss of potential seed. Ovule abortion is coped with in large part by adequate levels of pollen and (probably) by using fertile orchard sites. While to the extent embryo abortion results from embryos which are homozygous for lethal, sublethal, or defective genes, it is even desirable that it occurs.

In seed orchards it is possible because of the high unit cost of the product to contemplate expensive programs to increase seed yields; but this is less likely to prove justifiable in plantations or in natural stands. There is one situation, however, in which there is a strong interest in promoting abscission of reproductive structures. In species of *Pinus* with persistent cones, the pedicel of the strobilus is engulfed by the increasing diameter growth of the stem or branch which bears it until the bark of this reaches the base of the cone (Fielding, 1945). Subsequently further cambial growth exerts a pressure on the cone until the pedicel breaks at its connection with the stem tracheids. From this point on the cone is pushed outward with increasing branch diameter growth. Where (as frequently in *P. radiata*) the length of the pedicel is greater than the distance from the outer bark to the cambium, all wood laid down subsequently beneath the cone base contains a hole where the pedicel penetrates the cambium. In *P. radiata,* strobili occur frequently on the main stem of the tree and the cone–stem holes resulting from their presence cause a marked reduction in grade of timber. The commercial implications of this are important and preliminary attempts have been made to promote the shed of stem cones in plantations by chemical methods. The literature contains a number of references to forestry applications of compounds developed to promote flower and fruit drop in horticulture (see, e.g., Hong, 1965), but their main drawback is that they normally also restrict shoot growth. In seed orchards this may not be totally critical, but in production forestry stem growth is clearly of major importance and no sterilant which restricts this can be considered for practical use. I. R. Brown (unpublished data) has greatly accelerated the abscission of unpollinated strobili of *Pinus sylvestris* by applying 2 or 5% triiodobenzoic acid in a talc dust at anthesis, and to a lesser degree has achieved the same response with maleic hydrazide. The effect on growth of the trees is not known. A further compound, tested by the author with some success, is Ethrel (2-chloroethylphosphonic acid) applied as an aqueous spray at a concentration of 2000 ppm active ingredient. On one area of *Pinus radiata* this increased the level of drop by some 25% with no visible effect on tree growth. As with seed orchard treatments, however, there are problems in successfully carrying out large-scale sprayings on trees whose strobili cover a range of developmental stages. Currently it is probably not possible on a commercial scale to reduce the incidence of stem cones in plantations of *Pinus* by chemical means, but it seems likely that the possibility of doing so will be the subject of future research.

Acknowledgments

The author is very greatly indebted to Dr. D. L. Bramlett, Southeastern Forest Experiment Station, Blacksburg, Virginia; Dr. I. R. Brown, Department of Forestry, University of Aberdeen, Scotland; Dr. M. Hagman, The Finnish Forest Research Institute, Helsinki, Finland; Miss B. S. Henderson, Botany Department, Canterbury University, New Zealand; and Dr. S. L. Krugman, U. S. Forest Service, Washington, D. C., for making available much of their unpublished data for use in this chapter. Without their willing cooperation it would not have been possible to present up-to-date information on many aspects of the subject.

References

Abbott, D. L. (1960). The bourse shoot as a factor in the growth of apple fruits. *Ann. Appl. Biol.* **48**, 434–438.

Addicott, F. T. (1965). Physiology of abscission. *In* "Handbuch der Pflanzen-physiologie" (W. Ruhland, ed.), Vol. 15, Part 2, pp. 1094–1126. Springer-Verlag, Berlin and New York.

Allen, G. S. (1942). Parthenocarpy, parthenogenesis and self-sterility of Douglas fir. *J. Forest.* **40**, 642–644.

Andersson, E. (1965). Cone and seed studies in Norway spruce. *Stud. Forest. Suec.* **23**, 1–214.

Barnes, B. V., Bingham, R. T., and Squillace, A. E. (1962). Selective fertilisation in *Pinus monticola* Dougl. II. Results of additional tests. *Silvae Genet.* **11**, 103–111.

Bergman, A. (1968). Variation in flowering and its effect on seed cost. *N. C. State Univ., Sch. Forest Resour., Tech. Rep.* **38**.

Bingham, R. T., and Rehfeldt, G. E. (1970). Cone and seed yields in young western white pines. *U. S., Forest Serv. Res. Pap.* **INT-79**.

Bochurova, N. V. (1970). On geotropic bending of female strobiles in Scotch pine. *Sov. Plant Physiol.* **17**, 505–507.

Bramlett, D. L. (1972). Cone crop development records for six years in shortleaf pine. *Forest Sci.* **18**, 31–33.

Bray, J. R., and Gorham, E. (1964). Litter production in forests of the world. *Advan. Ecol. Res.* **2**, 101–157.

Brown, I. R. (1970a). Seed production in Scots pine. *In* "Physiology of Tree Crops" (C. V. Cutting and L. C. Luckwill, eds.), pp. 55–63. Academic Press, New York.

Brown, I. R. (1970b). Premature cone loss in grafted clones of Scots pine. *In* "Sexual Reproduction of Forest Trees," Vol. 1, 15 pp. Finnish Forest Res. Inst., Helsinki.

Bryndum, K., and Hedegart, T. (1969). Pollination of teak. *Silvae Genet.* **18**, 77–80.

Burdon, R. D., and Low, C. B. (1973). Seed production in *Pinus radiata* clones on four different sites. *N. Z. J. Forest Sci.* **3**, 211–219.

Ching, K., and Simak, M. (1971). Competition among embryos in polyembryonic seeds of *Pinus silvestris* L. and *Picea abies* (L.) Karst. *Inst. Skogsforyngring Res. Note* **30**, 1–12.

Davis, L. S. (1967). Investments in loblolly pine clonal seed orchards. *J. Forest.* **65**, 882–887.

Dickmann, D. I., and Kozlowski, T. T. (1969a). Seasonal growth patterns of ovulate strobili of *Pinus resinosa* in central Wisconsin. *Can. J. Bot.* **47**, 839–848.

Dickmann, D. I., and Kozlowski, T. T. (1969b). Seasonal changes in the macro- and micro-nutrient composition of ovulate strobili and seeds of *Pinus resinosa*. *Can. J. Bot.* **47**, 1547–1554.

Dogra, P. D. (1966). Observations on *Abies pindrow* with a discussion on the question of occurrence of apomixis in gymnosperms. *Silvae Genet.* **15**, 11–20.

Duff, G. H., and Nolan, N. J. (1958). Growth and morphogenesis in the Canadian forest species. III. The time scale of morphogenesis at the stem apex of *Pinus resinosa* Ait. *Can. J. Bot.* **36**, 687–706.

Ebel, B. H. (1964). The occurrence of *Ernobius granulatus* Le Conte in aborted first year cones of longleaf pine. *J. Forest.* **62**, 404–405.

Ferguson, M. C. (1904). Contributions to the knowledge of the life history of *Pinus* with special reference to sporogenesis, the development of the gametophytes and fertilisation. *Proc. Wash. Acad. Sci.* **6**, 1–202.

Fielding, J. M. (1945). Cone holes in Monterey pine. *Aust. Commonw. Forest. Bur. Leafl.* **55**.

Fielding, J. M. (1960). Branching and flowering characteristics of Monterey pine. *Aust. Forest. Timber Bur. Bull.* **37**.

Florence, R. G. (1964). A comparative study of flowering and seed production in six blackbutt (*Eucalyptus pilularis* Sm) forest stands. *Aust. Forest.* **28**, 23–33.

Forbes, D. C. (1971). Loss of shortleaf pine flowers under natural conditions—a probable explanation for high losses in controlled pollinations. *Tree Planters' Notes* **22**, 30.

Gabriel, W. J. (1967). Reproductive behaviour in sugar maple: Self compatibility, cross compatibility, agamospermy and agamocarpy. *Silvae Genet.* **16**, 165–168.

Grose, R. J. (1960). Effective seed supply for the natural regeneration of *Eucalyptus delegatensis* Hook f. APPITA **13**, 141–148.

Hagman, M. (1967). Serological studies of pollen and the incompatibility in forest trees. *Proc., Int. Union Forest Res. Organ, 14th, 1967* Sect. 22, No. 3, 60–71.

Hagman, M. (1970). Observations on the incompatibility in *Alnus*. In "Sexual Reproduction of Forest Trees," Vol. 1, 17 pp. Finnish Forest Res. Inst., Helsinki.

Hagman, M., and Mikkola, L. (1963). Observations on cross-, self-, and interspecific pollinations in *Pinus peuce* Griseb. *Silvae Genet.* **12**, 73–79.

Hard, J. S. (1963). Frost damage to red pine conelets. *U. S., Forest Serv. Res. Note* **LS-5**.

Hashizume, H. (1963). Initiation and development of flower buds in *Chamaecyparis obtusa*. *J. Jap. Forest. Soc.* **45**, 135–141.

Hillman, W. S. (1963). "The Physiology of Flowering." Holt, New York.

Hong, S. O. (1965). The effect of some growth regulators upon the development of male gametophyte of pitch pine. *Res. Rep. Office Rural Develop., Suwon, Korea* **6**, 45–60.

Hutchinson, J. G., and Bramlett, D. L. (1964). Frost damage to shortleaf pine flowers. *J. Forest.* **62**, 343.

Ivonis, I. Y. (1969). Gibberellin-like substance in young cones and pollen of Scotch pine trees. *Sov. Plant Physiol.* **16**, 782–784.

Jackson, D. I., and Sweet, G. B. (1972). Flower determination in temperate woody plants. *Hort. Abstr.* **42**, 9–24.

Katsuta, M. (1970). Cone development of *Pinus thunbergii* Parl. in response to chilling and day length. *In* "Sexual Reproduction of Forest Trees," Vol. 1, 7 pp. Finnish Forest Res. Inst., Helsinki.

Katsuta, M. (1971). Cone drop and pollination in *Pinus thunbergii* Parl. and *P. densiflora* Sieb. et Zucc. *Bull. Tokyo Univ. Forests* **65**, 87–106.

Katsuta, M., and Satoo, T. (1964). Cone development in *Pinus thunbergii*. *J. Jap. Forest. Soc.* **46**, 166–170.

Katsuta, M., and Satoo, T. (1965). Cone drop in *Pinus thunbergii*. *J. Jap. Forest. Soc.* **47**, 101–112.

King, J. P., Jeffers, R. M., and Nienstaedt, H. (1970). Effects of varying proportions of self-pollen on seed yield, seed quality and seedling development in *Picea glauca*. *In* "Sexual Reproduction of Forest Trees," Vol. 1, 15 pp. Finnish Forest Res. Inst., Helsinki.

Knox, R. B., and Heslop-Harrison, J. (1970). Pollen-wall proteins: Localization and enzymic activity. *J. Cell Sci.* **6**, 1–27.

Knox, R. B., and Heslop-Harrison, J. (1971). Pollen-wall proteins: The fate of intine-held antigens on the stigma in compatible and incompatible pollinations of *Phalaris tuberosa*. *J. Cell Sci.* **9**, 239–251.

Konar, R. N. (1958). Effect of IAA and kinetin on the pollen-tube growth of *Pinus roxburghii* Sar. *Curr. Sci.* **27**, 216–217.

Kopcewicz, J. (1969). The dynamics of growth substances during the development of the pine inflorescences. Part II. Female inflorescences. *Rocz. Nauk Roln., Ser. A* **95**, 231–238.

Kriebel, H. B. (1970). Emrbyo development and hybridity barriers in the white pines. *In* "Sexual Reproduction of Forest Trees," Vol. 1, 6 pp. Finnish Forest Res. Institute, Helsinki.

Krugman, S. L. (1966). Freezing spring temperatures damage knobcone pine conelets. *U. S., Forest Serv. Res. Pap.* **PSW-37**.

Krugman, S. L. (1970). Incompatibility and inviability systems among some western North American pines. *In* "Sexual Reproduction of Forest Trees," Vol. 2, 12 pp. Finnish Forest Res. Inst., Helsinki.

Luckwill, L. C. (1959). Fruit growth in relation to internal and external chemical stimuli. *In* "Cell, Organism and Milieu" (D. Rudnick, ed.), pp. 233–251. Ronald Press, New York.

Luckwill, L. C. (1970). The control of growth and fruitfulness of apple trees. *In* "Physiology of Tree Crops" (C. V. Cutting and L. C. Luckwill, eds.), pp. 237–253. Academic Press, New York.

Lyons, L. A. (1956). The seed production capacity and efficiency of red pine cones (*Pinus resinosa* Ait). *Can. J. Bot.* **34**, 27–36.

McWilliam, J. R. (1959). Interspecific incompatibility in *Pinus*. *Amer. J. Bot.* **46**, 425–433.

Michalski, L. (1967). Growth regulators in the pollen of pine (*Pinus sylvestris* L.). *Acta Soc. Bot. Pol.* **36**, 475–481.

Mikkola, L. (1969). Observations on interspecific sterility in *Picea*. *Ann. Bot. Fenn.* **6**, 285–339.

Mirov, N. T. (1967). "The Genus *Pinus*." Ronald Press, New York.

Mulcahy, D. L. (1971). A correlation between gametophytic and sporophytic characteristics in *Zea mays* L. *Science* **171**, 1155–1156.

Nielsen, P. C., and Schaffalitzky de Muckadell, M. (1954). Flower observations and controlled pollinations in *Fagus*. *Silvae Genet.* **3**, 6–24.

Nienstaedt, H. (1956). Receptivity of the pistillate flowers and pollen germination tests in genus *Castanea*. *Silvae Genet.* **5**, 40–45.

Orr-Ewing, A. L. (1957a). Possible occurrence of viable unfertilised seeds in Douglas fir. *Forest Sci.* **3**, 243–248.

Orr-Ewing, A. L. (1957b). A cytological study of the effects of self-pollination on *Pseudotsuga menziesii* (Mirb.) Franco. *Silvae Genet.* **6**, 179–185.

Owens, J. N. (1969). The relative importance of initiation and early development on cone production in Douglas fir. *Can. J. Bot.* **47**, 1039–1049.

Owens, J. N., and Smith, F. H. (1965). Development of the seed cone of Douglas fir following dormancy. *Can. J. Bot.* **43**, 317–332.

Pawsey, C. K. (1960). Cone production reduced, apparently by drought, in the south east of South Australia. *Aust. Forest.* **24**, 74–75.

Rehfeldt, G. E., Stage, A. R., and Bingham, R. T. (1971). Strobili development in western white pine: Periodicity, prediction and association with weather. *Forest Sci.* **17**, 454–461.

Rodin, L. E., and Bazilevich, N. I. (1967). "Production and Mineral Cycling in Terrestrial Vegetation." Oliver & Boyd, Edinburgh.

Rook, D. A., and Sweet, G. B. (1971). Photosynthesis and photosynthate distribution in Douglas fir strobili grafted to young seedlings. *Can. J. Bot.* **49**, 13–17.

Sarvas, R. (1962). Investigations on the flowering and seed crop of *Pinus sylvestris*. *Commun. Inst. Forest. Fenn.* **53**, 1–198.

Sarvas, R. (1968). Investigations on the flowering and seed crop of *Picea abies*. *Commun. Inst. Forest. Fenn.* **67**, 1–84.

Silen, R. R. (1967). Earlier forecasting of Douglas fir cone crop using male buds. *J. Forest.* **65**, 888–892.

Snyder, E. B., and Squillace, A. E. (1966). Cone and seed yields from controlled breeding of southern pines. *U. S. Forest Serv. Res. Pap.* **50**, 22.

Stanley, R. G. (1967). Factors affecting germination of the pollen grain. *Proc., Int. Union Forest Res. Organ., 14th, 1967* Sect. 22, No. 3, pp. 38–59.

Stettler, R. F., and Bawa, K. S. (1971). Experimental induction of haploid parthenogenesis in Black cottonwood. *Silvae Genet.* **20**, 42–46.

Sweet, G. B., and Bollmann, M. P. (1970). Investigations into the causes of conelet drop in *Pinus radiata* in New Zealand. *In* "Sexual Reproduction of Forest Trees," Vol. 2, 12 pp. Finnish Forest Res. Inst., Helsinki.

Sweet, G. B., and Bollmann, M. P. (1971a). Auxin sprays affect development of *Pinus radiata* strobili. *Forest Sci.* **17**, 14–15.

Sweet, G. B., and Bollmann, M. P. (1971b). Seasonal growth of the female strobilus in *Pinus radiata*. *N. Z. J. Forest. Sci.* **1**, 15–21.

Sweet, G. B., and Bollmann, M. P. (1972). Regional variation in Douglas fir seed yields. *N. Z. J. Forest.* **17**, 74–80.

Sweet, G. B., and Lewis, P. N. (1969). A diffusible auxin from *Pinus radiata* pollen and its possible role in stimulating ovule development. *Planta* **89**, 380–384.

Sweet, G. B., and Lewis, P. N. (1971). Plant growth substances in pollen of *Pinus radiata* at different levels of germination. *N. Z. J. Bot.* **9**, 146–156.

Sweet, G. B., and Thulin, I. J. (1969). The abortion of conelets in *Pinus radiata*. *N. Z. J. Forest.* **14**, 59–67.

Wang, C. W. (1970). Cone and seed production in controlled pollination of ponderosa pine. *Idaho, Forest Exp. Sta., Sta. Pap.* **7**, 1–15.

Williamson, M. J. (1966). Premature abscissions and white oak acorn crops. *Forest Sci.* **12**, 19–21.

. 10 .

Anatomical Changes
in Abscission of
Reproductive Structures

Roy K. Simons

I. Introduction

Evidence of anatomical changes in abscission of reproductive structures can be observed in the dissolution of cells, cell walls, and middle lamella, breakage of tissues, senescence of tissues that are separating, and formation of the abscission layer.

The abscission of reproductive structures in plants is characterized by a definite zone that is located at the base of an organ or associated plant parts. The process of detachment of plant organs has been defined by Esau (1960) as the *abscission zone* which is "the zone at base of leaf, or fruit, or flower, or other plant part that contains the abscission layer and the protective layer, both of which play a role in the separation of the plant part from the plant." The *abscission layer* is "In abscission zone. Layer of cells, the disjunction or breakdown of which separates a plant part, such as leaf, fruit, flower, etc. from the plant. Syn. separation layer."

Some examples of reproductive structures exhibiting abscission zones and layers in relation to development of the plant are separation of pedicels from fruits, petals from flowers, style abscission, abortion in ovulary development, and abscission of fruit from the pedicel. Specific studies on structural changes during bean leaf abscission (Scott *et al.,* 1967) and a morphogenetic study of leaf abscission in *Phaseolus* (Webster, 1970) have been reported.

Lott and Simons (1964, 1966, 1968a,b) reported on floral tube and style abscission for four species of *Prunus* which included peach, Montmorency cherry, sweet cherry, and apricot. Excerpts from these studies are included in this chapter.

Development of abscission zones and layers occurs at a very rapid rate for some plant parts, such as style abscission (Lott and Simons, 1968a,b), or may occur over a long period of time as in the maturation of apples with subsequent separation of the fruit pedicel from the spur (Simons, 1963). An understanding of the morphology and anatomy of the sequential stages of plant growth is critical for plant manipulation in order to maintain successful production practices such as fruit thinning and mechanical harvesting operations. The three successive waves of "drops in apples" provide an example that is visually observed in *Malus.* Seed and fruit enlargement cease with the ultimate change in color from a light green to a yellowish brown. The pedicel of the fruit becomes pigmented changing to a reddish brown or yellowish brown color depending on the cultivar. Abscission zones of drop fruit with abortive ovules have short developmental stages as compared with normal maturing fruit.

Early experimentation with plant growth substances containing an auxin such as α-naphthaleneacetic acid (NAA) for the effective control of harvest drop was reported by Gardner *et al.* (1939). Currently, some other growth regulators have been used for specific production controls and these include 2-(2,4,5-trichlorophenoxy)propionic acid (fenoprop) and 2,4,5-trichlorophenoxyacetic acid (2,4,5-T) and the growth retardant succinic acid—2,2-dimethylhydrazide (SADH).

A recent symposium on fruit abscission included topics such as: (1) abscission–regulation of senescence, protein synthesis, and enzyme secretion by ethylene (Abeles *et al.*, 1971); (2) physiological processes involved in abscission (Leopold, 1971); (3) apple abscission (Edgerton, 1971); (4) the correlative effect of competition on abscission in apricot and pistachio (Crane, 1971); (5) the nature and chemical promotion of abscission in maturing cherry fruit (Bukovac, 1971); and (6) citrus abscission (Biggs, 1971). Most of these reviews were concerned with abscission of mature fruits and not with individual flower parts of young developing fruits. However, Leopold (1971) defined 5 morphological stages in abscission which usually follows a set sequence and these were listed as follows: (1) differentiation of an abscission zone; (2) abscission zone in stage I (actual weakening of the break strength); (3) abscission zone in stage II (the ability of auxin to inhibit separation weakens and ordinarily causes an enhancement rather than a deferral of abscission); (4) separation—accomplished by changes in cell walls, by weakening of the cementing layers of the middle lamella, partial dissolution of the cellulosic walls, and rupture of the vascular cells; (5) healing: this may take the form of cell divisions to form a protective layer and/or the suberization of the exposed cells and plugging of the exposed vascular strands.

Addicott and Lynch (1955) stated that involved in the abscission of most organs there is a discrete abscission zone located at the base of the organ. The abscission zone is often constricted in relation to contiguous regions. Its cell walls are thin, completely or nearly lacking in lignin and suberin (Scott *et al.*, 1948). It is considered a region of arrested maturation, and its principal activities begin with the senescence or injury of the organ it subtends.

In many species, pollination, fertilization, and embryo development affect the abscission of flower parts. In *Clarkia*, for example, the petals abscise shortly after fertilization, the latter occurring about 40 hr after pollination. If fertilization does not occur, the petals remain attached for many days. If embryos fail to develop, the ovulary tissues are abscised; if they do develop, the ovulary remains attached until maturity (Lewis, 1946).

Chandler (1957) characterized the successive stages of abscission in apples. The first drop is of flowers without fertilized eggs. In flowers of the second wave there has been slight growth of the ovaries since the time of pollination and some growth of the endosperm, but very little embryo growth. In fruits that are retained to the third wave, endosperms and embryos have grown a little. In the fourth wave, endosperms are fully developed and embryos have grown considerably but are smaller than embryos in fruits that do not fall, and there are fewer seeds per fruit.

Barlow (1953) stated that abscission is a vital process and that ultimately the separation of the cells concerned is due to a softening of the pectic middle lamella and sometimes part of the cellulose cell wall.

Heinicke (1919) reported that pedicels of flowers or fruits often show constriction at the base where they unite with the stem. The abscission zone appears as a water-soaked line extending across the base of the lower or fruit stem, near, but not always at, the constricted portion. It is in a straight line but varies where it crosses the conducting tissues. There is no apparent difference between cells in the abscission zone region and those of adjoining tissues, although cortical cells in the constricted portion are usually somewhat smaller than those in the broader portion of the pedicel.

Yager (1959) presented anatomical differences between Lizard's Tail and Little Turkish varieties of *Nicotiana tabacum* which exist during development of the separation zone and as floral abscission occurs. Cells are larger in Little Turkish by an average of 5 to 6 μm in the abscission zone and by 10 μm in the developing fruit. The separation layer in Little Turkish is located 5 to 7 tiers distal to the groove and is usually composed of only 4 to 6 tiers.

In studies on the anatomy and morphology of *Solanum tuberosum* Weinheimer and Woodbury (1967) found that a protective layer is formed much earlier in Russet Burbank than in Menominee. Actual abscission results from formation of a separation layer formed from a few to several cells distant from the protective layer. The separation layer results from cell collapse and disintegration, first between the epidermis and the vascular system and, finally, over the entire cross section of the pedicel.

Landa and Kummerow (1963) reported the presence of 3 abscission zones in the peach varieties Philips Cling and Reina Elena. The first abscission zone is located at the union of the peduncle with the twig; the second, at the insertion point of the flower receptacle with the peduncle; and the third, at the union of the fruit to the disc.

In the first zone of abscission, no special structural changes were found to indicate the course of separation. In the second zone, cellular divisions and suberized cells exist in different degrees forming a separative tissue. In the third zone, a notable increase in the amounts of calcium oxalate, abundant mucilage, and an increase of cellular spaces were observed.

McCown (1943) reported on the anatomy of immature apple pedicels. At the pink stage of flower development, the pedicel was relatively soft. The cells of various tissues were immature and mostly parenchymatous. A very small amount of secondary xylem and phloem was evident. A constriction of the pedicel base and apex of peduncle was evident in immature stems. In this constricted zone, cells of all tissues, except those of the pith, were small when compared to adjacent tissues. This zone was usually 20–30 cells in width in cortical tissue with less width in the xylem and phloem areas.

Flowers and immature fruits abscised following differentiation of an abscission layer in the basal portion of the pedicel (McCown, 1943). This layer resulted from cell division and was differentiated within the limits of the constriction zone. In occasional pedicels, the layer appeared in the pedicel distal to the constriction zone. The cells in all tissues underwent division in the differentiation of the abscission layer, and the resulting layer was 6–8 cells in width across the pedicel. McCown (1938) reported that by the time of the completion of the June drop, the tissues of the pedicels of growing fruits could be considered to be mature. Maturity was evidenced by the modification of a varying percentage of the pith cells to stone cells, the presence of well-developed stone cells and fibers in the pericycle, very definite bands of secondary phloem and xylem, and rather extreme thickening by secondary cellulose lamellae of the walls of cells of the cortex and of the specialized cells of the abscission zone.

MacDaniels (1936) found that in the early abscission of McIntosh apple, the abscission layer transverses the parenchymatous tissues and the pedicel is cut off. The vascular tissue of the pedicel and cluster base at this stage is undifferentiated procambium with occasional protoxylem strands, but sclerenchyma tissue has not yet formed. The abscission layer ordinarily cuts through the pedicel base where it joins the cluster base. In this early season abscission, there is evidently cell division in the layer of cells through which separation takes place.

In the abscission zone, there is a reduction in diameter of the pedicel at the point of junction with the cluster base. MacDaniels also found abundant specialized collenchyma immediately underneath this constriction, extending through the area occupied by fibers and stone cells in a normal pedicel; a reduction in the amount or absence of sclerenchyma in what would be the cortical region of the pedicel; a reduction in the number of fibers and vessels of the vascular cylinder as compared with that found in the pedicel and their replacement by parenchyma; a modification of vessels from the normal porous type with round pits to scalariform type with scalariform pits; and a modification of epidermal cells in constriction about the abscission zone to form cushions of elongate cells which apparently separate readily in abscission.

Jensen and Valdovinos (1967, 1968) have studied the fine structure of abscission zones of the pedicels of tobacco and tomato flowers at anthesis. The indentation or groove which delineates the abscission zone extends some distance into the pedicel with branchings off the main groove. These branches are approximately 200 nm in width. Invaginations of the plasmalemma were observed with considerable frequency. Within these invaginations a material of about the same density as the cell wall was found except that it was more fibrillar. Plasmodesmata were found with considerable branching into the middle lamellae of cells comprising the abscission zones. Microbodies with crystalloid cores appear with considerable frequency in cells of the abscission zone. The crystalloids appear to be cubical in shape and composed of parallel sheets of osmiophilic material. Chloroplasts contain a granular component which is membrane-enclosed. The inner membrane of the chloroplast is highly invaginated, and DNA- and phytoferritinlike materials are observed within the plastids. Microtubules with an average diameter of 20 nm were observed adjacent and parallel to the plasmalemma.

Valdovinos and Jensen (1968) have shown cell wall changes in abscising pedicels of *Lycopersicon esculentum* and *Nicotiana tabacum* flowers occurring by separation of cells that appeared to be initiated primarily in the middle lamella region of the cell walls. Disintegration of the primary wall, which usually followed breakdown of the middle lamella region, also occurred concurrently with lysis of the middle lamella region. During cell wall degradation, the walls appeared to swell and become highly flexible. The walls of at least some cells in the zone of separation invaginated during the advanced stages of cell wall disintegration, and they ultimately collapsed.

When Leuty and Bukovac (1968) examined abscising, potential drops and persisting Redhaven peach fruits, they found a relationship between endosperm and embryo development and fruit abscission. Fruits about to abscise and those considered potential drops were characterized by a smaller pericarp, seeds lacking endosperm, and/or no embryo or an aborting embryo. Persisting fruits contained seeds with a rapidly developing embryo and endosperm.

Simons (1965) found tissue development in the apple associated with embryo abortion as follows: (1) an abscission zone appears to form first on the side of the funicular base toward the cavity end of the fruit; (2) ovule abortion results in an overall degenerative effect on the central axis or pith and placental tissue adjacent to the ovarian locules of the fruit; (3) of the composited fruit tissues, pith constitutes the major portion in dropped fruit; (4) less cortical tissue and cuticle existed in dropped fruit than in normal fruit; (5) abscission zone formation in the suspensor may be phase I of the sequential abscission development of the fruit; (6) ovule abortion causes an overall degenerative effect upon growth of the cortical tissues which results in morphological diversity of structures that are not characteristic of the variety.

In Starking apple (Simons and Chu, 1968) there was less placental tissue to which the ovule was attached than in the cultivars, Golden Delicious and Jonared. Also, in Starking, the "calyx tube" was open and continuous throughout the central axis of the fruit which extended to the base of the ovule attachment at the bottom of the carpel.

Simons *et al.* (1970) found in apple an abscission zone at the base of the style and it had developed to the ovarian locule region.

Biggs (1971) found 2 abscission zones in citrus—one at the base of the pedicel and the other at the base of the fruit.

Wittenbach and Bukovac (1972) found that abscission of maturing sweet cherry fruit (*Prunus avium cv.* Windsor) occurred at 2 different abscission zones depending on the stage of fruit development. Immature fruit abscised at the upper zone between the pedicel and peduncle; mature fruit abscised at the lower zone between the fruit and receptacle. Separation in the abscission layer began directly above the stony pericarp and resulted in formation of a cavity. Later separation occurred at the fruit–pedicel indentation and extended through the abscission layer toward the vascular bundles. Abscission involved the fracturing of cell walls as well as wall separation. There was no evidence of change in pectins, cellulose, or other polysaccharides in the cell walls of the abscission layer prior to or during fruit separation. No starch accumulation in the abscission zone or lignification of tissue adjacent to the abscission layer was observed through fruit maturity. Conversely, Wilson and Hendershott (1968) reported in anatomical studies of citrus fruit abscission that large quantities of starch were present in the separation layer at maturity. As starch increased during the abscission process, methylated pectins and total pectins disappeared from the separation layer. Preharvest application of 2,4-D prevented loss of pectic materials. Starch was not directly associated with abscission. The separation layer of immature fruits contained no starch. Abscission induced by iodoacetic acid (IOAC) was not distinguishable anatomically and histochemically from normal abscission even though IOAC accelerated the process considerably.

II. Peach—*Prunus persica* L. (Lott and Simons, 1964)

Development of the flower or the fruit of the peach is not characterized by easily determined physiological stages. The visual appearance of the abscission zone in the floral cup of the peach flower meets these requirements; it appears as a narrow, pale greenish yellow line around the outside of the floral cup about one-quarter of the distance from its base to its rim and it is specific in that the interval between the preceding and succeeding developmental stages occupies only a few days.

The use of the visual appearance of this abscission zone as a developmental reference point is much more nearly specific than the ambiguous terms "shuck-off," "shuck-split," and "shuck-fall" that have been used by various investigators (Blake, 1926; Kelley, 1955a,b; Powell *et al.*, 1963) none of whom have specifically defined the term used. If "shuck-off" and "shuck-fall" mean, as seems logical, the complete detachment of the "shuck" from the fruit, this would be associated with a quite variable interval in fruit development because, once the "shuck" is cut off from the floral cup base by the abscission zone, its subsequent detachment from the fruit is dependent on fruit enlargement and the vagaries of weathering agents. This is illustrated by the statement of Blake (1926) that "the shuck or dried calyx may remain on the tip of an occasional green fruit for many days beyond the normal period of shedding."

The existence of a floral cup abscission zone in the peach seems to have been recognized by Robbins (1931) who, in describing the Drupaceae, states "After fertilization, the receptacle, with its attached sepals, petals, and stamens is cut off by a circular abscission layer near its base"; by Robbins and Ramaley (1933) who make essentially the same statement; and by Lott (1942) who referred to it as a shuck abscission zone in Halehaven peach.

A second relatively specific reference point in the development of the peach fruit is the formation of the abscission zone at the base of the style which is followed by abscission of the style from 1 to 3 days later.

The presence of this style abscission zone has been mentioned by Lott (1942) and implied by Robbins and Ramaley (1933) who state in discussing the flowers of what they call the plum family, "the style and stigma do not persist in the fruit."

Because of the potential use of formation of these two abscission zones as developmental reference points their morphological and anatomical characteristics were investigated preceding, during, and subsequent to their appearance.

A. TERMINOLOGY

Flower, pistil, ovary, ovule, style, stigma, stamen, filament, anther, sepal, petal—the meaning commonly used in botanical and systematic pomology texts (Haupt, 1946; Zielinski, 1955).

Receptacle—the floral axis, on the distal end of which the floral cup and pistil are borne.

Floral cup—the cup-shaped organ borne on the distal end of the receptacle; on or near the rim of this cup the sepals, petals, and stamens are borne.

Floral cup abscission zone—the layers of abscission cells which separate the approximately distal three-quarters of the floral cup from the basal one-quarter. It appears macroscopically as a narrow, pale greenish yellow line around the outside of the floral cup.

Floral cup base—the basal part of the floral cup left when the distal part abscises. It persists to fruit maturity.

Floral tube—that part of the floral cup distal to the abscission zone which, together with the adherent dried remains of the sepals, petals, and stamens abscises; the petals nearly always abscise before the visual appearance of the abscission zone.

Shuck—same as floral tube, the use of which is preferred.

Color—the descriptive color terms used conform to the ISCC-NBS method of designating colors, as reported by Kelly and Judd (1955).

B. Morphological Development

Characteristics of the development of the flower parts and floral cup and style abscission zones were similar in 12 cultivars.

Two cultivars were chosen because Redhaven is small-petaled and early maturing and Redskin is large-petaled and matures in late mid-season.

The morphological characteristics of the flower parts and young fruits at each of 10 visually specific developmental states from full bloom to style abscission are described below.

Stage I—full bloom (Figs. 1A and C). Petals at maximum size and divergence; anthers dehisced and liberating pollen; apparently the stigma became receptive soon after the start of pollen liberation; in flowers with anthers nearly dry but still liberating pollen the stigmas were post-receptive and dry.

Stage II—petal fall (Figs. 1B and D). Petals abscised from the floral cup, anthers dry, pollen absent or dry, filaments purplish red, erect but drying at tip; stigma dry, style erect and purplish red one-third from tip to ovary.

Stage III (Figs. 1B, D, and 2A). This stage can be described only as embracing the interval from petal fall to the appearance of the floral cup abscission zone, since there were no externally apparent morphological changes in the floral cup during this interval, other than some increase in size.

FIG. 1. Developmental stages I and II. (A) Redhaven, stage I, nearest to full bloom, note small petals; (B) Redhaven, stage II, petal fall or just after; (C) Redskin, stage I, nearest to full bloom, note large petals; (D) Redskin, stage II, petal fall or just after. From Lott and Simons (1964).

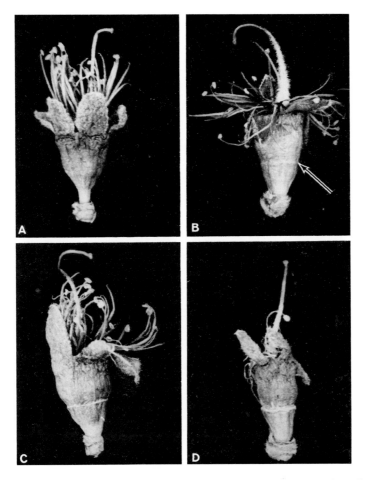

FIG. 2. Developmental stages III to VI. (A) Stage III, no visual evidence of floral cup abscission zone; (B) stage IV, floral cup abscission zone visible as a pale greenish yellow line at arrow; (C) stage V, floral tube abscised from base of floral cup but no vertical separation, apparently turgid; (D) stage VI, floral tube dry, shrunken tightly around ovary, separated from floral cup base up to $\frac{1}{16}$ inch, an occasional short vertical split. From Lott and Simons (1964).

Outside of floral cup uniform purplish red; inside brilliant orange except in the white-fleshed varieties Belle of Georgia and Early-Red-Fre in which it was pale greenish yellow to light greenish yellow. Anthers dry, pollen absent, filaments erect, dry down from tip $\frac{1}{8}$ to $\frac{1}{4}$ inch, remainder purplish red; stigma dry, style erect, dry $\frac{1}{8}$ inch below stigma, then purplish red two-thirds of distance to the pubescence.

Stage IV (Fig. 2B). Abscission zone evident as a pale greenish yellow

line around the outside of the floral cup about 25% of the distance from
the bottom of the floral cup to its rim; no visible separation of floral
tube from floral cup base; purplish red the dominant external color in
the floral tube, the floral cup base predominantly green externally with
red absent or nearly so; anthers dry, filaments erect, dry ¼ inch below
tip, remainder purplish red; stigma dry, style erect and dry ⅛ inch below
stigma, remainder purplish red to pubescence.

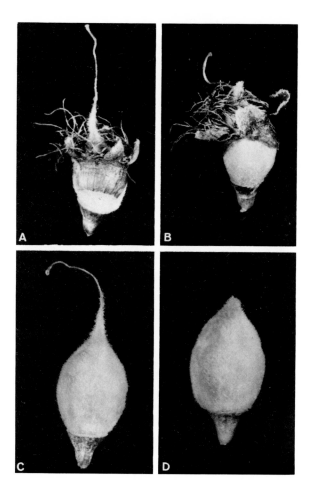

FIG. 3. Developmental stages VII to X. (A) Stage VII, floral tube separated from
floral cup base up to ⅜ inch vertically, one or more vertical splits; (B) stage VIII,
base of floral tube near point of greatest transverse diameter of fruit or beyond
toward tip; (C) stage IX, floral tube absent from fruit, style persistent; (D) stage X,
style abscised. From Lott and Simons (1964).

Stage V (Fig. 2C). Floral tube barely separate from the floral cup base, still apparently turgid, diameter less than that of floral cup base due primarily to continued growth in the latter, whereas the floral tube could no longer grow and had begun to shrink, an occasional vertical crack just started at base of the floral tube; the floral cup base predominantly green externally but some red still present, floral tube predominantly dull purplish red externally; anthers and upper one-fifth of filaments dry, basal four-fifths purplish red but flaccid; stigma and upper ¼ to ⅜ inch of style dry, remainder purplish red down to pubescence.

Stage VI (Fig. 2D). Floral tube dry, shrunken against ovary, separated vertically from floral cup base up to $\frac{1}{16}$ inch, about one-third had 1 or 2 vertical splits just started from their base; anthers and filaments dry; stigma and style dry ¼ to ⅜ inch below tip, remainder purplish red down to pubescence.

Stage VII (Fig. 3A). Floral tube dry, separated vertically from floral cup base ¼ to ⅜ inch, split vertically in 2 or 3 places; style dry half way to pubescence, lower part purplish red to pubescence.

Stage VIII (Fig. 3B). Base of floral tube near point of greatest transverse diameter of fruit or beyond toward tip, held on primarily by pubescence; style as in stage VII.

Stage IX (Fig. 3C). Floral tube absent from fruit, top of floral cup base slightly greater in diameter than base of fruit immediately distal to it, as in all previous stages; style persistent, dry nearly down to pubescence.

Stage X (Fig. 3D). Style abscised leaving suberized tip of fruit, if not abscised it fell off when touched; top of floral cup base less in diameter than base of fruit immediately distal to it, showing rapid enlargement in fruit base; compare shape of fruit bases in Figs. 3C and D.

Since fruit development is necessarily a progressive phenomenon it is obvious that stages exist which are intermediate between those described. However, from stage IV onward the interval between successive stages occupies only a few days and it is not feasible to attempt to define intermediate stages.

C. ANATOMICAL CHARACTERISTICS

Floral Cup Abscission Zone

In all cases the cell divisions by which the abscission zone was formed were initiated at the inner epidermis of the floral cup and progressed in the shape of a sigmoid curve to the outer epidermis. Consequently, when the abscission zone became visually evident around the outside of the floral cup (Fig. 2B), the floral tube had already been cut off from the floral cup base.

Figures 4A–D show the progressive development of the abscission zone. In Fig. 4A the arrow points to the area where cell divisions had just begun in the formation of the abscission zone; the ovary is to the right.

The arrow in Fig. 4B points toward a mass of recently divided cells in a stage of abscission zone development occurring soon after that in Fig. 4A. In Fig. 4C the abscission zone had progressed to a major vas-

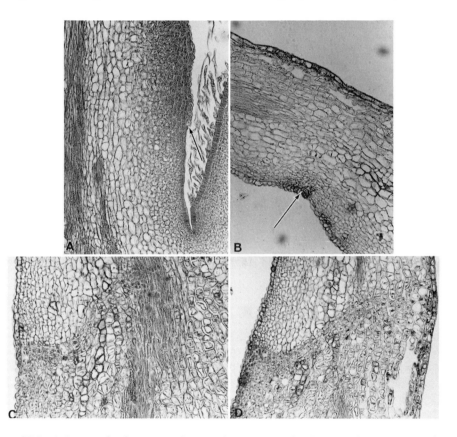

FIG. 4. Longitudinal sections through floral cup in developmental stages III and IV. (A) Stage III, initiation of cell division in floral cup abscission zone at arrow, ovary to the right, ×70; (B) stage III, mass of divided cells initiating floral cup abscission zone (arrow) in the inner side of floral cup, ×55; (C) later stage III, abscission zone formed from inner part of floral cup to a major vascular bundle, ×70; (D) stage IV, abscission zone formed completely through floral cup, ×55. Inner epidermis and ovary to left in B, C, and D. From Lott and Simons (1964).

cular bundle in the approximate center of the longitudinal section of the cup. Figures 4A–C are all in morphological stage III.

Figure 4D shows a typical abscission zone extending from the inner epidermis of the floral cup at the left to its outer epidermis at the right. This is a characteristic stage IV abscission zone, shown morphologically in Fig. 2B.

Figures 5A and B illustrate the fact that abscission occurred in the large vascular bundles of the floral cup after it was complete in the tissues on either side. These illustrate stage IV floral cups; the floral tube could be separated from the base by a very slight pull.

A stage V floral cup is seen in Figs. 5C and D. The floral tube is com-

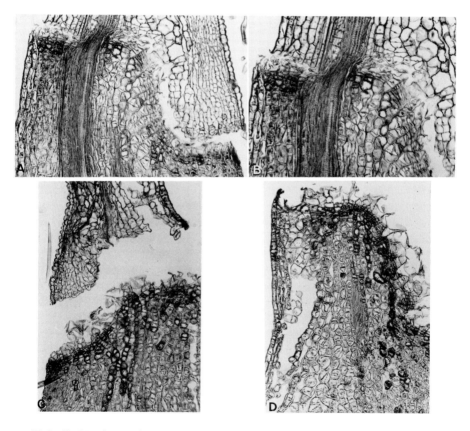

FIG. 5. (A,B) Developmental stage IV, abscission zone formed in main vascular bundles of floral cup later in other tissues, ovary to the right, ×70; (C,D) developmental stage V, floral tube abscised, separated slightly from floral cup base and beginning to shrink, ovary to the left, ×55. Floral cup base also represents that of stage VI. From Lott and Simons (1964).

pletely free and apparently pulled away from the base. However, when one considers that this is magnified 100 times it is obvious that the external appearance at this stage would be as in Fig. 2C.

A typical stage VI floral cup base produced a more pronounced sigmoid shape of the abscission zone as compared with stage IV (Fig. 4D). This, presumably, is the result of more rapid growth in the outer portion of the cup base than in the inner part.

Numerous attempts to obtain sections through the style abscission zone suitable for photographing have been unsuccessful. The pubescence makes sectioning very difficult, and attempts to remove it without damage to the fruit tissues have been only partly successful. However, indisputable evidence has been found of an abscission zone several cell layers wide formed at approximately right angles to the style. The effect of this can be seen in the fruit with abscised style in Fig. 3D.

D. Discussion

Examination of the literature applicable to the structure of the flowers of the genus *Prunus*, which are perigynous, shows the need for specific terminology concerning the nonpistil part of the flower. Terms are needed not only for this structure but also for the tissues that abscise and the part that remains after abscission. According to Marshall (1954), "The hollow or concave cuplike structure of the cherry flower that supports the calyx, corolla, and stamens on its rim is known as the receptacle. The ovary is located at the bottom of this cuplike receptacle." This interpretation of the *Prunus* flower agrees with others (Haupt, 1946; Holman and Robbins, 1947; Robbins, 1931; Robbins and Ramaley, 1933; Swingle, 1947). The cuplike part of the flower has also been referred to as "calyx-cup" (Hedrick, 1922) and "calyx-tube" (Swingle, 1947).

Jackson (1934) concluded, from an extensive study of the vascular structure of perigynous rosaceous flowers, that the cuplike portion of the flower was neither calyx, implied by the term calyx tube, nor receptacular, but was appendicular in nature and consisted of the fused basal portions of the sepals, petals, and stamens. She suggests the term floral tube for this cuplike part of the flower.

Since the shape of the structure outside of the pistil is more nearly that of a cup than of a tube the term floral cup has been adopted for it. The part that abscises is approximately tubular in shape and is, therefore, designated as floral tube which consists of the distal 70 to 75% of the floral cup, together with the dried remains of sepals, stamens, and occasionally petals. The portion of the floral cup remaining after abscission of the floral tube is referred to as the calyx-cup base.

III. Montmorency Cherry—*Prunus cerasus* L.
(Lott and Simons, 1966)

The developmental pattern of floral tube and style abscission was similar to that of the peach (Lott and Simons, 1964), but there were differences in the rate both chronologically and relatively.

A. MORPHOLOGICAL DEVELOPMENT

The morphological characteristics of the flower parts and young fruits at each of 8 visually specific developmental stages are described below.

Stage I—full bloom. Petals at maximum size and divergence; filaments erect, anthers mostly dehisced and liberating pollen; style erect, the stigma apparently became receptive coincident with the beginning of pollen liberation.

Stage II—petal fall. Petals abscised from the floral cup; filaments erect, anthers dry, pollen absent or dry; style erect, stigma dry.

Stage III (Figs. 6A and 7A). This includes the interval from petal fall to the external appearance of the floral cup abscission zone; filaments usually erect, tips and anthers dry; style erect, stigma dry.

Stage IV (Figs. 6B and 7B). Floral cup abscission zone evident as a pale greenish yellow line around the outside of the floral cup, about 20% of the distance from the bottom of the cup to its rim, no visible separation of floral tube from floral cup base; floral tube pale greenish yellow but floral cup base light yellowish green; leaflike sepals still green and apparently turgid; style erect but tip $3/16$ inch dry, style abscission zone not evident macroscopically but present at base of style (Fig. 7B).

Stage V (Figs. 6C and 7C). Floral tube barely separated vertically from floral cup base and beginning to shrink, diameter less than that of the floral cup base due mostly to rapid growth of the latter, floral tube in Fig. 7C midway between stages IV and V to show that abscission started at inner epidermis; style abscised.

Stage VI (Figs. 6D and 7D). Floral tube dry, shrunken against ovary, separated vertically from floral cup base up to $3/16$ inch; style abscission scar usually irregularly concave.

Stage VII (Fig. 6E). Floral tube dry, separated vertically from floral cup base $3/16$ inch or more, split vertically at one or more points; top of floral cup base greater in diameter than base of ovary immediately distal to it, as in all previous stages.

Stage VIII (Fig. 6F). Floral tube separated from fruit; a protuberance at stylar end of ovary formed by fleshy pericarp development adjacent

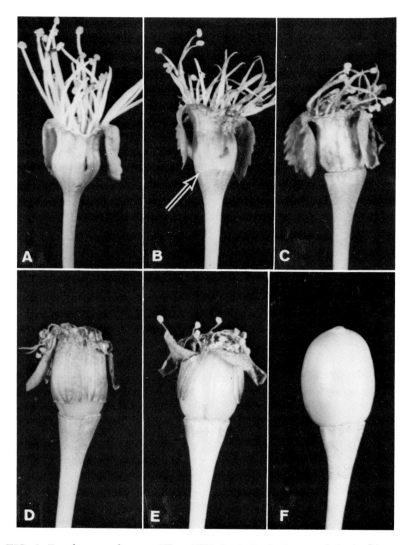

FIG. 6. Developmental stages III to VIII. In A, B, C, D, one of the leaflike sepals was excised to show the floral cup. (A) Stage III, floral cup abscission zone not evident; (B) stage IV, floral cup abscission zone evident as a pale greenish yellow line at arrow, floral tube lighter color than floral cup base; (C) stage V, floral tube abscised from floral cup base, barely separated vertically, and beginning to shrink as shown by edges of sepals in comparison with those of A and B; style abscised; (D) stage VI, floral tube dry, shrunken around ovary, separated from floral cup base up to $\frac{3}{16}$ inch; (E) stage VII, floral tube separated from floral cup base $\frac{3}{16}$ inch or more, 1 or 2 wide vertical splits; (F) stage VIII, floral tube separated from fruit, note fleshy pericarp development adjacent to style abscission scar. From Lott and Simons (1966).

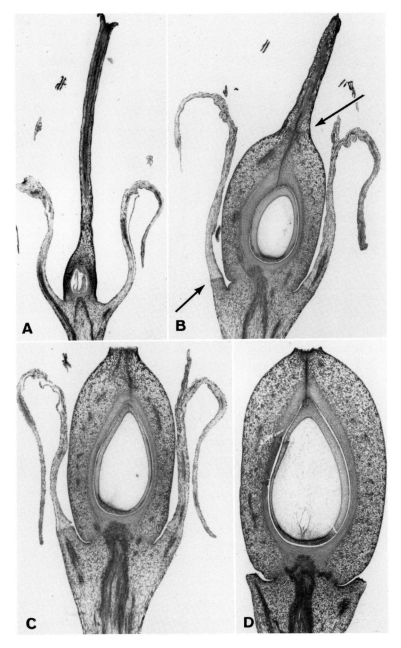

FIG. 7. Longitudinal sections of fruits in developmental stages III to VI. (A) Stage III, floral cup abscission zone not evident; (B) stage IV, floral cup abscission zone at lower arrow, and style abscission zone at upper arrow; (C) early stage V, floral tube separating from floral cup base and beginning to shrink; style abscised; note that diameter of ovary is less than that of floral cup base, compare with D; (D) stage VI, floral tube separated from floral cup base and lost in handling, see Fig. 6D. Photographed from slides. All ×8. From Lott and Simons (1966).

to the style abscission scar at a greater rate than in the tissue surrounding it; fruit free of pubescence; top of floral cup base less in diameter than that of fruit immediately distal to it, resulting from rapid enlargement of ovary.

B. ANATOMICAL CHARACTERISTICS

1. *Floral Cup Abscission Zone*

The formation of the abscission zone was initiated by cell divisions at the inner epidermis of the floral cup and progressed in an approximately straight line to the outer epidermis (Figs. 8A and B). Therefore, when the abscission zone became visible around the outside of the floral cup (Fig. 6B), the floral tube had, in effect, already been cut off from the base of the floral cup (Fig. 7B).

The developmental path of the abscission zone is shown in Figs. 8A and B of typical floral cups in stages III and IV, respectively. A stage V abscission zone is illustrated in Fig. 9A in which the floral tube is separated from the floral cup base except in the main vascular bundles; in such cases a very slight pull completed the separation. Developmental stage VI is seen in Fig. 9B; the floral tube is completely free from the floral cup base and, in this particular case, a secondary abscission zone has formed and a third one has been initiated at the inner epidermis. Only a few examples of such zones were found.

2. *Style Abscission Zone*

The formation of the abscission zone by which the style is cut off at its base from the ovary can be seen in the stage IV fruit in Fig. 10A. It developed across the style in an irregularly concave line.

The tip of a typical stage V (Fig. 6C) fruit after style abscission is shown in Fig. 10B. Such an irregularly concave surface was common. Apparently, there is associated with style abscission a stimulatory effect on development of the fruit tissue immediately adjacent to the abscission zone which results in a more rapid rate of growth than in the remainder of the fruit and the protuberance at the stylar end shown in Fig. 6F. This differential growth rate seems to be reversed later, since in the mature fruit the tip is usually depressed.

The cell divisions by which the abscission zone was formed originated on either side of the central vascular bundle of the style and progressed outward to the epidermis. At the same time it developed through the main vascular bundle. The style abscission zone originated 1 or 2 days

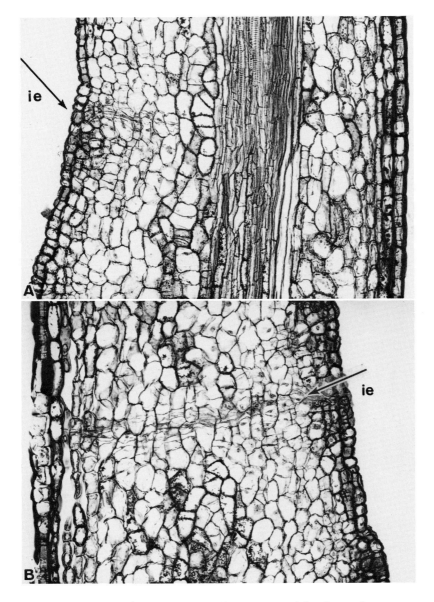

FIG. 8. Longitudinal sections showing development of floral cup abscission zone. (A) Stage III, floral cup abscission zone at arrow initiated at inner epidermis and progressing outward, ovary to the left; (B) stage IV, abscission zone at arrow developed across floral cup, ovary to the right. (ie denotes inner epidermis.) Both ×128. From Lott and Simons (1966).

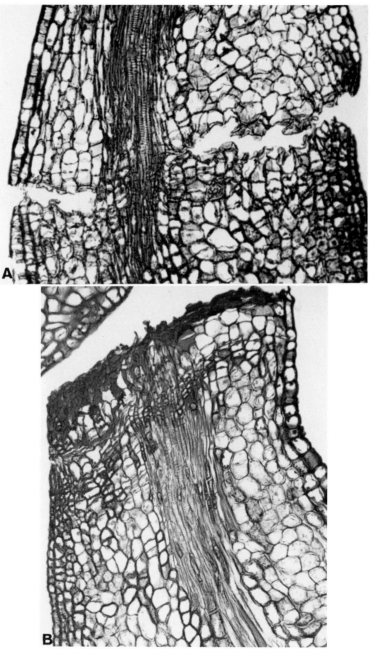

FIG. 9. Longitudinal sections of development stages V and VI. (A) Stage V, floral tube separated from floral cup base except in main vascular bundle, ovary to the right, ×128; (B) stage VI, floral tube (upper left) separated from floral cup base up to $\frac{3}{16}$ inch, ovary to the left; note secondary abscission zone in floral cup base, ×100. From Lott and Simons (1966).

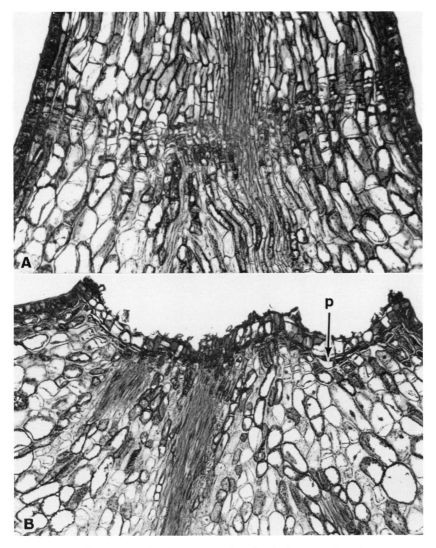

FIG. 10. Median longitudinal sections of style. (A) Stage IV, abscission zone formed across style at base, ×80; (B) stage V, style abscised; phellem forming in abscission zone at p, ×60. From Lott and Simons (1966).

later than the floral cup abscission zone, but it developed so rapidly that it had formed completely across the style before the floral tube had begun to separate from the floral cup base in progressing from stage IV to stage V. Thus, the style abscission zone in Fig. 10A is in a stage IV fruit. Both

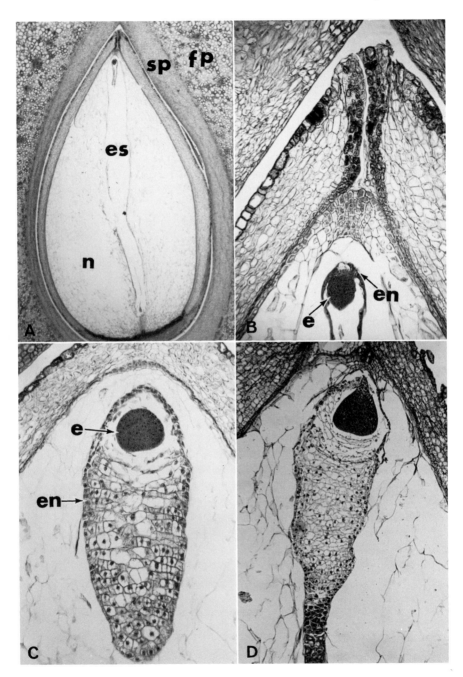

the floral cup and style abscission zones developed so rapidly that the time elapsed from their initiation to complete development could have been determined only by examining the same specimen throughout the interval, which was obviously impossible. However, it seems reasonably definite that the development of each required only 2 or 3 days at most, with initiation of the style abscission zone occurring 1 or 2 days after that of the floral cup abscission zone but developing more rapidly since, as shown in Fig. 7C, the style abscised before actual abscission of the floral tube.

3. *Relation of Morphological Stages to Ovule Development*

A median longitudinal section of the ovule and adjacent stony pericarp and fleshy pericarp of a stage VI fruit (Fig. 6D) is shown in Fig. 11A. The embryo sac extends through the nucellus nearly to the chalaza, essentially as described by others (Marshall, 1954; Pechoutre, 1902; Ruehle, 1924; Tukey, 1933, 1934; Tulasne, 1855; Went, 1887). There is a large endosperm nucleus near the chalazal end of the embryo sac; this is characteristic of sweet cherries also, according to Tukey (1933). Endosperm nuclei are also present in the cytoplasm surrounding the embryo near the tip of the embryo sac; these are readily evident in the higher magnification of the same ovule tip in Fig. 11B, as is also the multinucleate stage of the embryo.

The developmental status of the endosperm and embryo in a late stage VII fruit (Fig. 6E) is shown in Fig. 11C of longitudinal section through the ovule. Many nuclei have been formed in the endosperm and it has become cellular. There also has been a large increase in the number of nuclei and the size of the embryo. The comparison of the developmental status of the endosperm and embryo in the stage VI fruit in Fig. 11B (at 96×) with that in the stage VII fruit in Fig. 11C (at 75×) shows that there is quite active development in both during the early life of the fruit, even though their development is much more rapid after the beginning of stone hardening, as reported (Tukey, 1933, 1934).

That continued development in the endosperm and embryo is associated with morphological development of the fruit is shown by the

FIG. 11. Longitudinal sections through approximate center of ovule. (A) Stage VI, nucellus (n), embryo sac (es) extends through nucellus nearly to chalaza at base of ovule; stony pericarp (sp) and fleshy pericarp (fp) outside ovule, ×12; (B) higher magnification of tip of ovule in A, endosperm nuclei (en) and embryo (e), ×96; (C) stage VII, cellular structure of endosperm (en) and embryo (e), ×75; (D) stage VIII, further development of endosperm and embryo, ×45. From Lott and Simons (1966).

longitudinal section of the ovule of a stage VIII fruit (Fig. 6F) and in Fig. 11D where further development of both is apparent. The magnification of Fig. 11D is only 60% of that of Fig. 11C. Bradbury (1929) stated that "cellular endosperm is formed in the micropylar portion of the embryo sac," implying that the endosperm of the chalazal region remains noncellular. However, she also indicated that "The cellular endosperm, by continued division, gradually replaces the nucellar tissue." In sweet cherry, according to Tukey (1933), "In the region of the chalaza no cell walls are formed in the endosperm." The endosperm in Fig. 11D is cellular but it occupies only about one-third of the distance from the micropyle to the chalaza.

C. Discussion

Development of the floral cup and style abscission zones followed the same general pattern as reported for peach (Lott and Simons, 1964), but there were these important differences: (1) A more rapid rate of development from one stage to the next, particularly beyond stage IV. Thus, stage VIII in the cherry was reached in 10 to 12 days after bloom, but comparable developmental stage VII in the peach required about 20 days (Lott and Simons, 1964). Such differences may be expected because of the shorter interval from bloom to fruit maturity in cherry. (2) In cherry development of floral cup and style abscission zones was nearly coincident, as shown by the stage IV fruit in Fig. 7B, whereas in peach the style abscission zone did not develop until stage IX, after the floral tube was separated from the fruit (Lott and Simons, 1964); this is a definite genetic difference. (3) Actual abscission of the cherry style was coincident with or preceded that of the floral tube (Fig. 7C) instead of occurring 5 stages later after the floral tube was free of the fruit, as it did in peach (Lott and Simons, 1964). (4) The floral cup abscission zone in cherry developed in a nearly straight line, whereas in peach it developed as a sigmoid curve; this may result from the difference in rate of development.

It is possible to relate the morphological stages described here to development of the reproductive tissues as reported in the literature.

Stage I—full bloom. Pollen grains mature and being liberated, stigma receptive, the macrogametophyte mature (embryo sac) (Bradbury, 1929), initiation of rapid size increase in nucellus and integuments (Tukey, 1934) which continues through the 8 stages described herein.

Stage II—petal fall. Two to four days after full bloom; embryo sac extends about one-fifth the length of the nucellus at the time of double fertilization 2 to 4 days after full bloom (Bradbury, 1929).

Stages III to VI. Division of primary endosperm nucleus followed by division of zygote; elongation of embryo sac nearly to the chalaza, lined with cytoplasm containing endosperm nuclei; embryo becomes multi-nucleate and increases in size (Bradbury, 1929; Tukey, 1933).

Stages VII and VIII. Endosperm becomes cellular in micropylar end of ovule, embryo nuclei continue to divide (Bradbury, 1929; Tukey, 1933).

IV. Starking Hardy Giant Cherry—*Prunus avium* L.
(Lott and Simons, 1968b)

The morphological and anatomical characteristics of development of floral cup and style abscission were basically similar to those in peach, Montmorency cherry, and apricot (Lott and Simons, 1964, 1966, 1968b), but there was a marked increase in the rate of development of the style abscission zone.

A. MORPHOLOGICAL DEVELOPMENT

The morphological characteristics of flower parts and young fruits at each of 8 visually specific development stages are described below.

Stage I—anthesis (full bloom) (Fig. 12A). Petals at maximum size and divergence, white, roundish, 15 mm long by 15 mm wide; floral cup 6–7 mm long by 5–6 mm wide; exterior light yellow-green, interior light yellow-green at rim to moderate yellow-green at base; sepals recurved, with tips reaching nearly to base of floral cup, 5–6 mm long by 4–5 mm wide, moderate yellow-green; anthers light yellow, inner ones (at about 1 mm lower level than stigma) dehiscing just as stigmatic fluid became present, outer ones (at same level as stigmas) dehiscing during the next few hours while the stigmatic fluid was present, all dehiscing during the day, filaments erect, yellowish white; stigmatic fluid usually present one day only, stigma light to moderate yellow-green, 1 mm wide, styles erect, light yellow-green, 10–12 mm long; ovaries strong yellow-green, 3 mm long by 1.5 mm wide.

Stage II—petal fall. Petals abscised; floral cup as in stage I except central area of exterior washed 50 to 75% with light reddish purple; sepals as in stage I except both dorsal and ventral surfaces washed 50 to 75% with light reddish purple; anthers dry, light brown, filaments erect, yellowish white except tip dry and light brown 1–3 mm; stigmas dry, dark brown, styles variously persistent. In most styles the abscission zone was plainly evident at the base and styles abscised when touched; in such

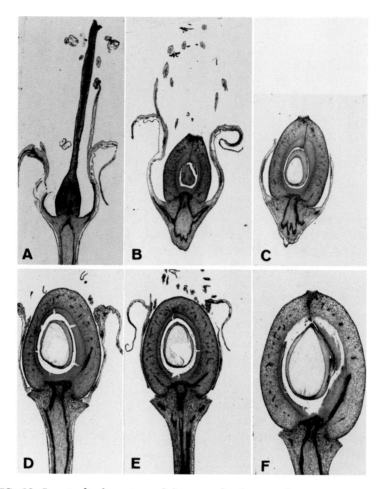

FIG. 12. Longitudinal sections of fruits in developmental stages I to VIII. (A) Stage I, full bloom, no evidence of style or floral cup abscission zone; (B) stage III, style abscised in stage II, floral cup abscission zone not evident; (C) stage IV, floral cup abscission zone evident around base of floral tube; (D) stage V, floral tube abscised and shrinking, just visually separated from floral cup base; (E) stage VI, floral tube dry, shrunken against ovary, separated from floral cup base; (F) stage VIII, floral tube absent, base of ovary greater in diameter than floral cup base (compare with E). All ×4. From Lott and Simons (1968b).

styles the tip 1–2 mm was dry and the remainder light greenish yellow with 4–5 mm of light reddish purple wash near the center; many styles were dry, moderate to dark brown and actually abscised but held erect by the stamens. Apparently the style abscission zone began rapid development coincident with the first appearance of the reddish purple color in the exterior of the cup, in the sepals, and in the style, the style had abscised when this color was readily noticed. The ovaries were strong yellow-green, 5 mm long by 3 mm wide.

Stage III (Fig. 12B). The interval between petal fall and appearance of the abscission zone around the outside of the floral cup; color of cup and sepals as in stage II; anthers dry, light brown, filaments variously erect to curved, yellowish white except tip 1–3 mm dry and light brown in early part of the interval but distal half dry and light to moderate brown in the latter part; styles abscised preceding start of this stage.

Stage IV (Fig. 12C). Abscission zone evident as a thin, light yellow-green line around the exterior of the floral cup, slightly above the base of the ovary, position of abscission zone emphasized by pale yellow-green band 1–2 mm wide immediately above it, remainder of floral tube exterior light to moderate yellow-green, interior light yellow-green at the rim to moderate yellow-green at the base; floral cup base moderate yellow-green; sepals moderate yellow-green with both surfaces washed 50 to 75% with light to moderate reddish purple; anthers and filaments dry and light to moderate brown; style abscission scars covered with phellem; ovaries strong yellow-green, 6 × 4 mm.

Stage V (Fig. 12D). Floral tube abscised and flaccid, just visually separated from floral cup base, floral cup in Fig. 17A midway between stages IV and V to show that actual abscission began in inner epidermis; anthers and filaments dry and light to moderate brown.

Stage VI (Fig. 12E). Floral tube dry, shrunken against ovary, separated from floral cup base up to 5 mm, usually less; style abscission scar irregularly straight across tip of ovary.

Stage VII. Floral tube dry, separated vertically from floral cup base 5 mm or more, vertical splits at one or more points, 3–4 mm wide at base and extending nearly to rim of floral tube; top of floral cup base greater in diameter than base of ovary, as in all previous stages.

Stage VIII (Fig. 12F). Floral tube separated from fruit; top of floral cup base less in diameter than base of ovary because of rapid enlargement of the latter; ovary light to moderate yellow-green, nonpubescent, as in all previous stages.

B. Anatomical Characteristics

1. *Associated Floral Organs*

Figure 13 shows the anatomical relationships of these floral organs of sweet cherry in a nearly median longitudinal section of a flower in the

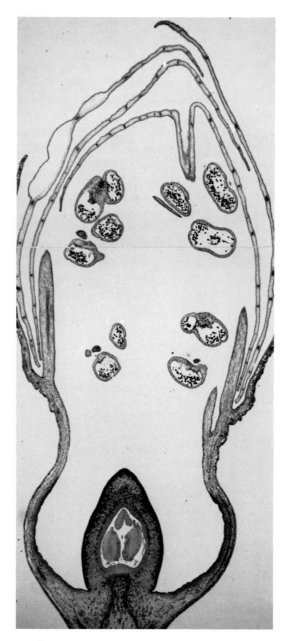

FIG. 13. Longitudinal section of flower, 1 or 2 days before anthesis. Note shape of floral cup, the sessile ovary, attachment of sepals and filaments, petals, and sections of anthers with pollen grains. ×9. From Lott and Simons (1968b).

"balloon" stage. This illustrates the shape and size of the floral cup in relation to other floral organs, the enfolded petals, the filaments arising at different levels on and near the rim of the floral cup, the sepals borne on its rim, the sessile ovary, and the cross sections of anthers with the enclosed pollen grains.

2. Style Abscission

A median longitudinal section through a stigma and adjacent style portion of a stage I flower are shown in Fig. 14A. Many of the stigmatic epidermal cells have developed into short hairs; this is characteristic of cherry (Esau, 1965). On the basis of reaction to staining it is evident that the central region of this part of the style is physiologically the same as the stigma. Esau (1965) prefers the term stigmatoid tissue for this area and states that "it is interpreted as a medium facilitating the progress of the pollen tube through the style and supplying the developing pollen tube with food."

The basal part of the style of a stage I flower is illustrated in Fig. 14B. Style abscission occurred in this region. There is no evidence of the initiation of the cell divisions by which the abscission zone is formed.

In the early stage II style of Fig. 15A the abscission layers have formed completely across the base of the style. In flowers having such styles only 1 or 2 petals per flower had fallen. When the petals had all fallen the style was abscised. In styles such as Fig. 15A the position of the abscission zone was plainly evident because the style was light greenish yellow above it and moderate yellow-green just below it.

The tip of a stage II fruit is illustrated in Fig. 15B, showing the cells left at the tip of the fruit by abscission of the style, and the phellogen and phellem formed under them. The cells above the phellem soon sloughed off to leave a corky scar surface.

3. Floral Tube Abscission

The cell divisions by which the abscission zone in the floral cup was formed originated at the inner epidermis and developed in an approximately straight line to the outer epidermis. In the stage II floral cup of Fig. 16A this cell division had not been definitely initiated. It always began with cell divisions in a mass of small cells adjacent to the inner epidermis, similar to those in Fig. 16A.

A floral cup abscission zone developed nearly across the cup is illustrated in the stage III fruit in Fig. 16B. In the stage IV floral cup of Fig. 17A the abscission zone had developed completely across the cup and abscission of the floral tube had begun in the inner epidermis and

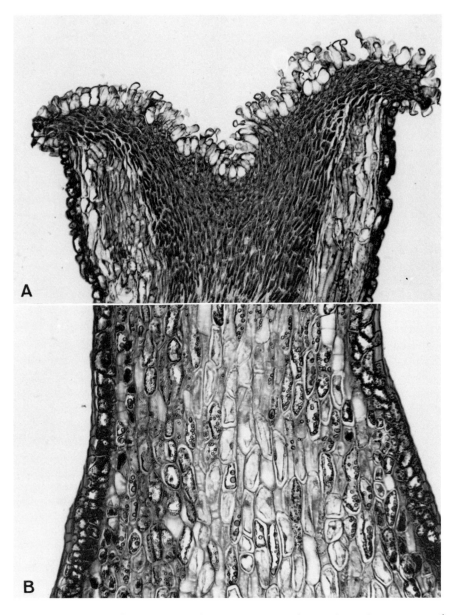

FIG. 14. Longitudinal sections through stage I styles. (A) Median section of stigma; note epidermal hairs, and stigmatoid tissue extending down style; (B) base of style; no evidence of formation of abscission zone. Both ×128. From Lott and Simons (1968b).

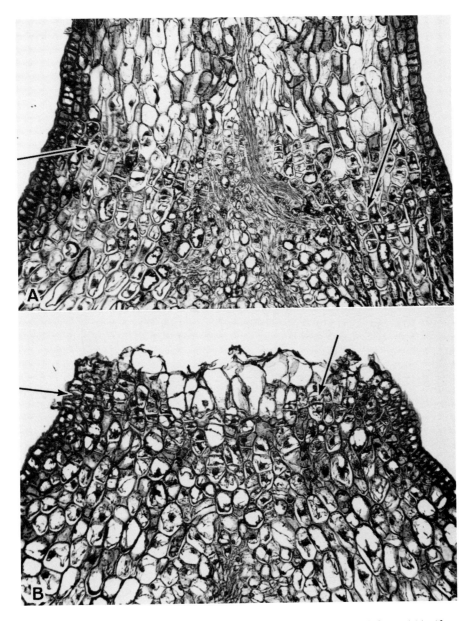

FIG. 15. Longitudinal sections through base of style and tip of fruit. (A) Abscission layers at arrows formed nearly completely across base of style, early stage II; (B) tip of fruit after style abscission, stage II. Note that abscission occurred distal to new cell layers at arrows. Both ×100. From Lott and Simons (1968b).

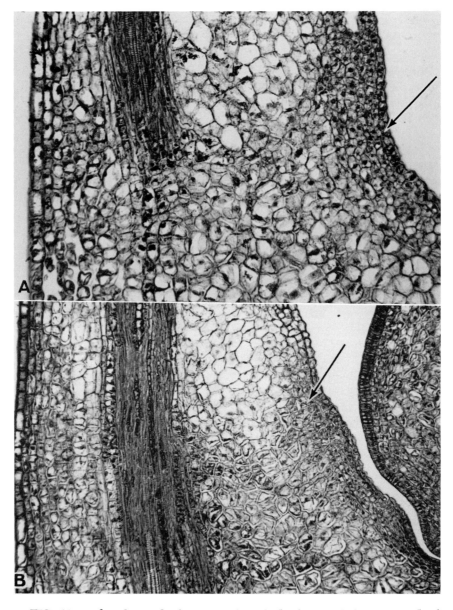

FIG. 16. Median longitudinal sections through floral cups. (A) Stage II, floral cup abscission cell layers not yet definitely initiated. They originate in region shown by arrows, ×128; (B) stage III, floral cup abscission cell layers at arrows established from inner epidermis nearly across the cup. ×100. Ovary to right in both. From Lott and Simons (1968b).

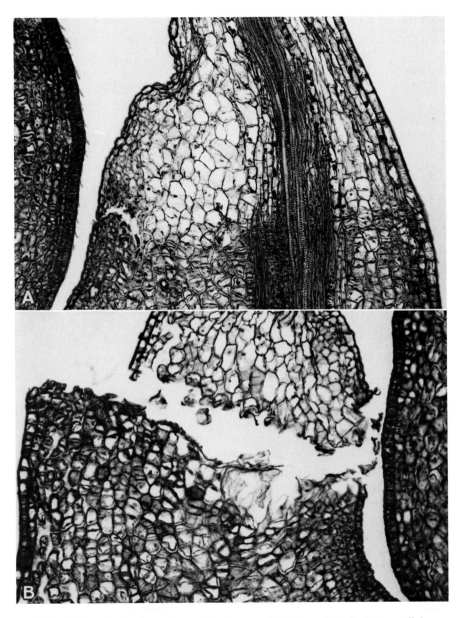

FIG. 17. Longitudinal sections of floral cups. (A) Stage IV, abscission cell layers formed across floral cup; floral tube abscising and shrinking at interior; (B) stage V, floral tube abscised, shrunken toward ovary, barely separated vertically from floral cup base. Both ×100. From Lott and Simons (1968b).

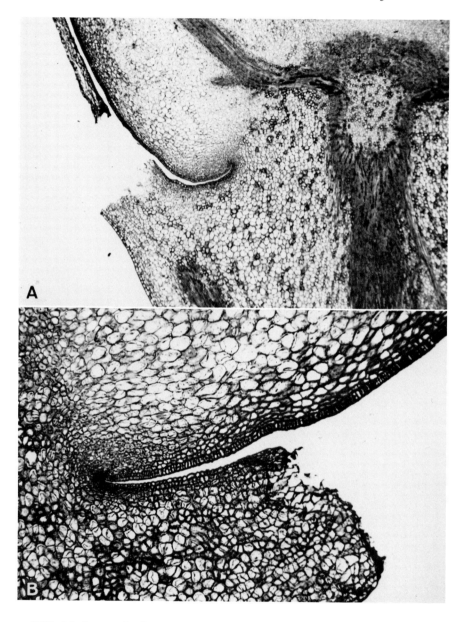

FIG. 18. Longitudinal sections through base of ovary and floral cup base. (A) Stage VI, floral tube shrivelled and separated vertically from floral cup base, diameter of floral cup base greater than base of ovary, ×25; (B) stage VIII, ovary base greater in diameter than floral cup base, ×63. From Lott and Simons (1968b).

adjacent cell layers, accompanied by shrinkage of the inner part of the floral tube. In such floral cups the abscission zone was plainly evident around the outside of the cup.

The typical stage V floral cup of Fig. 17B illustrates shrinkage of the floral tube that immediately followed its abscission from the floral cup base. Without magnification such floral tubes appeared to be shrunken inward from the outer epidermis of the floral cup base, and polar separation of the floral tube from it was generally just barely visible.

In stage VI fruits such as that of Fig. 18A the floral tube had become dry, shrunken against the ovule, and separated vertically from the floral cup base at various distances, depending on the rate of growth of the ovary. Note that the floral cup base still, in effect, enclosed the base of the ovary. By contrast, in the stage VIII fruit of Fig. 18B this was not the case because the rapid enlargement of the base of the ovary during this interval had resulted in a diameter greater than that of the floral cup base.

C. Discussion

Style abscission in Starking Hardy Giant sweet cherry developed more rapidly than in other *Prunus* species studied (Lott and Simons, 1964, 1966, 1968a). The comparative rates of development of floral tube and style abscission in 4 *Prunus* species are listed in Table I. The number of days shown in each case is for average weather conditions during these intervals. Unusually warm weather would shorten the intervals and unseasonably cool weather would lengthen them. However, the same

TABLE I

RELATIONSHIPS BETWEEN *Prunus* SPECIES IN DEVELOPMENT
OF FLORAL TUBE AND STYLE ABSCISSION

		Days from anthesis to:		
Species	Variety	Petal fall (stage II)	Floral tube abscission zone evident (stage IV)	Style abscission
P. persica L.	Twelve varieties[a]	3–6	7–10	20–24
P. armeniaca L.	Wilson Delicious	3–5	6–8	18–20
P. cerasus L.	Montmorency	3–4	5–6	5–6
P. avium L.	Starking Hardy Giant	3–4	5–6	3–4
P. avium L.	Stark Gold	3–4	5–6	5–6

[a] See Lott and Simons (1964).

relationships should exist between the species and varieties in any particular environment.

The Stark Gold sweet cherry is included in Table I to show that rapid development of style abscission in Starking Hardy Giant is not necessarily characteristic of sweet cherries. Stark Gold was included in the 1966 study but a freeze on May 10 made it impossible to obtain all the developmental stages. Collections had been made through the floral tube and style abscission stages of Stark Gold and these showed it to have the same developmental pattern as the Montmorency cherry (Table I and Lott and Simons, 1966). Observations in 1967 showed that Starking Hardy Giant and Stark Gold had the same developmental pattern as in 1966.

The short interval between anthesis and style abscission in Starking Hardy Giant introduces the question of its possible effect on fertilization. The stigmatic fluid was present coincident with anthesis and beginning of pollen liberation, and usually was evident during only one day, which should provide ample time for adequate pollination. But growth of the pollen tube must be rapid for it to progress down through the style below its point of abscission during the few days between anthesis and visible formation of the abscission zone completely across the style. As shown in Table I, this interval was only 3 to 4 days. The abscission zone was formed nearly across the style at least 1 day before it became visible on the exterior. This left only 2 to 3 days for the pollen grain to germinate and for the tube to grow through the 10–12 mm of the style distal to the abscission zone. Whether pollen tubes grow at this rate in this variety remains to be determined.

There seems to be little information on the rate of pollen-tube growth in the sweet cherry. It is known that there are wide differences between plant species in the rate of pollen-tube growth (Fuller, 1955). It is possible that differences occur also in rate of pollen tube growth between varieties within a species. Roy (1939) reported that, in the Noir de Schmidt sweet cherry pollinated by the cross-compatible variety Red Cluster, 4 days after pollination the longest pollen tubes had grown 6.1 mm in the styles, which had an average length of 10.4 mm. In Starking Hardy Giant the styles would have abscised by this time. He does not mention any later examination nor the time of style abscission, but does state that he obtained a 19% set of mature fruits. Afify (1933) pollinated Bigarreau Frogmore with Governor Wood and reported that "Flowers were examined for pollen-tube growth from 3 to 15 days after pollination," and that some pollen tubes grew through the style and reached the ovary, but he did not present data on the relation of pollen-tube penetration 2 days after pollination. In view of the present results it is surprising that he was able to examine styles 15 days after pollination; it does not seem likely that the fact the trees he used were grown in pots in a greenhouse could account for such prolonged retention of styles. He did not mention the time of style abscission.

It is possible that the type of style abscission that is characteristic of Starking Hardy Giant is present in other sweet cherry varieties and contributed to the self- and cross-unfruitfulness that has been reported widely in sweet cherry varieties (Marshall, 1954). Differences in the degree of compatibility between a female parent and various pollen parents could result in enough differences in rate of pollen-tube growth for the tubes of some combinations to fail to progress through the style before its abscission. Afify (1933) reported that in the styles of selfed Bedford Prolific and Bigarreau Frogmore hand-pollinated with Governor Wood there were pollen grains which failed to germinate, and those which germinated and penetrated the style for a short distance only then bent upward and ceased to grow "due to incompatibility." In the cross-pollination there were also pollen tubes which grew through the style and reached the ovary, resulting in a set of 19% mature fruits.

In any case, the time of style abscission should be determined in all compatibility investigations and in all varieties and potential varieties. Unfortunately, the literature on sweet cherry compatibility indicates that the investigators failed to notice that the styles abscise or have given no consideration to its potential significance in relation to fruitfulness. Typical of this situation are the summaries on the subject in reference works (e.g., Marshall, 1954) and research papers (e.g., Shoemaker, 1928).

The terminology and concept of the anatomy of abscission need clarification. We have used the term abscission zone to refer to the cell layers that are formed across the floral cup and the style (Lott and Simons, 1964, 1966, 1968a,b). Strictly speaking this is incorrect because the zone in which abscission occurs is present before any of the cell divisions by which this layer is formed have been initiated. Also, abscission has not occurred in this layer of new cells, but in apparently weakened cells 1 or 2 cell layers distal to it. Actually, the layers of cells that develop across the floral cup and style preceding abscission appear to be phellem formed from a phellogen initiated across each of these organs. The phellem thus presumably formed interrupts the flow of water and other nutrients to the part distal to the layer, causing senescence and abscission (Fuller, 1955). Gawadi and Avery (1950) stated, "That natural abscission is the ultimate result of senescence of cells of the abscising organ is fairly evident from the literature on the subject." They object to the term abscission layer because abscission occurs in some species before the formation of such a layer. They seem to prefer "separation layer" and correctly define it as being "immediately adjacent to the abscission layer, on its distal side" if one is willing to call the layer of cells that form across the organ an abscission layer. Pending the determination of whether this layer is actually phellem, as it appears to be, we have for practical purposes referred to it as the abscission zone but have no objection to referring to it as an abscission layer, even though abscission of floral tube and style occur adjacent and distal to it.

Investigation of the floral cup and style abscission zones in *Prunus* species has shown the same general pattern of development in 12 peach varieties, the Wilson Delicious apricot, the Montmorency sour cherry, and the Starking Hardy Giant and Stark Gold sweet cherries. There were significant differences in the rate of their development, being more rapid in cherries than in peach and apricot, and especially rapid in the styles of Starking Hardy Giant cherry.

V. Wilson Delicious Apricot—*Prunus armeniaca* L.
(Lott and Simons, 1968a)

The specificity and potential use of the visual appearance of the floral cup and style abscission zones as developmental reference points in peach and Montmorency cherry have been reported (Lott and Simons, 1964, 1966). In Montmorency cherry, these abscission zones differed from those of peach in their appearance at approximately the same time, in more rapid development from one stage to the next, and in style abscission occurring coincident with or slightly before floral tube abscission rather than about 2 weeks later. Consequently, a study of a third member of the *Prunus* genus, the apricot, was conducted to determine whether the development of these zones and associated floral organs differed from those in peach and cherry.

The only references to these abscission zones in the apricot seem to be by Robbins (1931) and Robbins and Ramaley (1933) who stated "In the maturing of the fruit, the parts of the flower are cut off by a basal ring of growth, as in the plum." Their statement regarding the plum is: "After fertilization, the cup-shaped receptacle withers with its attached sepals, petals, and stamens, being cut off by a circular abscission layer near its base."

A. Morphological Development

Development of the floral cup and style abscission zones in apricot was generally similar to those of peach (Lott and Simons, 1964), both chronologically and relatively. There were certain morphological differences in associated organs.

The morphological attributes of flower parts and young fruits at each of 10 visually specific developmental stages from full bloom to style abscission are given.

Stage I—full bloom. Petals at maximum size and divergence, white to pinkish white, 10 mm long by 9 mm wide; floral cup 5–6 mm long from base to rim, basal one-fifth to one-fourth of exterior light yellow-green, dotted and washed 50% with light to moderate reddish purple, remainder to rim moderate yellow-green overlaid 90% with moderate reddish purple; interior brilliant orange-yellow from base up 2–3 mm, remainder to rim yellowish white; sepals leafy, 3–4 mm in length and width, reflexed with tips extending down to lower one-third of floral cup, both surfaces 90% covered with moderate reddish purple over yellow-green background; numerous scales on pedicel, covering lower one-fifth to one-fourth of floral cup; anthers light to moderate yellow, mostly dehisced, and liberating pollen, filaments erect, yellowish white except basal 1–2 mm usually pale reddish purple; stigmas receptive, apparently coincident with pollen liberation, light to moderate orange-yellow, styles erect, pale greenish yellow 2–3 mm below stigma, then yellowish white down to 2–3 mm above base, remainder light yellow-green, pubescence absent 4–5 mm below stigma then increasing in amount to the thickly pubescent base and ovary.

Stage II—petal fall. Petals abscised; floral cup similar to stage I except basal interior changed to light orange-yellow; and upper exterior slightly darker and higher chroma reddish purple, no evidence of floral cup abscission zone after removal of the scales which covered the base of the cup; sepals as in stage I except color slightly darker and higher chroma reddish purple; anthers dry and moderate to dark brown, no pollen evident, filaments as in stage I except tip 1 mm dry and moderate brown; stigmas dry, moderate brown, styles as in stage I except tip 1 mm dry and moderate brown.

Stage III (Fig. 19A). This includes the interval from petal fall to just before appearance of the floral cup abscission zone, since there were no other specific morphological changes which could be construed as the termination of this stage; floral cup 5–6 mm long, basal 1 mm of exterior light yellow-green, then increasingly spotted, mottled, and splashed with light to moderate reddish purple halfway to the rim, upper half predominantly moderate reddish purple, basal 2 mm of cup interior light yellow. Scales persistent on pedicel as in previous stages; sepals as in stage II; anthers as in stage II, filaments erect to recurved, dry and moderate brown usually one-half to two-thirds below anther, then pale reddish purple to base; stigmas dry, dark brown, styles erect, dry and dark

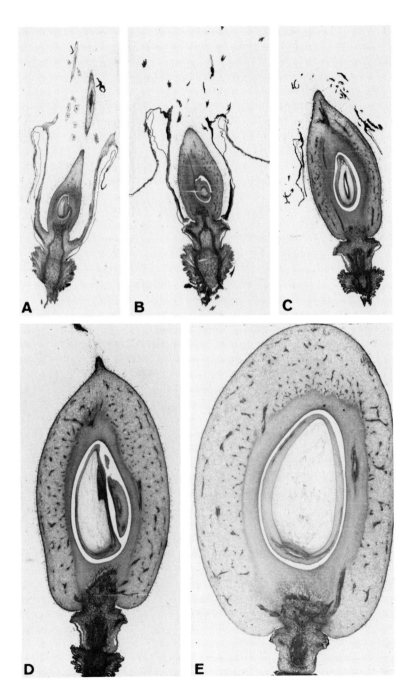

brown 1–3 mm below stigma, then pale reddish purple nearly to the greenish yellow base, nonpubescent 3–4 mm below stigma, then increasing in amount to the base, which was densely pubescent, as was also the yellow-green ovary.

Stage IV. Floral cup abscission zone evident around the outside of the cup, after removing the scales which extended above it, at or slightly above the level of the base of the ovary; floral cup moderate yellow-green below the abscission zone and pale greenish yellow immediately above it, color of remainder as in stage III; sepals increased to 5–6 mm long by 4–5 mm wide, color as in stage III; anthers dry, dark brown, filaments variously erect or recurved, dry to base, light to moderate brown; stigmas dry, dark brown, styles erect, moderate to dark brown below stigma to pubescence (3–4 mm), then as in stage III.

Stage V (Fig. 19B). Floral tube abscised from floral cup base but no visible vertical separation after scales removed, appeared to be shrunken against ovary, upper half somewhat shriveled, remainder flaccid, color of floral cup base as in stage IV, floral tube darker because of desiccation; sepals shriveling, color as in stage IV except darker; anthers dry, dark brown, filaments dry and dark brown at tip to light brown at base; stigmas dry, dark brown, styles dry to pubescence, moderate to dark brown.

Stage VI. Floral tube dry, shrunken against ovary, moderate to dark brown, separated from floral cup base up to 1.5 mm, an occasional vertical split up to 1 mm wide and 1 mm long, scales still extending above base of floral tube; sepals, anthers, and filaments dry and moderate to dark brown; stigmas dry, dark brown, styles persistent, semierect, dry nearly to base, dark brown at tip to light brown near the light yellow-green base.

Stage VII (Fig. 19C). Floral tube dry, shrunken, moderate to dark brown, usually separated vertically from floral cup base up to 2 mm, with the scales still nearly covering this gap, 1 or 2 vertical splits up to 2 mm wide by 4–5 mm long, indicating the expansion of the ovary, transverse diameter of floral cup base greater than that of ovary base;

FIG. 19. Longitudinal sections of fruits in developmental stages III to X. (A) Stage III, floral cup abscission zone not evident; note bases of scales broken off to observe base of floral cup; (B) stage V, floral tube abscised and shrivelling, abscission zone approximately at level of ovary base; (C) stage VII, note bases of broken-off scales, receptacle, and ovary base less in diameter than floral cup base; (D) stage IX, style persistent; ovary base much greater in diameter than floral cup base; (E) stage X, style abscised, but scar not evident in this slightly oblique section; ovary enlarging rapidly, base growing down around floral cup base. All ×3.6. From Lott and Simons (1968a).

styles persistent, dry, dark brown at tip to moderate brown near base. In Fig. 19C only approximately the upper half of the floral tube has been retained in the section; this accounts for the floral tube being separated from the floral cup base further than as described for stage VII.

Stage VIII. Floral tube separate from floral cup base 2–3 mm, usually 1 or 2 vertical splits 3–4 mm wide by 4–5 mm long; scales partly abscised; styles persistent, variously recurved, dark brown at tip and moderate brown down to pubescence. Rapid enlargement in ovary, transverse diameter of ovary base equal to or slightly greater than that of floral cup base.

Stage IX (Fig. 19D). Base of floral tube near point of greatest transverse diameter of ovary, or beyond toward tip, usually held on by pubescence; scales usually abscised, width of ovary base greater than that of floral cup base; styles persistent, dry and dark brown except basal 1 mm, which was light yellow overlaid with varying amounts of moderate reddish purple. There was a sharp line of demarcation between the light yellow color of the base of the style and the moderate greenish yellow tip of the ovary; this was the style abscission zone, shown by the fact that some styles would break off there with a slight to moderate pull.

Most ovaries had developed moderate reddish purple just below the tip on the dorsal suture (occasionally nearly around the tip) and extending varying distances down it toward the base. In some ovaries there were also light to moderate reddish purple blotches of various sizes extending down to the base; areas of moderate reddish purple on the cheeks at base of ovary and occasionally completely around it.

Stage X (Fig. 19E). Floral tube absent; styles abscised, leaving phellem around edge of style-abscission scar and mottled on the remainder of its surface. Rapid enlargement of ovary with its base extending down to or below top of floral cup base and as much as two and one-half times as wide; compare shape of ovary bases in Fig. 19D and E in relation to floral cup bases.

B. Anatomical Characteristics

1. *Floral Cup Abscission Zone*

This developed from cell divisions which were initiated in the region of the inner epidermis, at approximately the level of the ovary base, and proceeded in an irregular straight line to the outer epidermis. A stage I floral cup is shown in Fig. 20A; this was also representative of stage II, since the cell divisions that initiated the development of the abscission zone did not occur until stage III. A late stage III abscission zone is

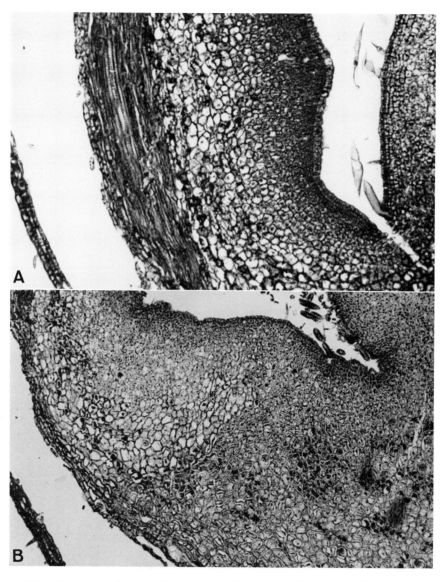

FIG. 20. Longitudinal sections through floral cups. (A) Stage I, initiation of abscission zone not evident; ovary to the right, ×125; (B) late stage III, abscission zone developed from inner epidermis nearly across floral cup; ovary to the right, ×80. From Lott and Simons (1968a).

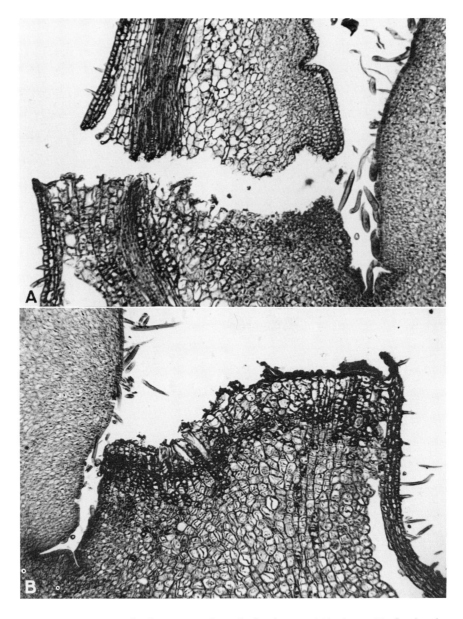

FIG. 21. Longitudinal sections through floral cup. (A) Stage V, floral tube abscised and shrivelling; ovary to the right, ×100; (B) stage VII, phellem layer formed in floral cup base; ovary to the left, ×80. From Lott and Simons (1968a).

illustrated in Fig. 20B, the abscission zone having progressed nearly across the floral cup. In cases where the abscission zone had developed completely across the floral cup it became evident around the outside of the cup, as described for stage IV.

A stage V abscission zone is seen in Fig. 21A. The floral tube has abscised and begun to shrink. Although the floral tube seems to be separated an appreciable distance from the floral cup base, the fact that the figure is magnified 100 times shows that no separation would have been visible without magnification.

The formation of phellogen and phellem layers in stage VII floral cup base is shown in Fig. 21B. The tissue distal to this zone eventually abscises to expose the phellem that is present on the surface of the floral cup base throughout fruit development.

2. *Style Abscission Zone*

The apricot style abscised in a manner similar to that of peach and sour cherry (Lott and Simons, 1964, 1966). In the stage VII style of Fig. 22A, development of the abscission zone has not been initiated. A style in the latter part of stage VIII is illustrated in Fig. 22B. The abscission zone had progressed halfway or more across the style. Because of the difficulty of sectioning the style, occasioned by the dense pubescence, it was not established that the abscission zone was initiated in the region of the central vascular bundle and developed radially from it as in Montmorency cherry (Lott and Simons, 1966), but it seems probable that this was the case. A stage IX style is shown in Fig. 22C. The abscission zone had developed completely across the style. Such styles broke off easily from wind or a slight to moderate pull. At style abscission, phellem was present around the edge of the style-abscission scar and progressed to cover it completely in 1 or 2 days.

C. DISCUSSION

Morphological characteristics of Wilson Delicious apricot flowers and young fruits differed from those of peach and Montmorency cherry (Lott and Simons, 1964, 1966) in several major respects: (1) The floral cup abscission zone occurred approximately at the level of the base of the ovary rather than definitely above it. This, combined with shrinkage of the floral tube against the ovary after abscission, resulted in the base of the floral tube being, in effect, shrunken under the base of the ovary. This was a major reason for the slow movement of the abscised floral tube away from the floral cup base as described in stages VI, VII, and

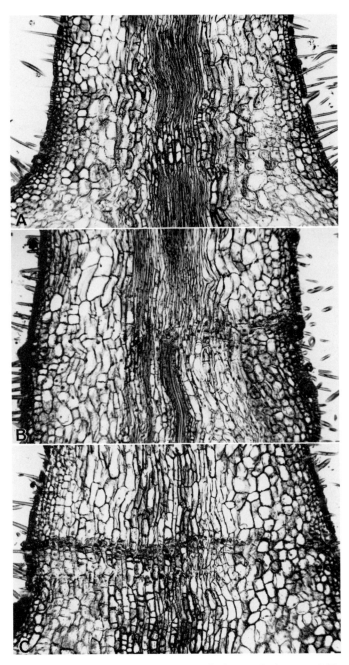

FIG. 22. Median longitudinal sections of style bases. (A) Stage VII, abscission zone not initiated, ×60; (B) stage VIII, abscission zone developed halfway across style, ×75; (C) stage IX, the abscission zone formed across base of style, but style not abscised, ×75. From Lott and Simons (1968a).

VIII. Appreciable separation occurred only after significant splits appeared in the floral tube as a result of increase in transverse diameter of the ovary; polar enlargement of the ovary was also a factor. (2) The ovary was attached to a short length of receptacle extending above the base of the floral cup rather than being sessile or nearly so in the base of the cup. This can be seen in Figs. 19A–E. (3) The scales surrounding the base of the floral cup extended up beyond the base of the floral tube as long as they persisted, which was usually to stage VIII, and occasionally into stage X, rather than reaching only to about the abscission zone and abscising in stage II or early stage III. (4) The section of pedicel to which the scales were attached was longer and thicker.

Anatomically, the floral cup and style abscission zones developed essentially as in peach (Lott and Simons, 1964) both chronologically and relatively, except that the floral cup abscission zone developed in a nearly straight line rather than as a sigmoid curve.

The development of a phellogen and a phellem layer in the floral cup base, shown in Fig. 21B, occurred also in peach and Montmorency cherry. It was not determined whether such layers are formed later and deeper in the floral cup base. At fruit maturity its surface is covered with phellem.

It is not assumed that all apricot varieties will show the same morphological and anatomical patterns as described for Wilson Delicious. Some differences may well occur, particularly in the kind and distribution of colors.

The morphological characteristics of 10 developmental stages in apricot flowers and young fruits are described, extending from anthesis to style abscission. Stage IV was characterized by the appearance of the floral cup abscission zone as a narrow, pale greenish yellow line around the cup at the approximate level of the ovary base. Anatomically, this abscission zone was initiated by cell divisions in the inner epidermis of the cup and progressed in an irregularly straight line to the outer epidermis. Stage X was characterized by abscission of the style at its base. Stages IV and X each occupied only a few days, were specific, and, therefore, useful as developmental reference points; stage X is preferred for this purpose because a larger percentage of the fruits reached it at the same time. These stages were similar to those found in peach, with these major differences: (1) Floral cup abscission occurred at approximately the level of the ovary base rather than somewhat above it. (2) The bud scales extended beyond the base of the floral tube and did not begin to abscise until in stage VIII, necessitating their removal to determine the abscission status of the floral tube; in the peach they were shorter and abscised in stages II and III.

References

Abeles, F. B., Leather, G. R., Forrence, L. E., and Craker, L. E. (1971). Abscission: Regulation of senescence, protein synthesis, and enzyme secretion by ethylene. *HortScience* 6, 371–376.

Addicott, F. T., and Lynch, R. S. (1955). Physiology of abscission. *Annu. Rev. Plant Physiol.* 6, 211–238.

Afify, A. (1933). Pollen tube growth in diploid and polyploid fruits. *J. Pomol.* 11, 11–19.

Barlow, H. W. B. (1953). The importance of abscission in fruit production. *Rep. Int. Hort. Congr., 13th, 1952* Vol. 1, pp. 145–152.

Biggs, R. H. (1971). Citrus abscission. *HortScience* 6, 388–392.

Blake, M. A. (1926). The growth of the Elberta peach from blossom bud to maturity. *Proc. Amer. Soc. Hort. Sci.* 22, 29–39.

Bradbury, D. (1929). A comparative study of the developing and aborting fruits of *Prunus cerasus. Amer. J. Bot.* 16, 525–542.

Bukovac, M. J. (1971). The nature and chemical promotion of abscission in maturing cherry fruit. *HortScience* 6, 385–388.

Chandler, W. H. (1957). "Deciduous Orchards." Lea & Febiger, Philadelphia, Pennsylvania.

Crane, J. C. (1971). The correlative effect of competition on abscission in apricot and pistachio. *HortScience* 6, 382–385.

Edgerton, L. J. (1971). Apple abscission. *HortScience* 6, 378–382.

Esau, K. (1960). "Anatomy of Seed Plants." Wiley, New York.

Esau, K. (1965). "Plant Anatomy," 2nd ed. Wiley, New York.

Fuller, H. J. (1955). "The Plant World," 3rd ed. Holt, New York.

Gardner, F. E., Marth, P. C., and Batjer, L. P. (1939). Spraying with plant growth substances for control of pre-harvest drop of apples. *Proc. Amer. Soc. Hort. Sci.* 37, 415–428.

Gawadi, A. G., and Avery, G. S., Jr. (1950). Leaf abscission and the so-called abscission layer. *Amer. J. Bot.* 37, 172–180.

Haupt, A. W. (1946). "An Introduction to Botany," 2nd ed. McGraw-Hill, New York.

Hedrick, U. P. (1922). "Cyclopedia of Hardy Fruits." Macmillan, New York.

Heinicke, A. J. (1919). Concerning the shedding of flowers and fruits and other abscission phenomena in apples and pears. *Proc. Amer. Soc. Hort. Sci.* 16, 76–83.

Holman, R. M., and Robbins, W. W. (1947). "Textbook of General Botany," 4th ed. Wiley, New York.

Jackson, G. (1934). The morphology of the flowers of *Rosa* and certain closely related genera. *Amer. J. Bot.* 21, 453–466.

Jensen, T. E., and Valdovinos, J. G. (1967). Fine structure of abscission zones. I. Abscission zones of the pedicels of tobacco and tomato flowers at anthesis. *Planta* 77, 298–318.

Jensen, T. E., and Valdovinos, J. G. (1968). Fine structure of abscission zones. III. Cytoplasmic changes in abscising pedicels of tobacco and tomato flowers. *Planta* 83, 303–313.

Kelley, V. W. (1955a). Effect of certain thinning materials on number of fruit buds formed in Elberta and Halehaven peaches. *Proc. Amer. Soc. Hort. Sci.* 66, 67–69.

Kelley, V. W. (1955b). Time of application of naphthaleneacetic acid for fruit thinning of the peach in relation to the June drop. *Proc. Amer. Soc. Hort. Sci.* **66**, 70–72.

Kelly, K. L., and Judd, D. B. (1955). The ISCC-NBS method of designating colors and a dictionary of color names. *Nat. Bur. Stand.* (*U. S.*), *Circ.* **533**, 1–158.

Landa, A., and Kummerow, J. (1963). Anatomical description of the abscission zone in the peach. *Maipu Chile ESTAC Exp. Agron.* **18**, 9–17.

Leopold, A. C. (1971). Physiological processes involved in abscission. *HortScience* **6**, 376–378.

Leuty, S. J., and Bukovac, M. J. (1968). A comparison of the growth and anatomical development of naturally abscising, non-abscising and naphthaleneacetic acid-treated peach fruits. *Phytomorphology* **18**, 372–379.

Lewis, D. (1946). Chemical control of fruit formation. *J. Pomol. Hort. Sci.* **22**, 175–183.

Lott, R. V. (1942). Effect of nitrate of soda on development of the Halehaven peach. *Ill., Agr. Exp. Sta., Bull.* **493**.

Lott, R. V., and Simons, R. K. (1964). Floral tube and style abscission in the peach and their use as physiological reference points. *Proc. Amer. Soc. Hort. Sci.* **85**, 141–153.

Lott, R. V., and Simons, R. K. (1966). Sequential development of floral-tube and style abscission in the Montmorency cherry (*Prunus cerasus* L.). *Proc. Amer. Soc. Hort. Sci.* **88**, 208–218.

Lott, R. V., and Simons, R. K. (1968a). The developmental morphology and anatomy of floral-tube and style abscission in the 'Wilson Delicious' apricot (*Prunus armeniaca* L.). *Hort. Res.* **8**, 67–73.

Lott, R. V., and Simons, R. K. (1968b). The morphology and anatomy of floral tube and style abscission and of associated floral organs in the Starking Hardy Giant cherry (*Prunus avium* L.). *Hort. Res.* **8**, 74–82.

McCown, M. (1938). Abscission of flowers and fruits of the apple. *Proc. Amer. Soc. Hort. Sci.* **36**, 320.

McCown, M. (1943). Anatomical and chemical aspects of abscission of fruits of the apple. *Bot. Gaz.* (*Chicago*) **105**, 212–220.

MacDaniels, L. H. (1936). Some anatomical aspects of apple flower and fruit abscission. *Proc. Amer. Soc. Hort. Sci.* **34**, 122–129.

Marshall, R. E. (1954). "Cherries and Cherry Products." Wiley (Interscience), New York.

Pechoutre, F. (1902). Contribution à l'étude du développement de l'ovule et de la graine des *Rosacées. Ann. Sci. Natur.: Bot. Biol. Veg.* [8] **16**, 1–158.

Powell, D., Meyer, R. H., and Owen, F. W. (1963). Pest control in commercial fruit plantings. *Ill. Ext. Circ.* **864**.

Robbins, W. W. (1931). "The Botany of Crop Plants," 3rd ed. McGraw-Hill (Blakiston), New York.

Robbins, W. W., and Ramaley, F. (1933). "Plants Useful to Man." McGraw-Hill (Blakiston), New York.

Roy, B. (1939). Studies on pollen tube growth in *Prunus. J. Pomol.* **16**, 320–328.

Ruehle, K. (1924). Beiträge zur Kenntmis der Gattung. *Prunus. Bot. Arch.* **7**, 224–229.

Scott, F. M., Schroeder, M. R., and Turrell, F. M. (1948). Development, cell shape, suberization of internal surface, and abscission in the leaf of the Valencia orange, *Citrus sinensis. Bot. Gaz.* (*Chicago*) **109**, 381–411.

Scott, P. C., Miller, L. W., Webster, B. D., and Leopold, A. C. (1967). Structural changes during bean leaf abscission. *Amer. J. Bot.* **54**, 730–734.

Shoemaker, J. S. (1928). Cherry pollination studies. *Ohio, Agr. Exp. Sta., Res. Bull.* **422.**

Simons, R. K. (1963). Anatomical studies of apple fruit abscission in relation to irrigation. *Proc. Amer. Soc. Hort. Sci.* **83**, 77–87.

Simons, R. K. (1965). Tissue development in the apple associated with embryo abortion. *Proc. Amer. Soc. Hort. Sci.* **87**, 55–65.

Simons, R. K., and Chu, M. C. (1968). Ovule development in the apple as related to morphological and anatomical variation in supporting tissues. *Proc. Amer. Soc. Hort. Sci.* **92**, 37–49.

Simons, R. K., Hewetson, F. N., and Chu, M. C. (1970). Fruit tissue injury by frost to 'York Imperial' apples. *J. Amer. Soc. Hort. Sci.* **95**, 821–827.

Swingle, D. B. (1947). "A Textbook of Systematic Botany." McGraw-Hill, New York.

Tukey, H. B. (1933). Embryo abortion in early-ripening varieties of *Prunus avium*. *Bot. Gaz. (Chicago)* **94**, 433–468.

Tukey, H. B. (1934). Growth of the embryo, seed, and pericarp of the sour cherry (*Prunus cerasus*) in relation to season of fruit ripening. *Proc. Amer. Soc. Hort. Sci.* **31**, 125–144.

Tukey, H. B., and Young, J. O. (1939). Histological study of the developing fruit of the sour cherry. *Bot. Gaz. (Chicago)* **100**, 723–749.

Tulasne, L. R. (1855). Nouvelles études d'embroyogenie végétale. *Ann. Sci. Natur.: Bot. Biol. Veg.* [4] **3**, 65–74.

Valdovinos, J. G., and Jensen, T. E. (1968). Fine structure of abscission zones. II. Cell-wall changes in abscising pedicels of tobacco and tomato flowers. *Planta* **83**, 295–302.

Webster, B. D. (1970). A morphogenetic study of leaf abscission in *Phaseolus*. *Amer. J. Bot.* **57**, 443–451.

Weinheimer, W. H., and Woodbury, G. W. (1967). The anatomy and morphology of the abscission layer in two cultivars of *Solanum tuberosum* L. *Amer. Potato J.* **44**, 402–408.

Went, F. (1887). Étude sur la forme du sac embryonnaire des *Rosacées*. *Ann. Sci. Natur.: Bot. Biol. Veg.* [7] **6**, 331–341.

Wilson, W. C., and Hendershott, C. H. (1968). Anatomical and histochemical studies of abscission of oranges. *Proc. Amer. Soc. Hort. Sci.* **92**, 203–210.

Wittenbach, V. A., and Bukovac, M. J. (1972). An anatomical and histochemical study of abscission in maturing sweet cherry fruit. *J. Amer. Soc. Hort. Sci.* **97**, 214–219.

Yager, R. E. (1959). Abscission zone anatomy, development, and separation in *Nicotiana tabacum*. *Iowa Acad. Sci.* **66**, 86–90.

Zielinski, Q. B. (1955). "Modern Systematic Pomology." W. C. Brown, Dubuque, Iowa.

. 11 .

Chemical Thinning
of Flowers and Fruits

L. J. Edgerton

I. Introduction

The benefits to the grower of regulating the crops on fruit bearing trees and vines have been recognized by horticulturists since the first treatises were written on fruit culture. In 1729 Langley wrote: "If our Apricots and Peaches are very numerous (on the tree), we must ease Nature of

435

her Burden. . . . About the 20th of June . . . as then the Fruits are beginning to swell . . . we should now thin them for good, taking away
the least promising ones, and preserving the best."

Nearly two and a half centuries later, in a feature article for the *New
York Times* about the modern fruit industry in New York's Hudson River
Valley, Kovach (1969) wrote: ". . . sprays (are applied that) contain
hormone which, at blossom time thin the production of the tree so a
controlled number of large, well-formed apples are produced rather than
a profusion of stunted fruit." Here the author speaks of thinning in the
same objective terms of assisting nature in producing fruits of better size
and appeal. The one significant difference noted in these two quotations
is that implied in the time and method of achieving the thinning; our
contemporary writer identifies a "hormone" spray at blossom time, while
his earlier counterpart implies hand thinning in late June. This distinction will become the principal theme of the present chapter.

Various reasons and theories were advanced by the early writers for
the improved fruit size and quality at harvest following such thinning or
adjustment of the crop load which formed on the tree following bloom.
However, it was generally recognized by even the earliest writers that
the improved leaf–fruit ratio following thinning permitted an obvious
increase in supply of elaborated food materials to the remaining fruits.

II. History of Thinning Sprays

Numerous methods have been utilized for removal of excess flowers
and/or developing fruits. However, through the centuries from the earliest
reports of thinning cited above until the late 1930's this was essentially a
hand operation, involving either removal of the unwanted specimens
directly with the fingers or use of a small hand shear to snip the pedicel
of the flower or fruit. In the 1900's a variety of tools came into use, particularly for thinning peaches, to speed up this hand elimination of
flowers and fruits. Wire brushes or combs were used to pull through the
branches just as the first flowers were opening to remove a large proportion of the bloom, leaving just enough buds and flowers to produce
the desired individual fruits through the fruiting surface of the tree.

In other instances the postbloom thinning was done with a light weight
club about 2 to 3 ft long, often with a short length of rubber hose on the
end. This club was used to strike the branches on the 2- or 3-year-old
wood with just the right force to dislodge a number of small fruitlets at
once. With such a tool, an experienced operator could greatly increase
his efficiency as compared with hand removal of the individual fruits,

and could at the same time obtain a reasonably good spacing and distribution of fruits that remained.

Many growers have successfully used a conventional high-pressure orchard sprayer directing a stream of water from a hand gun at the branches during or soon after full bloom. This is another form of mechanical removal of excess bloom, and has been employed principally with peaches.

While tracing the history of such forms of mechanical thinning and the examination of possible innovations would be an interesting exercise, it is beyond the scope and objectives of this chapter.

Although the early horticulturists recognized that one of the objectives of thinning was to prevent "exhaustion" of the tree and promote annual bearing, they realized that the conventional hand thinning of apples practiced after the early fruit drop periods were over failed to bring about complete recovery from the biennial cycle or alternate bearing habit of many cultivars. Russell and Pickering (1919) were among the first to show convincingly that this alternate bearing habit could be controlled if the "thinning" were practiced at the bloom stage instead of waiting 6 to 8 weeks after bloom to do the thinning as was the usual commercial practice. But of course the hand removal of the myriad of excess flowers in the typical "on-year" mature tree did not lend itself to practical use on a large scale.

Thus, it is not surprising that the first attempts to use chemical sprays for thinning apples were designed to eliminate a major portion of the flowers and permit the tree to differentiate flower buds for bloom the succeeding year. The idea was that growers with large blocks of such alternating varieties might have half of their trees flowering and fruiting one year while the others were barren, with the cycle reversing itself the following year.

This approach would of course be untenable for the contemporary orchardist with his high costs of production and overhead, where each producing unit in the orchard has to carry its full bearing potential each year rather than every other year.

Bagenal et al. (1925) were apparently the first to recognize that the drop of immature fruit from a healthy tree was increased by chemical sprays applied for control of pests when they observed that "lime-sulphur" (calcium polysulfide) induced excessive drop of young apples. The first conscious attempts at flower elimination were conducted 40 years ago by Auchter and Roberts (1934). Several of the common spray materials of that era were used in their tests including calicum and sodium polysulfide, copper sulfate, oil emulsion, and zinc sulfate. Prebloom applications of these materials not only killed the flowers in many instances but also caused injury to the spur tissues and foliage. Sub-

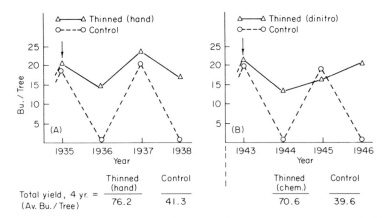

FIG. 1. Comparison between hand blossom thinning (A) and chemical thinning with dinitro (B) on the bearing habit of Wealthy apple trees over a 4-year period. Arrows indicate years the treatments were started. (A) 19-Year-old trees; averages of two trees in 1935 and 1936. (Adapted from Bobb and Blake, *Proc. Amer. Soc. Hort. Sci.* 36, 324, Table 1, 1938.) (B) 22-Year-old trees; averages of 7 to 10 trees per treatment. (Adapted from Hoffman, 1947.)

sequent tests by Auchter and Roberts (1935) included the use of tar oil distillate which proved effective in eliminating the flowers and caused less injury to the vegetative tissues.

As an outgrowth of these investigations, it became apparent that the proper choice of materials and concentrations might provide a selective removal of flowers which would permit some fruits to be borne in the year of application and still allow flower bud induction for a crop the following year. This concept was spurred on by the work of Bobb and Blake (1938) who were able to obtain annual cropping of the decidedly biennial Wealthy cultivar by hand thinning the flowers starting in the "on" year, and repeating the operation on the same trees in successive years (Fig. 1). Finally, the work of Gardner *et al.* (1940) and Magness *et al.* (1940) showed that dinitro might accomplish this desired objective.

III. Chemical Thinning of Apples (*Malus pumila* Mill.)

A. OBJECTIVES

1. *Quality Fruit Production*

Flowering and fruit production are exhaustive processes in the apple tree; if a heavy fruit set develops, the competition between developing fruits and the vegetative growth can deplete the vitality of the tree. The

TABLE I

EFFECT OF HAND THINNING JONATHAN APPLE FRUITS ON COLOR, SIZE, AND YIELD[a]

Thinning date and distance (leaves/fruit)	Color % of fruit		Size (diameter) % of fruit		Yield (bu/100 cm² cross section trunk area)
	0–25	Above 75	Under 2½ inches	Above 2¾ inches	
Unthinned	17.2	28.5	30.3	18.0	3.2
June 4—50 leaves	6.8	43.5	13.6	53.1	2.4
June 4—100 leaves	0	76.8	7.6	50.1	1.3

[a] Adapted from Fletcher (1932).

most obvious effect of this competition between individual fruits and between fruits and vegetative tissues is the small, unprofitable fruit size at harvest which is often coupled with delayed maturity of the fruit, as well as inferior color and quality (Table I).

Such excessive cropping frequently causes limb breakage varying from the small, lateral branches on younger trees and forked branches at the end of main scaffold limbs to the loss of large, main limbs on older trees. Such breakage may further weaken the tree, making it subject to decay and deterioration. An alternative to this predictable loss, even when some conscious adjustment in fruit set has been made, is the expensive practice of propping the branches.

Another effect of competition and depletion of reserve carbohydrates from woody tissues is increased susceptibility to winter injury. This has been observed both under orchard conditions following test winters and in controlled freezing tests in the laboratory (Way, 1954).

2. Annual Bearing

While these visible effects are taking place during the cropping season, an equally serious but unobserved effect is the failure of buds to differentiate flower primordia for the following season. Thus, not only is the current crop likely to be inferior in size and quality with limited sale value, but the tree is likely to produce inadequate bloom for a crop the following year. Thus, the alternate or biennial bearing habit of many cultivars is perpetuated. The early attempts with chemical sprays were directed at breaking up this cycle to get at least some of the trees out of step and to produce fruit in the off years.

3. Reduction in Cost of Hand Thinning

The process of hand removal of excess flowers or fruits from a heavily laden tree is costly. Hand thinning even with the aid of some of the mechanical devices mentioned earlier is a menial, time-consuming operation. Thus, one of the principal objectives of the search for and development of chemical thinning sprays has been to reduce this critical cost factor.

B. MATERIALS AND TIMING

1. *Dinitro*

As indicated earlier, it was discovered that one of the dinitros, dinitro-o-cyclohexyl phenol, applied at the prebloom or cluster bud stage could selectively eliminate flowers with less danger of injury to the young leaves and spur tissues than was caused by the tar oil distillates (Magness *et al.*, 1940). Experiments at about the same time with a related material, sodium dinitro-o-cresylate, indicated that such materials applied at bloom killed pollen grains adhering to the anthers and prevented germination of pollen on the stigmatic surface of the pistil (MacDaniels and Hildebrand, 1940). Thus, the mechanism for the thinning action of the dinitros was established. The work of Hoffman (1947) demonstrated that annual application of a dinitro starting in the "on" year could accomplish the annual bearing of the Wealthy cultivar similar to that obtained by Bobb and Blake (1938) with their hand flower thinning experiments (Fig. 1).

Other forms of dinitro were evaluated for thinning apples following these first experiments. A dry formulation of dinitro-o-cresol and the commercial sodium dinitro-o-cresylate or Elgetol gave comparable results in many trials when used at the same concentration of active ingredient (Batjer and Thompson, 1948; Batjer and Hoffman, 1951). On the other hand, the phenol forms of the dinitros caused some fruit russeting and leaf injury (Batjer and Hoffman, 1951).

Cautiously fruit growers took up the use of dinitros as thinning sprays for apples. The difficult-to-thin, strongly biennial cultivars such as Wealthy, Baldwin, and Golden Delicious were usually the first to receive attention by the grower as he embarked on this new practice. The counsel of local extension agents, experiment station personnel, and commercial fieldmen was generally involved in these initial attempts, and from these grower experiences came many of the guidelines upon which this practice was founded. For example, it was necessary to give two applications of dinitro on varieties like Wealthy to assure adequate thinning, the first when spur flowers were in bloom, and a follow-up application, usually at a lower concentration, a few days later when the lateral flowers on the terminal shoots opened.

Other apple varieties such as Red Delicious, Jonathan, and R. I. Greening which require cross-pollination for fruit set and are considered easier to thin also came under commercial thinning practice with dinitros (Batjer and Hoffman, 1951). Because of the need for more thorough pollination, i.e., extended periods of bee flight, in these cultivars the application of the dinitro was usually delayed until the second or third day after full bloom and a lower concentration was utilized. This delayed timing permitted cross-pollination to be completed in the first flowers opening in the clusters, then the dinitro application prevented pollen germination and fertilization in the later opening flowers.

Obviously, timing the spray application with respect to stage of bloom and extent of pollination was crucial to the success of this thinning spray. Inadequate or no thinning resulted if the spray application was delayed and pollen germination took place in a majority of the flowers. Likewise excessive thinning occurred if the dinitro spray was applied too early before sufficient bee activity and pollen germination took place. Often only a few hours to a day at most separated incorrect from ideal timing. Also the need for this early timing required the grower to make a decision about applying the thinning spray before actual need for thinning could be appraised and often well before the threat of killing frosts was over.

2. Growth Regulators

A few years after the initial work with dinitros, it was found that certain plant growth regulators or auxins would reduce the set of apples (Burkholder and McCown, 1941). Those first experiments were conducted with α-naphthaleneacetic acid (NAA) and its amide (NAD) and actually initiated with the idea that NAA, which delays fruit abscission at harvest, might likewise reduce the postbloom drop of developing fruitlets. Instead, Burkholder and McCown (1941) observed and recorded an *increased* drop of fruitlets after application of these materials. The set of Delicious apples was reduced from 15.1 to 77.7% by the NAA spray, depending on the concentration applied. The NAD treatments also reduced set but were less effective at a given concentration than the acid. Typical leaf epinastic effects also resulted from the NAA application along with some leaf injury.

While these original trials with NAA and NAD were made at the bloom stage similar to the dinitro applications, it was observed by Davidson *et al.* (1945) that thinning could be accomplished by postbloom applications of NAA and that the timing with respect to pollination was not critical. This immediately suggested an important benefit of the

TABLE II

RESPONSE OF YELLOW TRANSPARENT TO CHEMICAL THINNING
AT PETAL FALL AND 17 DAYS LATER[a,b]

Treatment	Concentration (ppm)	Date applied[c]	Fruits/100 blossoming clusters July 10	Thinning degree	Foliage condition
Control	—	—	57	—	—
NAD	20	May 27	25	Satisfactory	Good
NAD	40	May 27	18	Very good	Good
NAA	15	May 27	3	Excessive	Severe epinasty
NAA	15	June 13	51	None	Slight flagging
LSD at 5%	—	—	9.4	—	—

[a] Adapted from Hoffman et al., 1955.
[b] New York, 1952.
[c] May 27—petal fall.

auxin materials as compared with the dinitros. Extensive orchard trials and grower experience with NAA followed these initial experiments.

Although NAD had been included in the early tests by Burkholder and McCown (1941) and Schneider and Enzie (1943), it was not until the early 1950's that Thompson (1953) and Hoffman et al. (1955) conducted extensive trials with NAD in comparison with the other chemical thinners of that era. Similar to the earlier observations the NAD proved to be a milder thinning agent than NAA with less leaf epinasty and leaf dwarfing (Table II).

For many years following this work NAD was the preferred thinning agent for most of the late summer and early fall apple cultivars such as Yellow Transparent, Oldenburg, and Early McIntosh (Fig. 2). NAA, if applied at the 7- to 10-day interval after bloom, which was proving so successful on most cultivars, generally failed to thin these early cultivars properly and restricted growth rate of the remaining fruit. In addition the delayed postbloom application sometimes triggered early maturity of some of the fruits, with cracking and deterioration of the affected fruits (Fig. 3). This could be largely overcome by applying the NAA during or immediately after bloom as in the initial experiments by Burkholder and McCown (1941) but this negated the benefits of the postbloom timing in appraising the need for thinning, and the relatively high rates of NAA required to thin these early maturing cultivars caused undesirable leaf dwarfing when applied at the bloom stage.

NAD was also used successfully for thinning some of the relatively easy to thin late fall and winter cultivars such as McIntosh and R. I. Greening

Amid   Transparent   Control

FIG. 2. Yellow Transparent chemically thinned with a foliar spray of 50 ppm NAD at petal fall stage (left) and control branch (right). (Photo, August 10, 1972, Ithaca, N. Y.)

in seasons when the degree of pollination was uncertain and the need for thinning in doubt at the time a decision for a chemical thinning treatment had to be made.

Along with successful use of these two growth regulators on many apple cultivars, dissatisfaction developed with both NAA and NAD as thinning materials for Red Delicious and its color sports. With NAA a very narrow and often unpredictable concentration range frequently occurred between insufficient thinning and overthinning or virtual elimination of the crop. If the milder material, NAD, were used in the de-

FIG. 3. Red Astrachan apples photographed 1 week before normal maturity (August 10, 1972). This tree had received a chemical thinning spray containing 15 ppm NAA about 1 week after petal fall. Note the split fruits (A) and immature fruits (B).

sirable postbloom period some of the young fruits in the clusters, although being suppressed in growth and destined to drop as in other varieties, failed to abscise. The affected fruit continued to develop slowly but final size at harvest time was greatly reduced and the fruits were worthless (Fig. 4). For this reason many growers, particularly in the State of Washington, continued to use one of the dinitro materials for thinning Red Delicious although they had shifted to one of the auxins for most other cultivars.

3. Carbaryl

The discovery by Batjer and Westwood (1960) that 1-naphthyl N-methylcarbamate (carbaryl) would safely thin Delicious when applied in the postbloom stage ushered in a new era in chemical thinning. Carbaryl was originally developed as an insecticide, and the thinning action

FIG. 4. Clusters of Red Delicious showing normal fruits and small "pigmy" fruits which were affected by a thinning spray of 25 ppm NAD about 1 week after petal fall, but which failed to drop. (Photo, August 17, 1972.)

was first observed in Delicious orchards where insecticide trials were being conducted. Extensive experiments were initiated in many of the principal fruit-growing areas. Out of these evaluations the following features of carbaryl emerged: (1) Thinning action could be obtained with applications during an extended period of about 3 weeks after bloom; (2) some thinning could be obtained over a wide range of concentrations from as little as ¼ to 2 lb of the 50% formulation per 100 gal of water[1]; (3) over-thinning or complete elimination of the crop seldom occurred even at the higher concentrations; and (4) little, if any, objectionable foliage injury or fruit abnormalities could be attributed to the use of the compound. Thus, in contrast with some of the disadvantages of the auxins on Red Delicious as just pointed out, carbaryl proved to be particularly well adapted to this variety, and its use has become the standard recommendation for this cultivar throughout the United States and most parts of the world where chemical fruit thinning is practiced.

Many other cultivars have responded favorably to the use of carbaryl. These include Jonathan, McIntosh, Cortland, R. I. Greening, Northern Spy, Rome Beauty, and Winesap. In Washington, carbaryl at the higher

[1] Concentrations of carbaryl referred to in text and tables are based on amounts of a commercial formulation (Sevin) which contains 50% active ingredient.

concentrations is considered to be sufficiently effective on Golden Delicious to be acceptable for commercial practice, but in the northeastern United States inadequate thinning is usually obtained on this variety. The major concern with carbaryl as a thinning spray on Golden Delicious has been its failure, under eastern United States conditions, to affect flower initiation for the following year's crop. For this reason NAA, and more recently combinations of carbaryl and NAA, have been preferred for Golden Delicious.

In spite of the inadequate thinning generally obtained with carbaryl on Golden Delicious and its ineffectiveness on Macoun and Baldwin, carbaryl has proved to be a remarkably versatile chemical thinner. Certainly for most apple cultivars it could be considered an ideal thinner, one that always thins to some degree, but never overthins.

4. Combination of Carbaryl and Auxin

Because of the latitude in timing which is possible with carbaryl, some growers adopted the practice of applying NAA or NAD shortly after bloom using a conservative concentration. Then careful inspection of the trees was made 1 or 2 weeks later. If it appeared that insufficient thinning was taking place, an application of carbaryl was made. Often some additional thinning was accomplished with this follow-up treatment, provided the carbaryl was applied not later than 21 to 25 days after full bloom.

Horsfall and Moore (1963) tried a combination of carbaryl and NAA on a few varieties and found that this was more effective for thinning Golden Delicious when applied 19 days after full bloom than were two applications of carbaryl alone. Combinations of carbaryl with NAA or NAD have been used by growers in New York State for several years with good results. Such a combination has been especially useful on some varieties such as Golden Delicious and Early McIntosh on which carbaryl alone has given inadequate thinning and the high concentrations of NAA or NAD, when likewise used alone, frequently resulted in overthinning, leaf dwarfing, or other complications (Figs. 5 and 6). In these combinations carbaryl is generally used at 1 lb per 100 gal and NAA at 5 to 10 ppm, with the application made 1 to 2 weeks after full bloom. By using NAA at approximately half strength in this combination the danger of overthinning is reduced and the detrimental effects of the higher concentrations on the foliage and upon growth rate of the surviving fruits are lessened.

Flower induction for bloom the following year has also been a desirable feature of this combination when used on the heavy setting varieties such as Golden Delicious and Wealthy that have a strong

Sevin + NAA   Control

FIG. 5. Early McIntosh chemically thinned with a foliar spray of carbaryl at 1 lb/100 gal + 10 ppm NAA 6 days after petal fall (left), and control (right). (Photo, August 10, 1972.)

biennial tendency (Table III). Several research workers have suggested that NAA or NAD, when used as postbloom thinning sprays even at low concentrations which account for relatively little direct thinning, can promote flower initiation (Thompson, 1957; Harley *et al.*, 1958).

C. MODE OF ACTION

1. *Dinitros*

The mode of action of the dinitros, referred to previously, is primarily related to the action of the chemical on the pollen and on the stigmatic

Sevin + NAA   Control

FIG. 6. Golden Delicious chemically thinned with a foliar spray of carbaryl at 1 lb/100 gal + 10 ppm NAA 8 days after petal fall (left) and control (right). (Photo, August 10, 1972.)

surface of the flower (MacDaniels and Hildebrand, 1940). In fact, Hildebrand (1944) refers to one of the dinitros as a "pollenicide," and attributes the thinning action of the dinitros to their effect on pollen, both before and after transfer to the stigma. Obviously, if such a material is to be effective in preventing set of some of the flowers, it would have to be applied at a precise stage during the bloom period during which fertilization would be permitted in certain flowers and prevented in the remaining flowers.

TABLE III

EFFECT OF CHEMICAL THINNING TREATMENTS ON FRUIT SET AND
REPEAT BLOOM OF 13-YEAR-OLD GOLDEN DELICIOUS TREES[a]

Treatment	Date applied, 1970	Fruits/100 blossom clusters	Average yield (bu)	Bloom density, 1971[b]
Carbaryl				
2 lb/100 gal	June 2	31	12.5	55
2 lb/100 gal + NAA (10 ppm)	June 2	19	10.2	75
Control	—	62	13.0	10

[a] Ithaca, N. Y., 1970–1971.
[b] Average percent of spurs flowering.

2. Auxins

The mode of action of the auxins in inducing the shedding of flowers and small immature fruitlets has been the subject of extensive speculation and research. After all, the discovery that NAA and NAD would thin apple flowers was the outgrowth of a test designed to find treatments that might *increase* set of apples (Burkholder and McCown, 1941). This approach followed the earlier work of Gardner *et al.* (1939) which demonstrated that these auxins would delay *preharvest* drop of apples. Instead of increasing the number of flowers that remained on the low yielding Starking trees, Burkholder and McCown (1941) found that NAA at 0.001% reduced set by 15.1%, and 0.005% NAD reduced set by 34.0%. With this apparent anomaly in the effect of the auxins when applied at two different stages in fruit development, many investigations have been made as to the possible cause of the thinning action. The theories, and the research and observations supporting them, can be roughly divided into four categories.

a. INCREASED NUTRITIONAL COMPETITION BETWEEN FRUITLETS, AND SUBSEQUENT ABSCISSION OF THE WEAKER FRUITS. Struckmeyer and Roberts (1950) found a temporary delay in abscission of apple flowers and developing fruits following application of NAA. However, after the June drop was completed the set of fruit on several treated cultivars was less than on control plants, and thinning had indeed been accomplished. The authors proposed that the thinning action was the result of temporary nutritional competition among the developing fruits whose normal abscission was delayed by the auxin. The additional drop, precipitated by this competition, resulted in the thinning action with the desired improvement in fruit size and quality at harvest. In this concept the NAA is not exerting a direct thinning effect. Teubner and Murneek (1955) disagreed with this theory because they found that potential "drops" ceased to grow soon after NAA was applied and hence could not be competing sufficiently with other fruits to be responsible for the thinning.

b. WEAKENING IN THE CONDUCTIVE TISSUES AT THE BASE OF THE PEDICEL. Van Overbeek (1952) suggested that NAA caused a breakdown of pectic substances in the middle lamella of cells at the base of the pedicel, with concomitant enlargement and elongation of cells in this region. This weakening resulted in a mechanical fracture of the conductive tissues. This theory has not been substantiated by other investigators.

c. STIMULATION OF ABSCISSION BY ETHYLENE. It has been shown by Gawadi and Avery (1950) that plants treated with auxin produce ethylene. These workers concluded that the NAA application stimulated ethylene production and caused many of the fruitlets to abscise. However, following an extensive investigation on ethylene levels in apple fruits, Teubner and Murneek (1955) concluded that there did not seem to be a significant relationship between NAA treatment and increased ethylene production in the increased fruit drop for this to form the basis for the thinning action.

d. DESTRUCTION OF THE EMBRYO. Evidence accumulated in the early 1950's that the embryo and developing seed produced auxins which served to mobilize food materials for the growing fruitlet and prevent its abscission. Thus, it was natural to consider a possible role for NAA in embryo development. Luckwill (1953) found that increased drop of the young fruitlets following application of NAA was associated with seed abortion presumably induced by the applied NAA (Table IV). His data

TABLE IV

EFFECT OF NAA AT 20 ppm APPLIED AT DIFFERENT STAGES ON BRANCHES OF MILLER'S SEEDLING APPLE[a]

Stage	Fruit set/100 blossom clusters	Mean fruit wt (gm)	% Fruits less than 2 inches diam	% Abortive seeds at harvest
Control (unsprayed)	61 ± 3.6	38.6 ± 0.68	90.5	27.2
Pink bud	59 ± 12.0	49.3 ± 1.21	60.5	20.1
Full bloom	24 ± 7.7	49.5 ± 1.08	57.5	17.5
Petal fall	24 ± 6.6	53.4 ± 0.91	41.0	30.1
Petal fall + 7 days	27 ± 5.4	51.4 ± 1.10	50.0	36.0
Petal fall + 25 days	22 ± 7.2	29.5 ± 0.65	100.0	50.5
Petal fall + 41 days	53 ± 6.1	37.9 ± 0.75	88.0	26.6

[a] Adapted from Luckwill, 1953.

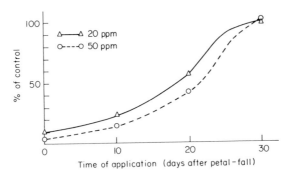

FIG. 7. The effect of time of application of NAA on fruit set of Crawley Beauty apples. (Adapted from Luckwill, 1953.)

further indicated that after cell wall formation occurred in the endosperm, NAA sprays were no longer effective in thinning. This work supported earlier observations that the effectiveness of NAA was limited to a period of two to four weeks after full bloom. This marks the start of cytokinesis in most varieties (Fig. 7). In more recent work with [14C]NAA Luckwill and Lloyd-Jones (1962) traced the movement of foliar applied auxin to the developing seeds. Only minute amounts could be recovered from the seeds after several days and none of this was in the original unmetabolized form. This appears to preclude the possibility of NAA exerting a direct toxic effect on the embryo or other seed tissues. Furthermore, Dennis (1970) was able to thin both seeded and seedless fruits of one apple cultivar with NAA. It thus appears that seed abortion is not the direct cause of the thinning action of the auxins.

In summarizing the evidence for the role of NAA and NAD it appears that the auxin affects the flow of endogenous hormones which control continued nourishment of the developing fruit. Embryo abortion may precede or accompany this restriction in mobilizing ability of the affected fruits. The auxin may temporarily delay abscission of certain fruitlets at the time of abortion as part of the overall interaction between applied and endogenous auxin. But it is only a short-lived effect, and subsequently the fruit drops when the flow of hormones down the pedicel from the seed is interrupted (Luckwill, 1953; Abbott, 1954).

3. Carbaryl

Batjer and Thomson (1961) found that carbaryl applied to Red Delicious usually resulted in reduced seed content of persisting fruit but had no consistent effect on seed numbers in other cultivars. In a study involving several cultivars, Bukovac and Mitchell (1962) concluded that embryo abortion was not the mechanism involved in the thinning action of carbaryl.

In a later investigation involving [14C]-labeled carbaryl, Williams and Batjer (1964) found that only trace amounts of the material could be detected in the seed and that extended periods of time elapsed before it could be detected. They proposed that carbaryl interferes with the translocation of necessary growth factors through vascular tissues in the developing fruit. This prevents the continued growth of the weaker fruits followed by their abscission. Seed abortion might or might not precede abscission.

D. Factors Influencing Results

1. *Concentration of Chemical*

With both the dinitros and the auxins a relationship between degree of thinning and concentration of the chemical was recognized with the first tests that were conducted (MacDaniels and Hoffman, 1941; Burkholder and McCown, 1941).

In several tests with the cultivar Wealthy Hoffman (1942) reported that Elgetol at 0.2% consistently reduced fruit set more than a concentration of 0.1% applied at the same time. The choice of concentration of dinitro to be applied, together with slight variations in time of application, became the basis on which adjustment could be made for the degree of thinning needed under the particular set of conditions being encountered.

Hartman and Howlett (1962) reported that in their tests with Golden Delicious in Ohio, increased thinning was obtained with NAA on a given date of application by increasing the concentration of the auxin. The range of useful concentrations with NAA on apples has been from 5 to 20 ppm (Batjer and Hoffman, 1951). Since NAD is a milder material, a higher range of concentrations of 15 to 50 ppm and even up to 75 ppm has been established for effective thinning on most cultivars. In many of the experiments that have been conducted with this material small differences in concentration of around 50% have not produced significant differences in fruit set. However, wider ranges in concentration have usually produced appreciable differences in fruit set. Inability to control or identify all the variables in an orchard experiment with large perennial plants such as apple trees will undoubtedly account for these inconsistencies in the more narrow concentration ranges.

Throughout the wide scope of experimental work and orchard obser-

TABLE V

EFFECT OF CARBARYL APPLIED AT DIFFERENT CONCENTRATIONS 17 DAYS
AFTER FULL BLOOM ON FRUIT SET OF 4 APPLE CULTIVARS[a]

			Fruits/100 blossom clusters				
		Pounds of 50% carbaryl/100 gal					LSD (0.05 level)
Cultivar	Control	0.75	1.0	1.5	2.0	3.0	
Delicious	62	42	48	42	46	38	13
Golden Delicious	80	—	—	47	—	45	15
Winesap	32	—	—	19	—	17	8
Jonathan	54	—	42	—	32	31	14

[a] Adapted from Batjer and Thomson, 1961.

vations there is general agreement that high concentrations of either material are needed to effectively thin heavy setting cultivars, whereas lower rates are applicable for cultivars easier to thin. As previously mentioned, the heavy setting, biennial varieties such as Baldwin and Wealthy require about twice the concentration for sufficient thinning as do the cultivars less prone to overbear.

Experience with different concentrations of carbaryl, on the other hand, indicates a slightly different relationship. In one of the earliest trials with this thinner it was found that 3 lb per 100 gal gave about the same degree of thinning as 1½ lb (Table V). This provided a safety factor with carbaryl for preventing overthinning which did not exist with the other chemical thinners. It is likely that the low solubility of carbaryl in water is the key factor in this relationship.

2. Method of Application

While concentration of the thinner in the spray is a factor in degree of thinning, and forms the basis for recommendations to growers, methods of application and spray coverage have to be considered. The uniformity of coverage of the leaf surface of all parts of the tree and, hence, the amount of chemical deposited on the leaf surface are important. Orchard trials with power sprayers have usually been conducted at the dilute rate. The basis of a dilute application is that all parts of the target are thoroughly wetted to the point of runoff. If the volume of spray applied is significantly less than this, the thinning action will be reduced since less active ingredient is applied to each unit.

With increased popularity in the application of pesticides with air blast sprayers in the concentrate range of 2 to 8× and higher, it was inevitable that some adventuresome growers would develop a pattern for applying the auxins for thinning at the concentrate range. This was done in spite of the admonition of many research workers who felt that the use of the dilute range eliminated at least one variable in the practice of chemical spray thinning. Undoubtedly many of the more observant and astute growers have been able to master this technique, and by carefully monitoring the rate of travel and calibration of equipment and by choosing favorable conditions for application have been able to use concentrate application successfully.

The use of carbaryl for thinning is probably more adaptable to concentrate application than the auxins because of the general lack of overthinning with this material. In tests with carbaryl applied at various concentrate levels with an Econ-O-Mist machine, Rogers and Thompson (1969) found no consistent difference in degree of thinning at concentration levels from 1 to 33×. In one test they even obtained about as much thinning when the application rate was reduced by one-half. This is probably related to the low solubility of carbaryl in water and is an additional example of the slight effect of increasing concentration from ¾ to 3 lb per 100 gal dilute application reported by Batjer and Thomson (1961).

In their study on concentrate applications Rogers and Thompson (1969) included NAA in a test on Golden Delicious. While the dilute spray was more effective, significant thinning was obtained at 3 and 6× concentrations. The research evidence and experience of growers seems to support the comment of Forshey and Hoffman (1966) that most consistent and successful results can be obtained with applications of the auxins at dilute or moderate concentrations. Carbaryl would seem to be an exception to this.

Another factor related to method of application is that of chemical thinning in orchards where the trees are closely planted and the lower fruiting wood is shaded and low in vigor. In such situations the more vigorous, heavy fruiting wood occurs through the upper canopy of the trees. In applying the thinning spray with an air blast machine in such an orchard, it is virtually impossible to obtain the proper coverage and spray deposit of a desirable concentration of the material through the tops of the trees where thinning will be needed without overthinning or even defruiting the lower, weaker wood which intercepts the full blast of the machine. To solve this problem many growers utilize high pressure sprayers with an operator standing on top of the machine spraying with a hand gun or spray broom. This method assures adequate coverage of the more vigorous wood in the upper parts of trees without overloading the lower, less vigorous wood.

3. *Variety and Blossom Density*

Varieties differ markedly in their fruit setting potential, depending upon their degree of self-fruitfulness and other factors. The research efforts of horticulturists and the practical experience of growers have continually demonstrated varietal differences in natural fruit set and response to thinning treatments. The late summer and early fall varieties such as Yellow Transparent, Duchess, and Early McIntosh are seldom thinned satisfactorily with NAA; if applied close to the blooming period it causes dwarfing and distortion of the foliage, and if applied about 2 weeks after bloom, the growth rate of the fruit may be checked to the point where size at harvest is not as good as with hand thinned fruits (Thompson, 1952). The delayed postbloom application may also cause premature ripening and splitting of many of the small fruits (Hoffman, 1954). For these reasons NAD has been the preferred material for these summer varieties as indicated previously.

Postbloom application of NAA on the fall and winter varieties does not ordinarily result in these abnormalities of fruit thinning and development, and its use is dictated by the fruit setting habits of the variety. A few varieties such as Baldwin and Macoun have not responded well to carbaryl.

Blossom density is a related factor which has also been noted to influence thinning response. As a result of several tests in Washington, Batjer and Billingsley (1964) reported that thinning sprays were less effective in reducing fruit set on trees carrying a light bloom than on those which flowered heavily. On the former the normal fruit set per 100 blossoming spurs is usually greater than on trees carrying a full complement of flowers. Veinbrandts (1972) has also reported that it is more difficult to reduce the number of fruit clusters with trees of certain varieties in Australia carrying a moderate bloom density than with trees of the same variety carrying a full bloom. In attempting to characterize a bloom density for certain varieties in determining the need for chemical thinning, Veinbrandts (1972) suggested that trees carrying 8 or fewer blossom clusters per centimeter of limb circumference would not need chemical thinning. These reports also suggest that if an occasional tree with light bloom density were sprayed, excessive thinning would not occur.

4. *Age of Trees*

Young apple trees just coming into full bearing may bloom heavily and appear to be setting an excessive crop of fruit. Probably because of the vigorous vegetative tissues which such trees are still developing,

FIG. 8. Effect of chemical thinning on alternate bearing. Tree on left was hand thinned in July, 1970; tree on right was chemically thinned with carbaryl + NAA 10 days after bloom in 1970. Only scattered bloom appeared on the hand thinned tree in 1971. (Photo, September, 1971.)

reserve food supplies are low during the early postbloom period and heavy June drops are the rule. Thus, chemical thinning is not usually needed with such trees. This is particularly true with cultivars such as Delicious, Winesap, and McIntosh which require thorough cross-pollination and are normally easy to thin. Such young trees often do not set fruit in proportion to bloom intensity. However, with each advancing year the tendency to overbear increases and the first thinning application will eventually be needed.

The first few thinning applications are usually made at about one-half the concentration recommended for mature trees of the same variety. Young Golden Delicious trees, particularly those on size-controlling stock, will benefit from chemical thinning at an earlier age than most cultivars, as such trees can quickly attain an alternate bearing pattern (Fig. 8). A somewhat reduced concentration is in order in this case also for the first few applications.

5. *Factors Affecting Absorption of Chemical*

It was recognized early in the work with the auxins that conditions affecting absorption of the chemical by leaves influenced the consistency and uniformity of the thinning response. This was an outgrowth of earlier work with these materials for controlling preharvest fruit drop. With only about ½ gm of the growth regulator being distributed over the surface of a mature, vigorous apple tree, the conditions influencing movement of these traces of chemical into the leaf for translocation to the abscission zone of the fruit stem or to the developing embryo were critical.

Humidity and air temperature are two environmental factors which were first identified as important in the absorption complex. Delicious and Winesap apples thinned more easily when NAA was applied during relatively cloudy, moist conditions one year than when similar applications were made on the same varieties in another season when clear, dry conditions prevailed (Batjer and Thompson, 1948).

Greenhouse and growth chamber experiments have shed some light on these conditions. Westwood and Batjer (1960) obtained increased absorption of NAA under conditions of high humidity and with a pretreatment temperature of 70°F than with 50°F. However, Batjer (1965) reported that under field conditions the humidity and the length of time the leaves remained wet might be more important than air temperature.

Edgerton and Haeseler (1959) working with [14C]-labeled NAA and NAD found that apple leaves preconditioned under high light intensity did not absorb the auxins as rapidly as leaves under low light. Temperature exerted a marked effect on rate of absorption of both materials with greater absorption at 70°F than at 50° to 55°F. These variations in light, temperature, and humidity are known to affect the thickness and composition of leaf cuticle (Skoss, 1955) and in this way might be expected to influence absorption of chemicals applied to leaf surfaces.

Several workers have reported that addition of spray surfactants such as Tween 20 may increase the absorption rate of the thinning materials, and perhaps more significantly tend to reduce the variabilities or variations in absorption rate under differing environmental conditions (Edgerton and Haeseler, 1959; Westwood and Batjer, 1958, 1960; Harley *et al.*, 1957).

6. *Physiological Factors (Tree Vigor)*

In early reports on the thinning action of dinitros, auxins, and carbaryl it was recognized that physiological characteristics or vigor of trees accounted for some of the variations observed. Southwick and Weeks (1949) were among the first to elucidate this point. The tagged and measured spurs of Wealthy trees which were then sprayed with NAA 2 weeks after calyx. Fruit set counts made following the June drop indicated that bloom from the smaller, weaker buds 4.7 mm or less was heavily thinned, while bloom from the larger buds was more resistant to the spray.

There is a temptation to explain this relationship between spur vigor and ease of thinning on the basis of carbohydrate reserve and nitrogen levels in the respective spurs. However, Hennerty and Forshey (1971) found that while treatments such as scoring, defoliating, and shading caused significant differences in flowering and fruit set of Golden Delicious trees, the differences did not appear to be related to static carbohydrate levels in shoots and spurs. When a thinning spray of NAA was superimposed on these treatments, a wide range of response to the spray was obtained, but this was related to fruit set potential rather than to differences in carbohydrate level (Hennerty and Forshey, 1972). Priestley (1970) also reported there was no consistent difference in carbohydrate level in young apple trees following ringing.

Probably the reason for failure of researchers to find a significant relationship between carbohydrate levels in the fruit-bearing tissues of spurs and twigs is the continual dynamic adjustment of the photosynthetic activity within the tree to "sinks" provided by vegetative growth, crop load, and other factors. Priestley (1970) supported this view: "Perhaps, in fact, the carbohydrates extracted during shoot extension form an integral part of the cells containing them. . . . an equilibrium exists within the bounds of the hypothesis that healthy apple trees tend to maintain within their living cells a fairly constant level of available carbohydrate."

With regard to nitrogen levels, Forshey and Hoffman (1966) reported that limited nitrogen availability in trees of only moderate vigor significantly increased the thinning action of the auxins. Thus, any condition such as inadequate nitrogen fertilization, drought, and impaired root activity from poor soil drainage that restricts nitrogen uptake and availability at the time of fruit setting will influence response to the thinning spray.

7. Time of Application

The choice of a specific time during the bloom or postbloom period seems to depend on the material being used, the cultivar involved, and the specialist making the recommendation. Many experienced growers have developed their own policies or practices and may diverge somewhat from published recommendations for a district. Nevertheless, certain principles with respect to time of application are well established.

In using the dinitros the choice of time is limited to 1 to 3 days during the full bloom to petal fall stage because of the nature of the thinning action as discussed earlier. The exact time would be determined by such conditions as the cultivar involved and its requirement for cross-pollination and temperature and weather conditions as they influence bee activity and fruit set.

Much less rigid limitations prevail as to timing in using the auxins. Reduction in set has been obtained with application of the auxins over a period extending from full bloom to 3 or 4 weeks after bloom (Fig. 7). In Ohio, Hartman and Howlett (1962) found that somewhat higher concentrations were needed for a given amount of thinning as the interval after full bloom was extended. When application of auxin was delayed to 35 days after petal fall no thinning was obtained. This limitation in effective period agrees with that cited by Luckwill (1953) (Fig. 7). In extensive tests with NAA and NAD in Maryland, Thompson (1957) obtained more consistent thinning when the applications were made between 10 and 21 days after full bloom. Experience with carbaryl also indicates considerable latitude in timing. Batjer and Thomson (1961) obtained about the same degree of thinning with applications made between 2 and 4 weeks after bloom.

The principal restrictions would seem to be the special cases with NAD on Delicious and both NAD and NAA on the early ripening cultivars. If NAD is applied later than about one week after petal fall on Delicious, the thinning action is often incomplete. Many fruitlets may be stunted by the treatment but fail to abscise in a normal fashion. Instead they persist throughout the season, growing little, if any, in size and represent a nuisance at harvest. Batjer (1965) referred to these as "pigmy" fruits. Figure 4 illustrates this phenomenon. NAA or carbaryl applied at the same time or even later does not appear to have this effect. A similar timing complication exists with early cultivars such as Transparent, Astrachan, Lodi, and Duchess where the delayed application of NAA or NAD likewise stunts or stops the development of many fruits which fail to abscise. In addition, it may set off premature ripening of partially grown fruit in advance of normal maturity dates (Fig. 3).

Beyond these peculiarities and limitations there is a rather wide range of time during which thinning can be accomplished with NAA, NAD, or carbaryl. While these materials may thin the fall and winter varieties effectively during the bloom to petal fall stage, as do the dinitros, little use is made of this timing. The advantages of a postbloom application which enable the grower to more accurately assess the need for thinning have already been referred to and have dominated commercial practice.

Some research workers (Tukey, 1965; Donoho, 1968) reported more effective and consistent thinning from sprays of NAA applied at a specific

TABLE VI

EFFECT OF NAA APPLICATION DATE ON CHEMICAL THINNING OF
JONATHAN APPLES IN 1961[a]

Date[b]	Fruits/100 blossom clusters[c]	Fruit length (mm)	Environmental conditions			
			°F	Relative humidity	Wind (mph)	Light[d]
May 24 (P.F.)	25*	5	70	30	5	726
May 29 (P.F. + 5)	27*	7	65	60	3	432
June 3 (P.F. + 10)	30*	12	73	48	2	524
June 9 (P.F. + 16)	13**	17	74	40	4	316
Control	42	—	—	—	—	—

[a] Adapted from Donoho, 1968.
[b] P.F. = petal fall.
[c] Significant difference at 5% level = *; significant difference at 1% level = **.
[d] Light expressed as cal/cm²/day.

stage of fruit development which could be identified by fruit measurements. Tukey (1965) suggested that this stage of effectiveness occurred when the fruits were 10 to 11 mm diameter. Donoho (1968) used fruit length as a criterion and suggested that best results were obtained when the largest fruits were 15 to 18 mm long (Table VI). These stages of fruit development are generally reached from 1 to 2 weeks after petal fall. Batjer *et al.* (1968), however, found thinning response from ap-

TABLE VII

EFFECT OF APPLYING CHEMICAL THINNERS AT DIFFERENT TIMES
ON FRUIT SET OF GOLDEN DELICIOUS[a,b]

Date of application	Days after petal fall	Fruit diameter (mm)	Fruit/100 blossom clusters	Weather conditions (temp. in °F)
June 3	5	4.2	19.7	78, sunny, humid
June 5	7	6.4	16.3	70, sunny, light breeze
June 7	9	8.3	25.2	60, sunny, 8 mph wind
June 9	11	9.6	29.1	65, cloudy, rained early A.M.
June 11	13	10.5	30.1	57, sunny, 10 mph wind
June 13	15	11.8	14.9	72, sunny, no wind
June 14	16	12.3	22.3	69, sunny, no wind
Control	—	—	69.4	—

[a] Ithaca, N. Y., 1972.
[b] Thinning treatment was carbaryl 1½ lb 50% formulation/100 gal + NAA 7½ ppm; 5 trees per treatment; 12-year-old trees on EM II rootstock; full bloom May 26; petal fall May 29.

plications over a period of 3 weeks from bloom to be more closely associated with weather conditions at time of application than with any specific stage of fruit development. A test conducted at Ithaca, New York on Golden Delicious in 1972 confirmed the report of Batjer *et al.* (1968). Reasonably uniform thinning was obtained with a combination of carbaryl and NAA applied periodically to separate trees over a period of 11 days (Table VII). The minor variations could be explained readily on weather conditions prevailing at the time of application similar to that reported by Batjer *et al.* (1968).

There is no obvious explanation for these differences which appear in the literature with respect to responsiveness of the fruit during a finite period of its development. However, the commercial practice of chemical thinning of apples is well established, and experience of the individual grower with concentrations and timings under his conditions remains the most vital link in the success of this practice. An observation made by Hoffman (1954) about chemical thinning of apples summarizes this problem: "Only diligence and continued experience will bring the practice closer to perfection—a worthy goal, but a difficult one to attain."

IV. Chemical Thinning of Pears (*Pyrus communis* L.)

A. OBJECTIVES

While the problem of oversetting and alternate bearing is less widespread in pears than in apples, some thinning is often needed to assure adequate fruit size at harvest for fresh fruit or processing outlets. The varieties Bartlett and Bosc may frequently set excess crops and require thinning to obtain desirable size. In some fruit districts such as California, Bartlett will set fruit without cross-pollination. Grown under these conditions, often without another pollinating variety, this cultivar seldom needs chemical thinning.

B. MATERIALS AND TIMING

1. *Dinitro*

The dinitros have been used successfully at bloom time to reduce the set of pears. Timing the application is similar to that suggested for apples. The dinitros have not been popular for thinning pears, particularly in northeastern United States where rain or high humidity occur frequently during bloom. Such conditions contribute to excessive leaf injury and unpredictable thinning. Also, it is impossible to assess accurately the full range of pollination and weather conditions at bloom time which will influence ultimate set and hence the need for chemical thinning.

2. Other Materials

The insecticide, carbaryl, does not thin pears effectively (Batjer, 1965). Both NAA and NAD are being used successfully for this purpose, with applications in the postbloom stage. If NAD is used, it must be applied at the petal fall stage or soon thereafter to avoid creating the "pigmy" effect as mentioned for Delicious apples (Fig. 4). This restriction in timing has limited the use of NAD on pears, although it was recommended for several years before NAA was registered for this purpose.

NAA applied 5 to 10 days after bloom is the preferred chemical for most cultivars in the northeastern United States because of this flexibility in timing. Rates vary from as little as 2 ppm on Bartlett to as much as 10 to 15 ppm when a heavy set is in prospect on Bosc. Seckel has not responded as successfully to NAA as most other cultivars.

C. Factors Influencing Results

1. Tree Vigor

Many pear orchards, particularly in the northeastern United States, have been grown under conditions of low nitrogen levels for many years. This was done intentionally to reduce the likelihood of infection and loss from the fire blight disease caused by Erwinia amylovora. With better understanding of the relationship of tree vigor and climatic conditions to epidemiology of the disease and with improved spray practices to contain it, growers are tending to raise the nutritional level and vigor of Bartlett and other cultivars. This has increased the potential for fruit set and hence the need for chemical thinning. By 1972 postbloom applications of NAA were widely used by pear growers in New York State.

2. Pollination

Other factors which determine the need for thinning and the concentrations of chemical to be applied are similar to those discussed for apple. These include provision for cross-pollination, weather conditions during and immediately following the blossom period, and intensity of bloom, as well as crop load and growing conditions during the previous year. In assessing these factors and arriving at a specific concentration to apply, the experience of the successful grower is an essential ingredient.

V. Chemical Thinning of Stone Fruits

A. INTRODUCTION

Application of a thinning spray to any species is based on the assumption that the final set will be excessive. Peaches and many of the other stone fruits generally present a thinning problem when the flower buds come through the winter without injury and where damage by spring frosts has not occurred.

Research on methods and materials for chemical thinning of stone fruits followed much the same procedures as were used on apples. In some cases materials that were found promising on apples were tested on peaches and other stone fruits, while in other cases extensive screening and orchard evaluation have been conducted explicitly with peaches (Thompson and Rogers, 1959). Some modest successes have been achieved, although at present the program for chemical thinning of stone fruits is less well refined and less widely practiced than with apples.

Several factors may have contributed to this. The basic physiology of fruit setting in peach and other stone fruits has not been detailed to the extent it has in apple. Furthermore, stone fruits contain but a single seed. The carpel contains two ovules but usually only one develops into a mature seed. On the other hand, flowers of pome fruits such as apple and pear have 10 or more ovules, all of which may be fertilized and produce seeds. Whatever the interpretation or explanation for fruit setting in apple may be, the number and quality of developing seeds is certainly involved (Luckwill, 1953). Thus, application of a growth regulating material such as NAA which serves to alter the level and the mobility of endogenous materials during the fruit setting period would appear to have a greater latitude in which final set could be realized in the apple than in the single seeded stone fruits.

B. MATERIALS AND TIMING

1. *Peach* (*Prunus persica* Sieb.)

a. DINITRO. The dinitros were the first chemicals to reduce set of flowers on peaches successfully (Batjer and Moon, 1943). Many peach growers adopted the practice of applying one of the dinitro materials during the blossom stage to prevent set of unwanted flowers. The usual procedure with many of the heavy setting cultivars was to apply the spray promptly when 95% of the flowers were open (Southwick et al., 1947). Pollination and fertilization of those flowers which had just opened were prevented, presumably by the action of the dinitro on pollen as well as on the stigma and style of the flower as discussed previously for apple. The supply of viable pollen available for subsequent distribution to the stigmatic surface of unopened flowers was also reduced, although the effect of this as part of the thinning action in peaches is not clear. Flowers which had previously been pollinated, and in which pollen tube growth through the style had occurred, continued to set fruit. As other materials came along which could be applied after bloom, the use of the dinitros decreased. At the present time the dinitros are not registered for use on peaches as a thinning spray in the United States.

b. N-1-Naphthylphthalamic Acid. One of the first successful materials that came out of the screening programs to find a postbloom thinner for peaches was N-1-naphthylphthalamic acid (NPA) (Edgerton and Hoffman, 1955). Application of NPA 3 to 5 days after full bloom at concentrations ranging from 100 to 400 ppm, depending on the cultivar and the degree of thinning needed, came into limited commercial practice. Aitken et al. (1972) reported that NPA was especially useful in thinning early maturing or "short-cycle" peaches in Florida. This timing offered some advantage over the more precise and earlier full bloom application necessary with the dinitros. However, in many peach areas, a still later timing of 2 to 4 weeks after full bloom is desirable in order to avoid the frost periods, and to more accurately assess the need for thinning. Sprays of NPA up to a month after bloom have reduced fruit set on peaches but these later applications generally result in leaf injury and stunted growth when used at concentrations necessary to reduce set. While NPA was registered for many years for use on peaches it has recently been dropped, and thus is not available at this time (1972) as a chemical thinner on peaches in the United States.

c. 3-Chlorophenoxypropionic Acid. Probably the most widely used chemical thinner on peaches in recent years has been 3-chlorophenoxypropionic acid (3CP). It was first reported to be effective as a peach thinner by Thompson and Rogers (1959).

Subsequently, a commercial formulation was developed containing a mixture of the acid and the amide of 3CP. This has been used with varying success on several cultivars in most of the principal peach growing districts when applied at concentrations of 100 to 300 ppm 2 to 4 weeks after bloom.

TABLE VIII

Effect of Ethephon on Fruit Set, Fruit Size, and
Shoot Growth of Redhaven Peaches[a,b]

Treatment,[c] ethephon, June 21 (ppm)	Fruit set[d] (%)	Fruit wt[e] (gm)	Shoot length (cm)
50	29.7	147.3	30.4
150	7.5	152.8	23.6
450	0	—	16.6
Control	67.2	126.0	30.8

[a] Hamlin, N. Y., 1967.
[b] Adapted from Edgerton and Greenhalgh, 1969.
[c] Full bloom May 20; 5 replicate branches/treatment.
[d] Percent of fruits present June 21 remaining July 19.
[e] Harvest August 21.

d. OTHER MATERIALS. A number of other chemicals have been reported effective in reducing set on peaches or in regulating the number of flower buds. These include NAA (Hibbard and Murneek, 1950), gibberellin (Hull and Lewis, 1959; Edgerton, 1965), 3-chloroisopropyl-N-phenyl-carbamate (CIPC) (Marth and Prince, 1953) and maleic hydrazide (Langer, 1952). None have come into commercial use because of excessive variability in thinning responses from year to year or undesirable side effects on the fruit or vegetative growth.

Edgerton and Greenhalgh (1969) noted that (2-chloroethyl)phosphonic acid (ethephon) thinned peach flowers when applied at concentrations of 50 to 450 ppm about 4 weeks after full bloom. Shoot growth was retarded on the Redhaven cultivar with application rates of 150 and 450 ppm (Table VIII). Some leaf yellowing and leaf abscission also occurred with all the cultivars treated at the high rate. However, satisfactory reduction in fruit set was obtained with ethephon at concentrations which avoided objectionable foliage effects. Stembridge and Gambrell (1971) conducted more extensive studies with ethephon on several peach cultivars and likewise found this growth regulator to be a promising chemical thinner for peaches. In addition to the effect of ethephon on set reduction, they found that fruit maturity was advanced depending on time of application.

2. Apricots, Nectarines, Plums, Prunes, and Cherries (*Prunus* sp.)

As with peaches, these stone fruits have been successfully thinned with applications of dinitro during the bloom stage. Proper timing and appropriate concentration have varied from year to year with different species and varieties. Fruit set on plums, for example, was reduced with a lower concentration than that required for peaches and apricots. Prunes were similar to peach in concentration of dinitro required for effective thinning (Batjer, 1965). The best thinning with apricots was obtained with an application when 60 to 75% of the blossoms are open.

Tests with NPA applied from full bloom to 6 days after bloom on nectarines and plums demonstrated that this growth regulator could reduce fruit set on several cultivars (Beutel *et al.*, 1969). However, phytotoxicity and overthinning were experienced in the trials, and the material has not come into commercial use on these fruits.

As discussed for peaches, 3CP has been used to a limited extent for thinning plums. A commercial formulation which was used for this purpose contained only the 3CP acid rather than a mixture of the acid and amide as used on peaches. The material was applied at 50 to 150 ppm 2 to 3 weeks after bloom (Beutel *et al.*, 1969).

Sweet cherries often benefit from thinning to obtain early ripening and large fruit size. Although fruit set can be reduced with an application of dinitro at full bloom, leaf injury is likely to result and a reduction in total yield is common.

While the dinitros have been used commercially on several stone fruits for blossom thinning in the past, they are not registered for this use on any of these crops at the present time in the United States.

VI. Chemical Thinning of Other Fruits

The discussion of chemical thinning thus far has been concerned with deciduous tree fruit species. The major worldwide interest in application of thinning sprays has been concerned with these fruits. Some attention has been given to thinning tropical and subtropical fruit species, although with many of these the concern in regulating fruit set has been the need for *increasing* fruit set rather than thinning. A brief review of the thinning trials with some of these fruits is presented here.

A. Avocado (*Persea americana* Mill.)

Differences in productivity exist among the varieties and selections of avocado in most areas where this species is grown. In studying this problem some attention has been given to regulating set and thus providing for more uniform cropping from year to year. Lahav and Tomer (1969) found that application of NAD at 20 to 50 ppm at the bloom stage resulted in satisfactory fruit thinning and improved size at harvest on the Hass cultivar. This variety is subject to overbearing and small fruit size in some seasons.

B. COCONUT (*Cocos nucifera* L.)

There appears to be little need for reducing set of coconut fruits in order to obtain the commercial benefits usually associated with thinning. However, some research has been conducted on regulation of set primarily on coconut palms located in parks and other areas where falling of mature fruits presents either a hazard or nuisance. Criley (1972) found that both ethephon and chlorflurenol applied to the pistillate inflorescence after the flowers were receptive induced abscission of the young fruitlets. However, this treatment also promoted abscission of any maturing fruit present on the tree at the time the spray was applied. Since coconuts continue to flower much of the year it would be necessary to repeat the sprays at frequent intervals on a long-range program to reduce the number of fruits reaching full size and falling to the ground when a once-over spray was applied.

C. MANGO (*Mangifera indica* L.)

Extensive experiments have been conducted on methods of inducing flowering and regulating fruit set in mango. In many major varieties a heavy crop year is followed by a year or more of light crops. By removing some of the panicles from a flowering tree, Singh and Khan (1939) obtained an increase in flowering the following year. However, this has not been developed into a commercial practice through the use of chemical thinning sprays. Defective fruit set with small seedless fruits is also a common problem. One approach to solving this problem in the Haden cultivar is to destroy the early inflorescences which are mainly responsible for producing small, imperfect fruits. Tests with growth regulators have been made in an effort to find an effective material that can be applied for this purpose (Gazit and Presman, 1969).

D. OLIVE (*Olea europaea* L.)

When a heavy crop develops on olive trees, fruit size at harvest is reduced. With most varieties such overproduction or overbearing also reduces the size of the crop in the following year, giving rise to an alternate bearing tendency typical of many apple cultivars. Hartmann (1952) found that relatively high concentrations of NAA applied as an aqueous spray with 1.5% summer oil 3 to 4 weeks after bloom was effective in reducing fruit set. This thinning resulted in improvement in fruit size and regularity of bearing. Reduction in fruit set was also accomplished by spraying during bloom with NAA at 40 to 50 ppm. However, this blossom thinning application was not recommended because the need for thinning could not be predicted at that time.

In some experiments with several olive cultivars under irrigated conditions Lavee and Spiegel-Roy (1967) obtained satisfactory thinning with applications of both NAA and NAD from 4 to 12 days after full bloom. The reduced fruit set was associated with earlier fruit maturation. On a concentration basis the NAA was from 1½ to 2 times more active in reducing fruit set than the NAD.

E. GRAPES (Vitis sp.)

In some preliminary trials with a number of growth regulators Weaver (1954) found that sodium monochloroacetate applied at 0.5% on Muscat of Alexandria at the bloom stage thinned the clusters and resulted in increased berry weight at harvest. Yields were reduced by the treatment and some injury occurred to the foliage. However, Weaver concluded from these tests that flower cluster and cluster thinning of grapes with chemical sprays offered some promise. Weaver and Pool (1969) evaluated 3 abscission promoting compounds on flower and berry abscission of 4 Vitis vinifera cultivars. The morphactin, methyl-2-chloro-9-hydroxy-fluorene-9-carboxylate, was very effective in promoting berry abscission in all cultivars when the clusters were dipped in the solution at 4 developmental stages; spray treatments were effective on 2 of the 4 cultivars. Ethephon and abscisic acid were less consistent in their thinning action on these cultivars, whether by dipping or spray application.

More recently Weaver and Pool (1971) found that GA applications at bloom successfully thinned the compact clusters of the seeded grape cultivars Tokay and Zinfandel. Such thinning is desirable to reduce the incidence of rotting of fruit in the cluster, although the total yield on the treated vines is generally reduced. GA applications have come into extensive use for regulating the set and cluster size of several grape cultivars, particularly Thompson Seedless for table use.

F. PECAN (Carya illinoinensis Koch.)

Mature pecan trees often set more nuts following a heavy bloom than will ripen with well-developed kernels. This heavy set with inferior nut quality at harvest also limits flowering and crop potential for the following year. The application of thinning sprays containing maleic hydrazide, 2,4,5-T, or CIPC about 3 weeks after bloom effectively thinned the Moore variety (Harris and Smith, 1957). Maleic hydrazide and 2,4,5-T were also effective on several other varieties. However, these applications resulted in some injury to young leaves and fruits and have not been extensively used commercially.

G. Mandarin (*Citrus reticulata* Blanco)

Fruit set on the Dancy mandarin or "tangerine" was reduced approximately 30% with an application of maleic hydrazide at 400 ppm during the bloom period (Gardner *et al.*, 1961). However, fruit size at harvest was not significantly improved and in some cases the application even reduced the total number of large sized fruits.

The Wilking mandarin is a typical alternate or biennial bearing cultivar. Fruits during the "on-year" cycle are small, while in the following light crop year the fruits may be too large and uneconomical to handle (Hield *et al.*, 1962). Research in California demonstrated that this mandarin could be successfully thinned with applications of NAA at 350 to 1000 ppm in late May when the young fruits averaged from 8.5 to 21.5 mm in diameter (Hield *et al.*, 1962). Fruit color, shape, and quality were normal on the treated trees; because of the larger fruits on the thinned trees the total yield was not reduced except in the thinning treatments where larger numbers of fruits were removed. Regarding the mode of action of the NAA in the mandarin, Lewis *et al.* (1964) suggested that the control of alternate bearing was through NAA-sensitive mechanisms and not directly related to carbohydrate–nitrogen levels.

VII. Conclusions

The objective of this chapter has been to characterize the use of various chemical sprays to induce shedding of flowers or developing fruits, and thereby regulate fruit set. It has been shown that by the proper choice of materials, concentration, and timing the desired reduction in set can be achieved on many fruit crops. This adjustment in fruit set will provide the benefits of traditional hand fruit thinning with the important additional advantage of regular cropping which was first achieved only by hand flower thinning.

The attempt to characterize the mode of action of growth regulators in reducing set has left some deficiencies. This is understandable since much of the research with chemical thinners has been biased toward description and identification of the factors that influence the response to a given chemical rather than coming to grips with the basic mechanisms involved.

To a large extent fruit growers are riding the crest of commercial practice where guidelines come more from experience and shrewd analysis of existing conditions than from comprehensive understanding of the physiological processes involved.

However, some progress has been recorded and, in the case of peaches at least, a concerted effort is being made to learn more basic details about the physiology of fruit setting as a base on which to build a more successful chemical thinning program. With modern techniques for analysis of endogenous hormones it will be possible to better assess the interplay between endogenous growth regulators and applied chemicals. In the meantime, the experience and intuition of the grower and field research of fruit specialists will have to provide the bench marks for expanded use and refinement of the practice.

References

Abbott, D. L. (1954). Recent application of growth substances in fruit growing. *Ann. Appl. Biol.* **41,** 215–220.

Aitken, J. B., Buchanan, D. W., and Sauls, J. W. (1972). Thinning short-cycle peaches with N-1-naphthyl-phthalamic acid. *HortScience* **7,** 255–256.

Auchter, E. C., and Roberts, J. W. (1934). Experiments in the spraying of apples for the prevention of fruit set. *Proc. Amer. Soc. Hort. Sci.* **30,** 22–25.

Auchter, E. C., and Roberts, J. W. (1935). Spraying apples for the prevention of fruit set. *Proc. Amer. Soc. Hort. Sci.* **32,** 208–212.

Bagenal, N. B., Goodwin, W., Salmon, E. S., and Ware, W. M. (1925). Spraying experiments against apple scab. *J. Min. Agr. (Gr. Brit.)* **32,** 137–150.

Batjer, L. P. (1965). Fruit thinning with chemicals. *U. S., Dep. Agr., Inform. Bull.* **289,** 1–27.

Batjer, L. P., and Billingsley, H. D. (1964). Apple thinning with chemical sprays. *Wash., Agr. Exp. Sta., Bull.* **651,** 1–25.

Batjer, L. P., and Hoffman, M. B. (1951). Fruit thinning with chemical sprays. *U. S., Dep. Agr., Circ.* **867,** 1–46.

Batjer, L. P., and Moon, H. H. (1943). Thinning apples and peaches with blossom-removal sprays. *Proc. Amer. Soc. Hort. Sci.* **43,** 43–46.

Batjer, L. P., and Thompson, A. H. (1948). Three years' results with chemical thinning of apples in the Northwest. *Proc. Amer. Soc. Hort. Sci.* **52,** 164–172.

Batjer, L. P., and Thomson, B. J. (1961). Effect of 1-naphthyl N-methylcarbamate (Sevin) on thinning apples. *Proc. Amer. Soc. Hort. Sci.* **77,** 1–8.

Batjer, L. P., and Westwood, M. N. (1960). 1-Naphthyl N-methylcarbamate, a new chemical for thinning apples. *Proc. Amer. Soc. Hort. Sci.* **75,** 1–4.

Batjer, L. P., Forshey, C. G., and Hoffman, M. B. (1968). Effectiveness of thinning sprays as related to fruit size at time of application. *Proc. Amer. Soc. Hort. Sci.* **92**, 50–54.

Beutel, J., Gerdts, M., LaRue, J., and Carlson, C. (1969). Chemical thinning for shipping peaches, nectarines and plums. *Calif. Agr.* **23**, 6–8.

Bobb, A. C., and Blake, M. A. (1938). Annual bearing of the Wealthy apple as induced by blossom thinning. *Proc. Amer. Soc. Hort. Sci.* **36**, 321–327.

Bukovac, M. J., and Mitchell, A. E. (1962). Biological evaluation of 1-naphthyl N-methyl-carbamate with special reference to the abscission of apple fruits. *Proc. Amer. Soc. Hort. Sci.* **80**, 1–10.

Burkholder, C. L., and McCown, M. (1941). Effect of scoring and of α-naphthyl acetic acid and amide spray upon fruit set and of the spray upon pre-harvest fruit drop. *Proc. Amer. Soc. Hort. Sci.* **38**, 117–120.

Criley, R. A. (1972). Coconut fruit drop induced by ethephon and chlorflurenol. *HortScience* **7**, 176.

Davidson, J. H., Hammer, O. H., Reimer, C. A., and Dutton, W. C. (1945). Thinning apples with the sodium salt of naphthyl acetic acid. *Mich., Agr. Exp. Sta., Quart. Bull.* **27**, 352–356.

Dennis, F. G., Jr. (1970). Effects of gibberellin and naphthaleneacetic acid on fruit development in seedless apple clones. *J. Amer. Soc. Hort. Sci.* **95**, 125–128.

Donoho, C. W., Jr. (1968). The relationship of date of application and size of fruit to the effectiveness of NAA for thinning apples. *Proc. Amer. Soc. Hort. Sci.* **92**, 55–62.

Edgerton, L. J. (1965). Gibberellin sprays for peach flower buds. *Farm Res.* **30**, 8–9.

Edgerton, L. J., and Greenhalgh, W. J. (1969). Regulation of growth, flowering and fruit abscission with 2-chloroethanephosphonic acid. *J. Amer. Soc. Hort. Sci.* **94**, 11–13.

Edgerton, L. J., and Haeseler, C. W. (1959). Some factors influencing the absorption of naphthaleneacetic acid and naphthaleneacetamide by apple leaves. *Proc. Amer. Soc. Hort. Sci.* **74**, 54–60.

Edgerton, L. J., and Hoffman, M. B. (1955). Effect of N-1-naphthyl phthalamic acid on fruit set of peaches. *Science* **121**, 467–468.

Fletcher, L. A. (1932). Effect of thinning on size and color of apples. *Proc. Amer. Soc. Hort. Sci.* **29**, 51–55.

Forshey, C. G., and Hoffman, M. B. (1966). Factors affecting chemical thinning of apples. *N. Y., Agr. Exp. Sta., Geneva, Res. Circ.* **4**, 1–12.

Gardner, F. E., Marth, P. C., and Batjer, L. P. (1939). Spraying with plant growth substances to prevent apple fruit dropping. *Science* **90**, 208–209.

Gardner, F. E., Reece, P. C., and Horanic, G. E. (1961). Thinning of Dancy tangerines at blossom time with maleic hydrazide. *Proc. Amer. Soc. Hort. Sci.* **77**, 188–193.

Gardner, V. R., Merrill, T. A., and Petering, H. G. (1940). Thinning the apple crop by spray at blooming: A preliminary report. *Proc. Amer. Soc. Hort. Sci.* **37**, 147–149.

Gawadi, A. G., and Avery, G. S., Jr. (1950). Leaf abscission and the so-called "abscission layer." *Amer. J. Bot.* **37**, 172–180.

Gazit, S., and Presman, E. (1969). Defective fruit set and the problem of small seedless fruits in the Haden variety. *Volcani Inst. Agr. Res. 1960–1969* p. 98.

Harley, C. P., Moon, H. H., and Regeimbal, L. O. (1957). Effects of the additive

Tween 20 and relatively low temperatures on apple thinning by naphthaleneacetic acid sprays. *Proc. Amer. Soc. Hort. Sci.* **69**, 21–27.

Harley, C. P., Moon, H. H., and Regeimbal, L. O. (1958). Evidence that post-bloom apple-thinning sprays of naphthaleneacetic acid increase blossom-bud formation. *Proc. Amer. Soc. Hort. Sci.* **72**, 52–56.

Harris, O. W., and Smith, C. L. (1957). Chemical thinning of pecan nut crops. *Proc. Amer. Soc. Hort. Sci.* **70**, 204–208.

Hartman, F. O., and Howlett, F. S. (1962). Effects of naphthaleneacetic acid on fruit setting and development in the apple. *Ohio, Agr. Exp. Sta., Res. Bull.* **920**, 1–66.

Hartmann, H. T. (1952). Spray thinning of olives with naphthaleneacetic acid. *Proc. Amer. Soc. Hort. Sci.* **59**, 187–195.

Hennerty, M. J., and Forshey, C. G. (1971). Effects of defruiting, scoring, defoliation and shading on the carbohydrate content of "Golden Delicious" apple trees. *J. Hort. Sci.* **46**, 153–161.

Hennerty, M. J., and Forshey, C. G. (1972). Tree physiological conditions as a source of variation in chemical thinning of apple fruits. *HortScience* **7**, 259–260.

Hibbard, A. D., and Murneek, A. E. (1950). Thinning peaches with hormone sprays. *Proc. Amer. Soc. Hort. Sci.* **56**, 65–69.

Hield, H. Z., Burns, R. M., and Coggins, C. W., Jr. (1962). Some fruit thinning effects of naphthaleneacetic acid on Wilking mandarin. *Proc. Amer. Soc. Hort. Sci.* **81**, 218–222.

Hildebrand, E. M. (1944). The mode of action of the pollenicide, Elgetol. *Proc. Amer. Soc. Hort. Sci.* **45**, 53–58.

Hoffman, M. B. (1942). Thinning Wealthy apples at blossom time with a caustic spray. *Proc. Amer. Soc. Hort. Sci.* **40**, 95–98.

Hoffman, M. B. (1947). Further experience with the chemical thinning of Wealthy apples during bloom and its influence on annual production and fruit size. *Proc. Amer. Soc. Hort. Sci.* **49**, 21–25.

Hoffman, M. B. (1954). Thinning apples with hormone sprays. *Proc. N. Y. State Hort. Soc.* **99**, 137–141.

Hoffman, M. B., Edgerton, L. J., and Fisher, E. G. (1955). Comparisons of naphthaleneacetamide for thinning apples. *Proc. Amer. Soc. Hort. Sci.* **65**, 63–70.

Horsfall, F., Jr., and Moore, R. C. (1963). Effect of 1-naphthyl N-methylcarbamate (Sevin) on thinning four varieties of apples. *Proc. Amer. Soc. Hort. Sci.* **82**, 1–4.

Hull, J., Jr., and Lewis, L. N. (1959). Response of one-year-old cherry and mature bearing cherry, peach and apple trees to gibberellin. *Proc. Amer. Soc. Hort. Sci.* **74**, 93–100.

Kovach, B. (1969). "Appleknockers" upstate reaping fruit of labor. *New York Times* **67**, No. 40,084.

Lahav, E., and Tomer, E. (1969). Flowering, fruit set and fruit drop. *Volcani Inst. Agr. Res. 1960–1969* pp. 57–60.

Langer, C. A. (1952). Effect of maleic hydrazide on the thinning of peaches during three successive years. *Mich., Agr. Exp. Sta., Quart. Bull.* **35**, 209–213.

Langley, B. (1729). "Pomona: Or, The Fruit-Garden Illustrated." G. Strahan, London.

Lavee, S., and Spiegel-Roy, P. (1967). The effect of time of application of two growth substances on the thinning of olive fruit. *Proc. Amer. Soc. Hort. Sci.* **91**, 180–186.

Lewis, L. N., Coggins, C. W., Jr., and Hield, H. Z. (1964). The effect of biennial bearing and NAA on the carbohydrate and nitrogen composition of Wilking mandarin leaves. *Proc. Amer. Soc. Hort. Sci.* **84**, 147–151.

Luckwill, L. C. (1953). Studies of fruit development in relation to plant hormones. II. The effect of naphthalene acetic acid on fruit set and fruit development in apples. *J. Hort. Sci.* **28**, 25–40.

Luckwill, L. C., and Lloyd-Jones, C. P. (1962). The absorption translocation and metabolism of 1-naphthaleneacetic acid applied to apple leaves. *J. Hort. Sci.* **37**, 190–206.

MacDaniels, L. H., and Hildebrand, E. M. (1940). A study of pollen germination upon the stigmas of apple flowers treated with fungicides. *Proc. Amer. Soc. Hort. Sci.* **37**, 137–140.

MacDaniels, L. H., and Hoffman, M. B. (1941). Apple blossom removal with caustic sprays. *Proc. Amer. Soc. Hort. Sci.* **38**, 86–88.

Magness, J. R., Batjer, L. P., and Harley, C. P. (1940). Spraying apples for blossom removal. *Proc. Amer. Soc. Hort. Sci.* **37**, 141–146.

Marth, P. C., and Prince, V. R. (1953). Effect of 3-chloro-isopropyl-N-phenyl carbamate on abscission of young fruits of peach. *Science* **117**, 497–498.

Priestley, C. A. (1970). Carbohydrate storage and utilization. *In* "Physiology of Tree Crops" (C. V. Cutting and L. C. Luckwill, eds.), pp. 113–127. Academic Press, New York.

Rogers, B. L., and Thompson, A. H. (1969). Chemical thinning of apple trees using concentrate sprays. *J. Amer. Soc. Hort. Sci.* **94**, 23–25.

Russell, H. A., and Pickering, S. (1919). "Science and Fruit Growing." Macmillan, New York.

Schneider, G. W., and Enzie, J. V. (1943). The effect of certain chemicals on the fruit set of the apple. *Proc. Amer. Soc. Hort. Sci.* **42**, 167–176.

Singh, L., and Khan, A. A. (1939). Relation of growth to fruit bearing in mangoes. *Indian J. Agr. Sci.* **9**, 835–867.

Skoss, J. D. (1955). Structure and composition of plant cuticle in relation to environmental factors and permeability. *Bot. Gaz.* (*Chicago*) **117**, 55–72.

Southwick, F. W., and Weeks, W. D. (1949). Chemical thinning of apples at blossom time and up to four weeks from petal fall. *Proc. Amer. Soc. Hort. Sci.* **53**, 143–147.

Southwick, F. W., Edgerton, L. J., and Hoffman, M. B. (1947). Studies in thinning peaches with blossom removal sprays. *Proc. Amer. Soc. Hort. Sci.* **49**, 26–32.

Stembridge, G. E., and Gambrell, C. E., Jr. (1971). Thinning peaches with bloom and postbloom applications of 2-chloroethylphosphonic acid. *J. Amer. Soc. Hort. Sci.* **96**, 7–9.

Struckmeyer, B. E., and Roberts, R. H. (1950). A possible explanation of how naphthalene acetic acid thins apples. *Proc. Amer. Soc. Hort. Sci.* **56**, 76–78.

Teubner, F. G., and Murneek, A. E. (1955). Embryo abortion as mechanism of "hormone" thinning of fruit. *Mo., Agr. Exp. Sta., Res. Bull.* **590**, 1–27.

Thompson, A. H. (1952). The present status of hormone sprays for thinning of apples and peaches. *Trans. Peninsula Hort. Soc.* **42**, 56–61.

Thompson, A. H. (1953). A new hormone for the chemical thinning of apples. *Trans. Peninsula Hort. Soc.* **43**, 54–59.

Thompson, A. H. (1957). 6 years' experiments on chemical thinning of apples. *Md., Agr. Exp. Sta., Bull.* **A-88**, 1–46.

Thompson, A. H., and Rogers, B. L. (1959). Three years' results with new chemicals for thinning of peaches after the blossoming period. *Proc. Amer. Soc. Hort. Sci.* **73,** 112–119.

Tukey, L. D. (1965). When to apply thinning sprays. *Amer. Fruit Grow.* **85,** 14–15.

van Overbeek, J. (1952). Agricultural application of growth regulators and their physiological basis. *Annu. Rev. Plant Physiol.* **3,** 87–108.

Veinbrants, N. (1972). Comparison of thiram, carbaryl, N.A.A. and N.A.D. for fruit thinning sprays on Granny Smith apples. *Aust. J. Exp. Agr. Anim. Husb.* **12,** 83–88.

Way, R. D. (1954). The effect of some cultural practices and of size of crop on the subsequent winter hardiness of apple trees. *Proc. Amer. Soc. Hort. Sci.* **63,** 163–166.

Weaver, R. J. (1954). Preliminary report on thinning grapes with chemical sprays. *Proc. Amer. Soc. Hort. Sci.* **63,** 194–200.

Weaver, R. J., and Pool, R. M. (1969). Effect of Ethrel, abscisic acid, and a morphactin on flower and berry abscission and shoot growth in *Vitis* vinifera. *J. Amer. Soc. Hort. Sci.* **94,** 474–478.

Weaver, R. J., and Pool, R. M. (1971). Thinning "Tokay" and "Zinfandel" grapes by bloom sprays of gibberellin. *J. Amer. Soc. Hort. Sci.* **96,** 820–822.

Westwood, M. N., and Batjer, L. P. (1958). Factors influencing absorption of dinitro-ortho-cresol and naphthaleneacetic acid by apple leaves. *Proc. Amer. Soc. Hort. Sci.* **72,** 35–44.

Westwood, M. N., and Batjer, L. P. (1960). Effects of environment and chemical additives on absorption of naphthaleneacetic acid by apple leaves. *Proc. Amer. Soc. Hort. Sci.* **76,** 16–29.

Williams, M. W., and Batjer, L. P. (1964). Site and mode of action of 1-naphthyl *N*-methylcarbamate (Sevin) in thinning apples. *Proc. Amer. Soc. Hort. Sci.* **85,** 1–10.

. 12 .

Chemical Control
of Fruit Abscission

William C. Cooper and William H. Henry

I. Introduction

In many of the cultivated fruit crops, which reflect breeding and selection for high quality fruit, the endogenous abscission mechanisms are not conducive to efficient fruit production and harvesting. Insufficient fruit set, overset, premature fruit drop, and too strong a retention of mature fruit by the tree are all examples of problems involved with these crops. Auxinology provided a chemical solution to one of these problems in 1939, when Gardner *et al.* found that the auxin, naphthaleneacetic acid (NAA),[1] delayed premature (preharvest) fruit drop of apples (*Pyrus malus* L.).

This finding led to a decade of applied research on use of exogenous auxins on fruit drop control and resulted in the widespread use of 2,4-D and 2,4,5-TP for this purpose (Batjer *et al.*, 1948; Edgerton and Hoffman, 1951; Stewart and Parker, 1947; Gardner *et al.*, 1950). Luckwill (1953, 1957), in basic studies on the role of endogenous auxin in control of fruit set of apples, showed that apples have two active internal meristems: the growing embryo and endosperm of seed, which produce auxin which moves out of the seed to prevent abscission in the base of the pedicel. Only low auxin levels occur in apples after fruit set and during fruit enlargement and maturity. Luckwill (1957) concluded that the developing apple fruit, to remain attached to the twig, requires a continuous internal supply of auxin, and immature fruit drop is caused by a deficiency in the supply of auxin to the abscission zone in the pedicel. The application of exogenous auxin sprays raises the internal level of auxin and alleviates the deficiency in endogenous auxin.

In the late 1950's fruit growers in the United States became faced with the double-edged problem of a dwindling and unstable supply of manual labor and increasing wages for the labor that were available. This led to an interest in the possibility of machine harvesting of fruit. Research to mechanize the harvest of various fruit crops has been underway in the United States since 1958. Mechanical shaking of trees has been demonstrated to be a feasible means of harvesting cherries (*Prunus avium* L;

[1] The following abbreviations are used for chemical names in this chapter: ABA [3-methyl, 5-(hydroxy-4-oxo-2,6,6-trimethyl-2-cyclohexen-1-yl-*cis,trans*-4-pentadienoic acid], or abscisic acid; Amo-1618 [(2-isopropyl-4-trimethylammonium chloride)-5-methylphenylpiperidine carboxylate]; CCC [(2-chloroethyl)trimethylammonium chloride]; CHI {B[2-(3,5-dimethyl-2-oxocyclohexyl)-2-hydroxyethyl]glutarimide} or cycloheximide; 2,4-D (2,4-dichlorophenoxyacetic acid); GA (gibberellins in general); GA_3 (gibberellic acid); IAA (indoleacetic acid); NAA (1-naphthaleneacetic acid); SADH (succinic acid 2,2-dimethylhydrazide); 2,4,5-TP (2,4,5-trichlorophenoxypropionic acid).

P. cerasus L.) (Levin *et al.*, 1960), apples (Edgerton, 1968), olives (*Olea europaea* L.) (Fridley *et al.*, 1971), and early and midseason varieties of oranges (*Citrus sinensis* Osb.) for processing use (Coppock, 1967). In the work with oranges, three types of tree shakers have been developed at the University of Florida Agricultural Research and Education Center at Lake Alfred. These are the "limb shaker" (Coppock and Jutras, 1960; Heddon and Coppock, 1965), the inertia "shock-wave trunk shaker" (Coppock and Jutras, 1962), and the "air shaker" (Whitney, 1968).

The successful removal of fruit by shaking the trees mechanically depends to some extent on the size and weight of the fruit and the attachment or fruit removal force (FRF), as well as on the effect of foliage and branch characteristics of the trees. The relatively high FRF of citrus fruits (15 to 22 lb), as compared to 2 lb for cherries and 6 lb for apples, has proven to be a serious handicap to successful mechanical harvesting of this fruit. Yields of machine-harvested oranges are generally less than 90% of the total crop on the tree. It, therefore, seems self-evident that if we can chemically reduce the FRF, fruit can be removed with a lower mechanical force, and the chemicals will thereby aid mechanical harvesting.

An abscission-accelerating substance was introduced when Ohkuma *et al.* (1963) demonstrated the presence of ABA in rapidly abscising cotton (*Gossypium hirsutum* L.) fruits. The presence of ABA is also associated with abscission of young lupin (*Lupinus* sp.) fruits (Van Stevenick, 1959) and with ripening and abscising pear fruits (Wang *et al.*, 1972). Although exogenously applied ABA accelerates fruit abscission of cherries (Zucconi *et al.*, 1969) and apples (Edgerton, 1971a), it has no effect on the FRF of citrus fruit (Cooper *et al.*, 1968a).

Ethylene gas has been known to promote leaf abscission since the pioneer work of Neluibou in 1901. In 1939, Nelson showed that there was an increase in the rate of ethylene production during ripening and natural fruit fall of apples. Blanpied (1972) has now established that high levels of ethylene in the abscission zone of ripening apples may be the result of diffusion from the flesh of the ripening fruit, where ethylene levels are very high (Table I). Cooper *et al.* (1968a) and Cooper and Henry (1971) have shown that the internal levels of ethylene in fruit near the abscission zone are a reliable indicator of the abscission process in citrus. When citrus trees are sprayed with certain chemicals, ethylene levels in the fruit increase, and there is a concomitant reduction in the FRF. This close relation between internal levels of ethylene in the fruit and abscission appears to hold true, whether ethylene is induced in the respiration climacteric as with apples or is induced by the fruit as a result of chemical treatment as with citrus.

We, therefore, hypothesize that a partial solution to the harvesting

TABLE I

EXTRACTABLE ETHYLENE CONTENT OF MCINTOSH APPLE TISSUES DURING
FRUIT MATURATION AND RIPENING ON THE TREE IN 1969

Days past full bloom	Cumulative fruit (%)	Ethylene content (nl/gm fresh wt)		
		Flesh	Pedicles	Cluster bases
128	0	0.04	0.13	0.13
132	1	0.52	0.11	0.12
136	7	1.85	0.40	0.18
139	24	—	0.57	0.14
143	40	5.05	—	0.25
146	58	9.24	1.39	0.22
149	86	34.20	2.39	0.24

[a] Data from Blanpied (1972).

problem lies in making the fruit generate "hyperethyleneism" by the use of certain chemicals. Just as auxinology started the first major "gold rush" in chemical plant physiology, now "ethyleneology" is the stimulus behind the chemical industry's present efforts to synthesize compounds that make the fruit produce ethylene.

This interest in chemical control of fruit abscission is shared by growers of apples, pears (*Pyrus communis* L.), olives, cherries, plums (*P. domestica* L.), and many other fruit, nut, and vegetable crops (Edgerton, 1968; Griggs *et al.*, 1970; Hartmann *et al.*, 1970; Bukovac, 1970; Looney and McMechan, 1970; Anderson, 1969). The trend is also worldwide. Field trials with abscission-accelerating chemicals are now in progress on olives in Italy (Jacoboni *et al.*, 1971; Calabrese and Sotille, 1971a), citrus in Victoria (El-Zeftawi, 1970), and coffee (*Coffea arabica* L.) in Kenya (Browning and Cannell, 1970).

II. Screening Program for Abscission-Inducing Chemicals

A. FIELD PLOT TECHNIQUES

Although small fruiting calamondin (*Citrus reticulata* var. *austera* ? × *Fortunella* sp. ?) trees grown in the greenhouse have been used for screening potential citrus fruit abscission chemicals (Rasmussen and Jones, 1969), most of the screening work (Cooper *et al.*, 1968a, 1969d; Cooper and Henry, 1968a) has been done either on whole citrus trees growing in commercial orchards or on branches of orchard-grown citrus trees. The calamondin fruit do not adhere to the tree as tightly as oranges. Therefore, results obtained on field-grown orange trees have more practical application to commercial fruit harvesting than do results with calamondins.

Generally, the first screening of a chemical is done on three branches: one branch on each of three trees. Branches are selected that contain about 100 mature fruits to provide sufficient fruits for ethylene analysis and FRF measurements. The FRF is measured with a pull force gauge by applying a straight pull at a zero angle parallel to the axis of the fruit, in order to transmit the force equally to all sides of the juncture of the woody peduncle and exocarp of the fruit. The average value of the FRF on 10 fruits, selected at random on each of the three replicate limbs, provides a good index for the effect of a chemical on fruit loosening at any one observation. Usually several FRF tests were made during the week or 10 days following a chemical treatment. Maximum loosening generally occurs after 5 to 7 days. When comparing FRF values in the tables and figures, significant differences usually exist when the larger of two values exceeds the smaller by 2 lb.

Promising chemicals from the branch screening tests are next tested on whole trees, using three replicate trees for each concentration of a given chemical. Generally, 10 gal of a test solution is adequate for wetting the foliage and fruit on a 20-ft-high tree.

Several chemicals including ascorbic acid, CHI, ethephon, and others have been evaluated in randomized large-scale field plots of 10 trees or more per replicate. These plots were evaluated for fruit abscission characteristics by FRF measurements and mechanical harvesting (Cooper *et al.*, 1969d; Wilson and Coppock, 1968; Wilson, 1971). Similar field plot techniques were employed in tests on cherries (Bukovac, 1970), apples (Edgerton, 1968), and olives (Hartmann *et al.*, 1967; Fridley *et al.*, 1971).

B. ETHYLENE MEASUREMENTS

A simplified procedure, based on gas chromatography, was used in all of the citrus fruit abscission work. Internal levels of ethylene in the fruit were measured by taking a 1-ml sample from under the exocarp near the abscission zone with a hypodermic syringe (Rasmussen and Cooper, 1969a). This procedure for obtaining internal gas samples from fruit is similar to that used by Burg (1958) on apples. The air in the syringe was directly injected into a gas chromatograph with activated alumina columns, dual flame ionization detectors, nitrogen carrier gas, and oxygen substituted for air in the flame. The column dimensions were 6 ft × ⅛ (1.83 m × 3.2 mm) inches and the nitrogen flow rate was 40 ml/min. The machine is capable of detecting quantities of ethylene approaching 0.01 ppm. Figure 1 shows triplicate gas chromatograms for Valencia oranges treated with CHI and for untreated fruit.

Complications are introduced by the transient nature of the chemically induced climacteric (Section IV,B) in oranges. A slight increase in internal level of ethylene may occur within 4 hr after chemical treatment, but the maximum increase may occur any time during the second to fourth day; the peak may be followed by a rapid decline in ethylene levels. Frequently, fruit of one replicate branch or tree will reach a peak

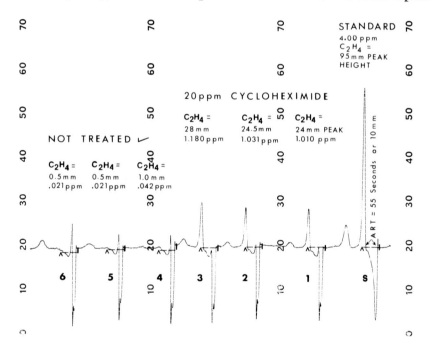

FIG. 1. Ethylene gas chromatograms of 1-ml samples of air withdrawn from fruits #1, 2, and 3 from a Valencia orange tree sprayed with 20 ppm CHI and untreated fruits 4, 5, and 6. Also shown is a chromatogram of a 4 ppm ethylene standard. The distance from the injection point to the ethylene standard peak was used to identify the ethylene peaks of the unknowns. The amount of ethylene present in the unknown peak was calculated from measurements of the height of the ethylene standard peak and the height of the unknown peak by the following equation:

$$\frac{\overset{\text{(Standard)}}{\text{95-mm peak height}}}{4 \text{ ppm}} \times \frac{\overset{\text{(Unknown \#1)}}{\text{24-mm peak height}}}{X} = 1.010 \text{ ppm ethylene in unknown \#1}$$

ethylene level a day earlier than the fruit of the other two replicates. Because of this transient nature of the ethylene response we measured ethylene on duplicate samples from each replicate branch several times during a 5-day period following treatment. The data given in Tables I, IV, VIII, and X and Figs. 4, 6, and 7 represent averages of the measurements on 6 fruits (duplicate fruits on 3 replicates). When comparing ethylene values, significant differences usually exist when the larger of two values exceeds the smaller by more than 150%, i.e., a value of 0.160 ppm is probably significantly higher than a value of 0.100 ppm.

Internal atmosphere of small green fruit, leaves, and stems could not be directly sampled with a hypodermic syringe. In such instances, Cooper *et al.* (1968a) measured the rate of ethylene emitted from the material during a 24-hr incubation period in a sealed 250-ml flask. A 1-ml sample of air was removed from the flask with a hypodermic syringe and was analyzed for ethylene as described above. The data were calculated in nl/g fresh weight/hr.

A vacuum apparatus for extraction of air from materials such as leaves and stems that cannot be directly sampled with a hypodermic syringe has recently been described by Blanpied (1971). The procedure requires about 1 hr per extraction. The Blanpied (1972) data for extractable ethylene content of flesh, pedicels, and cluster bases of apples given Table I (refer to Section I) were determined by the vacuum method. This procedure could be used to measure ethylene levels in clinically isolated abscission zones (distal end of peduncle, called the button, plus adjacent parenchyma tissue) of oranges. The apple data (Blanpied, 1972) support our thesis that increases in ethylene levels adjacent to the abscission zone as determined from hypodermic syringe extractions may have resulted from ethylene diffusion from the adjacent chemically treated exocarp.

C. Ethylene-Producing Chemicals as Abscission Agents

Abeles (1967) provided evidence that various abscission chemicals, including GA, ABA, and NAA, all have the ability to stimulate production of ethylene by plants. Other compounds that are now known to cause fruit to produce ethylene and abscise include ascorbic acid, citric acid, iodoacetic acid, maleic hydrazide, maleimides, 2-hydroxyethylhydrazine, ethephon, and cycloheximide (Cooper *et al.*, 1968a).

Among this broad spectrum of chemically diverse compounds, we now distinguish three groups of ethylene-producing chemicals as follows: (1) chemicals such as ethephon and 2-hydroxyethylhydrazine that decompose and release ethylene gas; (2) chemicals such as IAA, NAA, 2,4-D, ABA, and GA$_3$ which may stimulate the plant to produce ethylene from the endogenous precursor of ethylene, methionine (Burg and Clagett, 1967); and (3) chemicals such as cycloheximide, ascorbic acid, iodoacetic acid, maleimides, etc., that injure the fruit and cause production of wound ethylene from methionine (Abeles and Abeles, 1972).

Although CHI may stimulate ethylene generation by the fruit and thus promote abscission, it also inhibits or delays fruit abscission, even in the presence of ethylene. The site of placement of the chemical determines whether fruit abscission is accelerated or delayed (Cooper *et al.*, 1969a). (Refer to Section IV for further details on the action of CHI in abscission.)

IAA, NAA, and 2,4-D when sprayed on trees generally delay fruit abscission, even though they stimulate ethylene generation by the fruit. These auxins are abscission inhibitors and are translocated rapidly enough from leaf to fruit to delay fruit abscission, even in the presence of ethylene.

Ascorbic acid meets some of the requirements for an effective abscission chemical for citrus and olives (Gellini, 1965; Hartmann *et al.*, 1967; Cooper and Henry, 1967). Fruit on citrus trees sprayed with 20,000 ppm ascorbic acid contain adequate ethylene, usually from about 0.5 to 3 ppm, to promote abscission, while leaves usually do not contain enough ethylene to cause defoliation (Cooper *et al.*, 1968a; Palmer *et al.*, 1969). The addition of citric acid or Fe^{3+} or Cu^{2+} salts potentiates the effectiveness of ascorbic acid and enables the use of lower concentrations of ascorbic acid to promote abscission. These mixtures, however, cause fruit injury and are easily washed off by rain (Cooper *et al.*, 1968a). However, under dry weather conditions, the abscission response from ascorbic acid is good.

Iodoacetic acid, the first abscission chemical to be tested extensively on citrus (Hendershott, 1965), causes rind pitting and ethylene production by the fruit (Cooper *et al.*, 1968a). Levels of 500 to 1000 ppm are required to induce fruit abscission, and the response is generally less and more erratic than that induced by 20 ppm CHI. Also, attending leaf drop is excessive. Iron or copper chloride salts (Ben-Yehoshua and Biggs, 1970) and cyclamic acid (Wilson, 1969) were also found to promote ethylene production and fruit abscission in citrus.

Various substrates for ethylene production include methionine (Lieberman *et al.*, 1965; Burg and Clagett, 1967), *N*-benzoyl-DL-methionine (Demorest and Stahmann, 1971), ketomethylbutyrate (Yang, 1969), and linolenic acid (Lieberman and Mapson, 1964). Methionine is known to accelerate abscission of tobacco fruit (*Nicotiana tabacum* L.) (Yager and

Muir, 1958) and bean petioles (Rubinstein and Leopold, 1962). We sprayed all these substrates, at a 100-ppm level, on K-Early tangelo trees, and only N-benzoyl-DL-methionine showed abscission activity (W. C. Cooper and W. H. Henry, unpublished results, 1971). Likewise, these substrates were applied in combination with 20 ppm CHI, and none potentiated the CHI abscission effect.

In general, only two chemicals, namely, ethephon (refer to Section III) and CHI (refer to Section IV), are promising enough for consideration for commercial use at the present time. However, there are many other ethylene-producing chemicals similar to ethephon and CHI now under test by various chemical companies.

III. Effect of Ethephon on Fruit Abscission

Ethephon is the most effective abscission chemical of the group that is known to break down and release ethylene. 2-Hydroxyethylhydrazine releases copious quantities of ethylene on the surface of citrus fruit and leaves when sprayed on trees, but the ethylene escapes rapidly to the ambient air and is not translocated into the fruit (Rasmussen and Cooper, 1969a). Ethephon, on the other hand, not only releases ethylene on the fruit and leaf surfaces but enters the fruit and releases ethylene near the site of abscission (Edgerton and Hatch, 1972).

Ethephon has been tested extensively in citrus, olives, cherries, and apples. The initial attachment force of the fruit, the rate of penetration of the chemical into the fruit, and sensitivity of the leaves and fruit to the chemical are important factors in determining the concentration of the material that may be safely employed on various fruit crops.

A. Experiments with Citrus

Citrus fruit are attached much more tightly to the stem than are apples, cherries, or olives (Fig. 2). However, the rate of reduction in FRF is quite rapid for citrus and extremely slow for olives.

During cool weather [minimum temperatures of around 50°F (10°C)], 200 to 500 ppm ethephon is ineffective in loosening citrus. Likewise, rains following within a day after ethephon application tend to nullify the effect of treatment. Under ideal conditions of warm, dry weather, ethephon at 500 ppm will loosen fruits of tangerine and orange cultivars during the fall and winter. However, when 500 ppm of the material is used, considerable defoliation almost always occurs.

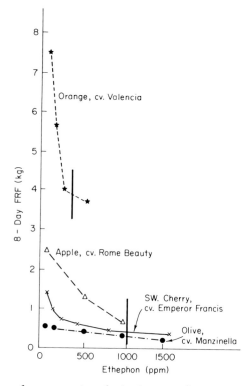

FIG. 2. Relation of concentration of ethephon to reduction in FRF of citrus, apples, sweet cherry, and olive. Perpendicular lines indicate approximate concentration beyond which excessive defoliation is likely to occur. From Bukovac *et al.* (1971); W. C. Cooper and W. H. Henry (unpublished results, 1971); Edgerton (1971a); and Hartmann *et al.* (1967).

Ethephon-induced defoliation in citrus is associated with high levels of ethylene production in both the leaf lamina and petiole (Table II). Ethylene production of both leaves and twigs is 50- to 100-fold that of the fruit. In certain sensitive cultivars, such as Robinson tangerine, these high levels of ethylene in the twigs induce gumming and dieback. (For a discussion of the data on CHI shown in Table II refer to Section IV,A.)

Low levels of ethephon (1 to 200 ppm) will loosen Robinson tangerines enough to aid hand picking and reduce plugging of the fruit (Young and Jahn, 1972). Pineapple oranges are highly sensitive to ethylene in February, and sometimes 100 ppm ethephon will loosen the fruit without defoliation (Cooper and Henry, 1968a). Also El-Zeftawi (1970) has had some success in loosening Valencia oranges in Mildura, Victoria, Australia, during December (comparable to June in Florida), but con-

TABLE II

ETHYLENE EMISSION FROM VARIOUS PARTS OF BRANCH OF
VALENCIA ORANGE TREE 24 hr AFTER SPRAYING[a]

Plant part	Water control (nl/gm/hr)	Ethephon (nl/gm/hr)	CHI (nl/gm/hr)
Leaf lamena	0.525	18.600	1.540
Leaf petiole	1.600	27.425	8.525
Stem	0.150	13.050	2.850
Fruit	0.010	0.200	0.250[b]

[a] Some trees were sprayed with ethephon (500 ppm) and others with CHI (20 ppm) and water. From Cooper and Henry (1970).

[b] Cycloheximide comes in contact only with the rind (see Section IV,B). To estimate the approximate rate of ethylene for the rind, multiply the whole fruit value by 7 as the rind constitutes about one-seventh of the weight of the whole fruit.

siderable defoliation resulted. Ethephon, when used at low concentrations, does not consistently loosen oranges enough to warrant its commercial use.

B. EXPERIMENTS WITH CHERRIES

Ethephon is readily absorbed by leaves and is rapidly translocated to the fruit where it breaks down into ethylene (Edgerton and Hatch, 1972). When radioactive ethephon was applied to leaf and fruit surfaces, the level of radioactive ethylene increased substantially in the fruit for 48 to 72 hr and then decreased (Edgerton and Hatch, 1972). The majority of the ethephon in the fruit moved there from the leaves.

Concentrations of 500 ppm of ethephon were effective in loosening Montmorency sour cherries, with little or no toxic effect on leaves or twigs (Bukovac *et al.*, 1969, 1971). There was no significant change in effectiveness of ethephon when applied over a 3-week period prior to maturity (Bukovac *et al.*, 1971). However, the greater the elapsed time between treatment and FRF measurement, the greater was the reduction in FRF. When applied earlier than July 2 (2 weeks before normal harvest), excessive premature abscission of fruits occurred with aborted embryos and pedicels attached (Bukovac *et al.*, 1971).

When used on Montmorency cherry at 500 ppm, ethephon lowers the FRF from 1000 to 200 gm and greatly potentiates mechanical harvesting (Bukovac *et al.*, 1969, 1971; Edgerton and Hatch, 1969). By relating the FRF to fruit removed by mechanical harvesting (Fig. 3), the distribution curves for treated trees have lower mean values and the individual

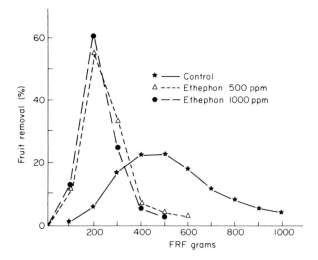

FIG. 3. Frequency polygon of FRF and fruit removal of control and ethephon-treated fruit population of Montmorency sour cherry. From Bukovac *et al.* (1969).

observations are clustered nearer the mean than for control trees (Bukovac *et al.*, 1971).

An Environmental Protective Agency (EPA) label permitting sale of cherries from experimental application of ethephon during 1972 will permit large-scale field evaluation in use of ethephon. Currently, it is recommended that ethephon be applied to Montmorency cherries at 250 ppm 7 to 14 days before harvest (Proebsting, 1972). However, based on the results of Bukovac *et al.* (1969), a concentration of 500 ppm would be safe to use on sour cherries.

FRF values for sweet cherry are generally 2 to 3 times higher than those for Montmorency sour cherry, and this is reflected in the need of an ethephon concentration of about 500 ppm for good fruit removal (Bukovac *et al.*, 1971). At this rate, there should be no gummosis induced nor delay in anthesis the following spring. Ethephon treatment advances the time that sweet cherries will be harvested without stems. Optimum response occurs about 8 days after application, but FRF continues to decrease the longer the fruit remains on the tree (Bukovac, 1970).

C. Experiments with Pome Fruits

Ethephon is highly effective in aiding fruit harvest of apples and pears without causing excessive defoliation (Edgerton, 1971a; Edgerton and Greenhalgh, 1969; Edgerton and Blanpied, 1968). The cultivar interaction with ethephon response, however, must be considered. Late maturing, firm fleshed apple cultivars, such as Granny Smith (Edgerton and Greenhalgh, 1969) and Rome Beauty (Edgerton and Hatch, 1969), appear to soften less in response to ethephon action than the early maturing cultivars. Loosening of the fruit is accomplished with concentrations of ethephon ranging from 250 to 1000 ppm, but the decision on what concentration to use must be based on field experience with specific cultivars and harvesting systems (Proebsting, 1972).

When radioactive ethephon was applied to fruit surfaces, more of it was absorbed through the rind by Cortland apples than by cherries (Edgerton and Hatch, 1972). This is because of the deep cavity around the stem of the fruit, which accumulates and holds appreciable quantities of the applied liquid.

D. Experiments with Olives

In California, olives are largely grown for table use and are harvested in October, when the green fruit show a slight pink coloration (Hartmann and Opitz, 1966). At this time, the fruits cling tightly to the fruit stems and are difficult to harvest mechanically.

During the table olive harvest season, ethephon loosens the fruit only when applied at concentrations of 2000 ppm or higher, but such treatments cause 20% or more of the leaves to drop (Hartmann *et al.*, 1970). The leaf drop is associated with high levels of ethylene production by the leaves.

Olives are grown in Italy primarily for oil and are generally not harvested until after the fruit ripens during the winter. At this time, the fruit is easily loosened at concentrations of 2500 and 3000 ppm, with a concomitant 20% leaf drop (Jacoboni *et al.*, 1971). Using a 2000 ppm treatment on trees of the Manzanilla cultivar, Tombesi (1970) reduced the FRF of fruit 50% and caused a 20% leaf drop and a 9-fold increase in ethylene production by leaves, as compared to that of fruit. At Palermo, Sicily, Calabrese and Sottile (1971b), employing a lower concentration (500 ppm of ethephon), obtained 30% leaf drop and 58% reduction in FRF of the Moraiola olive.

IV. Effect of Cycloheximide (CHI) on Fruit Abscission

A. The Ethylene Pathway to Abscission by CHI Treatment

CHI is an antibiotic produced by *Streptomyces griseus* (Whiffen *et al.*, 1946). It is widely known as an inhibitor of protein synthesis. Abeles and

Holm (1967) showed that in the bean petiole explant, CHI binds with DNA, blocking the transcription process and preventing abscission, but it has some side effects that result in ambiguous interpretation. The side effect that concerns abscission is that CHI increases ethylene production by the tissue. This freely diffusible ethylene induced by CHI may mask any effect of the inhibitor if the site of application is removed from the site of action (Abeles and Holm, 1967; Abeles, 1968).

Application of the CHI to the surface of Valencia oranges placed the inhibitor several centimeters distant from the abscission zone, and it failed to block abscission; instead, it accelerated abscission (Cooper *et al.*, 1969a). The combined effect of the CHI diffusion barrier and accelerated production of ethylene (which is freely diffusible) explains the ambiguity of the results. However, when CHI is injected into the fruit next to the abscission zone, it blocks abscission, even in the presence of ethylene (Cooper *et al.*, 1969a).

The rise in ethylene levels occurring just prior to loosening of the fruit suggests that ethylene may be the causal agent of the principal factor in CHI-induced fruit abscission. In order to obtain additional information relating to CHI action on abscission, we exposed CHI-treated trees to hypobar pressures to eliminate endogenous ethylene (Cooper and Henry, 1972). Burg and Burg (1966) and Imaseki (1970) employed a similar procedure to eliminate endogenous ethylene from stored bananas and sliced sweet potatoes, respectively.

Small fruiting calamondin, Valencia orange, and Persian lime (*C. aurantifolia* Swing.) trees were sprayed with 20 ppm CHI and incubated in the greenhouse at one-fifth atmospheric pressure in 99.7% O_2, plus 0.3% CO_2, which provided roughly the same O_2 and CO_2 levels as incubation in air at atmospheric pressure. The fruit showed the usual CHI-rind injury to the outer layer (flavedo) of the exocarp, but there was no accumulation of ethylene under the flavedo and only a small reduction in FRF. But, when 5 ppm exogenous ethylene was added under these reduced pressure conditions, a reduction in FRF occurred at a similar rate to that observed on CHI-treated trees on the greenhouse bench at atmospheric pressure. These results indicate that the inhibition of fruit loosening observed under reduced pressure is due to a lowered level of ethylene. Thus, CHI accelerates fruit abscission only under conditions in which ethylene is allowed to accumulate in the fruit.

B. WOUNDING AND ETHYLENE PRODUCTION

The peak in ethylene production by CHI-treated citrus fruit is associated with injury to the flavedo (Cooper *et al.*, 1968a; Cooper and

Henry, 1971). The large increase in ethylene production generally occurs within 1 to 3 days after CHI treatment and is followed by a rapid decline on the third day, at which time there is visible evidence of injury (Fig. 4). High ethylene levels always occur in advance of visible evidence of flavedo injury, and CHI must damage the cells of the flavedo sufficiently that injured tissue is visible.

In one lot of 50 fruits picked from a Valencia orange tree sprayed 3 days previously with 20 ppm CHI, we found 8 fruits with no evidence of injury. These fruits contained 0.15 ppm ethylene; whereas, slightly injured fruit contained 0.75 ppm and severely injured fruits, 1.3 ppm ethylene. Control fruits contained only 0.02 ppm ethylene. Although there is a 7.5-fold enhancement of ethylene level in the nonvisibly injured fruits, it is insignificant in relation to the 0.5 ppm required for fruit abscission.

Generally, fruit will show visible evidence of injury within 2 to 5 days after the trees are sprayed with 20 ppm CHI. Certain cells of the flavedo collapse in small sunken areas surrounding a protruding oil gland. When a substantial area of the fruit is affected, it takes on a marbled appearance. If the fruit is green, the green color is retained at the periphery of the injured areas, but the areas of the flavedo surrounding the injured areas become orange-yellow. In fully colored fruit, the sunken area develops a brownish color.

W. C. Cooper and W. H. Henry (unpublished results, 1971) dipped excised fruits of various species of fruit crops in a 20 ppm solution of CHI and found that injury occurred on Persian lime, Booth avocado (*Persea gratissima* Gaertn.), Red Delicious apple, and cantaloupe (*Cucumis melo* L.). With avocados and apples, the blemishes occurred

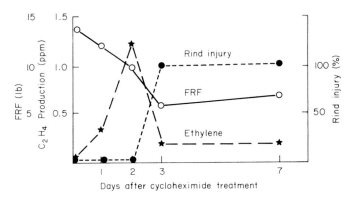

FIG. 4. Sequence of ethylene production, reduction in FRF, and rind injury to fruit of K-Early tangelo trees following spraying with 20 ppm CHI (W. C. Cooper and W. H. Henry, unpublished results, 1971).

TABLE III
EFFECT OF CHI ON ACCUMULATION OF ETHYLENE IN
EXCISED FRUIT 3 DAYS AFTER TREATMENT[a]

| Kind of fruit[b] | Maturity of fruit | C_2H_4 in fruit | | |
		Control (ppm)	CHI (ppm)	CHI injury
K-Early tangelo	Mature	0.030	1.050	Pitted
Persian lime	Yellow	0.002	0.500	Pitted
Lula avocado	Firm	0.450	27.180	Pitted
Deglet Noor date	Khalal	1.620	2.170	None
Delicious apple	Firm	4.630	45.080	Pitted
Plum	Red, firm	0.010	0.030	None
Bell pepper	Green	0.030	0.050	None
Tomato	Red	3.250	1.850	None
Cantaloupe	Yellow	151.800	207.000	Pitted

[a] From W. C. Cooper and W. H. Henry (unpublished results, 1971).

[b] Fruits were dipped in 20 ppm CHI and kept in laboratory for 3 days. K-Early tangelos and Persian limes were picked from trees at the U. S. Horticultural Station at Orlando; the Lula avocados were supplied by Harold Kendall, Coral Reef Nurseries, Goulds, Fla.; dates by John Carpenter, U. S. Date and Citrus Station, Indio, Calif.; Delicious apples by Max Williams, U. S. Horticultural Station, Wenatchee, Wash.; and plums, peppers, tomatoes, and cantaloupes (unknown cultivars) were purchased at the local market.

just beneath the exocarp in tiny isolated spots at the sites of the lenticels in the exocarp. Edgerton (1971a) also has reported that CHI causes some injury to apples, and Hartmann *et al.* (1970) reported CHI-induced injury on olives. On the other hand, fruits of date (*Phoenix dactilifera* L.) cv Deglet Noor, plum (*Prunus domestica* L.), pepper (*Cassicum annuum* var. *grossum* Sendt) cv Bell, and tomato (*Lycopersicum esculentum* L.) showed no evidence of injury. Significantly, only those fruits showing injury produced substantially more ethylene than the control fruits (Table III).

The production of wound ethylene in citrus fruits leads to a respiration climacteric similar to that of climacteric type fruit. Increases in CO_2 production, coloring of the flavedo, and fruit abscission occur (Table IV). There are also increases in ethylene and CO_2 production by the leaves.

C. PROMOTIVE EFFECT OF LEAVES

In addition to the effect of wound ethylene production by the fruit on reduction in FRF, there is also a promotive effect of leaves on reduction

TABLE IV

Effect of CHI on Ethylene Production Rate and Respiration Rate of Valencia Orange Fruit and Leaves in May[a,b]

CHI concentration (ppm)	C_2H_4 production (nl/gm/hr)		Respiratory rate (ml CO_2/kg/hr)	
	Fruit	Leaves	Fruit	Leaves
0 (Control)	0.003	0.014	7.4	7.4
5	0.016	0.051	8.0	7.8
20	0.150	0.130	12.2	18.4

[a] From W. C. Cooper and W. H. Henry (unpublished results, 1971).

[b] Measurements made 24 hr after spraying tree with CHI. Two fruits were sealed in each of 3 plastic containers (1.37 ml) and 5 leaves in each of 3 flasks (250-ml) and incubated 24 hr. Measurements made 3 days after treatment were essentially the same as those for the control after 24 hr.

in FRF. When leaves were trimmed from a branch and the leafless branch sprayed with 20 ppm CHI, fruit loosening was reduced by 50% compared with results from a leafy branch (Table V). When leaves only were sprayed with various concentrations of CHI, concentrations of 25 ppm and above caused substantial reduction in FRF of fruit on the branch (Table VI). In fact, the reduction in FRF from treated leaves (only) was comparable with that from treated fruit (only), but the maximum reduction in FRF occurred when both fruit and leaves were treated.

To explore further into the mechanism involved in the leaf effect, we incubated whole small fruiting calamondin trees in sealed bell jars after CHI treatment and measured levels of ethylene after various intervals.

TABLE V

Inhibitive Effect of Tree and Promotive Effect of Leaves on CHI-Induced Abscission of Valencia Oranges, January[a]

Branch treatment	Leaf treatment	FRF of control (lb)	FRF of 20 ppm CHI (lb)
On tree	Intact	21	5
	Trimmed off	21	11
Excised[b]	Intact	14	2
	Trimmed off	16	3

[a] Data from W. C. Cooper and W. H. Henry (unpublished results, 1971).

[b] Base of excised branches placed in water in a container in shade under the tree during the week of experiment.

TABLE VI

INFLUENCE OF CONCENTRATION AND SITE OF APPLICATION OF CHI ON
FRUIT REMOVAL FORCE (FRF) OF VALENCIA ORANGES IN MAY 1969[a]

CHI concentration (ppm)	FRF 5 days after treatment (lb)		
	Fruit only treated	Leaves only treated	Fruit + leaves treated
0	16	16	14
5	15	16	13
10	13	15	11
25	11	11	11
100	6	7	2

[a] From Cooper *et al.* (1969a).

Beginning on the first day and peaking on the second or third days, there was a high production of ethylene by trees on which the leaves only had been treated with CHI. When fruits only were treated, the ethylene levels were substantially lower than for leaves (W. C. Cooper and W. H. Henry, unpublished results, 1971). The peak level for leaves was similar to that for leaves plus fruits, but high ethylene levels were maintained longer for fruits plus leaves than for leaves only. Thus, it is apparent that leaves, when treated with CHI, produce considerable ethylene.

Leaves generally showed no evidence of visible injury, but occasionally leaf epinasty and defoliation occurred. Since leaf epinasty and abscission are responses to ethylene, it is obvious that CHI may induce ethylene production by leaves without visibly injuring them. Since it is highly unlikely that CHI breaks down chemically into ethylene, ethylene production by the leaves probably results from a stress effect on leaf metabolism by the CHI, referred to as "stress ethylene" by Abeles and Abeles (1972). It is likely stress is an invisible injury, and wound and stress-ethylene are the same.

The short diffusion path in leaves and open stomata on the abaxial side of the leaf permit ethylene to diffuse readily to the ambient air. For any substantial amount of this ethylene to be reabsorbed by the fruit from ambient air, a tent must enclose the tree.

Substantially more ethylene is produced by the leaf petiole than by the lamina (refer to Table II, Section III). Jackson and Osborne (1970) have shown that high rates of ethylene production by the petiole occur just prior to leaf abscission. However, in chemically treated plants the ethylene in petioles may well be caused, in part, by accumulation of gas in intercellular spaces in the restricted passageway of the petiole.

Zimmerman *et al.* (1931) showed that ethylene moved in all directions through tissues of tomato plants. The fact that the stems connecting the leaves to the fruit contain ethylene (refer to Table II, Section III,A) lends some credence to this view. Also, Cooper *et al.* (1969a) showed that there was an actual increase in ethylene content of untreated fruit adjacent to CHI-treated leaves. Although these observations suggest export of ethylene through the stem from leaf to fruit, they do not unequivocally prove it. There is also the possibility that other abscission-accelerating and retarding factors, as well as ethylene, may be exported to the fruit from the leaves (see Section V).

D. Penetration of CHI into the Exocarp

CHI penetrates the exocarp, but, once inside, its movement is more or less limited to exocarp tissue. Work with radioactive CHI showed that after 7 days the ^{14}C distribution stabilized with about 58% of the radio-carbon retained on the fruit surface, about 30% in the exocarp, and less than 2% in the endocarp (Fisher, 1971). Over 60% of the final exocarp uptake occurred within 48 hr (Table VII).

The first 4 to 8 hr after treating fruit with CHI are the most important for chemical uptake (L. G. Albrigo, unpublished data, 1971). The FRF of Valencia oranges in "on-the-tree" and "detached-fruit" tests was significantly reduced 5 days after treatment, even when the spray was washed off with water after 4 hr (Fig. 5). Some additional activity did occur from longer surface contact before washing, but major activity occurred during the first 4 hr. Increasing the early rate of uptake would apparently be most helpful.

The cuticle of the fruit, which lines the entire fruit surface and the

TABLE VII

DISTRIBUTION OF [^{14}C]CYCLOHEXIMIDE (RECOVERED) IN VALENCIA ORANGES[a]

Days after application	Water wash of peel surface (%)	Exocarp		Endocarp		Activity accounted for (%)
		Flavedo (%)	Albedo (%)	Rag (%)	Juice (%)	
1	83.3	4.5	2.2	1.4	0.2	91.7
2	72.0	12.7	5.0	1.4	0.7	92.0
3	59.3	17.7	9.9	0.9	0.7	88.6
5	58.0	19.1	10.6	0.7	0.8	89.4
7	58.2	19.7	10.8	0.9	0.7	90.5

[a] Data from Fisher (1971).

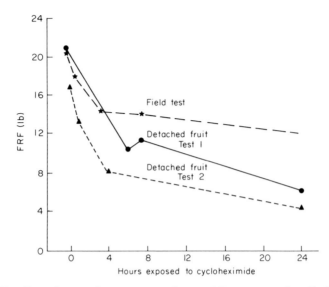

FIG. 5. Effect of time of exposure to 15 ppm CHI spray residue (before removal by washing surface) on the FRF of Valencia oranges 4 days after spraying (L. G. Albrigo, unpublished data, 1971).

substomatal cavities as a continuous membrane, is the primary barrier to the entry of CHI (Norris and Bukovac, 1968). This cuticle membrane consists of a hydrophilic cellulase matrix embedded with lipophilic wax platelets (Jansen, 1964). The wax on mature citrus fruit occurs in platelets (Albrigo and Brown, 1971; Albrigo, 1972a).

The wax on small Valencia orange fruitlets is soft and occurs as a continuous film with little surface structure (Albrigo, 1972b). CHI readily penetrates the young fruit surface, greatly enhances production of ethylene, and may cause excessive fruitlet drop.

The percentage of soft wax declines steadily during spring, with a general increase in hard wax occurring as a smooth continuous film that is quite resistant to penetration by CHI (Albrigo, 1972b). But in the fall when the fruit nears maturity, there is a progressive formation of overlapping wax platelet structures and a breaking of the continuous wax film (Albrigo, 1972b). The cuticle of Florida oranges generally develops the discontinuous overlapping waxed platelets much earlier than the cuticle on the same cultivars grown in California (L. G. Albrigo, unpublished data, 1971). Heavy rains, wind, and other climatic and cultural variables cause continual weathering of the outer layer of overlapping wax platelets and great variation in the thickness and structure of the wax in various parts of the fruit.

The large amount of diurnal shrinkage and swelling of fruits (Koz-lowski, 1971) probably contributes to the breaking up of the hard wax into platelets. Leaves generally have a continuous smooth hard wax layer that is more resistant than fruit to penetration of solutions and may account for the tendency of leaves to show less visible evidence of CHI toxicity.

Keeping CHI in solution greatly augments its rapid absorption. When 10 ppm CHI was applied as a mist for 1 hr and the fruit washed thoroughly with water, ethylene production 1 and 2 days later was as great as when the fruits were sprayed with 10 ppm CHI and the material left to dry on the fruit (Cooper, 1971).

A high relative humidity was found to favor the rapid decline in surface-extractable CHI on Orlando tangelos (Petzold and Chapman, 1970). This was reflected in increased ethylene production and a lower FRF than in fruit on trees held at a low relative humidity (Rasmussen and Cooper, 1969b). These results suggest that most of the material that enters probably penetrates the flavedo when in solution on the fruit surface.

On of the primary influences of a surfactant appears to be promotion of hydration of the cuticle under adverse humidity conditions. Water from a surfactant droplet is also more readily available to the cuticle as a result of reduced surface tension. In aqueous systems, most surfactants cause a very rapid drop in surface tension at extremely low concentrations. Within the range of 0.01 to 0.1%, a rapid fall in surface tension begins to level off, and little subsequent change is experienced with additional increases in concentration. However, surfactant enhancement of CHI fruit loosening activity on Valencia oranges takes place at higher concentrations (0.5%) than are needed for wetting (Table VIII). At these concentrations, the surfactants probably exist as strong colloidal solutions, which penetrate the cuticle causing even greater swelling of the hydrophilic channels (Jansen *et al.*, 1971). This may result in more young fruit abscission, as well as better mature fruit abscission.

In view of these results, it appears best to use the generally recommended 0.1% concentration of surfactant for citrus fruit. All six surfactants tested (Table VIII) at 0.1% concentration provided improved mature fruit abscission when used with 20 ppm CHI, as compared with no surfactant. In this test the best performance was obtained with Triton X-100 and Tween-20. In trials by L. G. Albrigo (unpublished data, 1971), X-77 surfactant gave superior results to Tween-20. In extensive field trials with citrus, Buttram (1970) obtained good results with Adsee-120. Hartmann *et al.* (1970) obtained good results on olives with X-77, and Edgerton (1971b) reported excellent results with Regulaid

TABLE VIII

EFFECT OF VARIOUS SURFACTANTS AT TWO CONCENTRATIONS ON
ABSCISSION PROPERTIES OF CHI WITH VALENCIA ORANGES
ON ROUGH LEMON ROOTSTOCK, MAY 1972[a]

Surfactant added to 20 ppm CHI	C_2H_4 under rind (ppm)	FRF (lb)	Degree of rind injury	Young leaf drop (%)	Old leaf drop (%)
No surfactant control	2.085	9	Slight	3	6
Triton X-100					
(0.1%)	3.170	4	Slight	7	5
(0.5%)	3.050	3	Severe	30	20
Tween-20					
(0.1%)	2.130	4	Slight	10	4
(0.5%)	2.690	3	Severe	20	12
Regulaid					
(0.1%)	1.510	8	Medium	2	4
(0.5%)	2.550	4	Severe	6	26
Adsee-120					
(0.1%)	1.125	7	Medium	2	5
(0.5%)	2.220	4	Severe	6	15
Carbowax-1500					
(0.1%)	1.480	6	Medium	4	1
(0.5%)	3.600	6	Severe	4	10
Vatsol O.T.					
(0.1%)	0.945	7	Slight	0	1
(0.5%)	1.030	7	Medium	0	1
Untreated control	0.010	19	None	0	1

[a] From W. C. Cooper and W. H. Henry (unpublished results, 1972).

(a mixture of polyoxyethane, propanol, and alkyl 2-ethoxyethanol) on apples.

When 1% solutions of a wax fraction from oil refining are applied to citrus fruit, it usually forms a film on the fruit which requires about 1 week to harden. One day after application, the film has a soft pliable structure readily penetrable by CHI. In fact, the film tends to promote uniform distribution of the chemical over the fruit surface, resulting in a fine marbled pitting in all parts of the fruit, rather than in localized areas of severe pitting where droplets of the solution accumulated. The soft structure of the film appeared to slow rapid evaporation of the CHI solution, and thus provided better penetration of the chemical. At the same time, the uniformly thick and continuous waxy film also appeared to slow the escape of ethylene and permitted accumulation of higher concentrations of the gas and enhanced fruit loosening, with no apparent deleterious effects on young fruits and leaves.

E. Retightening of Loosened Fruit

Generally, as long as the fruit contains 0.5 ppm or more ethylene, fruit loosening will continue but within a week after a decline in ethylene content of the fruit, retightening usually commences (Cooper and Henry, 1971). When the average FRF values at the end of fruit exposure to ethylene (5 to 7 days) are above 5 lb, retightening generally occurs. However, when the FRF is lowered to below 5 lb, fruit abscission usually takes place.

In examining microscopically the abscission zones of both CHI-treated and ethephon-treated fruit, we could find very little evidence of cell separation in the fruit with a FRF of 7 lb or higher. However, with fruit at an FRF of 5 lb or lower, some cell separation had usually occurred, and the fruit was easily separated by mechanical shakers.

Abeles *et al.* (1971a) interprets the retightening phenomenon to indicate that cell separation in the abscission zone is reversible and that enzyme systems capable of restructuring hydrolyzed walls must exist in the absicssion zone. Since up to 40 days are involved in the retightening process, there is ample time for such a restructuring of hydrolyzed walls. However, in instances where we have observed retightening of fruit, there is little evidence of visible cell wall separation prior to retightening. It may possibly occur where cell wall separation has taken place, but generally most of these fruits are heavy enough and exposed to enough wind to effect fruit fall. Where no wall separation has occurred, a plausible explanation for retightening lies in the known effect of ethylene on cell expansion (Burg *et al.*, 1971). Several workers have observed cell wall stretching of isolated cells in the abscission zone (Wilson, 1966; McCown, 1943; Barnell, 1939). The cell walls of an orange abscission zone are obviously stretched when a straight pull of 10 lb is made by the pull tester. When the limit of cell wall stretchability is reached, the cells in the abscission zone are torn asunder and the fruit breaks loose from the stem. When ethylene causes cells to expand, it alters the orientation of the newly deposited microfibrils, thus removing restraint on radial expansion (Apelbaum and Burg, 1971).

It may be that the stretching of cell walls results both from reorientation of microfibrils and by enzymatic dissolution of the cell walls. Both processes are triggered by ethylene. The retightening, occurring in the absence of high levels of ethylene, may be caused by a slow increase in juvenility factors moving to the fruit from the tree. When the phloem transport system is eliminated, as in fruit explants or in girdled branches, only those juvenility factors existing in the fruit are available for retightening. Since very little of these exist in the fruit, no retightening occurs.

TABLE IX

CHI AND CYCLAMIC ACID AS AN AID TO MACHINE HARVESTING
OF ORANGES IN FLORIDA[a]

Cultivar	Chemical and concentrate used	FRF (lb)	Fruit drop before harvest (%)	Fruit harvested by machine[b] (%)
Pineapple	Control	16	1	83
	CHI			
	(1 ppm)	8	2	86
	(5 ppm)	5	10	93
	(10 ppm)	4	15	96
Valencia	Control	15	0	78
	Cyclamic acid (1%)	10	0	79
	CHI (25 ppm)	7	5	96

[a] From Cooper *et al.* (1969c, 1969d).

[b] Used limb shaker for Pineapple oranges harvested in February and air shaker for Valencia oranges harvested in June.

The loosening process in these fruit explants is irreversible, even for those not treated with abscission chemicals.

As described in Section I, it seemed self-evident that any degree of chemical loosening of the fruit would aid mechanical harvesting. The data in Table IX and a large body of additional fruit harvesting experiments indicate that loosening the fruit to FRF values even as low as 8 lb does not greatly facilitate harvesting. A substantial reduction in FRF alone is not enough; fruit removal requires the initiation of cell wall separation. It appears that the loosening which occurs prior to development of the separation layer may be the effect of ethylene on cell wall stretching. The best indicator of the readiness for harvesting is when some of the treated fruits have dropped, or when the fruits will fall when a branch is shaken by hand.

F. CHI AS AN AID TO MECHANICAL HARVESTING

With an EPA label permitting sale of oranges from experimental applications of CHI since 1970, a good deal of experience in field use of CHI as an aid to mechanical harvesting has been gained. In field plots of an acre or more in size, Buttram (1970) demonstrated that early and mid-season orange cultivars in Florida can be readily harvested by machines following an application of 10 to 20 ppm CHI to the trees 5 days prior to harvest. In these tests, the CHI treatment reduced the FRF to near

4 lb. Pineapple orange trees (25-ft high), sprayed with 10 gal per tree of 10 ppm CHI, were harvested with an air shaker at the rate of 25 trees per hr, removing 98% of the crop. The chemical cost (10 ppm) on trees producing 500 boxes per acre is 4¢ per box (Wardowski and Wilson, 1971). The extremely low effective concentration makes CHI a highly competitive chemical to use on early and midseason orange cultivars in Florida.

The cell separation stage is more easily achieved by CHI treatment with these early and midseason orange cultivars which are harvested during the winter than it is with the Valencia orange cultivar, which is harvested in the late spring when the trees are carrying young fruit of the next season's crop.

During several weeks after petal fall, the mature Valencia orange fruit, now 12 months old and fully colored, are readily loosened by a 20 ppm CHI treatment. However, the small fruitlets contain a large amount of endogenous ethylene, and, when additional ethylene is imposed on these fruitlets by CHI treatment, fruitlet drop may be excessive (Cooper and Henry, 1970).

As the young fruitlets increase in size during late April and early May, they produce less ethylene and fruitlet drop declines. When these fruitlets reach golf ball size in late May, they do not produce detectable amounts of ethylene, and they are not readily loosened by CHI treatment (Cooper *et al.*, 1969d). The mature fruit, on the other hand, become more juvenile and are less responsive to CHI treatment than in early April (Cooper and Henry, 1971).

During June and July, the rejuvenated mature fruit revert to a senescent condition and are readily removed by CHI treatment (Cooper *et al.*, 1969d). Thus, there is considerable variability in responsiveness of mature Valencia oranges in Florida to CHI treatment.

Valencia oranges in California show even more variability in response to CHI treatment than those in Florida (Hield *et al.*, 1968, 1971a,b). Thus, the general use of CHI on Valencia oranges is not warranted. A possible solution to the Valencia orange abscission problem may lie in manipulation of the internal levels of juvenile and aging factors in the fruit by pretreatment with the growth retardant, SADH (see Section V,D).

The bulk of the world's crop of olives is also destined for processing, and it may be feasible to use cycloheximide to aid harvesting of ripe oil-type olives (Calabrese and Sottile, 1971a). When limbs treated with 50 ppm CHI were shaken by hand, 88% of the fruit was harvested as compared to 24% for control trees. Leaf drop, observed during 2½ months following treatment, was similar for treated and untreated trees.

V. Effect of Growth Regulators on Fruit Abscission

Our current understanding of growth regulator control of abscission is largely based on studies with explant abscission zones of leaves. In comparing the findings on growth regulator control of fruit abscission with that of leaf petiole explants, one should keep in mind certain differences as well as similarities in the anatomy of leaf and fruit abscission zones and in the accumulation of ethylene in these tissues. With leaves, cell division in a well-defined separation layer generally precedes cell separation. With fruits, no cell division occurs, and there is generally no well-defined separation layer. In the abscission of most fruits, cell separation occurs in an irregular path across a zone of parenchyma cells localized at a point where the fruit and receptacle tissues are contiguous (Barnell, 1939; McCown, 1943; Schneider, 1968; Stosser et al., 1969; Wittenbach and Buckovac, 1972).

In apples, which usually abscise after ripening, increased levels of ethylene occurring in the abscission zone tissues before abscission result from ethylene diffusion from adjacent flesh of ripening fruit containing high levels of ethylene (Blanpied, 1972). In citrus fruit, which hang on the tree for months after maturation, there is little to no ethylene production in either the flesh or abscission zone of the fruit. However, enhanced levels of ethylene are produced by fruit treated with CHI, and this production results in enhanced levels of ethylene in the abscission zone and in fruit abscission. Thus, in mature fruit as in leaves (Jackson and Osborne, 1970), hivh levels of ethylene occurring in abscission zone tissue are associated with abscission. However, in leaves, large quantities of ethylene are synthesized in the abscission zone tissues per se, just prior to abscission (Jackson and Osborne, 1970). The evidence with mature fruit suggests that the high levels of ethylene occurring in the abscission zone may result primarily from ethylene diffusion from adjacent flesh containing high levels of ethylene.

As Horton and Osborne (1967) reported for leaves and Pollard and Biggs (1970) and Rasmussen and Jones (1971) for citrus fruit, the cell separation protein synthesized as a result of ethylene treatment is cellulase. The cellulase is localized in the abscission zone and requires a 3-hr induction period after addition of ethylene (Abeles et al., 1971a).

A simple explanation of these findings is that ethylene controls abscission by turning on synthesis of the cell wall degrading enzyme, cellulase, in cells of the abscission zone. However, the ability of ethylene to act depends on the sensitivity of the leaves and fruit to ethylene. Presumably, the lowering of the levels of diffusible auxin in the tissues, a process referred to as aging (Abeles *et al.*, 1971a) or a certain stage of senescence (Jackson and Osborne, 1970), is a prerequisite of an ethylene effect on leaf abscission. Subsequently, ethylene initiates synthesis of cellulase and other biochemical sequences essential for leaf abscission (Burg, 1969; Jackson and Osborne, 1970; Abeles *et al.*, 1971a). Abeles *et al.* (1971a) consider ethylene to be the primary leaf-aging hormone, as well as the hormone that initiates cellulase synthesis. This explanation may be too simple for the fruit abscission process.

A. Effect of Auxin

Any plant tissue when exposed to exogenous auxin emits ethylene (Zimmerman and Hitchcock, 1939). McIntosh apples treated with 2,4,5-TP (an auxin) are no exception (Table X). Enhanced production by treated fruit was pronounced for immature fruit harvested on August 25 and September 8, but diminished with advancing maturity when climacteric-induced ethylene reached its peak. Concomitantly with the 2,4,5-TP hastening of ethylene production in immature fruit, there was a

TABLE X

Effect of Harvest Date and 2,4,5-T (20 ppm) and Ethephon (500 ppm) Treatments on Respiration, Ethylene Production, and Fruit Firmness of McIntosh Apples[a]

Harvest date	Treatment[b]	Respiration rate (ml CO_2/ kg/hr)	C_2H_4 (nl/gm/hr)	Fruit firmness (kg)	Fruit abscission (%)
Aug. 25	Control	7.3	14.3	7.48	—
	2,4,5-T	8.9	73.8	5.67	0
	Ethephon	11.4	135.6	5.62	100
Sept. 8	Control	8.4	78.9	5.56	—
	2,4,5-T	9.2	101.7	4.40	0
	Ethephon	11.0	108.1	4.84	100
Sept. 22	Control	9.2	106.3	4.35	—
	2,4,5-T	10.0	89.1	4.43	0
	Ethephon	12.9	126.7	3.94	100

[a] From Looney (1971).
[b] Trees were sprayed with various chemicals 2 weeks prior to each harvest date. Respiration rate recorded 24 hr after harvest, rate of ethylene production recorded after 1 week in respiration chambers. Firmness was evaluated after 1 week in respiration chambers by using a Ballauf pressure tester with an 11-mm tip.

hastening of the respiration climacteric, fruit softening, and coloring, but not of fruit abscission (Looney, 1971).

In the above experiment, the action of auxin consisted of inhibiting abscission, while acceleration of the respiration climacteric softening of the fruit and enhanced coloring (all manifestations of senescence) are mediated by a side effect of auxin in which the effecter molecule, ethylene gas, is formed. These results indicate that senescence in the apple fruit is not necessarily locked together with abscission, since auxin inhibited abscission while having no direct effect on senescence (the side effect was due to auxin-induced ethylene).

When NAA and 2,4-D are applied to citrus trees, mature fruit abscission is delayed and there is no effect on fruit maturity (Gardner *et al.*, 1950; Cooper and Henry, 1968b). In these instances, there is little to no auxin-induced ethylene production by the fruit (but there is some by the leaves).

Preharvest and postharvest treatment of lemons, oranges, and grapefruit have prolonged the storage life of these fruits, but the response primarily involved a delay in abscission of buttons which prevented entry of fungi into the fruit (Stewart, 1949; Erickson and Haas, 1956). There was some retardation of aging and coloring of the fruits (Erickson and Haas, 1956), but this was minor compared to that produced by GA (Coggins and Hield, 1968).

In most of the bean (*Phaseolus vulgaris* L.) petiole explant studies, auxin was found to retard the development of senescence in the pulvinal cells on the distal side of the abscission zone. This and other considerations prompted various workers to conclude that auxin retarded abscission in leaves through regulating senescence (Horton and Osborne, 1967; Abeles, 1968). This may possibly be the case in leaves, but auxin does not appear to be a major factor in retarding senescence of fruit, especially citrus fruit. As we show in Section V,B GA appears to retard senescence and not abscission in citrus fruits, while auxin retards abscission and not senescence. The observations by Horton and Osborne (1967) and Abeles *et al.* (1971a) that auxin prevents synthesis of cellulase seem to be in line with the known action of auxin as an abscission inhibitor.

The main metabolites of naturally occurring auxin and applied IAA found in apples are indoleacetylaspartic acid and indoleacetyl-D-glucose (Zenk, 1962). Also, the amino acid precursor of auxin, tryptophan, is conjugated in apples as N-malonyltryptophan (von Raussendorf-Bargen, 1962). Thus, only low levels of metabolically active auxin are expected to exist in apples at harvest time.

The 2,4,5-TP-induced ethylene in apple (as shown in Table IX) may probably exercise some feedback control over auxin transport in the tree. When radioactive NAA was applied to apple leaves, 75% was inactivated by ultraviolet light on the leaf surface, 15% was bound in the vascular (probably phloem) transport system, and only 10% was transported (acropetally) to the fruit (Luckwill and Lloyd-Jones, 1962). This acropetal movement of NAA, though relatively small, is highly significant in control of preharvest drop of apples. Possibly the bound NAA in the phloem occurs as naphthaleneacetylaspartic acid and NAA-induced ethylene is the conjugating agent. This hypothesis is in line with the observation of Craker *et al.* (1970) that ethylene is associated with conjugation of IAA and with the observation of Leep and Peel (1971) that rapid conjugation of considerable IAA to IAAsp occurs in the phloem.

Although young fruitlets of citrus produce substantial amounts of auxin (Gustafson, 1939; Cooper and Knowlton, 1939; Khalifah *et al.*, 1963), only traces are found in older fruit. The auxin in citrus has been identified as IAA (Goldschmidt *et al.*, 1971). Whether or not enough metabolically active auxin exists in mature citrus to account for fruit hanging on the tree long after maturity remains to be determined.

B. EFFECT OF GIBBERELLINS

As described in Section IV,F, small fruitlets of Valencia orange following petal fall are highly sensitive to abscission chemical treatment. At this time the fruitlets normally produce large amounts of ethylene and when additional ethylene is added by CHI sprays, fruitlet drop may be excessive (Cooper and Henry, 1970). This toxic effect of CHI on the young fruitlets is partly alleviated by a pretreatment spray of 25 ppm GA_3 made at petal fall. The GA_3 pretreatment resulted in decreased ethylene production, and increased growth rate of the young fruitlets (Fig. 6). When these GA_3-treated trees were sprayed with CHI 3 weeks after the GA_3 treatment, there was less fruitlet drop than on control-CHI trees.

The GA_3 pretreatment also suppresses ethylene production and abscission of CHI-treated mature fruit, but the suppression is slower developing than in the young fruitlets. In tests conducted with Valencia oranges, the suppressing effects of a GA_3 pretreatment on CHI-induced ethylene production and reduction in FRF of mature fruit were not apparent after 3 weeks but was very evident after 6 and 9 weeks (Fig. 7). Thus, the rapid effect of GA_3 on suppressing CHI-induced ethylene production of young fruitlets and slow effect on mature fruit makes it possible to use GA_3 to retard young fruitlet drop for several weeks after bloom, while not influencing the effectiveness of the CHI treatment on abscission of the mature fruits during this period.

However, the long-term effect of a GA_3 treatment on senescent mature

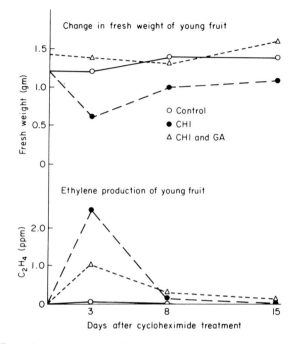

FIG. 6. Effect of pretreatment of Valencia orange trees with 25 ppm GA_3 1 week after anthesis (March 21) on ethylene production and growth rate of young immature fruit on trees treated with 20 ppm CHI on April 21.

fruits appears to be one of rejuvenation in which the fruits produce less ethylene when treated with CHI and the abscission response is reduced. These opposing effects of ethylene and GA_3 confirm earlier findings of these effects on abscission with other plants (Scott and Leopold, 1967). They also confirm the earlier findings of the effect of GA_3 on delaying and inhibiting color changes in the flavedo associated with maturation and ripening of citrus (Coggins *et al.*, 1960), tomato (Dostal and Leopold, 1967), and apricot (*Prunus armeniaca* Lam.) (Russo *et al.*, 1968).

Under natural conditions in Florida, senescent, mature Valencia oranges tend to regreen during May and become less responsive to CHI treatment. According to West and Fall (1972) shoot and root apices and young new leaves of plants may be sites for GA synthesis. In citrus GA's produced by the spring flush of new growth are probably exported to the mature fruit where they promote regreening. G. K. Rasmussen (unpublished data, 1971) has shown that GA activity of flavedo tissues of Valencia orange increases from March through May and then declines during June and July. The increase in GA activity precedes an increase in chlorophyll content of the flavedo of regreening oranges. We believe that the unresponsiveness of the mature regreened Valencia orange to

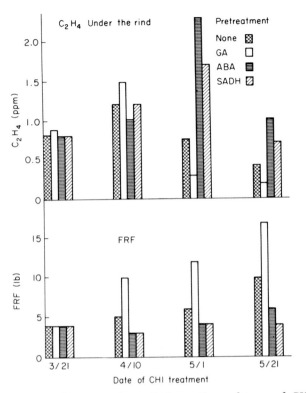

FIG. 7. Interaction of exogenously applied growth regulators and CHI (20 ppm) on ethylene production and FRF of Valencia oranges, Clermont, Fla., March 21 to May 21, 1972. GA, at 20 ppm, was applied only on March 21; ABA at 200 ppm, applied weekly beginning March 21; SADH at 4000 ppm, applied 1 week before each date of CHI treatment. All fruit with growth regulator treatment but not treated with CHI maintained a FRF of about 20 lb at all times and contained little to no ethylene.

CHI during May is caused, in part, by accumulation of endogenous GA in the fruit.

GA_3 is well known as a leaf-abscission accelerator (Bornman, 1965). This abscission effect is partially accounted for by enhanced ethylene production of GA_3-treated leaves (Abeles, 1967; Cooper *et al.*, 1968a). However, GA_3 may have abscission-accelerating activity other than that accounted for by GA_3-induced ethylene.

GA_3-treated citrus fruit do not produce ethylene, and fruit maturity and abscission are retarded as described above. Yet we have observed cell wall separation in the pith area of the abscission zone, but not in the portion of the abscission zone in the tissue adjacent to the flavedo. In GA_3-treated fruit the flavedo cells adjoining the abscission zone are generally green and appear to resist separation (Cooper and Henry, 1968b).

There is also considerable cambial activity in the peduncle of GA_3-treated fruit. A cambium layer covers the dome (button), as well as the stele of the woody peduncle (Schneider, 1968). The enhanced cambial activity is evidenced by (1) thicker peduncles, (2) bark slipping on the peduncle, and (3) production of enlarged cells in the cambium covering the button (Cooper and Henry, 1968b). At the same time, some cell separation occurred in the pith area of the abscission zone. Some of the adjacent enlarged cambial cells developed a callus which appeared to maintain contact between the peduncle and fruit. Bornman (1965) described a similar callus that appeared to maintain contact across the abscission zone of cotton petiole explants. The GA_3-treated fruit in our experiments did not abscise, and, instead, actually showed a slight increase in FRF, as compared with untreated controls. These observations suggest the primary action of GA_3 on fruit is retardation of maturity (senescence) and enhancement of cambial activity. The delay in abscission was merely a side effect resulting from retardation in senescence.

C. Effect of Abscisic Acid

The role of ABA in abscission is incompletely known. The substance was first isolated from young fruit of cotton (Ohkuma et al., 1963) and was demonstrated to be an abscission-accelerating hormone in young cotton fruit (Addicott et al., 1964; Davis and Addicott, 1972), young lupin (Lupinus sp.) fruit (Van Stevenick, 1959), and coleus (Coleus blumei Benth.) leaves (Chang, 1971). Exogenously applied ABA promotes abscission of cherries (Zucconi et al., 1969) and apples (Edgerton, 1971a), but not of citrus fruit (Cooper and Henry, 1971b). The abscission activity of ABA on apples and not on citrus fruit is associated with ABA stimulation of ethylene production in apples (Edgerton, 1971a) but not in citrus (W. C. Cooper and W. H. Henry, unpublished results, 1971).

Endogenous ABA is known to increase in debladed petioles of coleus during aging and abscission (Chang, 1971). These aging debladed petioles also produce ethylene (Cooper and Horanic, 1972). When debladed plants are held under hypobaric pressure, ethylene is eliminated and abscission is slowed but not prevented. The petioles turned yellow and abscised 3 days after abscission occurred on control plants held at atmospheric pressure. It seems likely that endogenous ABA promoted the yellowing of the petioles. Whether abscission, in this instance, was caused by ABA or by a senescence factor (Osborne et al., 1972) remains to be determined.

Endogenous ABA is known to increase during ripening of pears (Rudnicki *et al.*, 1968) and during coloring of citrus (Goldschmidt *et al.*, 1972; G. K. Rasmussen, unpublished data, 1971). Since ethylene production is also promoted in pears during ripening (Wang *et al.*, 1971), there is still some question whether ABA directly promotes ripening or whether it stimulates ethylene production, and the ethylene in turn stimulates ripening.

When ABA is taken up into citrus fruit through the cut end of stems on fruit explants, it promotes fruit abscission, but ethylene is also produced by such fruit (Cooper and Henry, 1971). By placing the explants with stems in ABA under hypobaric pressure, fruit abscission developed just as rapidly as when the fruit explants were held under atmospheric conditions (Cooper and Horanic, 1972). Thus, abscission is achieved in the absence of ethylene. However, to obtain the abscission-inducing effect, it is necessary to apply the ABA through cut stems or to inject the material into the fruit. When sprayed on the surface of citrus fruit it does not promote abscission (Cooper *et al.*, 1968b). However, when sprayed on the fruit ABA may promote senescence, making the fruit produce more ethylene when treated with CHI (see Fig. 7, Section V,B). This effect on Valencia oranges in April and May is achieved only by repeated application. El-Antably *et al.* (1967) reported this requirement for a variety of exogenous ABA effects on plant growth. However, it may be possible to achieve the ABA response by other means, as described in Section V,D.

We have indicated earlier that there is a promotive effect of leaves on fruit abscission (see Section IV,A). Some of this beneficial effect of nearby leaves may be caused by export of ABA from leaves to fruits. Mature leaves are known to inhibit growth of axillary buds, even when the trees have been sprayed with GA_3 (Cooper *et al.*, 1969b). Goren and Goldschmidt (1970) have demonstrated that mature citrus leaves contain large amounts of ABA. That high levels of ABA in mature leaves is a general phenomenon is indicated by high levels found in mature leaves of apple (Pieniazek and Rudnicki, 1967), *Acer* and *Betula* (Eagles and Wareing, 1964), and coleus (Chang, 1971). So it appears that though young leaves may retard mature fruit abscission by export of GA, mature leaves may promote mature fruit abscission by export of ABA.

D. THE ROLE OF THE GROWTH RETARDANT, SADH

The growth retardants, Amo-1618, CCC, and SADH inhibit GA biosynthesis in most plants, preventing formation of kaurene (the precursor of GA) from melvalonate (Dennis *et al.*, 1965). In some plants CCC interferes with the ability of the plant to respond to GA, while effects of CCC on GA biosynthesis are of little relevance (Reid and Crozier, 1970). SADH treatment interferes with GA metabolism and enhances senescence in peach (Byers and Emerson, 1969; Byers *et al.*, 1972), cherry (Unrath *et al.*, 1969), and citrus (W. C. Cooper and W. H. Henry, unpublished results, 1971). In apple, SADH does the opposite and retards senescence (Looney, 1968). In some instances, growth retardants are known to cause production of an altered complex of GA's, instead of preventing GA biosynthesis (Reid and Carr, 1967). If such abnormal GA's were produced in apple by application of SADH, they could conceivably retard senescence.

The senescence-promoting effects of SADH on peaches led Gambrell *et al.* (1972) to use SADH as an aid in mechanical harvesting of peaches (*Prunus persica* L.). SADH at 1000 to 2000 ppm, applied at the pit-hardening stage, advanced the maturity date by 1 week, caused fruit to abscise more readily and left less fruit remaining on trees that were harvested mechanically.

Consistent with the report that SADH treatment advances the maturity of cherries (Unrath *et al.*, 1969), 2000 ppm SADH, applied 2 weeks after bloom, advanced development of red color of Montmorency cherry (Looney and McMechan, 1970). However, the treatment did not reduce FRF enough to aid machine harvesting. When 500 ppm ethephon was applied to trees previously treated with 2000 ppm SADH, adequate fruit removal was achieved.

SADH, at 4000 ppm applied at 3 weekly intervals (beginning during bloom) to mature colored Valencia oranges, slowed regreening during April and May (Cooper and Henry, 1972b). The SADH treatments alone had no effect on ethylene production by the fruit, but SADH-treated fruits, like ABA-treated fruits, were very sensitive to CHI treatment. When 20 ppm CHI was applied 7 days after a SADH treatment, ethylene production by the fruit was greatly enhanced over that produced by fruit treated with CHI only. Also, the FRF of SADH + CHI-treated fruit was lowered to 4 lb on May 22, as compared to only 10 lb for fruit treated with CHI only (see Fig. 7, Section V,B).

Thus, SADH has the same effect as ABA in making the fruits produce more ethylene when treated with CHI. The effects of either SADH or ABA can be partially overcome by treatment with a simultaneous GA_3 treatment. Therefore, both ABA and SADH appear to accelerate senescence of citrus through antagonizing GA responses. Because SADH is already being used extensively on other fruit crops, it may be feasible to use it on Valencia oranges.

Because of auxin-induced ethylene production, auxins induce an early ripening climacteric in pome fruits. Therefore, the use of auxin chemicals as preharvest "stop-drop" sprays may be hazardous to pome fruit growers if conditions develop that prevent harvesting the crop on time. The combined use of SADH with auxin and ethephon now appears to be a promising solution to the preharvest stop-drop problem with apples (Edgerton and Blanpied, 1968; Looney, 1971). Further experimentation with SADH on apples by Edgerton and Blanpied (1970), Looney (1971), and Rhodes *et al.* (1969) has shown that SADH delays the onset of the respiration climacteric by several weeks with the length of delay depending on the SADH concentration used and age of the fruit. Further, SADH reduces respiration of the fruit at the climacteric peak by about 20%. The onset of the climacteric in ethylene production is similarly retarded by about 10 days, and, in early harvests, the climacteric rate of ethylene production is less than 20% of the controls.

Treatment with SADH greatly suppressed the effects of 2,4,5-TP and ethephon and all combinations of these treatments on apples. However, addition of ethephon overcame the abscission delay, hastened red color development, and slightly reduced the increased firmness imparted by SADH (Edgerton and Blanpied, 1970). The further addition of NAA counteracted the abscission stimulation of ethephon. Thus, SADH-treated fruits reached maturity earlier under the influence of ethephon, and with the addition of NAA, a satisfactory fruit removal force was retained until harvest.

VI. Side Effects of Abscission-Inducing Chemicals on Fruit Trees

A. DEFOLIATION

Since the use of ethephon or cycloheximide accelerates fruit abscission through stimulation of ethylene production, the leaves, as well as the fruit, may be affected (Cooper *et al.*, 1968a). Just prior to and at the time of the spring flush of new growth, old citrus leaves are extremely sensitive to ethylene, and heavy defoliation (10 to 50%) may result from the use of abscission chemicals. Since many of these old leaves will fall normally during March and April, chemically induced premature defoliation at this season may not be as harmful as it may appear. Yet, in some instances a decrease in fruit set of the next year's crop has been observed.

Since Hamlin and Pineapple orange fruit harvested during February and March are also quite sensitive to ethylene, it is possible to achieve good fruit abscission with the use of 5 and 10 ppm CHI, as compared to 20 ppm that is normally used during December and January. Defoliation at these low levels of CHI is greatly reduced or entirely absent.

Because ethephon-treated trees normally produce 100 to 200 times as much ethylene in the leaves as in the fruit, defoliation is likely to occur when this chemical is used on citrus and olive trees. Abscission chemicals are more effective on highly turgid leaves than on wilted leaves. Prior application of antitranspirant materials greatly increases chemically induced abscission (Hartmann *et al.*, 1972).

At the time of fruit harvest, leaves of citrus and olives are generally more sensitive to abscission chemicals than leaves of cherry and apples. However, cultivar interaction with ethylene response within each species must be considered (Cooper and Henry, 1970; Proebsting, 1972). Also, climatic conditions prior to and during fruit harvesting influence the ethylene response of the tree. The leaves on citrus trees, exposed to a frost and not showing local freeze lesions, are likely to be producing very high levels of ethylene (Cooper *et al.*, 1969e). When additional ethylene production is imposed on these leaves by application of abscission chemicals, severe defoliation may result (Cooper and Henry, 1970). Therefore, the decision of what concentration of the abscission chemical to use must be based on field experience with specific cultivars and climatic conditions. Even though leaves of cherry and apple are shown to tolerate ethephon concentrations above 1000 ppm, usually much lower concentrations are recommended (Proebsting, 1972).

B. Twig Dieback and Gumming

Bukovac *et al.* (1971) reported that high concentrations (above 2000 ppm) of ethephon may cause twig dieback, gumming, and enlarged lenticels of current season's growth of cherries. Enlarged lenticels also occur on olives as a result of either ethephon or CHI treatment (Hartmann *et al.*, 1970). Gummosis was observed around lateral ends and pruning wounds of cherry trees when high concentrations were used, but when 500 to 1000 ppm ethephon was used, the injury was minor (Bukovac *et al.*, 1971).

Similar twig dieback and gumming have been observed in citrus trees when either 500 ppm ethephon or 20 ppm CHI was applied to trees near the time of the spring growth flush (Cooper and Henry, 1970). These observations have led to the commercial practice of using a lower concentration at this season of the year.

Various tissues of citrus trees are capable of producing ethylene (see

Table II, Section III), and CHI sprays enhance these high rates of ethylene production. Stem tissue of Robinson tangerine is highly susceptible to ethylene-induced gummosis and dieback. This cultivar is also highly susceptible to a naturally occurring dieback and gummosis disorder associated with the presence of the fungus *Diplodia natalensis* Pole-Evans. Feder and Hutchins (1966) produced dieback in healthy Robinson trees by inoculation with *Diplodia* cultures. We have cultured *Diplodia* on stem pieces of Robinson tangerine and find that the infected stem piece produces both ethylene and gum (W. C. Cooper and W. H. Henry, unpublished results, 1971). Since ethylene is known to trigger the enzymatic degradation of the middle lamella in the cell walls of the abscission zone of citrus fruit (Pollard and Biggs, 1970; Rasmussen and Jones, 1971) and causes the production of a cell wall maceration factor in the albedo tissue of *Citrus grandis* (Rogers and Hurley, 1971), it seems likely it may play a role in the dieback of Robinson and tangerine stems.

Although degradation of the plant cell wall is of central importance in these phenomena, little is known about the reactions requisite to cell wall degradation. It seems unlikely that a wall-degradative enzyme is produced by the ethylene-treated plant, unless it was the ethylene produced by *Diplodia natalensis* that caused the reaction. However, attractive as this hypothesis may be, the pathogenesis of Robinson tangerine dieback is probably involved in a complex interaction of pathogen, hydrolytic enzymes, and ethylene.

C. Enhanced Fruit Growth

In fruits such as the fig and peach, which typically have a double sigmoid growth pattern, ethephon applied during the period of static growth triggers a rapid growth rate which normally does not occur until much later (Maxie and Crane, 1968; Byers *et al.*, 1969). Bukovac *et al.* (1971) observed enhancement in fruit enlargement of cherries when ethephon was applied 3 weeks before harvest.

A similar increase of fruit growth was obtained with both ethephon- and CHI-treated Valencia oranges (Cooper and Henry, 1972). Such enhancement of the rate of fruit growth was achieved both with rapidly growing green fruit during the summer and slow growing mature fruit during the winter. The largest effect was obtained with 5 and 10 ppm CHI treatments. These treatments enhanced ethylene production but did not reduce the FRF below 8 lb, and, consequently, the fruit retightened and no drop occurred.

Not all fruits show an enhanced growth rate from ethylene treatments.

Edgerton and Greenhalgh (1969) report that ethephon decreased the growth rate of immature apples. We obtained inhibition of fruit growth of oranges when abscission chemicals were applied during the month after anthesis, when fruit growth was by cell division.

The results with radioautography have confirmed that a reduced rate of DNA synthesis is associated with, and perhaps the cause of, inhibition of cell division in the growing point of etiolated peas (*Pisum sativum* L.) (Burg *et al.*, 1971). Thus, inhibition of cell division would account for the effect of ethylene on the reduced growth rate of small fruitlets. It is also plausible that later, when ethylene enhances fruit growth, the gas causes the cells to expand isodiametrically by altering the orientation of newly deposited microfibrils; thus, removing the restraint on radial expansion (Burg *et al.*, 1971). This would be the same phenomenon that permits stretching of the cells in the abscission zone during the fruit-loosening process.

D. Enhanced Coloring of Fruit

The endocarp of many early maturing citrus varieties is sufficiently mature for harvest, but the flavedo still carries enough chlorophyll to give the appearance of immaturity. When such fruits are treated with abscission chemicals, they cause the rind to degreen (Cooper *et al.*, 1968b; Young and Jahn, 1972). The degreening usually occurs concomitantly with loosening of the fruits, but degreening can be accomplished at lower concentrations of the chemicals than are needed for good fruit loosening. Thus 100 to 200 ppm ethephon provides preharvest coloring without attendant excessive defoliation (Young and Jahn, 1972).

Bukovac *et al.* (1971) reported that ethephon at 500 and 1000 ppm enhanced accumulation of anthocyanin pigments on the fruit. This beneficial effect of ethephon also occurs in apples (Looney, 1971; Edgerton, 1971a).

E. Effect on Air and Soil Pollution

If the million acres of citrus trees in Florida were sprayed with CHI, ethylene gas would be emitted to the air by the leaves and fruit. In this regard, it seems pertinent to emphasize that the gas is a naturally occurring product and is very widespread in plants. It is doubtful that a 20 ppm CHI application on a citrus tree will cause the emission of any more ethylene by the tree than if the tree were exposed to a freeze. Such a freeze may damage widespread areas of cultivated crops overnight.

CHI treatment, on the other hand, is regulated to give a specific level of response, its use is spread over a period of months, and it is limited to citrus. Perhaps ethylene emission from CHI-treated citrus is substantial in absolute terms but relative to an area-wide freeze, it is probably not very significant.

The ultimate sink for air and soil pollutants is unknown, but recent reports favor the idea that reaction with soil through microbial or chemical means can remove ethylene, as well as other hydrocarbons from the air and soil (Abeles *et al.*, 1971b).

We tested the fate of ethylene in Florida soils by incubating for 3 days various soils in flasks to which 5 ppm ethylene was added (W. C. Cooper and W. H. Henry, unpublished results, 1971). In control flasks without soil, there was no reduction in ethylene; whereas, in flasks containing nonsterilized soil taken from a healthy tree (with unidentified soil flora), there was a 58% reduction in ethylene. In flasks containing soil infested with *Phytophthora* sp., there was a 100% reduction in ethylene. With autoclaved soil, there was a 20 to 35% reduction in ethylene. We conclude from these observations that the soil may represent an important sink for ethylene and that some sort of organism functions to reduce levels of the gas.

Rapid breakdown of CHI also took place in Florida soils (pH 6.8), with degradation through three half-lives in less than 3 weeks (Petzold *et al.*, 1971). The first half-life was 3 days, while the second and third were 4 days. This fast decline was also demonstrated in a number of field tests with soil from under trees sprayed with 20 ppm CHI. Residues were usually below detectable levels (0.02 ppm) within a week after application in these tests. This rapid decline, however, was not observed in soils where soil moisture was low, but rapid decline did resume following rainfall.

VII. General Discussion and Conclusions

In addition to ethylene, ABA, GA, and auxin are shown to occur in fruits. It is quite apparent that hormonal control of development of senescence and abscission involves a complex interplay of these major plant hormones. Ethylene appears to have a synergistic effect on production of ABA, while ABA may enhance the production of ethylene by the fruits. Both of these hormones antagonize the action of GA in retarding senescence of fruits and antagonize the action of auxin in inhibiting fruit abscission.

Whereas auxin is the primary factor opposing the development of senescence in leaves, GA appears to be the primary factor opposing the development of a state of senescence in fruits in which they become sensitive to ethylene action. Although there is considerable endogenous auxin in young immature fruit, most of this appears to be conjugated (possibly by action of ABA or ethylene) before fruit maturity. Where a deficiency in endogenous auxin causes premature fruit drop, this problem is controlled by application of exogenous auxin.

In overview we interpret the complex interplay of the major plant hormones in orange trees as follows: With onset of cool winter weather callose deposits appear in the sieve tubes (Schneider, 1968) and phloem transport is slowed. ABA is produced in mature leaves and maturing fruit and accumulates in these organs, while GA levels decrease. As the fruits color, their capacity to produce ethylene increases and, when treated with CHI, they produce copious amounts of ethylene and readily abscise.

The mechanism of CHI action in accelerating fruit abscission is through injury to the flavedo, which causes it to produce ethylene. The more senescent the fruit, the more readily it is injured by CHI and the more ethylene is produced. The ethylene readily diffuses to the abscission zone and triggers cellulase synthesis. We, therefore, propose that a solution to the harvesting problem of early and midseason oranges (those that mature and are harvested during the winter) lies in making fruit generate ethylene by the use of appropriate chemicals. When CHI is used, wound ethylene is produced. Wound ethylene production leads to an imposed respiratory climacteric and fruit abscission results.

Recent investigations on biosynthesis of ethylene indicate that methionine is probably the precursor of auxin-induced ethylene (Burg and Clagett, 1967), fruit-ripening ethylene (Baur *et al.*, 1971), and CHI-induced ethylene by leaves (Abeles and Abeles, 1972). Therefore, it seems quite likely that methionine is a source of wound ethylene in citrus fruit.

Ethylene production associated with triggering the respiration climacteric (Burg and Burg, 1965) in climacteric-type fruits such as apples is dependent on new synthesis of enzyme, as well as methionine (Rhodes and Wooltorton, 1967; Frenkel *et al.*, 1968). However, the burst of ethylene production at the climacteric peak is less dependent on new enzyme synthesis (Frenkel *et al.*, 1968) and may be a product of the digestive processes occurring in the cells of the flesh of the ripening fruit, as a result of an increase in hydrolytic enzyme activity.

Electron micrographs of cells of ripening pears show that the climacteric rise in respiration of pear tissue is coincident with accumulation of electron-dense particles in the vacuoles, invaginations of the tonoplast, and extensive vacuolation (Bain and Mercer, 1964). Similar ultrastructural changes have been observed in the exocarp of senescent navel oranges (Thomson, 1972) and in the cells of senescent corollas of *Ipomoea* (Matile and Winkenbach, 1971). The vacuoles of the cells in these tissues may well serve as lysosomes (Matile and Winkenbach, 1971).

This progressive vesiculation of cytoplasm and organelles is by no means unique to aging cells. Webster (1972) observed similar cellular changes in the abscission zone of *Phaseolus*. Bain and Mercer (1964) observed it in tissues subjected to a wide range of abnormal conditions, including exposure to hypotonic solutions, freezing and thawing, fungal invasions, and from mechanical shearing. Shumway and Ryan (1972) reported the same changes in cells of tomato leaves as a response to wounding. It seems highly probable that in each of these conditions the ethylene substrate, methionine, is likely to be put in contact with lysosomal enzymes in the vacuoles and result in ethylene production. We know that freezing causes citrus leaves and fruit to produce ethylene (Cooper *et al.*, 1969e), and we show in Section VI,B that *Diplodia natalensis* infection of Robinson tangerine produces ethylene. Where the rind tissue of citrus is subject to mechanical or pathological or chemical wounding, the ethylene substrate and the lysosomal enzymes are likely to be mixed and result in ethylene production, which we know to occur.

Recent studies by Osborne *et al.* (1972) suggest that a senescence factor (SF), which is not ABA or methionine, regulates ethylene production. Such a factor could possibly exist in the vacuole and would diffuse or leak from senescent or damaged nonsenescent fruit in much the same manner as in the model described above for lysomal enzymes.

Fruit of the Valencia orange is not mature enough for optimum eating quality during the winter, even though the flavedo of the exocarp is fully colored and responds to CHI treatment by production of ethylene and abscission. Consequently, the fruit is left on the trees until April and May before harvest.

We theorize that, possibly as a result of a vernalization effect, GA or GA precursors accumulate in dormant buds during the winter and set the stage for bud break at the advent of warm weather. With the advent of the spring flush of new growth, GA from the young leaves moves to the mature fruit via the newly formed and functional sieve tubes of the phloem. This movement of GA from young leaves to mature fruit is probably more extensive than movement of ABA from mature leaves to the fruit. Also, during this period, many of the old mature leaves abscise and their ABA content is lost to the tree. Thus, during May, the GA in the mature fruit is dominant over ABA and regreening of the exocarp results. Consequently, the sensitivity of the fruit to CHI action declines.

Although the flesh of the fruit is more mature in May than March, the tissues of the exocarp including the abscission zone are less mature during May than March. The relative maturity or senescence of the exocarp appears to influence the amount of ethylene produced by CHI treatment, rather than by influencing sensitivity of the tissue to ethylene action. It is this reaction that is conditioned by the internal level of GA in relation to ABA. Auxin on the other hand directly inhibits ethylene action, and this effect is quite different from the GA/ABA effect on senescence and the capacity of the fruit to produce ethylene. GA and ABA may possibly also influence sensitivity of the tissue to ethylene action, but this is not clearly shown from the data presented.

A theoretical solution to the harvest of Valencia oranges with rejuvenated exocarp tissues during May is the application of exogenous ABA, which appears to be more effective than exogenous ethylene in inducing senescence. However, a more practical solution may lie in the use of growth retardants to lower GA levels in the exocarp, thereby restoring a higher rate of ABA to GA levels in the fruit.

References

Abeles, A. L., and Abeles, F. B. (1972). Biochemical pathway of stress-induced ethylene. *Plant Physiol.* **50**, 496–498.

Abeles, F. B. (1967). Mechanism of action of abscission accelerators. *Physiol. Plant.* **20**, 442–454.

Abeles, F. B. (1968). Role of RNA and protein synthesis in abscission. *Plant Physiol.* **43**, 1577–1586.

Abeles, F. B., and Holm, R. E. (1967). Abscission: Role of protein synthesis. *Ann. N. Y. Acad. Sci.* **144**, 367–373.

Abeles, F. B., Leather, G. R., Forrence, L. E., and Craker, L. E. (1971a). Abscission: Regulation of senescence, protein synthesis and enzyme secretion by ethylene. *HortScience* **6**, 371–376.

Abeles, F. B., Craker, L. E., Forrence, L. E., and Leather, G. R. (1971b). Fate of air pollutants: Removal of ethylene, sulfur dioxide and nitrogen dioxide by soil. *Science* **173**, 914–916.

Addicott, F. T., Carns, H. R., Lyon, J. L., Smith, O. E., and McMeans, J. L. (1964). On the physiology of abscission. *In* "Régulateurs naturels de la croissance végétale" (J. P. Nitsch, ed.), pp. 687–703. CNRS, Paris.

Albrigo, L. G. (1972a). Distribution of stomata and epicuticular wax on oranges as related to stem-end breakdown and water loss. *J. Amer. Soc. Hort. Sci.* **97**, 220–223.

Albrigo, L. G. (1972b). Appearance and persistence of pinolene anti-transpirant spray on Valencia orange leaves. *HortScience* **7**, 247–248.

Albrigo, L. G., and Brown, G. E. (1971). Orange peel topography as affected by a preharvest plastic spray. *HortScience* **5**, 470–472.

Anderson, J. L. (1969). The effect of ethrel on the ripening of Montmorency sour cherries. *HortScience* **4**, 92–93.

Apelbaum, A., and Burg, S. P. (1971). Altered cell microfilbrillar orientation in ethylene-treated *Pisum sativum* stems. *Plant Physiol.* **48**, 648–652.

Bain, J. M., and Mercer, F. V. (1964). Organization resistance and the respiration climacteric. *Aust. J. Biol. Sci.* **17**, 78–85.

Barnell, E. (1939). Studies in tropical fruits. V. Some anatomical aspects of fruit-fall in two arboreal plants. *Ann. Bot.* (*London*) **111**, 77–89.

Batjer, L. P., Thompson, A. H., and Gerhardt, F. (1948). A comparison of NAA and 2,4-D sprays for controlling preharvest drop of Bartlett pears. *Proc. Amer. Soc. Hort. Sci.* **51**, 71–76.

Baur, A. H., Yang, S. F., Pratt, H. K., and Biale, J. B. (1971). Ethylene biosynthesis in fruit tissues. *Plant Physiol.* **47**, 696–699.

Ben-Yehoshua, S., and Biggs, R. H. (1970). Effects of iron and copper ions in promotion of selecting abscission and ethylene production by citrus fruit and inactivation of indoleacetic acid. *Plant Physiol.* **45**, 604–607.

Blanpied, G. D. (1971). Apparatus for ethylene extraction from plant tissue. *HortScience* **6**, 132–134.

Blanpied, G. D. (1972). A study of ethylene in apple, red raspberry, and cherry. *Plant Physiol.* **49**, 627–630.

Bornman, C. H. (1965). Histological and histochemical effects of gibberellin and auxin in abscission. Ph.D. Thesis, University of California, Davis.

Browning, G., and Cannell, M. G. R. (1970). Use of 2-chloroethane-phosphonic acid to promote abscission and ripening of fruit of *Coffeae arabica* L. *J. Hort. Sci.* **45**, 223–232.

Bukovac, M. J. (1970). Promotion of cherry fruit abscission in relation to mechanical harvest. *Mich. State Univ. Hort. Rep.* No. 34, pp. 7–9.

Bukovac, M. J., Zuconni, F., Larsen, R. P., and Kesner, C. D. (1969). Chemical promotion of fruit abscission in cherries and plums with special reference to 2-chloroethylphosphonic acid. *J. Amer. Soc. Hort. Sci.* **94**, 226–230.

Bukovac, M. J., Zucconi, F., Wittenbach, V. A., Flore, J. A., and Inoue, H. (1971). Effects of (2-chloroethyl)phosphonic acid on the development and abscission of maturing sweet cherry fruit. *J. Amer. Soc. Hort. Sci.* **96**, 777–781.

Burg, S. P. (1958). Biogenesis of ethylene. Ph.D. Thesis, Harvard University, Cambridge, Massachusetts.

Burg, S. P. (1969). Ethylene, plant senescence and abscission. *Plant Physiol.* **43**, 1503–1511.

Burg, S. P., and Burg, E. A. (1965). Ethylene action and the ripening of fruits. *Science* **148**, 1190–1196.

Burg, S. P., and Burg, E. A. (1966). Fruit storage at subatmospheric pressure. *Science* **153**, 3733.

Burg, S. P., and Clagett, C. O. (1967). Conversion of methionine to ethylene in vegetative tissue and fruits. *Biochem. Biophy. Res. Commun.* **27**, 125–130.

Burg, S. P., Apelbaum, A., Eisinger, W., and Kang, B. G. (1971). Physiology and mode of action of ethylene. *HortScience* **6**, 359–364.

Buttram, J. R. (1970). Harvesting processing oranges with the use of cycloheximide. *Proc. Fla. State Hort. Soc.* **83**, 253–256.

Byers, R. E., and Emerson, F. H. (1969). Effect of succinamic acid, 2,2-dimethylhydrazide on peach fruit maturation and tree growth. *J. Amer. Soc. Hort. Sci.* **94**, 641–645.

Byers, R. E., Dostal, H. C., and Emerson, F. H. (1969). Regulation of fruit growth with 2-chloroethanephosphonic acid. *BioScience* **19**, 903–904.

Byers, R. E., Emerson, F. H., and Dostal, H. C. (1972). The effect of succinamic acid-2,2-dimethylhydrazide and other growth regulating chemicals on peach fruit maturation. *J. Amer. Soc. Hort. Sci.* **97**, 420–423.

Calabrese, F., and Sottile, I. (1971a). Prova di efficacia dialcuni prodotti chimica sul distacco delle drupedi olivo. *Estratto Tech. Agr.* **13**, 1–15.

Calabrese, F., and Sottile, I. (1971b). Effecto cascolanti della sulle drupe di olivo. *Estratto Tech. Agr.* **23**, 1–15.

Chang, Y. (1971). The movement of indoleacetic acid in coleus petioles as affected by abscisic acid. Ph.D. Thesis, Princeton University.

Coggins, C. W., and Hield, H. Z. (1968). *In* "The Citrus Industry" (W. Reuther, L. D. Batchelor, and H. J. Webber, eds.), Vol. II, pp. 371–389. Univ. of California Press, Berkeley.

Coggins, C. W., Hield, H. Z., and Garber, J. M. (1960). The influence of potassium gibberellate on Valencia orange trees and fruit. *Proc. Amer. Soc. Hort. Sci.* **76**, 199–207.

Cooper, W. C. (1971). Chemical control of citrus fruit abscission. *Indian Agr.* **15**, 9–17.

Cooper, W. C., and Henry, W. H. (1967). The effect of ascorbic acid on citrus fruit abscission. *Citrus Ind.* **48** (6), 5–7.

Cooper, W. C., and Henry, W. H. (1968a). Field trials with potential abscission chemicals as an aid to mechanical harvesting of citrus in Florida. *Proc. Fla. State Hort. Soc.* **81**, 62–68.

Cooper, W. C., and Henry, W. H. (1968b). Effect of growth regulators on the coloring and abscission of citrus fruit. *Isr. J. Agr. Res.* **18**, 161–174.

Cooper, W. C., and Henry, W. H. (1970). Ethylene production by the Valencia orange tree as related to the use of abscission chemicals. *Proc. Fla. State Hort. Soc.* **83**, 89–92.

Cooper, W. C., and Henry, W. H. (1971). Abscission chemicals in relation to citrus fruit harvest. *J. Agr. Food Chem.* **19**, 559–563.

Cooper, W. C., and Henry, W. H. (1972). Stimulation of fruit growth of Valencia oranges by treatment with cycloheximide. *HortScience* **7**, Sect. 2, 19 (abstr.).

Cooper, W. C., and Horanic, G. (1972). The requirement for ethylene in cycloheximide-induced citrus fruit abscission. *Plant Physiol.* **49s**, 17 (abstr.).

Cooper, W. C., and Knowlton, K. R. (1939). Distribution of auxin in subtropical fruit plants. *Amer. J. Bot.* **26**, 23s (abstr.).

Cooper, W. C., Rasmussen, G. K., Rogers, B. J., Reece, P. C., and Henry, W. H. (1968a). Control of abscission in agricultural crops and its physiological basis. *Plant Physiol.* **43**, 1560–1576.

Cooper, W. C., Rasmussen, G. K., and Smoot, J. J. (1968b). Induction of degreening of tangerines by preharvest applications of ascorbic acid, other ethylene releasing agents. *Citrus Ind.* **49** (10), 25–27.

Cooper, W. C., Rasmussen, G. K., and Hutchison, D. J. (1969a). Promotion of abscission of orange fruits by cycloheximide as related to the site of treatment. *BioScience* **19**, 443–444.

Cooper, W. C., Young, R., and Henry, W. H. (1969b). Effect of growth regulators on bud growth and dormancy in citrus as influenced by season of year and climate. *Proc. Int. Citrus Symp., 1st, 1968* Vol. 1, pp. 301–314.

Cooper, W. C., Henry, W. H., Reece, P. C., Rasmussen, G. K., and Rogers, B. J.

(1969c). Ethylene participation in natural and chemically induced senescence and abscission of citrus fruits and leaves. *Proc. Conf. Trop. Subtrop. Fruits, 1969* pp. 121–127.

Cooper, W. C., Henry, W. H., Rasmussen, G. K., and Hearn, C. J. (1969d). Cycloheximide: An effective abscission chemical for oranges in Florida. *Proc. Fla. State Hort. Soc.* **82**, 99–103.

Cooper, W. C., Rasmussen, G. K., and Waldon, E. S. (1969e). Ethylene production in freeze injured citrus trees. *J. Rio Grande Val. Hort. Soc.* **23**, 29–37.

Coppock, G. E. (1967). Harvesting early and mid-season citrus fruit with tree-shaker harvest systems. *Proc. Fla. State Hort. Soc.* **80**, 98–104.

Coppock, G. E., and Jutras, P. J. (1960). Mechanizing citrus fruit picking. *Trans. ASAE (Amer. Soc. Agr. Eng.)* **3**, 130–132.

Coppock, G. E., and Jutras, P. J. (1962). Harvesting citrus with an inertia shaker. *Proc. Fla. State Hort. Soc.* **75**, 297–301.

Craker, L. E., Chadwick, A. V., and Leather, G. R. (1970). Abscission: Movement and conjugation of auxin. *Plant Physiol.* **45**, 790–793.

Davis, L. A., and Addicott, F. T. (1972). Abscisic acid: Correlations with abscission and with development of cotton fruit. *Plant Physiol.* **49**, 644–648.

Demorest, D. M., and Stahmann, M. N. (1971). Ethylene production from peptides and protein containing methionine. *Plant Physiol.* **47**, 450–451.

Dennis, D. T., Upper, C. D., and West, C. A. (1965). An enzymatic site of inhibition of gibberellin synthesis by Amo-1618 and other growth retardants. *Plant Physiol.* **40**, 948–952.

Dostal, H. C., and Leopold, A. C. (1967). Gibberellin delays ripening of tomatoes. *Science* **185**, 1579–1580.

Eagles, C. F., and Wareing, P. F. (1964). The role of growth substances in the regulation of dormancy. *Physiol. Plant.* **17**, 697–709.

Edgerton, L. J. (1968). New materials to loosen fruit for mechanical harvesting. *Proc. N. Y. State Hort. Soc.* **113**, 99–102.

Edgerton, L. J. (1971a). Apple abscission. *HortScience* **6**, 378–382.

Edgerton, L. J. (1971b). Effect of some adjuvants on absorption and activity of 2-chloroethylphosphonic acid. *HortScience* **6**, Sect. 2, 286 (abstr.).

Edgerton, L. J., and Blanpied, G. D. (1968). Regulation of growth and fruit maturation with 2-chlorophosphonic acid. *Nature (London)* **219**, 1064–1065.

Edgerton, L. J., and Blanpied, G. D. (1970). Interaction of succinamic acid 2,2-dimethylhydrazide, 2-chloroethylphosphonic acid, and auxins on maturity, quality and abscission of apples. *J. Amer. Soc. Hort. Sci.* **95**, 664–666.

Edgerton, L. J., and Greenhalgh, W. J. (1969). Regulation of growth, flowering, and fruit abscission with 2-chloroethylphosphonic acid. *J. Amer. Soc. Hort. Sci.* **94**, 11–13.

Edgerton, L. J., and Hatch, A. H. (1969). Promoting abscission of cherries and apples for mechanical harvesting. *Proc. N. Y. State Hort. Soc.* **114**, 109–113.

Edgerton, L. J., and Hatch, A. H. (1972). Absorption and metabolism of [14]C (2-chloroethyl)phosphonic acid in apples and cherries. *J. Amer. Soc. Hort. Sci.* **97**, 112–114.

Edgerton, L. J., and Hoffman, M. B. (1951). The effectiveness of several growth regulating chemicals in delaying the harvest drop of the McIntosh apple. *Proc. Amer. Soc. Hort. Sci.* **57**, 120–124.

El-Antably, H. M. M., Wareing, P. F., and Hillman, J. (1967). Some physiological responses to D,L-abscision (Dormin). *Planta* **73**, 74–90.

El-Zeftawi, B. M. (1970). Chemical induction of mature citrus fruit abscission. *J. Aust. Inst. Agr. Sci.* 139–141.

Erickson, L. C., and Haas, A. R. C. (1956). Size, yield, and quality of fruit produced by Eureka lemon trees sprayed with 2,4-D and 2,4,5-T. *Proc. Amer. Soc. Hort. Sci.* **67,** 215–221.

Feder, W. A., and Hutchins, P. C. (1966). Twig gumming and dieback of the Robinson tangerine. *Plant Dis. Rep.* **50,** 429–430.

Fisher, J. F. (1971). Distribution of radiocarbon in Valencia orange after treatment with ^{14}C-cycloheximide. *J. Agr. Food Chem.* **19,** 1162–1164.

Frenkel, C., Klein, K., and Dilley, D. R. (1968). Protein synthesis in relation to ripening of pome fruits. *Plant Physiol.* **43,** 1146–1153.

Fridley, R. B., Hartmann, H. T., Mehlschau, J. J., Chen, P., and Whisler, J. (1971). Olive harvest mechanization in California. *Calif., Agr. Exp. Sta., Bull.* **855,** 26.

Gambrell, C. E., Sims, E. T., Stembridge, G. E., and Rhodes, W. H. (1972). Response of peaches to succinic acid-2-2-dimethylhydrazide as an aid in mechanical harvesting. *J. Amer. Soc. Hort. Sci.* **97,** 265–268.

Gardner, F. E., Marth, P. C., and Batjer, L. P. (1939). Spraying with plant growth substances for control of preharvest drop of apples. *Proc. Amer. Soc. Hort. Sci.* **37,** 415–428.

Gardner, F. E., Reece, P. C., and Horanic, G. E. (1950). The effect of 2,4-D on preharvest drop of citrus under Florida conditions. *Proc. Fla. State Hort. Soc.* **63,** 7–11.

Gellini, R. (1965). Results obtained with maleic hydrazide and ascorbic acid and report of experimentation in 1964. *Symp. Inter. Rac. Mec. Olive Res.* **1,** 3–24.

Goldschmidt, E. E., Monselise, S. P., and Goren, R. (1971). On identification of native auxins in citrus tissues. *Can. J. Bot.* **49,** 241–244.

Goldschmidt, E. E., Eilati, S. K., and Goren, R. (1972). Increase in ABA-like growth substances and decrease in gibberellin-like substances during ripening and senescence of citrus fruit. *Proc. Int. Conf. Plant Growth Substances, 7th, Canberra 1970* Vol. 7, pp. 611–617.

Goren, R., and Goldschmidt, E. E. (1970). Regulative systems in the developing citrus fruit. I. The hormonal balance in orange fruit tissues. *Physiol. Plant.* **23,** 937–947.

Griggs, W. H., Iwakiri, B. T., Fridley, R. B., and Mehlschau, J. (1970). Effect of 2-chloroethylphosphonic acid and cycloheximide on abscission and ripening of Bartlett pears. *HortScience* **5,** 264–266.

Gustafson, F. G. (1939). The cause of natural parthenocarpy. *Amer. J. Bot.* **26,** 135–138.

Hartmann, H. T., and Opitz, C. (1966). Olive production in California. *Calif., Agr. Exp. Sta., Circ.* **540,** 1–63.

Hartmann, H. T., Fadl, J., and Whisler, J. (1967). Inducing abscission of olive fruits by spraying with ascorbic acid and iodoacetic acid. *Calif. Agr.* **21,** 5–7.

Hartmann, H. T., Tombesi, A., and Whisler, J. (1970). Promotion of ethylene evolution and fruit abscission in the olive by 2-chloroethylphosphonic acid and cycloheximide. *J. Amer. Soc. Hort. Sci.* **95,** 635–639.

Hartmann, H. T., El-Hamady, M., and Whisler, J. (1972). Cycloheximide as an abscission-inducing chemical for olive fruits. *HortScience* **7,** Sect. 2, 19 (abstr.).

Heddon, S. L., and Coppock, G. E. (1965). A tree-shaker harvest system for citrus. *Proc. Fla. State Hort. Soc.* **78,** 302–306.

Hendershott, C. H. (1965). The effect of iodoacetic acid on citrus fruit abscission. *Proc. Fla. State Hort. Soc.* **78,** 36–41.

Hield, H. Z., Palmer, R. L., and Lewis, L. N. (1968). Progress report: Chemical aids to citrus fruit loosening. *Calif. Citrogr.* **53,** 386–392.

Hield, H. Z., Palmer, R. L., and Lewis, L. N. (1971a). Chemical abscission studies on California Valencias. *Calif. Citrogr.* **56,** 380–383.

Hield, H. Z., Palmer, R. L., and Lewis, L. N. (1971b). Abscission responses of Valencia oranges to January–February sprays of cycloheximide. *Calif. Citrogr.* **56,** 411, 414, and 415.

Horton, R. F., and Osborne, D. J. (1967). Senescence, abscission and cellulase activity in Phaseolus vulgaris. *Nature (London)* **214,** 1086–1088.

Imaseki, H. (1970). Induction of peroxidase activity by ethylene in sweet potato. *Plant Physiol.* **46,** 172–174.

Jackson, M. B., and Osborne, D. J. (1970). Ethylene, the natural regulator of leaf abscission. *Nature (London)* **225,** 1019–1022.

Jacoboni, N., Tombesi, A., and Cartechini, A. (1971). Ancora speranze sull'ausilio dei cascolanti nella racolta meccanica delle olive. *Inform. Agr.* **4,** 3–7.

Jansen, L. L. (1964). Surfactant enhancement of herbicide entry. *Weeds* **12,** 251–255.

Jansen, L. L., Gentner, W. A., and Shaw, W. C. (1971). Effects of surfactants on the herbicidal activity of several herbicides in aqueous spray systems. *Weeds* **9,** 381–405.

Khalifah, R. A., Lewis, L. N., and Coggins, C. W. (1963). New natural growth promoting substance in young citrus fruit. *Science* **142,** 399–400.

Kozlowski, T. T. (1971). "Growth and Development of Trees," Vol. 2. Academic Press, New York.

Leep, N. W., and Peel, A. J. (1971). Patterns of translocation and metabolism of ¹⁴C-labeled IAA in the phloem in willow. *Planta* **96,** 62–73.

Levin, J. H., Gaston, H. P., Heddon, S. L., and Whittenberger, R. T. (1960). Mechanizing the harvest of red tart cherries. *Mich., Agr. Exp. Sta., Quart. Bull.* **42,** 656–685.

Lieberman, M., and Mapson, L. W. (1964). Genesis and biogenesis of ethylene. *Nature (London)* **204,** 343–345.

Lieberman, M., Kunishi, A. T., Mapson, L. W., and Wardale, D. A. (1965). Ethylene production from methionine. *Biochem. J.* **97,** 449–459.

Looney, N. E. (1968). Inhibition of apple ripening by succinic acid 2,2-dimethylhydrazide and its reversal by ethylene. *Plant Physiol.* **43,** 1133–1137.

Looney, N. E. (1971). Interaction of ethylene, auxin, and succinic acid 2,2-dimethylhydrazide in apple fruit ripening control. *J. Amer. Soc. Hort. Sci.* **96,** 350–353.

Looney, N. E., and McMechan, A. D. (1970). The use of 2-chloroethylphosphonic acid and 2,2-dimethylhydrazide to aid mechanical shaking of sour cherries. *J. Amer. Soc. Hort. Sci.* **95,** 452–455.

Luckwill, L. C. (1953). Studies of fruit development in relation to plant hormones. 1. Hormone production by the developing apple seed in relation to fruit drop. *J. Hort. Sci.* **28,** 14–24.

Luckwill, L. C. (1957). Hormonal aspects of fruit development in higher plants. *Symp. Soc. Exp. Biol.* **11,** 63–85.

Luckwill, L. C., and Lloyd-Jones, C. P. (1962). The absorption, translocation, and metabolism of NAA applied to apple leaves. *J. Hort. Sci.* **37,** 190–206.

McCown, J. (1943). Anatomical and chemical aspects of abscission of the fruits of the apple. *Bot. Gaz. (Chicago)* **105**, 212–220.

Matile, P. H., and Winkenbach, F. (1971). Function of lysosomes and lysosomal enzymes in the senescing corolla of the morning glory (*Ipomoea purpurea*). *J. Exp. Bot.* **22**, 759–771.

Maxie, E. C., and Crane, J. C. (1968). Effect of ethylene on growth and maturation of the fig fruit. *Proc. Amer. Soc. Hort. Sci.* **92**, 255–267.

Nelson, R. C. (1939). Studies on the production of ethylene by ripening processes in apples and bananas. *Food Res.* **4**, 173–190.

Neluibow, D. (1901). Uber die horizontale nutation der stengel von *Pisum sativum* und einiger anderen. *Pflanzen. Beih. Bot. Zentralbl.* **10**, 128–139.

Norris, R. G., and Bukovac, M. J. (1968). Structure of the pear leaf cuticle with special reference to cuticular penetration. *Amer. J. Bot.* **55**, 975–983.

Ohkuma, K., Lyon, J. L., Addicott, F. T., and Smith, O. E. (1963). Abscisin II, an abscission accelerating from young cotton fruit. *Science* **142**, 1592–1593.

Osborne, D. J., Jackson, M. B., and Milborrow, B. V. (1972). Physiological properties of abscission accelerator from senescent leaves. *Nature (London) New Biol.* **240**, 98–101.

Palmer, R. L., Hield, H. Z., and Lewis, L. N. (1969). Chemical control of abscission. *Proc. Int. Citrus Symp., 1st, 1968* Vol. 3, pp. 1135–1144.

Petzold, E. N., and Chapman, D. D. (1970). "Effects of Relative Humidity and Temperature upon the Rate of Decline of Surface Extractable Cycloheximide," Rep. 120-9760-25. Upjohn Co., Kalamazoo, Michigan.

Petzold, E. N., Negg, A. W., and Chapman, D. D. (1971). "Effect of Repeated Applications of Water upon Migration and Persistence of Cycloheximide in a Treated Plot of Florida Soil," Rep. 120-9760-52. Upjohn Co., Kalamazoo, Michigan.

Pieniazek, J., and Rudnicki, R. (1967). The presence of Abscisin II in apple leaves and apple fruit juice. *Bull. Acad. Pol. Sci., Cl. V* **15**, 251–254.

Pollard, J. E., and Biggs, R. H. (1970). Role of cellulase in abscission of citrus fruits. *Proc. Amer. Soc. Hort. Sci.* **95**, 667–673.

Proebsting, E. L. (1972). Value of growth regulators as aids to mechanical harvesting of apples, cherries and grapes. *Int. Symp. Growth Regulators Fruit Production, 1972.* (In press.)

Rasmussen, G. K., and Cooper, W. C. (1969a). Abscission of citrus fruits induced by ethylene-producing chemicals. *Proc. Amer. Soc. Hort. Sci.* **93**, 191–198.

Rasmussen, G. K., and Cooper, W. C. (1969b). Influence of temperature and humidity on cycloheximide-induced abscission and ethylene content of citrus. *Proc. Fla. State Hort. Soc.* **81**, 81–84.

Rasmussen, G. K., and Jones, J. W. (1969). Evolving ethylene from calamondin fruits and seedlings treated with ascorbic acid. *HortScience* **4**, 1.

Rasmussen, G. K., and Jones, J. W. (1971). Cellulase activity and ethylene concentration in citrus fruit sprayed with ethylene-inducing chemicals. *HortScience* **6**, 402–403.

Reid, D. M., and Carr, D. J. (1967). The effects of the dwarfing compound CCC on the production and export of gibberellin-like substances by root systems. *Planta* **73**, 1–12.

Reid, D. M., and Crozier, A. (1970). CCC-induced increase of gibberellin levels. *Planta* **94**, 95–106.

Rhodes, M. J. C., and Wooltorton, L. S. C. (1967). The respiration climacteric in apple fruits: The action of hydrolytic enzymes in peel tissue during the climacteric period in fruit detached from the tree. *Phytochemistry* **6**, 1–12.

Rhodes, M. J. C., Harkett, P. J., Wooltorton, L. S. C., and Hulme, A. C. (1969). Studies on the effect of N-dimethylaminosuccinamic acid in ripening of apple fruits. *J. Food Technol.* **4**, 377–387.

Rogers, B. J., and Hurley, C. (1971). Ethylene and the appearance of an albedo macerating factor in citrus. *J. Amer. Soc. Hort. Sci.* **96**, 811–813.

Rubinstein, B., and Leopold, A. C. (1962). Effects of amino acids on bean leaf abscission. *Plant Physiol.* **37**, 398–401.

Rudnicki, R., Machnits, J., and Pieniazek, J. (1968). Accumulation of abscisic acid during ripening of pears (Clapp's Favourite) in various storage conditions. *Bull. Acad. Pol. Sci., Cl. V* **16**, 500–512.

Russo, L., Dostal, H. C., and Leopold, A. C. (1968). Chemical regulation of fruit ripening. *BioScience* **18**, 109.

Schneider, H. (1968). The anatomy of citrus. *In* "The Citrus Industry" (W. Reuther, L. D. Batchelor, and H. J. Webber, eds.), Vol. II, pp. 1–8. Univ. of California Press, Berkeley.

Scott, P. C., and Leopold, A. C. (1967). Opposing effects of gibberellin and ethylene. *Plant Physiol.* **43**, 1021–1022.

Shumway, L. K., and Ryan, C. A. (1972). Vacuolar protein synthesis in tomato leaves in response to wounding. *Plant Physiol.* **49s**, 42 (abstr.).

Stewart, W. S. (1949). Effects of 2,4-dichlorophenoxy acetic acid and 2,4,5-trichlorophenoxyacetic acid on citrus fruit storage. *Proc. Amer. Soc. Hort. Sci.* **54**, 109–117.

Stewart, W. S., and Parker, E. R. (1947). Preliminary studies on effects of 2,4-D sprays on preharvest drop, yield, and quality of grapefruit. *Proc. Amer. Soc. Hort. Sci.* **50**, 187–194.

Stosser, R., Rasmussen, H. P., and Bukovac, M. J. (1969). A histological study of abscission layer formation in cherry fruits during maturation. *J. Amer. Soc. Hort. Sci.* **94**, 239–243.

Thomson, W. W. (1972). Ultrastructural studies of the rind of senescent Washington navel orange. (By private communication with author.)

Tombesi, A. (1970). Meccanizzazione della raccolta delle olive. *Estac. Ann. Fac. Agr. Univ. Perugia* **25**, 37.

Unrath, C. R., Kenworthy, A. L., and Bedord, C. L. (1969). The effect of alar on fruit maturation, quality, and vegatative growth of sour cherries, *Prunus cerasus* cv. "Montmorency." *J. Amer. Soc. Hort. Sci.* **94**, 387–391.

Van Steveninck, R. F. M. (1959). Factors affecting the abscission of reproductive organs in yellow lupins (Lupinus luteus L.). III. Endogenous growth substances in virus infected and healthy plants and their effects on abscission. *J. Exp. Bot.* **10**, 367–376.

von Raussendorff-Bargen, G. (1962). Indolederivate in apfel. *Planta* **58**, 471–482.

Wang, C. Y., Mellenthin, W. M., and Hansen, E. (1971). Effect of temperature on development of premature ripening in Bartlett pears. *J. Amer. Soc. Hort. Sci.* **96**, 122.

Wang, C. Y., Wang, S. Y., and Mellethin, W. M. (1972). Identification of abscisic acid in Bartlett pears and its relationship to premature ripening. *J. Agr. Food Chem.* **20**, 451–453.

Wardowski, W. F., and Wilson, W. C. (1971). Observations on early and midseason orange abscission demonstrations using cycloheximide. *Proc. Fla. State Hort. Soc.* **84**, 81–83.

Webster, B. D. (1972). Cellus of the abscission zone of Phaseoulus. *Plant Physiol.* **49s**, 18 (abstr.).

West, C. A., and Fall, R. R. Gibberellin biosynthesis and its regulation. *Proc. Int. Conf. Plant Growth Substances, Canberra, 7th, 1970* p. 133–142.

Whiffen, A. J., Bohones, H., and Emerson, R. L. (1946). The production of an artificial antibiotic by *Streptomyces griseus. J. Bacteriol.* **52**, 610.

Whitney, J. D. (1968). Citrus fruit removal with an air harvester concept. *Proc. Fla. State Hort. Soc.* **81**, 43–48.

Wilson, W. C. (1966). The anatomy and physiology of citrus fruit abscission induced by iodoacetic acid. Ph.D. Thesis, University of Florida, Gainesville.

Wilson, W. C. (1969). Four years of abscission studies on oranges. *Proc. Fla. State Hort. Soc.* **82**, 75–81.

Wilson, W. C. (1971). Field testing of cycloheximide for abscission of oranges grown in the Indian River area. *Proc. Fla. State Hort. Soc.* **84**, 67–69.

Wilson, W. C., and Coppock, G. E. (1968). Chemical abscission studies and trials with mechanical harvesters. *Proc. Fla. State Hort. Soc.* **81**, 39–43.

Wittenbach, V. A., and Bukovac, M. J. (1972). An anatomical and histochemical study of abscission in maturing sweet cherry fruit. *J. Amer. Soc. Hort. Sci.* **97**, 214–219.

Yager, R. E., and Muir, R. M. (1958). Amino acid factor in control of abscission. *Science* **127**, 82–83.

Yang, S. F. (1969). Further studies on ethylene formation from α-keto-y-methyl-thiobutyric acid or β-methylthiopropionaldehyde by peroxidase in the presence of sulfite and oxygen. *J. Biol. Chem.* **244**, 4360–4365.

Young, R., and Jahn, O. (1972). Degreening and abscission of citrus fruit with pre-harvest applications of (2-chloroethyl)phosphonic acid (ethephon). *J. Amer. Soc. Hort. Sci.* **97**, 237–241.

Zenk, M. H. (1962). Aufnahme und stoffwechsel von a-napthylessigsaure durch erbsenepicotyle. *Planta* **58**, 75.

Zimmerman, P. W., and Hitchcock, A. E. (1939). Experiments with vapors and solutions of growth substances. *Contrib. Boyce Thompson Inst.* **10**, 481–518.

Zimmerman, P. W., Hitchcock, A. E., and Crocker, W. (1931). Movement of ethylene into and through plants. *Contrib. Boyce Thompson Inst.* **3**, 313–320.

Zucconi, F., Stosser, R., and Bukovac, M. J. (1969). Promotion of fruit abscission with abscisic acid. *BioScience* **19**, 815–817.

AUTHOR INDEX

Numbers in italics refer to the pages on which the complete references are listed.

SUBJECT INDEX

A

ABA (Abscisic acid), 50, 69, 77, 98, 100–104, 110, 111, 125, 140, 141, 161, 162, 323, 476, 477, 481, 482, 506–508, 513–516, *see also* Abscisin II, Dormin

Abortion, *see also* Shedding
anatomy of, 155
conelets, 355, 358, 370, 377
embryos, 373, 374, 376, 377, 388, 452, 485
factors affecting
age of plants, 156–158
external, 155–157
hormones, 158–160
internal, 157–162, 349–353, 357–370
flowers, 103–105, 109, 116, 117
fruits, 114, 116, 117
ovules, 103, 341, 342, 349–353, 355–358, 362, 365, 367, 369, 371, 372, 376, 377, 385, 388
pollen, 308
proembryos, 373
seeds, 450–452
shoots, 12, 94, 149–162, 169
strobili, 15

Abortive terminal, 162

Abscisic acid, *see* ABA

Abscisin II, 161

Abscission, *see also* Abscission layer, Abscission scar, Abscission zone, Autumnal abscission, Factors affecting abscission, Shedding, Summer abscission, Vernal abscission
anatomical changes, 45–79, 383–431
factors affecting
atmospheric, 85, 111–114
climatic, 85, 99–106
edaphic, 85, 106–111
internal, 125–143
histochemical changes in, 45–79

Abscission layer, 1, 2, 4, 59–61, 67, 76, 97, 116, 168, 182, 183, 321, 349, 354, 383, 384, 386, 387, 389, 390, 413, 421, 498, *see also* Abscission zone

Abscission scar, 168, 171, 173, 174, 186, 208, 400, 402, 411, 426, 429, *see also* Branch scar

Abscission zone, 3, 25, 45, 46, 49, 52–58, 65, 66, 68, 72–79, 86, 91, 94, 101, 102, 112, 116, 117, 126, 130, 133–135, 138, 151, 155, 169, 175, 182, 183, 209, 323, 375, 384–391, 393, 395–403, 405, 407, 408, 410, 422, 426–429, 457, 476, 488, 497, 500, 505, 506, 511, 512, 514–516, *see also* Abscission layer, Floral cup abscission zone, Style abscission zone

Absorption
minerals, 30, 100, 104
NAA, 457
NAD, 457
nitrogen, 104, 110
water, 111

Acetic acid, 368

Achene, 322, 329, 330, 333

Adaptation, 47, 93

Adenosine triphosphate, *see* ATP

Adenylate, 97

Adenylate pool, 98

Adsee-120, 496

Adult stage, 17, *see also* Aging

Adventitious roots, 106, 231, 239, 251, 266–268

Aeration, 19, 97, 106, 110, 213, 214, 253, 269, 270

Aerial roots, 239, 254

Agamocarpy, 353, 367

Agamospermy, 349, 367

Agar, 141, 160

Age class, 232, 233

Age of plants, 191, 312, *see also* Aging

Aging, 16, 17, 46, 57, 69, 128, 156, 189, 217, 219, 220, 224, 248, 358, 359, 499, 501, 502, 506, 515, *see also* Senescence

Air sac, 314

Air shaker, 477

Corky bark disease, 116
Corm, 239
Corolla, 398
Correlation, 86
Cortex, 12, 53, 57, 58, 61, 62, 65–68, 72, 73, 75, 174, 175, 182–186, 206, 207, 209, 210, 215, 217, 219, 224, 226, 229–231, 239, 240, 246, 262, 264, 265, 270, 277, 286, 386, 388
Cortical browning, 259, 261
Cotyledons, 24, 47, 64, 67, 77, *see also* Senescence of cotyledons, Shedding of cotyledons
  fusion of, 24
Coumarins, 19
Cristae, 78
Crossing, 371, 372, 374, *see also* Outcrossing
Cross-pollination, 320, *see also* Pollination
Crown class, 16, 27, 188, 193
Crown to height ratio, 193
Cryptogams, 208, 334
Crystalloids, 388
Culms, 266
Cultivar, 246, 389, 391, 437–442, 446, 449, 451–453, 456, 458, 459, 461, 462, 464–468, 484, 486, 487, 490, 498, 510, 511
Curing, 24, 25
Curling, 24
Cuticle, 388, 457, 493–495
Cutin, 61, 69
Cuttings, 251, 375
Cyanamid, 26
Cyanides, 19
Cycads, 302
Cyclamic acid, 498
Cycloheximide (CHI) β[2-(3,5-dimethyl-2-oxocyclohexyl)-2-hydroethyl] glutamide, 476, 479–485, 487–496, 498–500, 503–505, 508–516
Cytokinesis, 62, 451
Cytokinins (CK), 50, 64, 97, 98, 100, 103, 104, 106–108, 110, 127, 129, 131, 132, 139, 161, 162, 214, 323, 369, 372, *see also* individual cytokinins
Cytolysis, 1, 25, 75, 76
Cytoplasm, 60, 133, 155, 373, 407, 515

D

2,4-D, (2,4-dichlorophenoxyacetic acid), 23, 27, 129, 194, 214, 389, 476, 482, 502
Day length, *see* Photoperiod
Death of roots, 262–269
Debarking, 26–28
  chemical, 26–28
Debudding, 159
Decarboxylation, 111
Decomposition of litter, 19
Defoliants, 24, 25, 109
Defoliation, 13, 17, 22, 106, 107, 458, 482–484, 486, 492, 509, *see also* Abscission, Defoliants, Shedding of leaves, and individual chemical defoliants
  chemical, 26
  for military purposes, 1, 28–30
  of nursery stock, 26
Dehiscence, 3, 86, 299, 302, 303, 318, *see also* Shedding of anthers, 300, 301, 305–307, 320, 391, 399, 409, 423
  of fruits, 332–334
  of strobili, 306
Denaturation of protoplasm, 24, 25
Deoxyribonucelic acid, *see* DNA
Depolymerization, 74
Desiccants, 24 *see also* Desiccation, individual desiccants
Desiccation, 19, 22, 24, 105, 126, 185, 192, 206, 219, 300, 423, *see also* Water deficits
Dew, 25
2,4-Dichlorophenoxyacetic acid, *see* 2,4-D
Dictyosomes, 133
Dictyosome vesicles, 133
Dieback, 21, 24, 30, 114, 279, 484, 510, 511
Differentiation, 36, 51, 69, 133, 151, 157, 214, 224, 277, 385, 387
Diffusion, 141, 241, 270, 364, 365, 369, 481, 488, 492, 500, 514, 515
Dihydroxyphenols, 109
Dilation, 163, 185, 217
Dimorphism, 21, 313